A Walk Through Combinatorics

An Introduction to Enumeration and Graph Theory

A Walk Through Combinatorics

An Introduction to Enumeration and Graph Theory

Third Edition

Miklós Bóna
University of Florida, USA

World Scientific

NEW JERSEY • LONDON • SINGAPORE • BEIJING • SHANGHAI • HONG KONG • TAIPEI • CHENNAI

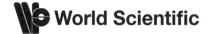

Published by

World Scientific Publishing Co. Pte. Ltd.
5 Toh Tuck Link, Singapore 596224
USA office: 27 Warren Street, Suite 401-402, Hackensack, NJ 07601
UK office: 57 Shelton Street, Covent Garden, London WC2H 9HE

Library of Congress Cataloging-in-Publication Data
Bóna, Miklós.
A walk through combinatorics : an introduction to enumeration and graph theory / by Miklós Bóna. -- 3rd ed.
p. cm.
Includes bibliographical references and index.
ISBN-13 978-981-4335-23-2 (hardcover : alk. paper)
ISBN-10 981-4335-23-1 (hardcover : alk. paper)
1. Combinatorial analysis. I. Title.
QA164.B66 2011
511'.6--dc22
2011001455

British Library Cataloguing-in-Publication Data
A catalogue record for this book is available from the British Library.

First published 2011
Reprinted 2012
Reprinted 2013 (paperback edition only)
ISBN 978-981-4460-00-2 (pbk)

Printed in Singapore.

To Linda
To Mikike, Benny and Vinnie

Foreword

The subject of combinatorics is so vast that the author of a textbook faces a difficult decision as to what topics to include. There is no more-or-less canonical corpus as in such other subjects as number theory and complex variable theory. Miklós Bóna has succeeded admirably in blending classic results that would be on anyone's list for inclusion in a textbook, a sprinkling of more advanced topics that are essential for further study of combinatorics, and a taste of recent work bringing the reader to the frontiers of current research. All three items are conveyed in an engaging style, with many interesting examples and exercises. A worthy feature of the book is the many exercises that come with complete solutions. There are also numerous exercises without solutions that can be assigned for homework.

Some relatively advanced topics covered by Bóna include permutations with restricted cycle structure, the Matrix-Tree theorem, Ramsey theory (going well beyond the classical Ramsey's theorem for graphs), the probabilistic method, and the Möbius function of a partially ordered set. Any of these topics could be a springboard for a subsequent course or reading project which will further convince the student of the extraordinary richness, variety, depth, and applicability of combinatorics. The most unusual topic covered by Bóna is pattern avoidance in permutations and the connection with stack sortable permutations. This is a relatively recent research area in which most of the work has been entirely elementary. An undergraduate student eager to do some original research has a good chance of making a worthwhile contribution in the area of pattern avoidance.

I only wish that when I was a student beginning to learn combinatorics there was a textbook available as attractive as Bóna's. Students today are fortunate to be able to sample the treasures available herein.

Richard Stanley
Cambridge, Massachusetts
February 6, 2002

Preface

The best way to get to know Yosemite National Park is to walk through it, on many different paths. In the optimal case, the gorgeous sights provide ample compensation for our sore muscles. In this book, we intend to explain the basics of Combinatorics while walking through its beautiful results. Starting from our very first chapter, we will show numerous examples of what may be the most attractive feature of this field: that *very* simple tools can be *very* powerful at the same time. We will also show the other side of the coin, that is, that sometimes totally elementary-looking problems turn out to be unexpectedly deep, or even unknown.

This book is meant to be a textbook for an introductory combinatorics course that can take one or two semesters. We included a very extensive list of exercises, ranging in difficulty from "routine" to "worthy of independent publication". In each section, we included exercises that contain material not explicitly discussed in the text before. We chose to do this to provide instructors with some extra choices if they want to shift the emphasis of their course.

It goes without saying that we covered the classics, that is, combinatorial choice problems, and graph theory. We included some more elaborate concepts, such as Ramsey theory, the Probabilistic Method, and Pattern Avoidance (the latter is probably a first of its kind). While we realize that we can only skim the surface of these areas, we believe they are interesting enough to catch the attention of some students, even at first sight. Most undergraduate students enroll in at most one Combinatorics course during their studies, therefore it is important that they see as many captivating examples as possible. It is in this spirit that we included two new chapters in the second edition, on Algorithms, and on Computational Complexity. We believe that the best undergraduate students, those who will get to the end of the book, should be acquainted with the extremely intriguing questions that abound in these two areas. The third edition has two challenging new chapters, one on Block Designs and codes obtained from designs, and the other one on counting unlabeled structures.

We wrote this book as we believe that combinatorics, researching it, teaching it, learning it, is always fun. We hope that at the end of the walk, readers will agree.

****

Exercises that are thought to be significantly harder than average are marked by one or more + signs. An exercise with a single + sign is probably at the level of a harder homework problem. The difficulty level of an exercise with more than one + sign may be comparable to an independent publication. An exercise that is thought to be significantly easier than average is marked by a - sign.

We provide Supplementary Exercises without solutions at the end of each chapter. These typically include, but are not limited to, the easiest exercises in that chapter. A solution manual for the Supplementary Exercises is available for Instructors.

Acknowledgments

This book has been written while I was teaching Combinatorics at the University of Florida, and during my sabbatical at the University of Pennsylvania in the Fall of 2005. I am certainly indebted to the books I used in my teaching during this time. These were "Introductory Combinatorics" by Kenneth Bogarth, "Enumerative Combinatorics I.-II" by Richard Stanley, "Matching Theory" by László Lovász and Michael D. Plummer, and "A Course in Combinatorics" by J. H. van Lint and R. M. Wilson. The two new chapters of the second edition were certainly influenced by the books of which I learned the theory of algorithms and computation, namely "Computational Complexity" by Christos Papadimitriou, "Introduction to the Theory of Computation" by Michael Sipser, who taught me the subject in person, "Algorithms and Complexity" by Herbert Wilf, and "Introduction to Algorithms" by Cormen, Leiserson, Rivest and Stein. Several exercises in the book come from my long history as a student mathematics competition participant. This includes various national and international contests, as well as the long-term contest run by the Hungarian student journal KÖMAL, and the Russian student journal Kvant.

I am grateful to my students who never stopped asking questions and showed which part of the material needed further explanation.

Some of the presented material was part of my own research, sometimes in collaboration. I would like to say thanks to my co-authors, Andrew MacLennan, Bruce Sagan, Rodica Simion, Daniel Spielman, Géza Tóth, and Dennis White. I am also indebted to my former advisor, Richard Stanley, who introduced me to the fascinating area of Pattern Avoidance, discussed in Chapter 14.

I am deeply appreciative for the constructive suggestions of my colleagues Vincent Vatter, Andrew Vince, Neil White, and Aleksandr Vayner.

A significant part of the first edition was written during the summer of 2001, when I enjoyed the hospitality of my parents, Miklós and Katalin Bóna, at the Lake Balaton in Hungary.

My gratitude is extended to Joseph Sciacca for the cover page. If you do not know why a book entitled "A Walk Through Combinatorics" has such a cover page, you may figure it out when reading Chapter 10.

After the publication of the first edition in 2002, several mathematicians contributed lists of typographical errors to be corrected. Particularly extensive lists were provided by Margaret Bayer, Richard Ehrenborg, John Hall, Hyeongkwan Ju, Sergey Kitaev, and Robert Robinson. I am thankful for their help in making the second edition better by communicating those lists to me, as well as for similar help from countless other contributors who will hopefully forgive that I do not list all of them here. The second edition was improved by a significant list of comments by Margaret Bayer, while the third edition was similarly helped by the remarks of Glenn Tesler. I am indebted to Thomas Zaslavsky for numerous conversations and suggestions that made the paperback version of the third edition better.

Most of all, I must thank my wife Linda, my first reader, who made it possible for me to spend long hours writing this book while she also had her hands full. See Exercise 3 of Chapter 15 for further explanation.

Contents

III. Graph Theory

IV. Horizons

Chapter 1

Seven Is More Than Six. The Pigeon-Hole Principle

1.1 The Basic Pigeon-Hole Principle

Seven is more than six. Four is more than three. Two is more than one. These statements do not seem to be too interesting, exciting, or deep. We will see, however, that the famous Pigeon-hole Principle makes excellent use of them. We choose to start our walk through combinatorics by discussing the Pigeon-hole Principle because it epitomizes one of the most attractive treats of this field: the possibility of obtaining very strong results by very simple means.

Theorem 1.1 (Pigeon-hole Principle). *Let n and k be positive integers, and let $n > k$. Suppose we have to place n identical balls into k identical boxes. Then there will be at least one box in which we place at least two balls.*

Proof. While the statement seems intuitively obvious, we are going to give a formal proof because proofs of this nature will be used throughout this book.

We prove our statement in an *indirect way*, that is, we assume its contrary is true, and deduce a contradiction from that assumption. This is a very common strategy in mathematics; in fact, if we have no idea how to prove something, we can always try an indirect proof.

Let us assume there is no box with at least two balls. Then each of the k boxes has either 0 or 1 ball in it. Denote by m the number of boxes that have zero balls in them; then certainly $m \geq 0$. Then, of course, there are $k - m$ boxes that have one. However, that would mean that the total number of balls placed into the k boxes is $k - m$ which is a contradiction because we had to place n balls into the boxes, and $k - m \leq k < n$.

Therefore, our assumption that there is no box with at least two balls must have been false. □

In what follows, we will present several applications that show that this innocuous statement is in fact a *very* powerful tool.

Example 1.2. There is an element in the sequence $7, 77, 777, 7777, \cdots$, that is divisible by 2003.

Solution. We prove that an even stronger statement is true, in fact, one of the first 2003 elements of the sequence is divisible by 2003. Let us assume that the contrary is true. Then take the first 2003 elements of the sequence and divide each of them by 2003. As none of them is divisible by 2003, they will all have a remainder that is at least 1 and at most 2002. As there are 2003 remainders (one for each of the first 2003 elements of the sequence), and only 2002 possible values for these remainders, it follows by the Pigeon-hole Principle that there are two elements out of the first 2003 that have the same remainder. Let us say that the ith and the jth elements of the sequence, a_i and a_j, have this property, and let $i < j$.

	7777777777777777777777777	j digits
−	777777777777777777	i digits
	7777777000000000000000000	j-i digits equal to 7, i digits equal to 0

Fig. 1.1 The difference of a_j and a_i.

As a_i and a_j have the same remainder when divided by 2003, there exist non-negative integers k_i, k_j, and r so that $r \leq 2002$, and $a_i = 2003k_i + r$, and $a_j = 2003k_j + r$. This shows that $a_j - a_i = 2003(k_j - k_i)$, so in particular, $a_j - a_i$ is divisible by 2003.

This is nice, but we need to show that there is an element in our sequence that is divisible by 2003, and $a_j - a_i$ is not an element in our sequence. Figure 1.1 helps understand why the information that $a_j - a_i$ is divisible by 2003 is nevertheless very useful.

Indeed, $a_j - a_i$ consists of $j - i$ digits equal to 7, then i digits equal to 0. In other words,

$$a_j - a_i = a_{j-i} \cdot 10^i,$$

and the proof follows as 10^i is relatively prime to 2003, so a_{j-i} must be divisible by 2003.

In this example, the possible values of the remainders were the boxes, all 2002 of them, while the first 2003 elements of the sequence played the role of the balls. There were more balls than boxes, so the Pigeon-hole Principle applied.

Example 1.3. A chess tournament has n participants, and any two players play one game against each other. Then it is true that in any given point of time, there are two players who have finished the same number of games.

Solution. First we could think that the Pigeon-hole Principle will not be applicable here as the number of players ("balls") is n, and the number of possibilities for the number of games finished by any one of them ("boxes") is also n. Indeed, a player could finish either no games, or one game, or two games, and so on, up to and including $n-1$ games.

The fact, however, that two players play their games *against* each other, provides the missing piece of our proof. If there is a player A who has completed all his $n-1$ games, then there cannot be any player who completed zero games because at the very least, everyone has played with A. Therefore, the values 0 and $n-1$ cannot both occur among the numbers of games finished by the players at any one time. So the number of possibilities for these numbers ("boxes") is at most $n-1$ at any given point of time, and the proof follows.

1.2 The Generalized Pigeon-Hole Principle

It is easy to generalize the Pigeon-hole Principle in the following way.

Theorem 1.4 (Pigeon-hole Principle, general version).
Let n, m and r be positive integers so that $n > rm$. Let us distribute n identical balls into m identical boxes. Then there will be at least one box into which we place at least $r+1$ balls.

Proof. Just as in the proof of Theorem 1.1, we assume the contrary statement. Then each of the m boxes can hold at most r balls, so all the boxes can hold at most $rm < n$ balls, which contradicts the requirement that we distribute n balls. □

It is certainly not only in number theory that the Pigeon-hole Principle proves to be very useful. The following example provides a geometric application.

Example 1.5. Ten points are given within a square of unit size. Then there are two of them that are closer to each other than 0.48, and there are three of them that can be covered by a disk of radius 0.5.

Solution. Let us split our unit square into nine equal squares by straight lines as shown in Figure 1.2. As there are ten points given inside the nine small squares, Theorem 1.1 implies that there will be at least one small square containing two of our ten points. The longest distance within a square of side length 1/3 is that of two opposite endpoints of a diagonal. By the Pythagorean theorem, that distance is $\frac{\sqrt{2}}{3} < 0.48$, so the first part of the statement follows.

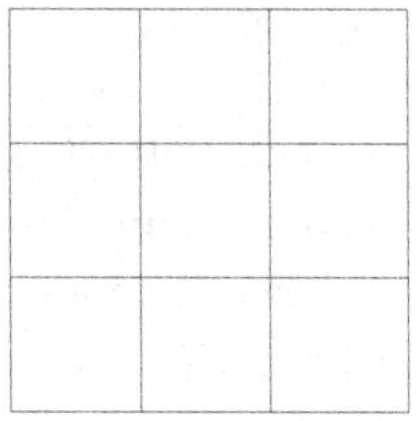

Fig. 1.2 Nine small squares for ten points.

To prove the second statement, divide our square into four equal parts by its two diagonals as shown in Figure 1.3. Theorem 1.4 then implies that at least one of these triangles will contain three of our points. The proof again follows as the radius of the circumcircle of these triangles is shorter than 0.5.

We finish our discussion of the Pigeon-hole Principle by two highly surprising applications. What is striking in our first example is that it is valid for *everybody*, not just say, the majority of people. So we might as well discuss our example choosing the reader herself for its subject.

Example 1.6. During the last 1000 years, the reader had an ancestor A such that there was a person P who was an ancestor of *both* the father and

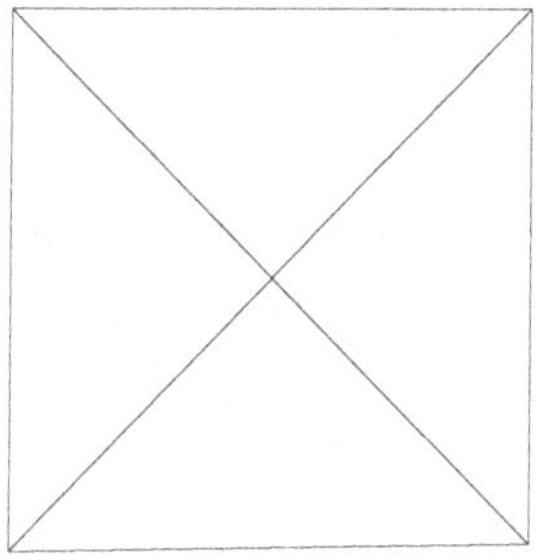

Fig. 1.3 Four triangles for ten points.

the mother of A.

Solution. Again, we prove our statement in an indirect way: we assume its contrary, and deduce a contradiction. We will use some rough estimates for the sake of shortness, but they will not make our argument any less valid.

Take the family tree of the reader. This tree is shown in Figure 1.4.

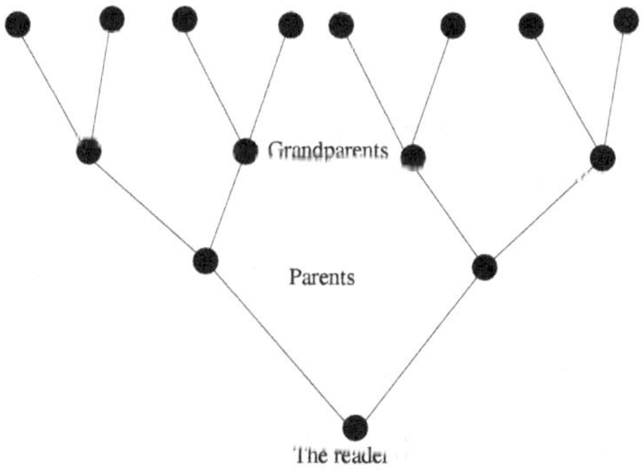

Fig. 1.4 The first few levels of the family tree of the reader.

The root of this tree is the reader herself. On the first level of the tree, we see the two parents of the reader, on the second level we find her four grandparents, and so on. Assume (for shortness) that one generation takes 25 years to produce offspring. That means that 1000 years was sufficient time for 40 generations to grow up, yielding $1 + 2 + 2^2 + \cdots + 2^{40} = 2^{41} - 1$

nodes in the family tree. If any two nodes of this tree are associated to the same person B, then we are done as B can play the role of P.

Now assume that no two nodes of the first 40 levels of the family tree coincide. Then all the $2^{41} - 1$ nodes of the family tree must be distinct. That would mean $2^{41} - 1$ distinct people, and that is a lot more than the number of all people who have lived in our planet during the last 1000 years. Indeed, the current population of our planet is less than 10^{10}, and was much less at any earlier point of time. Therefore, the cumulative population of the last 1000 years, or 40 generations, was less than $40 \cdot 10^{10} < 2^{41} - 1$, and the proof follows by contradiction.

Our assumption that every generation takes 25 years to produce offspring was a realistic one. Given that by all available data, the average life expectancy of humans is longer today than ever before, 25 seems to be a high-end estimate. The reader should spend a little time thinking about how (and if) the argument would have to be modified if 25 were to be replaced by a smaller or larger number.

Our last example comes from the theory of graphs, an extensive and important area of combinatorics to which we will devote several chapters later.

Example 1.7. Mr. and Mrs. Smith invited four couples to their home. Some guests were friends of Mr. Smith, and some others were friends of Mrs. Smith. When the guests arrived, people who knew each other beforehand shook hands, those who did not know each other just greeted each other. After all this took place, the observant Mr. Smith said "How interesting. If you disregard me, there are no two people present who shook hands the same number of times".

How many times did Mrs. Smith shake hands?

Solution. The reader may well think that this question cannot be answered from the given information any better than say, a question about the age of the second cousin of Mr. Smith. However, using the Pigeonhole Principle and a very handy model called a *graph*, this question *can* be answered.

To start, let us represent each person by a node, and let us write the number of handshakes carried out by each person except Mr. Smith next to the corresponding vertex. This way we must write down nine *different* nonnegative integers. All these integers must be smaller than nine as nobody shook hands with himself/herself or his/her spouse. So the numbers we wrote down are between 0 and 8, and since there are nine of them, we must have written down each of the numbers $0, 1, 2, 3, 4, 5, 6, 7, 8$ exactly once. The diagram we have constructed so far can be seen in Figure 1.5.

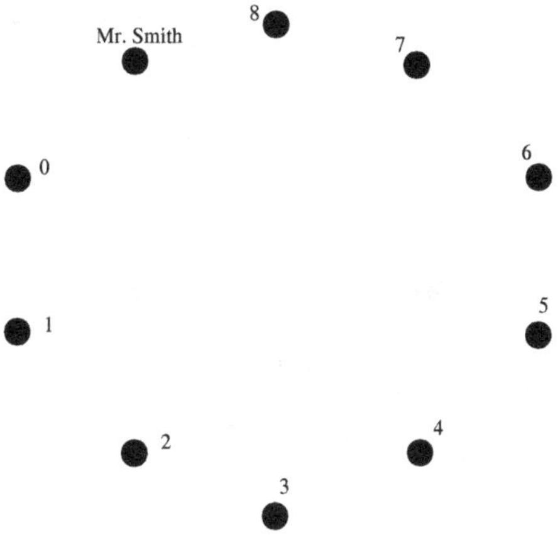

Fig. 1.5 The participants of the party.

Now let us join two nodes by a line if the corresponding two people shook each other's hands. Such a diagram is called a *graph*, the nodes are called the *vertices* of the graph, and the lines are called the *edges* of the graph. So our diagram will be a graph with ten vertices.

Let us denote the person with i handshakes by Y_i. (Mr. Smith is not assigned any additional notation.) Who can be the spouse of the person Y_8? We know that Y_8 did not shake the hand of only one other person, so that person must have been his or her spouse. On the other hand, Y_8 certainly did not shake the hand Y_0 as nobody did that. Therefore, Y_8 and Y_0 are married, and Y_8 shook everyone's hand except for Y_0. We represent this by joining his vertex to all vertices other than Y_0. We also encircle Y_8 and Y_0 together, to express that they are married.

Now try to find the spouse of Y_7, the person with seven handshakes. This person did not shake the hands of two people, one of whom was his/her spouse. Looking at Figure 1.6, we can tell who these two people are. One of them is Y_0 as he or she did not shake anyone's hand, and the other one is Y_1 as he or she had only one handshake, and that was with Y_8. As spouses do not shake hands, this implies that the spouse of Y_7 is either Y_0 or Y_1. However, Y_0 is married to Y_8, so Y_1 must be married to Y_7.

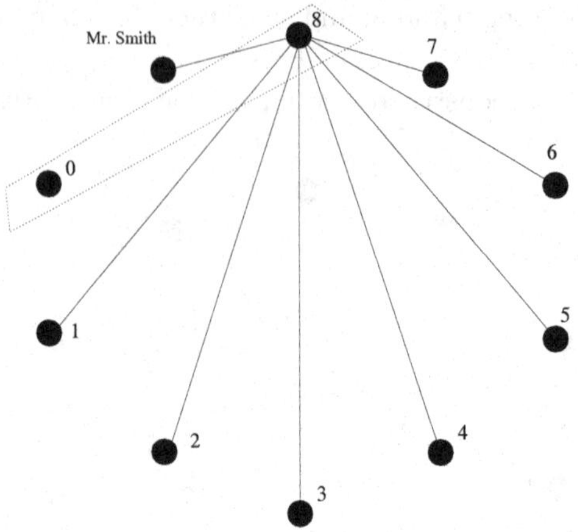

Fig. 1.6 Y_8 and Y_0 are married.

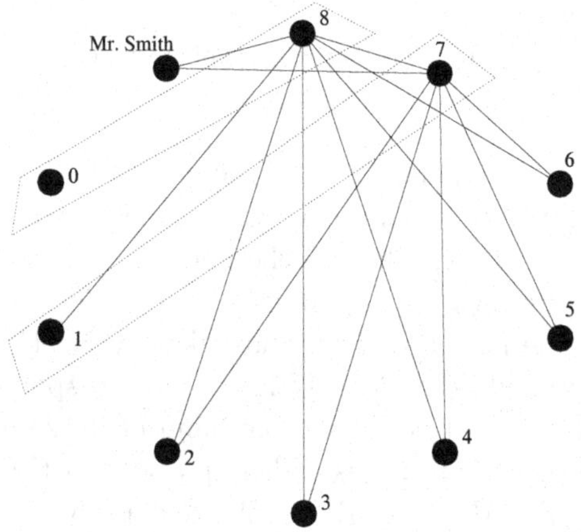

Fig. 1.7 Y_1 and Y_7 are married.

By a similar argument that the reader should be able to complete, Y_6 and Y_2 must be married, and also, Y_5 and Y_3 must be married. That implies that by exclusion, Y_4 is Mrs. Smith, therefore Mrs. Smith shook hands four times.

How did we obtain such a strong result from "almost no data"? The truth is that the data we had, that is, that all people except Mr. Smith

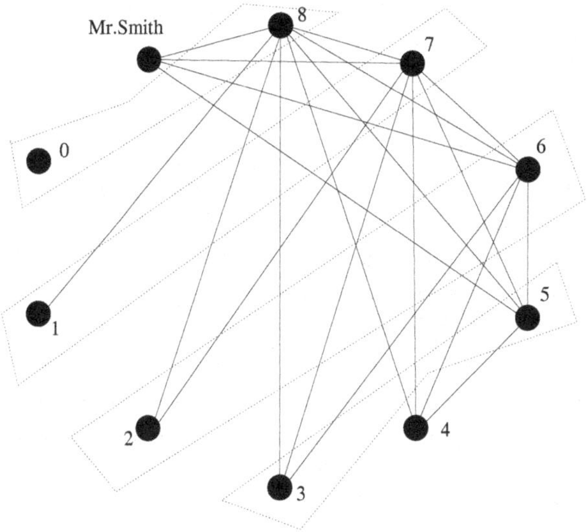

Fig. 1.8 Mrs. Smith shook hands four times.

shook hands a different number of times, is quite restrictive. Indeed, consider Example 1.3 again. An obvious reformulation of that Example shows that it is simply impossible to have a party at which no two people shake hands the same number of times (as long as no two people shake hands more than once). Example 1.7 relaxes the "all-different-numbers" requirement a little bit, by waiving it for Mr. Smith. Our argument then shows that with that extra level of freedom, we can indeed have a party satisfying the new, weaker conditions, but only in one way. That way is described by the graph shown in Figure 1.8.

Exercises

(1) A busy airport sees 1500 takeoffs per day. Prove that there are two planes that must take off within a minute of each other.
(2) Find all triples of positive integers $a < b < c$ for which

$$\frac{1}{a} + \frac{1}{b} + \frac{1}{c} = 1$$

holds.

(3) One hundred points are given inside a cube of side length one. Prove that there are four of them that span a tetrahedron whose volume is at most $1/99$.

(4) (+) We have distributed two hundred balls into one hundred boxes with the restrictions that no box got more than one hundred balls, and each box got at least one. Prove that it is possible to find some boxes that together contain exactly one hundred balls.

(5) (+) Last year, the Division One basketball teams played against an average of eighteen different opponents. Is it possible to find a group of teams so that each of them played against at least ten other teams of the group?

(6)(a) The set M consists of nine positive integers, none of which has a prime divisor larger than six. Prove that M has two elements whose product is the square of an integer.

(b) (+) (Some knowledge of linear algebra and abstract algebra required.) The set A consists of $n+1$ positive integers, none of which has a prime divisor that is larger than the nth smallest prime number. Prove that there exists a non-empty subset $B \subseteq A$ so that the product of the elements of B is a perfect square.

(7) (++) The set L consists of 2003 integers, none of which has a prime divisor larger than 24. Prove that L has four elements, the product of which is equal to the fourth power of an integer.

(8) (+) The sum of one hundred given real numbers is zero. Prove that at least 99 of the pairwise sums of these hundred numbers are non-negative. Is this result the best possible one?

(9) (+) We colored all points of R^2 with integer coordinates by one of six colors. Prove that there is a rectangle whose vertices are monochromatic. Can we make the statement stronger by limiting the size of the purported monochromatic rectangle?

(10) Prove that among 502 positive integers, there are always two integers so that either their sum or their difference is divisible by 1000.

(11) (+) We chose $n+2$ numbers from the set $1, 2, \cdots, 3n$. Prove that there are always two among the chosen numbers whose difference is more than n but less than $2n$.

(12) There are four heaps of stones in our backyard. We rearrange them into five heaps. Prove that at least two stones are placed into a smaller heap.

(13) There are infinitely many pieces of paper in a basket, and there is a positive integer written on each of them. We know that no matter how

we choose infinitely many pieces, there will always be two of them so that the difference of the numbers written on them is at most ten million. Prove that there is an integer that has been written on infinitely many pieces of paper.

(14) (+)

(a) A soccer team played 30 games this year, and scored a total of 53 goals, scoring at least one goal in each game. Prove that there was a sequence of consecutive games in which the team scored exactly six goals.

(b) Prove that the claim of part (a) does not hold for a team that scored 60 goals, with the other parameters unchanged.

(c) Prove that the claim of part (a) does hold for a team that scored 59 goals, with the other parameters unchanged.

(15) (+) The set of all positive integers is partitioned into several arithmetic progressions. Show that there is at least one among these arithmetic progressions whose initial term is divisible by its difference.

Supplementary Exercises

(16) (-) Prove that every year contains at least four and at most five months that contain five Sundays.

(17) (-) A soccer league features 17 games for today. Let us assume that no team will score more than three goals. Prove that there will be a result that will occur more than once. (A result consists of the number of goals scored by the home team, followed by the number of goals scored by the visiting team. So 3-2 and 2-3 are considered different scores.)

(18) (-) A group of seven co-workers are trying to predict the total number of points scored in a given basketball game. The first six people already took their guesses, and, curiously, they all picked distinct even numbers. Mr. Slow is the last person to guess, and he knows all previous guesses. Is there a strategy for him that assures that his guess will be better than the guesses of half of his colleagues?

(19) (-) A soccer team scored a total of 40 goals this season. Nine players scored at least one of those goals. Prove that there are two players among those nine who scored the same number of goals.

(20) (-)

(a) In April, Ms. Consistent went to the swimming pool 26 times. Is it true that there were six consecutive days when she went to the swimming pool?

(b) Same as (a), but for the month of May instead of April.

(21)(a) We select 11 positive integers that are less than 29 at random. Prove that there will always be two integers selected that have a common divisor larger than 1.

(b) Is the statement of part (a) true if we only select ten integers that are less than 29?

(22) Prove that there exists a positive integer n so that $44^n - 1$ is divisible by 7.

(23) The sum of five positive real numbers is 100. Prove that there are two numbers among them whose difference is at most 10.

(24) Find all 4-tuples (a, b, c, d) of distinct positive integers so that $a < b < c < d$ and

$$\frac{1}{a} + \frac{1}{b} + \frac{1}{c} + \frac{1}{d} = 1.$$

(25) Complete the following sentence, that is a generalization of the Pigeon-hole Principle to real numbers. "If the sum of k real numbers is n, then there must be one of them which is...". Prove your claim.

(26) We are given 17 points inside a regular triangle of side length one. Prove that there are two points among them whose distance is not more than $1/4$.

(27) Prove that the sequence 1967, 19671967, 196719671967, $\cdots$, contains an element that is divisible by 1969.

(28) A teacher receives a paycheck every two weeks, always the same day of the week. Is it true that in any six consecutive calendar months she receives exactly 13 paychecks?

(29) (+) Let T be a triangle with angles of 30, 60 and 90 degrees whose hypotenuse is of length 1. We choose ten points inside T at random. Prove that there will be four points among them that can be covered by a half-circle of radius 0.42.

(30) We select $n + 1$ different integers from the set $\{1, 2, \cdots, 2n\}$. Prove that there will always be two among the selected integers whose largest common divisor is 1.

(31)(a) Let $n \geq 2$. We select $n + 1$ different integers from the set $\{1, 2, \cdots, 2n\}$. Is it true that there will always be two among the selected integers so that one of them is equal to twice the other?

(b) Is it true that there will always be two among the selected integers so that one is a multiple of the other?

(32) One afternoon, a mathematics library had several visitors. A librarian noticed that it was impossible to find three visitors so that no two of them met in the library that afternoon. Prove that then it was possible to find two moments of time that afternoon so that each visitor was in the library at one of those two moments.

(33) (+) Let r be any irrational real number. Prove that there exists a positive integer n so that the distance of nr from the closest integer is less than 10^{-10}.

(34) Let p and q be two positive integers so that the largest common divisor of p and q is 1. Prove then for any non-negative integers $s \leq p-1$ and $t \leq q-1$, there exists a non-negative integer $m \leq pq$ so that if we divide m by p, the remainder is s, and if we divide m by q, the remainder is t.

Solutions to Exercises

(1) There are 1440 minutes per day. If our 1440 minutes are the boxes, and our 1500 planes are the balls, the Pigeon-hole Principle says that there are two balls in the same box, that is, there are two planes that take off within a minute of each other.

(2) It is clear that $a = 2$. Indeed, $a = 1$ is impossible because then the left hand side would be larger than 1, and $a \geq 3$ is impossible as $a < b < c$ implies $\frac{1}{a} > \frac{1}{b} > \frac{1}{c}$, so $a = 3$ would imply that the left-hand side is smaller than 1. Thus we only have to solve

$$\frac{1}{b} + \frac{1}{c} = \frac{1}{2},$$

with $3 \leq b < c$. We claim that b must take its smallest possible value, 3. Indeed, if $b \geq 4$, then $c \geq 5$, and so $\frac{1}{b} + \frac{1}{c} < \frac{1}{4} + \frac{1}{5} < \frac{1}{2}$. Thus $b = 3$, and therefore, $c = 6$.

(3) Split the cube into 33 prisms by planes that are parallel to its base and are at distance $1/33$ from each other. By Pigeon-hole Principle, one of these prisms must contain four of our points. The volume of the tetrahedron spanned by these four points is at most one third of that of the prism, and the statement follows.

(4) Arrange our boxes in a line so that the first two boxes do not have the same number of balls in them. We can always do this unless all boxes have two balls, in which case the statement is certainly true.

Let a_i denote the number of balls in box i, for all positive integers $1 \leq i \leq 100$. Now look at the following sums: a_1, a_1+a_2, $a_1+a_2+a_3$, $\cdots$, $a_1+a_2+\cdots+a_{100}$. If two of them yield the same remainder when divided by 100, then take the difference of those two sums. That will yield a sum of type $a_i+a_{i+1}+\cdots+a_j$ that is divisible by 100, is smaller than 200, and is positive. In other words, $a_i + a_{i+1} + \cdots + a_j = 100$, so the total content of boxes $i, i+1, \cdots, j$ is exactly 100 balls.
Now assume this does not happen, that is, all sums $a_1 + a_2 + \cdots + a_k$ yield different remainders when divided by 100. Attach the one-element sum a_2 to our list of sums. Now we have 101 sums, so by Theorem 1.1, two of them must have the same remainder when divided by 100. Since we assumed this did not happen before a_2 joined the list, we know that there is a sum S on our list that has the same remainder as a_2. As we know that $a_1 \neq a_2$, we also know that $S \neq a_1$, and we are done as in the previous paragraph, since $S - a_2 = a_1 + a_3 + \cdots + a_t = 100$.
We note that this argument works in general with $2n$ boxes and $4n$ balls. We also note that we in fact proved a stronger statement as our chosen boxes are almost consecutive.

(5) Yes. Take a team T that played against at most nine opponents. If there is no such team, then the group of all Division One teams has the required property, and we are done. Omit T; we claim that this will not decrease the average number of opponents. Indeed, as we are only interested in the number of opponents played (and not games), we can assume that any two teams played each other at most once. The 18-game-average means that all the m Division One teams together played $9m$ games as a game involves two teams. Omitting T, we are left with $m-1$ teams, who played a grand total of at least $9m-9$ games. This means that the remaining teams still played at least 18 games on average against other remaining teams.
Now iterate this procedure- look for a team from the remaining group that has only played nine games and omit it. As the number of teams is finite, this elimination procedure has to come to an end. The only way that can happen is that there will be a group of which we cannot eliminate any team, that is, in which every team has played at least ten games against the other teams of the group.

(6)(a) Each element of M can be written as $2^i 3^j 5^k$ for some non-negative integers i, j, k. Therefore, we can divide the elements of M into eight classes according to the *parity* of their exponents i, j, k. By the Pigeon-hole Principle, there will be two elements of M, say x and y, that are in the same class. As the sum of two integers of the same parity is even, this implies that $x \cdot y = 2^{2a}3^{2b}5^{2c}$ for some non-negative integers a, b, c, therefore, $xy = (2^a 3^b 5^c)^2$.

(b) The $n+1$ elements of A can be considered as elements of an n-dimensional vector space over the binary field. Let B be a linearly dependent subset of A, then the product of all elements of B is a perfect square since all prime factors must occur in that product an even number of times in that product.

(7) If we try to copy the exact method of the previous problem, we may run into difficulties. Indeed, the elements of L can have nine different prime divisors, $2, 3, 5, 7, 11, 13, 17, 19, 23$. If we classify them according to the remainder of the exponents of these prime divisors modulo four, we get a classification into $4^9 > 2003$ classes. So it seems that it is not even sure that there will be a class containing two elements of L, let alone four.

The reason this attempt did not work is that it tried to prove too much. For the product of four integers to be a fourth power, it is not necessary that the exponents of each prime divisor have the same remainder modulo four in each of the four integers. For example, 1,1,2,8 do not have that property, but their product is $16 = 2^4$.

A more gradual approach is more successful. Let us classify the elements of L again just by the parity of the exponents of the nine possible prime divisors in them. This classification creates just $2^9 = 512$ classes. Now pick two elements of L that are in the same class, and remove them from L. Put their product into a new set L'. This procedure clearly decreased the size of L by 2. Then repeat this same procedure, that is, pick two elements of L that are in the same class, remove them, and put their product into L'. Note that all elements of L' will be squares as they will contain all their prime divisors with even exponents. Do this until you can, that is, until there are no two elements of L in the same class. Stop when that happens. Then L has at most 511 elements left, so we have removed at least 1492 elements from L. Therefore L' has at least 746 elements, all of which are squares of integers.

Now classify the elements of L' according to the remainders of the exponents of their prime divisors modulo four. As the elements of L' are all squares, all these exponents are even numbers, so their remainders modulo four are either 0 or 2. So again, this classification creates only 512 classes, and therefore, there will be two elements of L' in the same class, say u and v. Then uv is the fourth power of an integer, and since both u and v are products of two integers in L, our claim is proved.

(8) **First Solution**: Let $a_1 \leq a_2 \leq \cdots \leq a_{100}$ denote our one hundred numbers. We will show 99 non-negative sums. We have to distinguish two cases, according to the sign of $a_{50} + a_{99}$. Assume first that $a_{50} + a_{99} \geq 0$. Then we have

$$0 \leq a_{50} + a_{99} \leq a_{51} + a_{99} \leq a_{52} + a_{99} \leq \cdots \leq a_{100} + a_{99},$$

providing 51 non-negative sums. On the other hand, for any i so that $50 \leq i \leq 100$, we now have

$$0 \leq a_i + a_{99} \leq a_i + a_{100},$$

providing the new non-negative sums $a_{50} + a_{100}, a_{51} + a_{100}, \cdots, a_{98} + a_{100}$, which is 49 new sums, so we have found 100 non-negative sums. Now assume that $a_{50} + a_{99} < 0$. Then necessarily

$$a_1 + a_2 + \cdots + a_{49} + a_{51} + \cdots + a_{98} + a_{100} > 0. \qquad (1.1)$$

In this case we claim that all sums $a_i + a_{100}$ are non-negative. To see this, it suffices to show that the smallest of them, $a_1 + a_{100}$ is non-negative. And that is true as

$$0 > a_{50} + a_{99} \geq a_{49} + a_{98} \geq a_{48} + a_{97} \geq \cdots \geq a_2 + a_{51},$$

and therefore the left-hand side of (1.1) can be decomposed as the sum of $a_1 + a_{100}$, and 48 negative numbers. So $a_1 + a_{100}$ is positive, and the proof follows.

Second Solution: It is well known from everyday life that one can organize a round robin tournament for $2n$ teams in $2n - 1$ rounds, so that each round consists of n games, and that each team plays a different team each round. A rigorous proof of this fact can be found in Chapter 2, Exercise 4. Now take such a round robin tournament, and replace the teams with the numbers $a_1, a_2, \cdots, a_{100}$. So the fifty games of each round are replaced by fifty pairs of type $a_i + a_j$. As each team plays in each round, the sum of the 100 numbers, or 50 pairs, in any given round is zero. Therefore, at least one pair must have a non-negative sum in any given row, otherwise that row would have a negative sum.

This result is the best possible one: if $a_{100} = 99$, and $a_i = -1$ if $1 \leq i \leq 99$, then there will be exactly 99 non-negative two-element sums.

(9) There is only a finite number of choices for the color of each point, so there is only a finite number F of choices to color the integer points of a 7×7 square. Now take a column built up from $F+1$ squares of size 7×7 that have the same x coordinates. (They are "above one another".) By the Pigeon-hole Principle, two of them must have the very same coloring. This means that if the first one has two points of the same color in the ith and jth positions, then so does the second, and a monochromatic rectangle is formed. The Pigeon-hole Principle ensures that such i and j always exist, and the proof follows. In fact, we also proved that there will always be a monochromatic rectangle whose shorter side contains at most 7 points with integer coordinates.

(10) Consider the remainders of each of the given integers modulo 1000, and the opposites of these remainders modulo 1000. Note that if an integer is not congruent to 0 or 500 modulo 1000, then its remainder and opposite remainder modulo 1000 are two different integers.
We distinguish two cases. First, if at least two of our integers are divisible by 1000, or if at least two of our integers have remainder 500 modulo 1000, then the difference and sum of these two integers are both divisible by 1000, and we are done.
If there is at most one among our integers that is divisible by 1000, and there is at most one among our integers that has remainder 500 modulo 1000, then we have at least 500 integers that do not fall into either category. Consider their remainders and opposite remainders modulo 1000, altogether 1000 numbers. They cannot be equal to 0 or 500, so there are only 998 possibilities for them. Therefore, the Pigeon-hole Principle implies that there must be two equal among them, and the proof follows.

(11) Denote $3n - a$ the largest chosen number (it could be that $a = 0$). Let us add a to all our chosen numbers; this clearly does not change their pairwise differences. So now $3n$ is the largest chosen number. Therefore, if any number from the interval $[n+1, 2n-1]$ is chosen, we are done. Otherwise, we had to choose a total of $n+1$ numbers from the intervals $[1, n]$ and $[2n, 3n-1]$. Consider the n pairs

$$(1, 2n); (2, 2n+1); \cdots ; (i, i+2n-1), \cdots ; (n, 3n-1).$$

As there are n such pairs, and we chose $n+1$ integers, there is one pair with two chosen elements. The difference of those two chosen elements is $2n-1$, and our claim is proved.

(12) Let the numbers of stones in the original four heaps be $a_1 \geq a_2 \geq a_3 \geq a_4$, and let the numbers of stones in the five new heaps be

$b_1 \geq b_2 \geq b_3 \geq b_4 \geq b_5$. Then $a_1 + a_2 + a_3 + a_4 > b_1 + b_2 + b_3 + b_4$. Let k be the smallest index so that $a_1 + \cdots + a_k > b_1 + \cdots + b_k$. (It follows from the previous sentence that there is such an index.) This implies that $a_k > b_k$. Then the stones from the k largest old heaps could not all go to the k largest new heaps. (Indeed, there are too many of them.) In fact, note that $a_1 + \cdots + a_k > b_1 + \cdots + b_{k-1} + 1$. So at least two of these stones had to go to a heap with b_k stones or less, and we are done as $a_1 \geq \cdots \geq a_k > b_k \geq b_{k+1} \geq \cdots \geq b_5$.

(13) Assume the contrary, that is, that each positive integer appears on a finite number of pieces only. As we have an infinite number of pieces, this means that there is an infinite sequence of different positive integers $a_1 < a_2 < a_3 < \cdots$ so that each a_i appears on at least one piece of paper. Then the subsequence $a_1, a_{10^7+1}, a_{2\cdot 10^7+1}, a_{3\cdot 10^7+1}, \cdots$, is an infinite set in which any two elements differ by at least ten million. As all elements of this subsequence appear on some pieces of paper, we have reached a contradiction.

(14)(a) Let a_i denote the number of goals the team scored in the ith game. Consider the 30 numbers $b_i = a_1 + a_2 + \cdots + a_i$ for all i satisfying $1 \leq i \leq 30$, and the 30 numbers $b_i + 6$ for $1 \leq i \leq 30$. This is a collection of 60 numbers, each of which is a positive integer, and none of which is larger than $53 + 6 = 59$. So by the Pigeon-hole Principle, two of these numbers are equal. One of them must be b_i and the other must be $b_j + 6$ for some $j < i$, since all the b_i are different. Then the team scored exactly six goals total in games $j+1, j+2, \cdots, i$.

(b) A counterexample is given by the sequence 2, 1, 2, 2, 3, 2, repeated four more times, for the numbers $a_1, a_2, \cdots$ as defined in the solution of part (a). Another counterexample is given by the sequence 1, 1, 1, 1, 1, 7 repeated four more times.

(c) Let the numbers a_i and b_i be defined as in the solution of part (a). Let us assume that our claim does not hold. Consider the sequence of the ten integers $F = \{1, 7, 13, \cdots, 55\}$. Let B denote the sequence $b_1, b_2, \cdots, b_{30}$.

At most five elements of F can be elements of B since no two *consecutive* elements of F can be part B. Similarly, at most five elements of the sequence $2, 8, \cdots, 56$ can be part of B. The same goes for the sequence $3, 9, \cdots, 57$, the seqeunce $4, 10, \cdots, 58$, and the sequence $5, 11, \cdots, 59$. Therefore, since B consists of 30 positive integers, the largest of which is 59, the sequence of the remaining positive integers not larger than 60, that is, the sequence $6, 12, \cdots, 54$ must contain at least five elements of B. If our claim does not hold, then $6 \notin B$, so the eight-element sequence $S = \{12, 18, \cdots, 54\}$ contains at least five numbers b_i. That means that the there are two *consecutive* elements of S that are part of B, which is a contradiction.

(15) Let $a_1, a_2, \cdots, a_k$ be the initial terms of our k progressions, and let $d_1, d_2, \cdots, d_k$ be their differences. The number $d_1 d_2 \cdots d_k$ is an element of one of these progressions, say, the ith one. Therefore, there is a positive integer m so that

$$d_1 d_2 \cdots d_k = a_i + m d_i,$$

$$d_1 d_2 \cdots d_k - m d_i = a_i.$$

So a_i is divisible by d_i. This problem had nothing to do with the Pigeon-hole Principle. We included it to warn the reader that not all that glitters is gold. Just because we have to prove that one of many objects has a given property, we cannot necessarily use the Pigeon-hole Principle.

Chapter 2

One Step at a Time. The Method of Mathematical Induction

2.1 Weak Induction

Let us assume it is almost midnight, and it has not rained all day today. If, from the fact that it does not rain on a given day, it followed that it will not rain the following day, it would then also follow that it would *never* rain again. Indeed, from the fact that it does not rain today, it would follow that it will not rain tomorrow, from which it would follow that it will not rain the day after tomorrow, and so on.

This simple logic leads to another very powerful tool in mathematics: the method of mathematical induction. We can try to apply this method any time we need to prove a statement for all natural numbers m. Our method then has two steps.

(1) The Initial Step. Prove that the statement is true for the smallest value of m for which it is defined, usually 0 or 1.
(2) The Induction Step. Prove that from the fact that the statement is true for n ("the induction hypothesis"), it follows that the statement is also true for $n + 1$.

If we can complete both of these steps, then we will have proved our statement for all natural values of m. Indeed, suppose not, that is, that we have completed the two steps described above, but still there are some positive integers for which our statement is not true. Let $m + 1$ be the smallest such integer. Then $m + 1$ is not the smallest integer for which our statement is defined, for that would contradict the fact that we completed the Initial Step. So our statement is defined, and therefore, true, for m as $m + 1$ was the smallest integer for which it was false. So our statement is true for m, but false for $m+1$, which contradicts the fact that we completed the Induction Step. Indeed, choosing $m = n$ in the Induction Step yields this contradiction.

Having seen that the method of mathematical induction is a valid one, let us survey some of its applications.

Example 2.1. For all positive integers n,

$$1^2 + 2^2 + \cdots + m^2 = \frac{m(m+1)(2m+1)}{6}. \tag{2.1}$$

Without the method of mathematical induction, we could be in trouble here. The left-hand side is a sum that is *not* an arithmetic series or a geometric series, so we could not use the known formulae for those series. Moreover, the right-hand side looks slightly counter-intuitive; for example, it is not clear how the number 6 will show up in the denominator. The method of mathematical induction, however, solves this problem effortlessly as we will see below.

Solution. (1) The Initial Step. If $m = 1$, then the left-hand side is 1, and so is the right-hand side, so the statement is true.

(2) The Induction Step. Now assume equation (2.1) is true for n, and prove it for $n+1$. The statement for $n+1$ can be obtained from (2.1) by replacing n by $n+1$ and is as follows.

$$1^2 + 2^2 + \cdots + n^2 + (n+1)^2 = \frac{(n+1)(n+2)(2n+3)}{6}. \tag{2.2}$$

To prove (2.2) from (2.1), note that these two equations look pretty much alike; in fact, their difference is a rather simple equation. We are going to prove that this difference is an equation that is in fact an identity. This is true as the difference of the two left-hand sides is clearly $(n+1)^2$, while that of the two right-hand sides is

$$\frac{(n+1)[(n+2)(2n+3) - n(2n+1)]}{6} = (n+1)^2.$$

Therefore, adding the true statements

$$1^2 + 2^2 + \cdots + n^2 = \frac{n(n+1)(2n+1)}{6}$$

and

$$(n+1)^2 = \frac{(n+1)[(n+2)(2n+3) - n(2n+1)]}{6}$$

we get that

$$1^2 + 2^2 + \cdots + n^2 + (n+1)^2 = \frac{(n+1)(n+2)(2n+3)}{6}.$$

Therefore, the statement holds for all positive integers m.

The previous example shows the one serious advantage and one serious disadvantage of the method of mathematical induction. The advantage is that instead of having to prove a general statement, we only have to prove two specific statements. That is, first, we have to complete the initial step, which is usually easy as the substitution $m = 0$ or $m = 1$ usually simplifies the expressions at hand significantly. Then we have to complete the induction step which only involves proving the statement for $n + 1$ *assuming that it is true for* n, which is again usually easier than proving the statement for $n + 1$ without the induction hypothesis.

The drawback will become more apparent after the next example.

Example 2.2. Let $f(m)$ be the maximum number of domains into which m straight lines can divide the plane. Then for all positive integers m, the equality $f(m) = \frac{m(m+1)}{2} + 1$ holds.

It is clear that one straight line always divides the plane into two domains, so $f(1) = 2$, and the initial step is complete. The reader can easily verify that the constructions below are optimal for $m = 2$ and $m = 3$, and therefore $f(2) = 4$, and $f(3) = 7$. This step is not a necessary part of our induction proof, but it helps the reader visualize the problem.

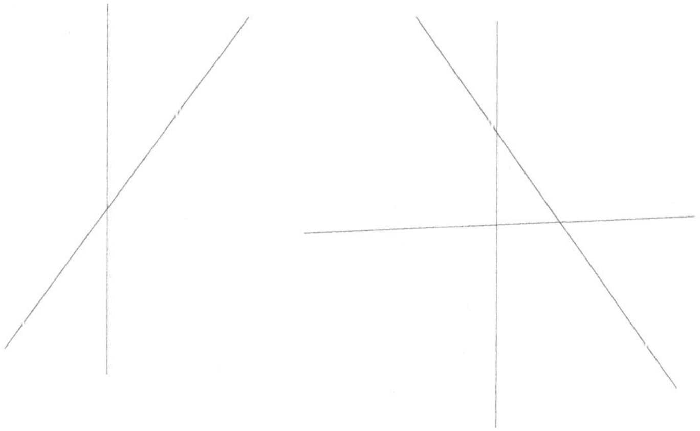

Fig. 2.1 Optimal constructions for $m = 2$ and $m = 3$.

Now let us assume the statement is true for an integer n, and let us prove that it is true for $n + 1$. Let s be one of our $n + 1$ straight lines; we may think of s as the straight line we added to our picture last. Then s

intersects at most n other straight lines, since there are only n other lines in the picture. Denote by $t_1, t_2, \cdots, t_k$ the straight lines that s crosses, in the order it crosses them, in some order. As we said, $k \leq n$ since there are $n+1$ lines altogether. This means that s passes through $k+1$ different domains formed by the other n lines, and cuts each of them into two new domains. Indeed, s cuts through a domain *before* crossing each t_i, and *after* crossing t_k. In other words, s increases the number of domains by $k+1 \leq n+1$. Therefore, we have just proved that $f(n+1) \leq f(n)+n+1$, and equality occurs if and only if s does intersect all the other n lines. Thus $f(n+1) = f(n) + n + 1$. However, the induction hypothesis claims that $f(n) = \frac{n(n+1)}{2} + 1$. Therefore,

$$f(n+1) = f(n) + n + 1 = \frac{n(n+1)}{2} + 1 + n + 1 = \frac{(n+1)(n+2)}{2} + 1,$$

completing the proof.

This proof was possible because we were given a formula for $f(m)$ to prove, just as we were given a formula to prove for the sum of squares in the previous example. Had we been not given these formulae beforehand, first we would have had to guess them, then we could have proved them by the method of mathematical induction. This is the disadvantage of the inductive method we were referring to. However, this guessing is not always hard to do, as the following example shows.

Example 2.3. Let $a_0 = 1$, and let $a_{n+1} = 3a_n + 1$, for all positive integers $n \geq 1$. Find an explicit formula for a_m.

We will learn techniques that enable us to solve problems like this without any guessing. For now, however, let us compute the first few values of the sequence. We get that they are 1, 4, 13, 40, 121. It is easy to conjecture that $a_m = (3^m - 1)/2$. Now we are going to prove our statement by induction. For $m = 1$, the statement is trivially true. Now assume that the statement holds for n. Then

$$a_{n+1} = 3a_n + 1 = \frac{3 \cdot (3^n - 1)}{2} + 1 = \frac{3^{n+1} - 1}{2},$$

so the statement also holds for $n+1$, and the proof follows.

We point out that the formula $a_{n+1} = 3a_n + 1$ in Example 2.3 is called a *recurrence relation*, since it provides a way to compute the value of an element of a sequence from another element of that sequence. The formula $a_{n+1} = \frac{3^{n+1}-1}{2}$ is called an *explicit* formula since it provides a way to compute a_{n+1} directly, without computing other elements of the sequence

$\{a_i\}$ first. It is also called a *closed* formula because it does not contain a $\sum$ or $\prod$ sign. Even more precisely, a closed formula can contain the sum or product of a *fixed number* of parts, but not the sum or product of a *changing number* of parts. The formulas $f(n) = \sum_{i=1}^{n} i$ and $f(n) = (n+1)n/2$ are both explicit, but the former is not closed, while the latter is closed. Induction is often a good way to turn a recurrence relation into an explicit formula, or to turn an explicit, but not closed formula into a closed formula.

Remark. Readers should have a basic understanding of the method of mathematical induction by now, and probably noticed that at the end of the induction proofs, we always choose $m = n$. Therefore, *we will no longer use different variables* for m and n.

For our purposes, a finite *set* is a finite unordered collection of different objects. That is, $\{1, 3, 2\}$ and $\{2, 1, 3\}$ are the same as sets, because they only differ in the order of their elements, and as we said, sets are unordered structures. If an element is allowed to appear more than one time in a collection, such as the element 1 in the collection $(1, 1, 2, 3)$, then that collection is called a *multiset*. We say that the set B is a *subset* of the set A, denoted $B \subseteq A$, if each element of B is also an element of A. In this case it is clear that B has at most as many elements as A.

In combinatorial enumeration, the most important property of a set is *the number of its elements*. Usually, if a statement of enumerative combinatorial nature is true for one set of size n, then it is true for all sets of size n. Therefore, it is permissible, and certainly convenient, to use one example of n-element sets for most of our discussion: that of the first n positive integers, that is, the set $\{1, 2, 3, \cdots, n\}$. As this set will be our canonical example, we introduce the notation $[n] = \{1, 2, 3, \cdots, n\}$ for this set.

Theorem 2.4. *For all positive integers n, the number of all subsets of $[n]$ is 2^n.*

Proof. For $n = 1$, the statement is true as $[1]$ has two subsets, the empty set, and $\{1\}$.

Now assume we know the statement for n, and prove it for $n + 1$. We divide the subsets of $[n + 1]$ into two classes: there will be those subsets that do not contain the element $n + 1$, and there will be those that do. Those that do not contain $n + 1$ are also subsets of $[n]$, so by the induction hypothesis, their number is 2^n. Those that contain $n+1$ consist of $n+1$ and a subset of $[n]$, however, that subset of $[n]$ can be any of the 2^n subsets of

$[n]$, so the number of these subsets of $[n+1]$ is once more 2^n. So altogether, $[n+1]$ has $2^n + 2^n = 2^{n+1}$ subsets, and the theorem is proved. □

With all its strength, the method of induction can also be dangerous if not applied carefully. One common pitfall is to omit a careful proof of the Initial Step, then "prove" a faulty statement by a correct Induction Step. For example, we could "prove" the faulty statement that all positive integers of the form $2n+1$ are divisible by 2, if we could start the induction somewhere, that is, if we could find just one positive integer n for which this property holds. The Induction Step would be easy to complete as $2(n+1)+1-(2n+1) = (2n+1)+2-(2n+1) = 2$ is certainly divisible by 2, the Initial Step, however, cannot be completed.

The following provides an example of a much more subtle fallacy.

We claim that all horses have the same color. As the number of all horses in the world is certainly finite, we can restate our claim as follows. *For any positive integer n, any n horses always have the same color.* And here is our "proof" by induction. For $n = 1$, the statement is obviously true: any one horse has the same color as itself. Now suppose that the statement is true for n, and prove it for $n+1$. Take $n+1$ horses, and line them up. Then the first n horses must have the same color, say black, by the induction hypothesis, but the last n horses also must have the same color, by the same induction hypothesis, so they too must be black as we already have seen that all the first n horses were black, and that included the second, third, fourth,$\cdots$, nth horses, which are also included among the last n horses. Therefore, all $n+1$ horses are black.

It is not so easy to catch the faulty step in this argument because this argument would indeed work for all values of n, except for $n = 1$. When, however, we want to apply this argument to prove that the statement holds for two horses using the fact that it holds for one horse, we encounter insurmountable difficulties. The reason for this is simple: in this case the "first n horses" simply means the first horse, while the "last n horses" means the last horse. These two sets have no intersection, so nothing forces the color of the horse in the first set to be the same as that of the horse in the second one!

This shows that we must be careful that our Induction Step is correct for all values of n at least as large as the value used in the Initial Step.

Our argument proves that if any two horses did have the same color, then all horses would have the same color, but that result would be a horse of a different color.

2.2 Strong Induction

Example 2.5. Let the sequence $\{a_n\}$ be defined by the relations $a_0 = 0$, and $a_{n+1} = a_0 + a_1 + a_2 + \cdots + a_n + n + 1$ if $n \geq 0$. Prove that for all positive integers n, the equality $a_n = 2^n - 1$ holds.

Here we certainly could not hope to prove our statement by our usual way of induction. Indeed, a_{n+1} depends not only on a_n, but also on $a_{n-1}, a_{n-2}, \cdots, a_1$, so simply using the fact that $a_{n-1} = 2^{n-1} - 1$ cannot be sufficient.

Solution. (of Example 2.5) As $a_0 = 0$, the initial case is true. Now let us assume that we know that the statement is true for all positive integers less than or equal to n. Then, by our recurrence relation,

$$a_{n+1} = a_0 + \cdots + a_n + n + 1 = (2^0 - 1) + \cdots + (2^n - 1) + n + 1$$
$$1 + 2 + 4 + \cdots + 2^n = 2^{n+1} - 1.$$

This shows that our explicit formula is correct for $n + 1$, and the proof is complete.

Note that if we remove a_0 from our sequence $\{a_n\}$, we get a geometric series.

Let us review the steps of this *strong induction* algorithm.

(1) The Initial Step. Prove that the statement is true for the smallest value of n for which it is defined, usually 0 or 1.
(2) The Induction Step. Prove that from the fact that the statement is true for all integers less than $n + 1$ ("the induction hypothesis"), it follows that the statement is also true for $n + 1$.

Just as in the case of weak induction, if we can complete both of these steps, then we will have proved our statement for all natural numbers n. Indeed, suppose not, that is, that we have completed the two steps described above, but still there are some positive integers for which our statement is not true. Let $n + 1$ be the smallest such integer. Then $n + 1$ is not the smallest integer for which our statement is defined, for that would contradict the fact that we completed the Initial Step. So our statement is defined, and therefore, true, for all integers less than or equal to n, because $n + 1$ was the smallest integer for which it was false. So our statement is true for all integers less than or equal to n, but false for $n + 1$, which contradicts the fact that we completed the Induction Step.

Let us see one more application of the strong induction algorithm. For the rest of this book, denote **N** the set of natural numbers, that is, the set of non-negative integers.

Example 2.6. Let $f(0) = 1$, let $f(1) = 2$, and let $f(n+1) = f(n-1) + 2f(n)$ if $n \geq 1$. Prove that then $f(n) \leq 3^n$ for all $n \in N$.

Solution. It follows from the conditions that the statement is true for $n = 0$ and $n = 1$. Now let us assume that the statement is true for all non-negative integers that are less than or equal to n, and prove it for $n + 1$. For $n \geq 1$, we have

$$\begin{aligned} f(n+1) &= f(n-1) + 2f(n) \\ &\leq 3^{n-1} + 2 \cdot 3^n \\ &= 7 \cdot 3^{n-1} \\ &< 3^{n+1}, \end{aligned}$$

and our induction proof is complete. Note that we have used the induction hypothesis when passing from the first line to the second. Also note that we did need the strong induction hypothesis in that we needed the *both* inequalities $f(n-1) \leq 3^{n-1}$ and $f(n) \leq 3^n$ in order to complete that step.

Notes

It is sometimes convenient to shift the parameters in an induction proof. This means that the Induction Step involves assuming the statement for $n - 1$, and proving it for n (in the weak case), or assuming the statement for all integers less than n, and proving it for n. It can also happen that we want to prove some property of *even* integers, or *odd* integers, in which case we would have to adjust our Induction Step accordingly. We will see some of these variations of the method of mathematical induction in the upcoming chapters of this book.

Exercises

(1) (+) Let $p(k)$ be a polynomial of degree d. Prove that $q(n) = \sum_{k=1}^{n} p(k)$ is a polynomial of degree $d + 1$. Prove that this polynomial q satisfies $q(0) = 0$.

(2) At a tennis tournament, every two players played against each other exactly one time. After all games were over, each player listed the names of those he defeated, and the names of those defeated by someone he defeated. Prove that at least one player listed the names of everybody else.

(3) At a tennis tournament, there were 2^n participants, and any two of them played against each other exactly one time. Prove that we can find $n+1$ players that can form a line in which everybody has defeated all the players who are behind him in the line.

(4) Prove that for all positive integers n, it is possible to organize a round robin tournament of n football teams in

   (a) $n-1$ rounds if n is even,
   (b) n rounds if n is odd.

   A round is a set of games in which each team plays one opponent if n is even, and there is only one idle team if n is odd. A round-robin tournament is a tournament in which any pair of teams meet exactly once.

(5) Let $a_0 = 1$, and let $a_{n+1} = 3a_n + 2$, for all non-negative integers n. Prove that $a_n = 2 \cdot 3^n - 1$.

(6) Let $a_0 = 1$, and let $a_{n+1} = 4a_n - 1$, for all non-negative integers n. Prove that $a_n = \frac{2 \cdot 4^n + 1}{3}$.

(7) Let $a_0 = 1$, and let $a_{n+1} = 2\sum_{i=0}^{n} a_i$ for all non-negative integers n. Find an explicit formula for a_n.

(8) There are n patients waiting in a doctor's office. Each of them took a number, from 1 to n. The patients are told that they will not necessarily be called in the order their numbers would indicate, but nobody will be preceded by more patients than he would be if the order of their numbers were strictly respected. That is, the patient holding number i will be preceded by at most $i-1$ patients.
When Mr. Jones heard this, he said, "This is just the same as respecting the order of the numbers." Was he right?

(9) Prove that for all natural numbers n, the number $a(n) = n^3 + 11n$ is divisible by 6.

(10) Prove that $3^n > n^4$ if $n \geq 8$.

(11) Prove that if n is a positive integer, then $8^n - 14n + 27$ is divisible by 7.

(12) We cut a square into four smaller squares, then we cut some of the obtained small squares into four smaller squares, and so on. Prove that

at any given point of time during this operation, the number of all squares we have is of the form $3m+1$.

(13) (Some calculus required.) Recall that $n! = 1 \cdot 2 \cdot \cdots \cdot n$. Prove that for all positive integers n, the inequality $n! > \frac{n^n}{3^n}$ holds.

(14) Prove that there exists a positive integer N so that if $n > N$, then the inequality

$$n! < \frac{n^n}{(2.5)^n}$$

holds.

(15) (+) Give an induction proof for the inequality between the geometric and the arithmetic mean, that is, prove that if $a_1, a_2, \cdots, a_n$ are non-negative numbers, then

$$\sqrt[n]{a_1 a_2 \cdots a_n} \leq \frac{a_1 + a_2 + \cdots + a_n}{n}. \tag{2.3}$$

(16) (+) Give an induction proof for the inequality between the harmonic mean and the geometric mean, that is, prove that if $a_1, a_2, \cdots, a_n$ are positive real numbers, then

$$\frac{n}{\frac{1}{a_1} + \frac{1}{a_2} + \cdots + \frac{1}{a_n}} \leq \sqrt[n]{a_1 a_2 \cdots a_n}.$$

Supplementary Exercises

(17) (-) Prove that for all positive integers n, we have

$$1 + 3 + \cdots + (2n-1) = n^2.$$

(18) (-) Let n be a positive integer. Prove that it is possible to cut up a cube into $7n+1$ smaller cubes.

(19) (-) Let $a_1 = 3$, and let $a_n = a_1 \cdot a_2 \cdot \cdots \cdot a_{n-1} + 2$ for $n \geq 2$. Prove that $a_n = 2^n + 1$.

(20) (-) Prove that

$$1 \cdot 2 + 2 \cdot 3 + \cdots + (n-1)n = \frac{(n-1)n(n+1)}{3}.$$

(21) (-) Prove by induction that the sum of the angles of a convex n-gon is $(n-2)180$ degrees.

(22) Prove that for all positive integers n,

$$1^3 + 2^3 + \cdots + n^3 = (1 + 2 + \cdots + n)^2. \tag{2.4}$$

(23) Prove that for all positive integers n,
$$2(1+2+\cdots+n)^4 = (1^5+2^5+\cdots+n^5)+(1^7+2^7+\cdots+n^7).$$
(24) Find a closed formula (no summation signs) for the expression $\sum_{i=1}^{n} i(i+1)$.
(25) Let $a_0 = 1$, and let $a_{n+1} = 10a_n - 1$. Prove that for all $n \geq 1$, the equality $a_n = (8 \cdot 10^n + 1)/9$ holds.
(26) Let $a_0 = 1$, and let $a_{n+1} = 10a_n - 3$. Find an explicit formula for a_n.
(27) Let $a_0 = 3$, and let $a_{n+1} = \sqrt{a_n + 7}$ if $n > 0$. Prove that $3 < a_n < 4$ for all $n > 0$.
(28) Let $a_0 = 0$, $a_1 = 1$, and let $a_{n+2} = 6a_{n+1} - 9a_n$ for $n \geq 0$. Prove that $a_n = n \cdot 3^{n-1}$ for all $n \geq 0$.
(29) Let $a_0 = a_1 = 1$, and let $a_{n+2} = a_{n+1} + 5a_n$ for $n \geq 0$. Prove that $a_n \leq 3^n$ for all $n \geq 0$.
(30) Let H be a ten-element set of two-digit positive integers. Prove that H has two disjoint subsets A and B so that the sum of the elements of A is equal to the sum of the elements of B.
(31) Prove that a positive integer is divisible by 3 if and only if the sum of its digits is divisible by 3.
(32) Let $a_1, a_2, \cdots, a_n$ be the digits of a positive integer m, from left to right. Prove that m is divisible by 11 if and only if $a_1 - a_2 + a_3 - \cdots + (-1)^{n-1}a_n$ is divisible by 11.
(33) Let $a_1 = 5$, and let $a_{n+1} = a_n^2$. Prove that the last n digits of a_n are the same as the last n digits of a_{n+1}.
(34) Prove that for any positive integer n, it is possible to partition any triangle T into $3n+1$ similar triangles.
(35) Let $n > 14$ be an integer. Prove that a square can be partitioned into n smaller squares.
(36) Prove that if $n > 2$ is a natural number, then n can be written as a product of primes.
(37) Define a function μ on the set of non-negative integers as follows. Let $\mu(1) = 1$, and let $\mu(n) = 0$ if $n > 1$ and n is divisible by the square of an integer $a > 1$. Otherwise, if $n = p_1p_2\cdots p_k$, where the p_i are all distinct primes, then let $\mu(n) = (-1)^k$. Use induction to prove that for all positive integers $m > 1$,
$$Z_n = \sum_{d|n} \mu(d) = 0.$$
The summation is taken over all positive *divisors* d of n. (This is what $d|n$ denotes.)

Solutions to Exercises

(1) We prove the statement by strong induction on d. If $d = 0$, then p is a constant polynomial, say $p = c$. Then $\sum_{i=1}^{n} p(i) = nc$, and the statement is true.
Now let us assume that we know the statement for all polynomials of degree less than d, and let p be a polynomial of degree d. First we claim that it suffices to prove our statement for the polynomial $p(d) = n^d$. Let $a_0, a_1, \cdots, a_d$ be real numbers, with $a_d \neq 0$. Then the statement is true for the polynomial n^d if and only if it is true for the polynomial $a_d n^d$. Moreover, the statement is true for the polynomial $a_d n^d$ if and only if it is true for the polynomial $h(n) = a_d n^d + a_{d-1}n^{d-1} + \cdots + a_1 d + a_0$. Indeed, $r(n) = a_{d-1}n^{d-1} + \cdots + a_1 d + a_0$ is a polynomial of degree $d-1$, so the induction hypothesis implies that $\sum_{i=1}^{n} r(i)$ is a polynomial of n of degree at most d. Therefore,

$$\sum_{i=1}^{n} h(i) - \sum_{i=1}^{n} a_d i^d = \sum_{i=1}^{n} r(i)$$

is a polynomial of degree at most d.
To prove that the statement is true for n^d, it suffices to show that there exists a polynomial $z(n)$ of degree $d+1$ so that $z(n+1) - z(n) = n^d$ for all positive integers n, and $z(0) = 0$. That will imply that

$$\begin{aligned} 1^d + 2^d + \cdots + n^d &= (z(1) - z(0)) + \cdots + (z(n+1) - z(n)) \\ &= z(n+1) - z(0) \\ &= z(n+1). \end{aligned}$$

Finally, in order to prove that such a polynomial $z(n)$ exists, let us recall that $(n+1)^{d+1} - n^{d+1}$ is a polynomial of degree d. This is not exactly what we want, that is, the polynomial n^d. However, using the induction hypothesis just as we did in the previous paragraph, it is easy to show that this implies the existence of $z(n)$.

(2) **First solution.** We claim that the winner of the tournament (or any winner, if there is a tie at the top) always lists the names of everyone else. Indeed, suppose W is a winner of the tournament, that is, he won k games, and nobody won more than k games. Now assume there is a player P whose name W did not list. That means that P defeated W, and P also defeated all the k players W defeated. So P won at least $k+1$ games, which is a contradiction.

Second solution. Induction on n, the number of players at the tournament. If $n = 2$, the statement is true, for the player who won the sole game lists the name of his opponent. Now assume the statement is true for n, and take a tournament with $n + 1$ players. Call the player with the smallest number of victories A. (If there is a tie at the bottom, any player from that tie will do.) If we temporarily disregard A, we have n players left, so by the induction hypothesis there will be one of them, say B, who will list the names of the other $n-1$ players. Now if B defeated A, or if anyone defeated by B defeated A, then B lists the name of A, too, and we are done. If not, then A has defeated B, and all the players defeated by B, so A won more games than B, a contradiction.

(3) Induction on n. For $n = 1$, the statement is trivially true. Now assume the statement is true for n and prove it for $n + 1$. The winner X of a tournament with 2^{n+1} games must have won at least 2^n games (why?). Take X, and 2^n people he defeated. By the induction hypothesis, we can find $n + 1$ people among the 2^n people defeated by X who can form a line with the required property. Then we put X to the front of this line and we have obtained a line of length $n + 2$ that has the required property.

(4) We are going to prove the statement by strong induction on n. For $n = 1, 2$, the statement is trivially true. Now assume that we know the statement for all positive integers less than $n + 1$, and prove it for $n + 1$.

First, we claim that we can assume that $n + 1$ is even. Indeed, if $n + 1$ is odd, then we can add one more player to the tournament, and have an even number of players. Once we have our round robin tournament, we can simply take away the extra player, and say that his opponent has a bye in each round.

Thus $n + 1$ is even. We distinguish two cases.

- First assume that $n+1 = 4k$. Let us split our group of players into two groups of size $2k$ each. Have both groups play a round-robin tournament. By the induction hypothesis, that is possible in $2k-1$ rounds. Then denote the players in the two groups $a_1, a_2, \cdots, a_{2k}$ and $b_1, b_2, \cdots, b_{2k}$. Have them play $2k$ rounds as follows. In the first round, a_i plays b_i. In the second round a_i plays b_{i+1}, modulo $2k$, that is, a_{2k} plays b_1. Continue this way, in round j, a_i will play b_{i+j}. This completes a round robin tournament in $4k - 1 = n$

rounds, as claimed.

- If $n+1 = 4k+2$, then again split the group of players into two groups of size $2k+1$ each. Proceed as before, except that when the groups play their tournaments, there will be an idle player in each of them, in each round. Have those two play each other.

(5) The statement is true for $n = 0$. Now assume it is true for n, and prove it for $n+1$. We know that $a_{n+1} = 3a_n + 2$. By our induction hypothesis, we have $a_n = 2 \cdot 3^n - 1$. Substituting this for a_n, we get $a_{n+1} = 3 \cdot (2 \cdot 3^n - 1) + 2 = 2 \cdot 3^{n+1} - 3 + 2 = 2 \cdot 3^{n+1} - 1$, and the statement is proved.

(6) The statement is true for $n = 0$. Now assume it is true for n, and prove it for $n+1$. We know that $a_{n+1} = 4a_n - 1$. By our induction hypothesis, $a_n = \frac{2 \cdot 4^n + 1}{3}$. Substituting this for a_n, we get

$$a_{n+1} = 4 \cdot \frac{2 \cdot 4^n + 1}{3} - 1 = \frac{2 \cdot 4^{n+1} + 4}{3} - 1 = \frac{2 \cdot 4^{n+1} + 1}{3},$$

which was to be proved.

(7) Computing the first few elements, we find that $a_0 = 1$, $a_1 = 2$, $a_2 = 6$, $a_3 = 18$, $a_4 = 54$, and so on. This seems to suggest that $a_n = 2 \cdot 3^{n-1}$ if $n \geq 1$. We prove this by strong induction on n. The initial case is true. Now assume we know the statement for all positive integers less than or equal to n. Then, by our recurrence relation,

$$\begin{aligned} a_{n+1} &= 2a_0 + 2a_1 + \cdots + 2a_n \\ &= 2 + 2(2 + 6 + \cdots + 2 \cdot 3^{n-1}) \\ &= 2 + 4 \cdot \frac{3^n - 1}{2} \\ &= 2 \cdot 3^n. \end{aligned}$$

This proves that our explicit formula is correct for $n+1$, and the proof is complete.

(8) Yes, he was. Let us identify the patients by their numbers, and let $f(i)$ be the function that tells when patient i is called. Then we have to prove that the only one-to-one function $f : \{1, 2, \cdots, n\} \rightarrow \{1, 2, \cdots, n\}$ that satisfies $f(i) \leq i$ for all i is the identity function. (That is, the function defined by $f(i) = i$ for all i.) Note that a one-to-one function between two sets of the same size is necessarily onto. A function that is both one-to-one and onto is called a bijection. We will use bijections often in later chapters. We will then explain these words, though we suspect you heard them before.

We prove our statement by induction on n. The statement is obviously true for $n = 1$. Now assume we know that the statement is true for n, and prove it for $n+1$. Let $f : \{1, 2, \cdots, n+1\} \rightarrow \{1, 2, \cdots, n+1\}$ be a bijection that satisfies $f(i) \leq i$ for all i. Then we must have $f(n+1) = n+1$. Indeed, there has to be an i so that $f(i) = n+1$, and if this i is not $n+1$, then the condition $f(i) \leq i$ is violated. So $f(n+1) = n+1$. This means that f maps the set $\{1, 2, \cdots, n\}$ onto the set $\{1, 2, \cdots, n\}$, and of course, satisfies $f(i) \leq i$. However, the induction hypothesis then says that $f(i) = i$ for all $i \leq n$, and the statement follows.

(9) As $a(0) = 0$, the initial step is complete. Now assume we know that the statement is true for n, and prove it for $n+1$. As $a(n)$ is divisible by six, it suffices to show that $a(n+1) - a(n)$ is divisible by six, and that will prove that so is $a(n+1)$. Indeed,

$$\begin{aligned} a(n+1) - a(n) &= (n+1)^3 + 11(n+1) - n^3 - 11n \\ &= 3n^2 + 3n + 1 + 11 \\ &= 3(n^2 + n + 4), \end{aligned}$$

and the statement follows as $n^2 + n$ is always an even number.

(10) The statement is true for $n = 8$. Indeed, $3^8 = 9^4 > 8^4$. This will be our initial step. Now assume that we know that the statement is true for n (where $n \geq 8$). We then have to prove that $b_{n+1} = \frac{3^{n+1}}{(n+1)^4} > 1$. We know that $b_n > 1$, and that

$$b_{n+1} = b_n \cdot 3 \cdot \left(\frac{n}{n+1}\right)^4.$$

Therefore, to show that $b_{n+1} > 1$, it suffices to show that $(\frac{n}{n+1})^4 > \frac{1}{3}$ when $n \geq 8$. As $(\frac{n}{n+1})^4 = (1 - \frac{1}{n+1})^4$ obviously grows when n grows, it suffices to show that this holds when $n = 8$. Indeed, $\left(\frac{8}{9}\right)^4 = 0.624 > \frac{1}{3}$.

(11) Let $a_n = 8^n - 14n + 27$. Then $a_1 = 21$ is divisible by seven. Now assume the statement is true for n, and prove it for $n+1$. To do that, it suffices to show that $a_{n+1} - a_n$ is divisible by seven. One verifies easily that $a_{n+1} - a_n = 8^{n+1} - 14(n+1) - 8^n - 14n = 7 \cdot 8^n - 14 = 7(8^n - 2)$, which is always divisible by seven.

(12) We prove the statement by induction on the number n of squares that have been cut up. When $n = 0$, then we have one square, and the statement is true. Now assume the statement is true for n, and prove it for $n+1$. At step $n+1$, we cut up one additional square. This increases the number of all squares by three, so if that number was

of the form $3m+1$, now it is of the form $3m+4=3(m+1)+1$. This proves our claim. A little additional thought shows that in fact, $n=m$, that is, after we cut up n squares, we have $3n+1$ squares.

(13) Let $a_n = \frac{n!}{(n/3)^n}$. We have to prove that $a_n > 1$. If $n=1$, then we have $a_1 = 3$, and the statement is true. Assume the statement is true for n. To prove it for $n+1$, we show that $a_{n+1}/a_n > 1$. Indeed,

$$\frac{a_{n+1}}{a_n} = \frac{3^{n+1}\cdot(n+1)!}{(n+1)^{n+1}} \cdot \frac{n^n}{3^n \cdot n!} = 3\cdot\left(\frac{n}{n+1}\right)^n.$$

It is a well-known fact in Calculus that the sequence $\left(\frac{n}{n+1}\right)^n$ is decreasing and converges to $1/e$. In particular, it is always larger than $1/e$, let alone $1/3$, and our statement is proved.

(14) Let $b_n = \frac{n!}{(n/2.5)^n}$. Then we compute

$$\frac{b_{n+1}}{b_n} = \frac{2.5^{n+1}\cdot(n+1)!}{(n+1)^{n+1}} \cdot \frac{n^n}{2.5^n \cdot n!} = 2.5\cdot\left(\frac{n}{n+1}\right)^n.$$

As the sequence $c_n = (\frac{n}{n+1})^n$ is decreasing, the ratio $\frac{b_{n+1}}{b_n} = 2.5c_n$ is decreasing with n. Moreover, $c_n \to 1/e$, so there exists an integer m such that if $n > m$, then $\frac{b_{n+1}}{b_n} < \frac{2.5}{2.6}$. As $(\frac{2.5}{2.6})^n$ converges to 0, it follows that eventually, we will have an N so that $b_N < 1$, and the proof follows by induction.

(15) We prove the statement by induction on n. For $n=1$, the statement is trivially true. Now assume we know that the statement is true for all integers less than n, and prove it for n.
Assume first that n is even, say $n=2k$. Then apply this same inequality for the numbers $a_1, \cdots, a_k$ and $a_{k+1}, \cdots, a_{2k}$. As $k<n$, we know by the induction hypothesis that for both sets of numbers, the geometric mean is at most as large as the arithmetic mean. Replace each of the numbers $a_1, \cdots, a_k$ by their arithmetic mean A, and replace each of the numbers $a_{k+1}, \cdots, a_{2k}$ by their arithmetic mean B. Then the left-hand side of (2.3) increases, while the right-hand side does not change. For our new sets of numbers, the inequality between the geometric and arithmetic means is the following.

$$\sqrt[2k]{A^k B^k} \leq \frac{k(A+B)}{2k}. \tag{2.5}$$

Note if we can prove (2.5), we will also get a proof of our original inequality (2.3). Indeed, (2.5) was obtained from (2.3) by increasing the left-hand side and leaving the right-hand side unchanged. Therefore (2.5) implies (2.3).

To see that (2.5) holds, note that (2.5) simplifies to

$$\sqrt{AB} \leq \frac{A+B}{2},$$

$$0 \leq (A-B)^2.$$

If n is odd, then assume without loss of generality that a_n is maximal among the a_i. Replace the numbers $a_1, a_2, \cdots, a_{n-1}$ with their arithmetic mean C. By the induction hypothesis, this is larger than their geometric mean. Therefore, this operation increases the left-hand side of (2.3) or leaves it the same, and leaves the right-hand side unchanged. Just as in the case of even n, we have turned our inequality into a sharper one, namely

$$\sqrt[n]{C^{n-1}a_n} \leq \frac{(n-1)C + a_n}{n}.$$

Again, it suffices to prove this inequality as it implies (2.3). Let us prove this inequality. As $a_n \geq C$, the arithmetic mean $\frac{(n-1)C+a_n}{n}$ is at distance d from C, and distance $(n-1)d$ from a_n. We will modify our numbers so that the left-hand side increases and the right-hand side does not change. We will do this in $n-1$ steps, and in each step, we will change two numbers, one of which will always be the maximal number. First we take one of our $n-1$ copies of C, add d to it, and subtract this d from a_n. Clearly, the sum, and therefore, the arithmetic mean of our numbers did not change. On the other hand, their geometric mean grew as $Ca_n \leq (C+d)(a_n - d)$. Then add d to another copy of C, and subtract d from $a_n - d$, and so on. After $n-1$ steps, all our entries are equal to $C+d$. So raising the geometric mean and keeping the arithmetic mean unchanged, we reach a point where these two are equal. This shows that the geometric mean could not be larger than the arithmetic mean.

Remark. In the second case, we have not used the fact that n was odd, so we could have done the whole proof with just that method. It would have been faster, but we wanted to show the nice trick of splitting the set of our numbers into two subsets. If n is not even, but not prime, the same method would have worked. We just would have had to split the set of our numbers into k equal parts, where k is a prime divisor of n.

(16) This is similar to the solution of the previous exercise. The only difference is that we substitute the relevant sets of numbers by their geometric means, not their arithmetic means.

Chapter 3

There Are A Lot Of Them. Elementary Counting Problems

In the first two chapters, we have explained how to use the Pigeon-hole Principle and the method of mathematical induction to draw conclusions from certain numbers. However, *to find* those numbers is not always easy. It is high time that we learned some fundamental counting techniques.

3.1 Permutations

Let us assume that n people arrived at a dentist's office at the same time. The dentist will treat them one by one, so they must first decide the order in which they will be served. How many different orders are possible?

This problem, that is, arranging different objects linearly, is so omnipresent in combinatorics that we will have a name for both the arrangements and the number of arrangements. However, we are going to answer the question first.

Certainly, there are n choices for the person who will indulge in dental pleasures first. How many choices are there for the person who goes second? There are only $n-1$ choices as the person who went first will not go second, but everybody else can.

The crucial observation now is that for *each of the* n *choices* for the patient to be seen first, we have $n-1$ *choices for the patient who will be second.* Therefore, we have $n(n-1)$ ways to select these two patients. If you do not believe this, try it out with four patients, called A, B, C, and D, and you will see that there are indeed 12 ways the first two lucky patients can be chosen.

We can then proceed in a similar manner: we have $n-2$ choices for the patient to be seen third as the first two patients no longer need to be seen. Then we have $n-3$ choices for the patient to be seen fourth, and so on, two choices for the patient to be seen next-to-last, and only one choice, the remaining, frightened patient, to be seen last. Therefore, the number of orders in which the patients can sit down in the dentist's chair is $n \cdot (n-1) \cdot (n-2) \cdots 2 \cdot 1$.

Definition 3.1. The arrangement of different objects into a linear order using each object exactly once is called a *permutation* of these objects. The number $n \cdot (n-1) \cdot (n-2) \cdots 2 \cdot 1$ of all permutations of n objects is called n *factorial*, and is denoted by $n!$.

So we have just proved the following basic theorem.

Theorem 3.2. *The number of all permutations of an n-element set is $n!$.*

We note that by convention, $0! = 1$. If you really want to know why we choose $0!$ to be 1, and not, say, 0, here is an answer. Assume there are n people in a room and m people in another room. How many ways are there for people in the first room to form a line and people in the second room to form a line? The answer is, of course, $n! \cdot m!$ as any line in the first room is possible with any line in the second room. Now look at the special case of $n = 0$. Then people in the second room can still form $m!$ different lines. Therefore, if we want our answer, $n!m!$ to be correct in this singular case too, we must choose $0! = 1$. You will soon see that there are plenty of other situations that show that $0! = 1$ is the good definition.

The number $n!$ is extremely important in combinatorial enumeration, as you will see throughout this book. You may wonder how large this number is, in terms of n. This question can be answered at various levels of precision. All answers that are at least somewhat precise require advanced calculus. Here we will just mention, without proof that

$$n! \sim \sqrt{2\pi n}\left(\frac{n}{e}\right)^n. \tag{3.1}$$

The symbol $n! \sim z(n)$ sign means that $\lim_{n\to\infty} \frac{n!}{z(n)} = 1$. Relation (3.1) is called *Stirling's formula*, and we will use it in several later chapters.

Example 3.3. How many different flags can we construct using colors red, white, and green if all flags must consist of three horizontal stripes of different colors?

Solution. By Theorem 3.2, the answer is $3! = 3 \cdot 2 \cdot 1 = 6$. It is easy to convince ourselves that this is indeed correct by listing all six flags: RWG, RGW, WRG, WGR and GWR, and GRW.

The simplicity of the answer to the previous question was due to several factors: we used each of our objects exactly once, the order of the objects mattered, and the objects were all different. In the rest of this section we will study problems *without* one or more of these simplifying factors.

Example 3.4. A gardener has five red flowers, three yellow flowers and two white flowers to plant in a row. In how many different ways can she do that?

This problem differs from the previous one in only one aspect: the objects are *not all different.* The collection of the five red, three yellow, and two white flowers is often called a *multiset.* A linear order that contains all the elements of a multiset exactly once is called a *multiset permutation.*

How many permutations does our multiset have? We are going to answer this question by reducing it to the previous one, in which all objects were different. Assume our gardener plants her flowers in a row, in any of A different ways, then sticks labels (say numbers 1 through 5 for the red flowers, 1 through 3 for the yellow ones, and 1 through 2 for the white ones) to her flowers so that she can distinguish them. Now she has ten different flowers, and therefore the row of flowers she has just finished working on can look in 10! different ways. We have to tell how many of these arrangements differ *only* because of these labels.

The five red flowers could be given five different labels in 5! different ways. The three yellow flowers could be given three different labels in 3! different ways. The two white flowers could be given two different labels in 2! different ways. Moreover, the labeling of flowers of different colors can be done independently of each other. Therefore, the labeling of all ten flowers can be done in $5! \cdot 3! \cdot 2!$ different ways once the flowers are planted in any of A different ways. Therefore, $A \cdot 5! \cdot 3! \cdot 2! = 10!$, or, in other words,

$$A = \frac{10!}{5! \cdot 3! \cdot 2!} = 2520.$$

This argument can easily be generalized to a general theorem. However, we will need a greater level of abstraction in our notations to achieve that. This is because we will take general variables for the number of objects, but also for the number of *different kinds* of objects. In other words, instead of saying that we have five red flowers, three yellow flowers, and two white flowers, we will allow flowers of k different colors, and we will say that there are a_1 flowers of the first color, a_2 flowers of the second color, a_3 flowers of the third color, and so on. We complete the set of these conditions by saying that we have a_k flowers of color k (or a_k flowers of the kth color).

This is a long set of conditions, so some shorter way of expressing it would certainly make it less painful. We will achieve this by saying that we have a_i flowers of color i, for all $i \in [k]$. Instead of saying that we plant our flowers in a line, we will often say that we *linearly order* our objects.

Now we are in a position to state our general theorem.

Theorem 3.5. *Let $n, k, a_1, a_2, \cdots, a_k$ be non-negative integers satisfying $a_1 + a_2 + \cdots + a_k = n$. Consider a multiset of n objects, in which a_i objects are of type i, for all $i \in [k]$. Then the number of ways to linearly order these objects is*

$$\frac{n!}{a_1! \cdot a_2! \cdot \cdots \cdot a_k!}.$$

Proof. This is a generalization of Example 3.4, and the same idea of proof works here. The reader should work out the details. □

3.2 Strings over a Finite Alphabet

Now we are going to study problems in which we are not simply arranging certain objects, knowing how many times we can use each object, but rather construct strings, or words, from a finite set of symbols, which we call a finite *alphabet*. We will *not* require that each symbol occur a specific number of times; though we may require that each symbol occur *at most once*.

Theorem 3.6. *The number of k-digit strings one can form over an n-element alphabet is n^k.*

Proof. We can choose the first digit in n different ways. Then, we can choose the second digit in n different ways as well since we are not forbidden to use the same digit again (unlike in case of permutations). Similarly, we can choose the third, fourth, etc., kth element in n different ways. We can make all these choices independently from each other, so the total number of choices is n^k. □

Example 3.7. The number of k-digit positive integers is $9 \cdot 10^{k-1}$.

Solution. There are two ways one can see this. From Theorem 3.6, we know that the number of k-digit strings that can be made up from the alphabet $\{0, 1, \cdots, 9\}$ is 10^k. However, not all these yield a k-digit positive integer. Indeed, those with first digit 0 do not. What is the number of these *bad* strings? Disregarding their first digit, these strings are $(k-1)$-digit strings over $\{0, 1, \cdots, 9\}$ with no restriction, so Theorem 3.6 shows that there are 10^{k-1} of them. Therefore, the number of k-digit strings that do not start with 0, in other words, the number of k-digit positive integers is $10^k - 10^{k-1} = 9 \cdot 10^{k-1}$ as claimed.

Alternatively, we could argue as follows. We have 9 choices for the first digit (everything but 0), and ten choices for each of the remaining $k-1$ digits. Therefore, the number of total choices is $9 \cdot 10 \cdot 10 \cdot \cdots \cdot 10 = 9 \cdot 10^{k-1}$, just as in the previous argument.

Before we discuss our next example, we mention a general technique in enumeration, the method of *bijections*. Suppose there are many men and many women in a huge ballroom. We do not know the number of men, but we know that the number of women is exactly 253. Suppose we think that the number of men is also 253, but we are not sure. What is a fast way to test this conjecture? We can ask the men and women to form man-woman pairs. If they succeed in doing this, that is, nobody is left without a match, and everyone has a match of the opposite gender, then we know that the number of men is 253 as well. If not, then there are two possibilities: if some man did not find a woman for himself, then the number of men is more than 253. If some woman did not find a man, then the number of men is less than 253.

This technique of matching two sets element-wise and then conclude (in case of success) that the sets are equinumerous is very often used in combinatorial enumeration. Let us put it in a more formal context.

Definition 3.8. Let X and Y be two finite sets, and let $f : X \to Y$ be a function so that

(1) if $f(a) = f(b)$, then $a = b$, and
(2) for all $y \in Y$ there is an $x \in X$ so that $f(x) = y$,

then we say that f is a *bijection* from X onto Y. Equivalently, f is a bijection if for all $y \in Y$, there exists a *unique* $x \in X$ so that $f(x) = y$.

In other words, a bijection matches the elements of X with the elements of Y, so that each element will have exactly one match.

The functions that have only one of the two defining properties of bijections also have their own names.

Definition 3.9. Let $f : X \to Y$ be a function. If f satisfies criterion (1) of Definition 3.8, then we say that f is *one-to-one* or *injective*, or is an *injection*. If f satisfies criterion (2) of Definition 3.8, then we say that f is *onto* or *surjective*, or is a *surjection*.

Proposition 3.10. *Let X and Y be two finite sets. If there exists a bijection f from X onto Y, then X and Y have the same number of elements.*

Proof. The bijection f matches elements of X to elements of Y, in other words it creates pairs with one element from X and one from Y in each pair. Say f created m pairs, then both X and Y have m elements. □

The advantages of the bijective method are significant. Instead of enumerating the elements of X, we can enumerate the elements of Y if that is easier. Then, we can find a bijection from X onto Y. Let us illustrate this by computing the number of all subsets of $[n]$ without resorting to induction.

Example 3.11. The number of all subsets of an n-element set is 2^n.

Solution. We construct a bijection from the set of all subsets of an n-element set into that of all n-digit strings over the binary alphabet $\{0, 1\}$. As this latter set has 2^n elements by Theorem 3.6, it will follow that so does the former.

To construct the bijection, let B be any subset of $[n]$. Now let $f(B)$ be the string whose ith digit is 1 if and only if $i \in B$ and 0 otherwise. This way $f(B)$ will indeed be an n-digit word over the binary alphabet. Moreover, it is clear that given any string s of length n containing digits equal to 0 and 1 only, we can find the unique subset $B \subseteq [n]$ for which $f(B) = s$. Indeed, B will precisely consist of the elements $i \in [n]$ so that the ith element of s is 1.

Example 3.12. A city has recently built ten intersections. Some of these will get traffic lights, and some of those that get traffic lights will also get a gas station. In how many different ways can this happen?

Solution. It is easy to construct a bijection from the set of all distributions of lights and gas stations onto that of ten-digit words over the alphabet A, B, C. Indeed, for each distribution of these objects, we define a word over $\{A, B, C\}$ as follows: if the ith intersection gets both a gas station and a traffic light, then let the ith digit of the word that we are constructing be A, if only a traffic light, then let the ith digit be B, and if neither, then let the ith digit be C.

Clearly, this is a bijection, for any ten-digit word can be obtained from exactly one distribution of gas stations and traffic lights this way. So the number we are looking for is, by Proposition 3.10, the number of all ten-digit words over a three-digit alphabet, that is, 3^{10}.

Theorem 3.13. *Let n and k be positive integers satisfying $n \geq k$. Then the number of k-digit strings over an n-element alphabet in which no letter is used more than once is*

$$n(n-1)\cdots(n-k+1) = \frac{n!}{(n-k)!}.$$

Proof. Indeed, we have n choices for the first digit, $n-1$ choices for the second digit, and so on, just as we did in the case of factorials. The only difference is that here we do not necessarily use all our n objects, we stop after choosing k of them. □

The number $n(n-1)\cdots(n-k+1)$ is sometimes denoted $(n)_k$.

Example 3.14. A president must choose five politicians from a pool of 20 candidates to fill five different cabinet positions. In how many different ways can she do that?

Solution. We can directly apply Theorem 3.13. We have a 20-element alphabet (the politicians) and we need to count the number of 5-letter words with no repeated letters. Therefore, the answer is $(20)_5 = 20 \cdot 19 \cdot 18 \cdot 17 \cdot 16$. If the candidates are all equally qualified, it may take a while...

3.3 Choice Problems

At the national lottery drawings in Hungary, five numbers are selected at random from the set [90]. To win the main prize, one must guess all five numbers correctly. How many lottery tickets does one need in order to secure the main prize?

This problem is an example of the last and most interesting kind of elementary enumeration problems, the choice problems. In these problems, we have to choose *subsets* of a given set. We will often require that the subsets have a specific size. The important difference from the previous two sections is that the order of the elements of the subset will not matter;

for example, $\{1, 43, 52, 8, 3\}$ and $\{52, 1, 8, 43, 3\}$ are identical as subsets of $[90]$.

The number of k-element subsets of $[n]$ is of pivotal importance in enumerative combinatorics. Therefore, we have a symbol and name for this number.

Definition 3.15. The number of k-element subsets of $[n]$ is denoted $\binom{n}{k}$ and is read "n choose k".

The numbers $\binom{n}{k}$ are often called *binomial coefficients*, for reasons that will become clear in Chapter 4.

Theorem 3.16. *For all non-negative integers $k \leq n$,*

$$\binom{n}{k} = \frac{n!}{k!(n-k)!} = \frac{(n)_k}{k!}.$$

Proof. To select a k-element subset of $[n]$, we first select a k-element string in which the digits are elements of $[n]$. By Theorem 3.6, we can do it in $n!/(n-k)!$ different ways. However, in these strings the order of the elements *does* matter. In fact, each k-element subset occurs $k!$ times among these strings as its elements can be permuted in $k!$ different ways. Therefore, the number of k-element subsets is $1/k!$ times the number of k-element strings, and the proof follows. □

Therefore, if we want to be absolutely sure to win at the Hungarian lottery, we have to buy $\binom{90}{5} = \frac{90 \cdot 89 \cdot 88 \cdot 87 \cdot 86}{1 \cdot 2 \cdot 3 \cdot 4 \cdot 5} = 43949268$ tickets. If you do that, make sure you fill them out right...

Definition 3.17. Let $S \subseteq [n]$. Then the *complement* of S, denoted S^c is the subset of $[n]$ that consists precisely of the elements that are not in S. In other words, S^c is the unique subset of $[n]$ that for all $i \in [n]$ satisfies the following statement: $i \in S^c$ if and only if $i \notin S$.

The following proposition summarizes some straightforward properties of the numbers $\binom{n}{k}$. We choose to announce these easy statements as a proposition since they will be used incessantly in the coming sections.

Proposition 3.18. *For all non-negative integers $k \leq n$, the following hold.*

(1)

$$\binom{n}{k} = \binom{n}{n-k}.$$

(2)

$$\binom{n}{0} = \binom{n}{n} = 1.$$

Proof.

(1) We set up a bijection f from the set of all k-element subsets of $[n]$ onto that of all $n - k$-element subsets of n. This f will be simplicity itself: it will map any given k-element subset $S \subseteq [n]$ into its complement S^c. Then for any $n-k$-element subset $T \subseteq [n]$, there is exactly one S so that $f(S) = T$, namely $S = T^c$. So f is indeed a bijection, proving that the number of k-element subsets of $[n]$ is the same as that of $n - k$-element subsets of $[n]$, which, by definition, means that $\binom{n}{k} = \binom{n}{n-k}$.

(2) The first equality is a special case of the claim of part 1, with $k = 0$. To see that $\binom{n}{0} = 1$, note that the only 0-element subset of $[n]$ is the empty set.

$\square$

We note in particular that $\binom{0}{0} = 1$, and that sometimes it is convenient to define $\binom{n}{k}$ even in the case when $n < k$. It goes without saying that in that case, we define $\binom{n}{k} = 0$ as no set has a subset that is larger than the set itself.

Example 3.19. A medical student has to work in a hospital for five days in January. However, he is not allowed to work two consecutive days in the hospital. In how many different ways can he choose the five days he will work in the hospital?

Solution. The difficulty here is to make sure that we do not choose two consecutive days. This can be assured by the following trick. Let a_1, a_2, a_3, a_4, a_5 be the dates of the five days of January that the student will spend in the hospital, in increasing order. Note that the requirement that there are no two consecutive numbers among the a_i, and $1 \leq a_i \leq 31$ for all i is equivalent to the requirement that $1 \leq a_1 < a_2 - 1 < a_3 - 2 < a_4 - 3 < a_5 - 4 \leq 27$. In other words, there is an obvious bijection between the set of 5-element subsets of $[31]$ containing no two consecutive elements and the set of 5-element subsets of $[27]$.

Instead of choosing the numbers a_i, we can choose the numbers $1 \leq a_1 < a_2 - 1 < a_3 - 2 < a_4 - 3 < a_5 - 4 \leq 27$, that is, we can simply choose a five-element subset of $[27]$, and we know that there are $\binom{27}{5}$ ways to do that.

The trick we used here is also useful when instead of requiring that the chosen elements are far apart, we even allow them to be identical.

Example 3.20. Now assume that we play a lottery game where five numbers are drawn out of [90], but the numbers drawn are put back into the basket right after being selected. To win the jackpot, one must have played the same multiset of numbers as the one drawn (regardless of the order in which the numbers were drawn). How many lottery tickets do we have to buy to make sure that we win the jackpot?

Solution. We are going to apply the same trick as in the previous example, just backwards. We claim there is a bijection from the set of 5-element multisets

$$1 \leq b_1 \leq b_2 \leq b_3 \leq b_4 \leq b_5 \leq 90 \tag{3.2}$$

onto the set of 5-elements subsets of [94]. Indeed, such a bijection f is given by $f(b_1, b_2, b_3, b_4, b_5) = (b_1, b_2 + 1, b_3 + 2, b_4 + 3, b_5 + 4)$. It is obvious that the numbers b_i satisfy the requirements given by (3.2) if and only if $f(b_1, b_2, b_3, b_4, b_5) = (b_1, b_2 + 1, b_3 + 2, b_4 + 3, b_5 + 4)$ is a subset of [94]. Therefore, we need to buy $\binom{94}{5}$ lottery tickets to secure a jackpot.

There is nothing magic about the numbers 90 and 5 here. In fact, the same argument can be repeated in a general setup, to yield the following Theorem.

Theorem 3.21. *The number of k-element multisets whose elements all belong to $[n]$ is*

$$\binom{n+k-1}{k}.$$

The following table summarizes our enumeration theorems proved in this chapter.

	parameters	formula
Permutations	n distinct objects	$n!$
	a_i objects of type i, $\sum a_i = n$	$\frac{n!}{a_1!a_2!\cdots a_k!}$
Lists	n distinct objects list of length k	$(n)_k = \frac{n!}{(n-k)!}$
	n distinct letters words of length k	n^k
Subsets	k-element subsets of $[n]$	$\binom{n}{k}$
	k-element multisets with elements from $[n]$	$\binom{n+k-1}{k}$

Notes

One of the most difficult exercises of this chapter is Exercise 24. The first one to prove the formula given in that exercise was probably P. A. MacMahon [29], in 1916. The proof presented here is due to the present author [12]. A high-level survey (using commutative algebra) of results concerning magic squares can be found in "Combinatorics and Commutative Algebra" [39] by Richard Stanley, while a survey intended for undergraduate and starting graduate students is presented in Chapter 9 of "Introduction to Enumerative Combinatorics" [7] by the present author.

Exercises

(1) How many functions are there from $[n]$ to $[n]$ that are not one-to-one?
(2) Prove that the number of subsets of $[n]$ that have an odd number of elements is 2^{n-1}.
(3) A company has 20 employees, 12 males and eight females. How many

ways are there to form a committee of 5 employees that contains at least one male and at least one female?

(4) A track and field championship had participants from 49 countries. The flag of each participating country consisted of three horizontal stripes of different colors. However, no flag contained colors other than red, white, blue, and green. Is it true that there were three participating countries with identical flags?

(5) In countries that currently belong to the European Union, 17 languages are spoken by at least ten million people. For any two of these languages, the European Commission employs an interpreter who can translate documents from one language to the other, and vice versa. One journalist has recently noted that when the soon-to-be admitted countries bring the number of languages spoken by at least ten million people in the Union to 22, more than a hundred new interpreters will be needed. Was she right? (No interpreter works two jobs.)

(6) How many five-digit positive integers are there with middle digit 6 that are divisible by three?

(7) How many five-digit positive integers are there that contain the digit 9 and are divisible by three?

(8) How many ways are there to list the digits $\{1, 2, 2, 3, 4, 5, 6\}$ so that identical digits are not in consecutive positions?

(9) How many ways are there to list the digits $\{1, 1, 2, 2, 3, 4, 5\}$ so that the two 1s *are* in consecutive positions?

(10) A cashier wants to work five days a week, but he wants to have at least one of Saturday and Sunday off. In how many ways can he choose the days he will work?

(11) A car dealership employs five salespeople. A salesperson receives a 100-dollar bonus for each car he or she sells. Yesterday the dealership sold seven cars. In how many different ways could this happen? (Let us consider two scenarios different if they result in different bonus payments.)

(12) A traveling agent has to visit four cities, each of them five times. In how many different ways can he do this if he is not allowed to start and finish in the same city?

(13) A college professor has been working for the same department for 30 years. He taught two courses in each semester. The department offers 15 different courses. Is it sure that there were at least two semesters when this professor had identical teaching programs? (A year has two semesters.)

(14) A restaurant offers five different soups, ten main courses, and six desserts. Joe decided to order at most one soup, at most one main course, and at most one dessert. In how many ways can he do this?

(15) A student in physics needs to spend five days in a laboratory during her last semester of studies. After each day in the lab, she needs to spend at least six days in her office to analyze the data before she can return to the lab. After the last day in the lab, she needs ten days to complete her report that is due at the end of the last day of the semester. In how many ways can she choose her lab days if we assume that the semester is 105 days long?

(16)(a) Three friends, having the nice names A, B, and C played a ping-pong tournament each day of a given week. There were no ties at the end of the tournament. Prove that there were two days when the final ranking of the three people was the same.

(b) A fourth person, called D, joined the company of the mentioned three. These four friends played a tennis competition each day for five weeks. When the five weeks were over, one of them noticed that none of their one-day tournaments resulted in a tie at the first place, or in a tie at the last place. Is it true that there were two contests with the same final ranking of players?

(c) Now A, B and, C are playing a round-robin chess tournament each day starting January 1. Each player plays against each other player once playing the white pieces, and once playing the black pieces. The three friends agreed that they will stop when there will be two days with completely identical results. (That is, if on the earlier day, A beat B when playing the whites, but played a draw with him when playing the blacks, then, on the last day the friends play, A has to beat B when playing the whites, and has to play a draw with him when playing the blacks, and the same coinciding results must occur for the pair (B, C), and for the pair (A, C).)

When their left-out friend, D, heard about their plan, she said "are you sure you want to do this? You might be playing chess for two years!" Was she exaggerating?

(17) Let $k \geq 1$, and let $b_1, b_2, \cdots, b_k$ be positive integers with sum less than n, where n is a positive integer. Prove that then

$$b_1!b_2! \cdots b_k! < n!$$

holds. Can you make that statement stronger?

(18) How many 6-digit positive integers are there in which the sum of the digits is at most 51?

(19) How many ways are there to select an 11-member soccer team and a 5-member basketball team from a class of 30 students if

(a) nobody can be on two teams
(b) any number of students can be on both teams
(c) at most one student can be on both teams?

(20) On the island of Combinatoria, all cars have license plates consisting of six numerical digits only. A witness to a crime could only give a partial description of the getaway car. In particular, she noticed that the license plate was from Combinatoria, there was only one digit that occurred more than once, and that digit occurred three times. A police officer estimated that this information will exclude more than 90 percent of all cars as suspects. Was his estimate correct?

(21) (+) A round robin chess tournament had $2n$ participants from two countries, n from each country. There were no two players with the same number of points at the end. Prove that there was at least one player who scored at least as many points against his compatriots as against the players of the other country. (In chess, a player gets one point for a win and one half of a point for a draw.)

(22) (+)

(a) At a round robin chess tournament, at least $3/4$ of the games ended by a draw. Prove that there were two players who had the same final score.
(b) Now assume the tournament has been interrupted after t rounds, that is, after each player has finished t games. (We assume, for simplicity, that the number of players is even.) Is it still true that if at least $3/4$ of the games played ended by a draw, then there were two players with the same total score?
(c) Prove that if the games of the tournament are played in a random order (there are no rounds; one player can finish many games before another player starts), and the tournament is interrupted at some point. Could it happen that three $3/4$ of the finished games ended by a draw, but there were no two players with the same total score?
(d) Is there a constant $K < 1$ such that if we organize the tournament as in the preceding case, and we interrupt the tournament at a point when at least K of the finished games ended by a draw, then there will always be two players with the same total score?

(23) In how many different ways can we place 8 identical rooks on a chess board so that no two of them attack each other?

(24) (++) A *magic square* is a square matrix with non-negative integer entries in which all row sums and column sums are equal. Let $H_3(r)$ be the number of magic squares of size 3×3 in which each row and column have sum r. Prove that

$$H_3(r) = \binom{r+4}{4} + \binom{r+3}{4} + \binom{r+2}{4}, \tag{3.3}$$

where $H_3(r)$ is the number of 3×3 magic squares of line sum r. We will return to formula (3.3) in Chapter 11. The material covered in that chapter will allow us to give a simpler proof to this result.

Supplementary Exercises

(25) (-) How many three-digit positive integers contain two (but not three) different digits?

(26) (-) How many ways are there to list the letters of the word ALABAMA?

(27) (-) How many subsets does $[n]$ have that contain exactly one of the elements 1 and 2?

(28) (-) How many subsets does $[n]$ have that contain at least one of the elements 1 and 2?

(29) (-) How many three-digit positive integers start and end with an even digit?

(30) How many four-digit positive integers are there in which all digits are different?

(31) How many four-digit positive integers are there that contain the digit 1?

(32) How many three-digit numbers are there in which the sum of the digits is even? (We do not allow the first digit to be zero.)

(33) In this exercise, the words *precede* does not mean *immediately precede.*

(a) In how many ways can the elements of $[n]$ be permuted if 1 is to precede 2 and 3 is to precede 4?

(b) In how many ways can the elements of $[n]$ be permuted if 1 is to precede both 2 and 3?

(34) In how many ways can the elements of $[n]$ be permuted so that the sum of every two consecutive elements in the permutation is odd?

(35) Let $n = p_1^{a_1} p_2^{a_2} \cdots p_k^{a_k}$, where the p_i are distinct primes, and the a_i are positive integers. How many positive divisors does n have?

(36)(a) Let $d(n)$ be the number of positive divisors of n. For what numbers n will $d(n)$ be a power of 2?

(b) Is it true that for all positive integers n, the inequality $d(n) \leq 1 + \log_2 n$ holds?

(37) A student needs to work five days in January. He does not want to work on more than one Sunday. In how many ways can he select his five working days? (Assume that in the year in question, January has five Sundays.)

(38) (+) A host invites n couples to a party. She wants to ask a subset of the $2n$ guests to give a speech, but she does not want to ask *both* members of any couple to give speeches. In how many ways can she proceed?

(39) We want to select as many subsets of $[n]$ as possible so that any two selected subsets have at least one element in common. What is the largest number of subsets we can select?

(40) We want to select an ordered pair (A, B) of subsets of $[n]$ so that $A \cap B \neq \emptyset$. In how many different ways can we do this?

(41) We want to select three subsets A, B, and C of $[n]$ so that $A \subseteq C$, $B \subseteq C$, and $A \cap B \neq \emptyset$. In how many different ways can we do this?

(42) A two-day mathematics conference has n participants. Some of the participants give a talk on Saturday, some others give a talk on Sunday. Nobody gives more than one talk, and there may be some people who do not give a talk at all. At the end of the conference, a few talks are selected to be included in a book. In how many different ways is this all possible if we assume that there is at least one talk selected for inclusion in the book?

(43) A group organizing a faculty-student tennis match must match four faculty volunteers to four of the 13 students who volunteered to be in the match. In how many ways can they do this?

(44) Let P be a convex n-gon in which no three diagonals intersect in one point. How many intersection points do the diagonals of P have?

(45) A student will study 26 hours in preparation for an exam. She will due this in the course of six consecutive days. On each of these days, she will study either four hours, or five hours, or six hours. In how many different ways is this possible?

(46) (+) Andy and Brenda play with dice. They throw four dice at the same time. If at least one of the four dice shows a six, then Andy wins, if not, then Brenda. Who has a greater chance of winning?

(47) (+) A store has n different products for sale. Each of them has a

different price that is at least one dollar, at most n dollars, and is a whole dollar. A customer only has the time to inspect k different products. After doing so, she buys the product that has the lowest price among the k products she inspected. Prove that on average, she will pay $\frac{n+1}{k+1}$ dollars.

(48) In how many ways can we place n non-attacking rooks on an $n \times n$ chess board?

(49) A class is attended by n sophomores, n juniors, and n seniors. In how many ways can these students form n groups of three people each if each group is to contain a sophomore, a junior, and a senior?

(50) The National Football League consists of 32 teams. These teams are first divided into two conferences, the American Conference and the National Conference, each of which consists of sixteen teams. Then each conference is divided into four divisions of four teams each. Each division has a distinct name. In how many ways can this be done?

(51) Answer the question of the previous exercise if there are two teams from New York City in the National Football League, and they cannot be assigned to the same conference.

(52) Let $P_3(r)$ be the number of 3×3 magic squares that are symmetric to their main diagonal. Prove that $P_3(r) \leq (r+1)^3$. (Magic squares are defined in Exercise 24.)

(53) How many $n \times n$ square matrices are there whose entries are 0 or 1 and in which each row and column has an even sum?

Solutions to Exercises

(1) The number of all functions from $[n]$ to $[n]$ is n^n by Theorem 3.6. Indeed, such a function f is defined by the array $(f(1), f(2), f(3), \cdots, f(n))$, and any entry in this array can be any element of $[n]$. If f is a one-to-one function, then the array $(f(1), f(2), f(3), \cdots, f(n))$ is a permutation of the elements $1, 2, \cdots, n$ as it contains each of them exactly once. So the number of one-to-one functions from $[n]$ to $[n]$ is $n!$, by Theorem 3.2. Therefore, the number of functions from $[n]$ to $[n]$ that are *not* one-to-one is $n^n - n!$.

Remark: Note that we were asked to compute the number of functions that were *not* one-to-one, and we obtained that number in an indirect way. We first computed the number of all functions from $[n]$

to $[n]$, then we computed the number of all functions from $[n]$ to $[n]$ that *were* one-to-one, and then we subtracted the second number from the first.

This technique of "number of good objects is equal to that of all objects minus that of bad objects" is very often used in combinatorial enumeration. Several exercises in this chapter can be solved this way.

(2) As in the proof of Example 3.11, we can bijectively encode all subsets of $[n]$ by 0-1 sequences consisting of n digits. If we want this sequence to contain an odd number of ones, then we can choose the first $n-1$ digits any way we want. The last digit can be used to make sure that the number of all ones is odd. That is, if there were an odd number of ones among the first $n-1$ digits, then the last digit has to be a zero, otherwise it has to be a one. Therefore, we make a choice $n-1$ times, and each time we have two possibilities. So the total number of possibilities is 2^{n-1}.

(3) There are $\binom{20}{5}$ ways to choose five people out of our twenty employees. However, $\binom{12}{5}$ of these choices will result in male-only committees, and $\binom{8}{5}$ will result in female-only committees. Therefore, the number of good choices is $\binom{20}{5}-\binom{12}{5}-\binom{8}{5}$.

(4) There are $4 \cdot 3 \cdot 2 = 24$ different 3-color flags that can be made from our four colors. As $2 \cdot 24 = 48 < 49$, it follows from the general version of the Pigeon-hole Principle that there are three identical flags among any 49 such flags.

(5) There are $\binom{17}{2} = \frac{17 \cdot 16}{2} = 136$ pairs that can be formed of the 17 languages currently spoken by at least ten million people in the European Union. When the number of these languages grows to 22, the number of pairs of languages will be $\binom{22}{2} = \frac{22 \cdot 21}{2} = 231$, so 95 new interpreters will be needed. Therefore, the journalist was wrong.

(6) It is well-known (see Exercise 31 of Chapter 2) that a positive integer is divisible by three if and only if the sum of its digits is divisible by three. Therefore, a five-digit a integer with middle digit six is divisible by three if and only if the four-digit integer obtained by deleting the middle digit of a is divisible by three. There are 9000 four-digit positive integers, and the third, sixth, ninth,....9000th of them are divisible by 3 (these are the integers 1002, 1005, 1008,...,9999). In other words, there are 3000 four-digit positive integers divisible by three, so there are 3000 five-digit positive integers divisible by three and having middle digit 6.

(7) The number of all five-digit positive integers is 90000, and one third of

them, 30000, are divisible by three. Let us count how many of these 30000 numbers do not contain the digit nine. Such a number can start with one of eight digits $(1, 2, \cdots, 8)$, then can have any of nine digits $(0, 1, 2, \cdots, 8)$ in the second, third, and fourth positions. For the fifth digit, we have more limited choice. We have to pick the fifth digit so that the sum of all five digits is divisible by three. Depending on the first four digits, we can either choose one of 0,3,6, or one of 1,4,7, or one of 2,5,8. Either way, this means three choices. The total number of choices we have is $8 \cdot 9^3 \cdot 3 = 17496$, so this is the number of 5-digit positive integers that are divisible by three, but do not contain the digit 9. Therefore, there are $30000 - 17496 = 12504$ 5-digit positive integers that are divisible by three and do contain the digit 9.

(8) The number of all permutations of this multiset is given by Theorem 3.5, and is equal to $\frac{7!}{2!} = 2520$. However, we have to subtract the number of those permutations in which the two identical digits are in consecutive positions. To count these, let us *glue* the two identical digits together. Then we have six digits, which are all different, and therefore Theorem 3.2 shows that they have $6! = 720$ permutations. Therefore, the number of all permutations of our multiset in which the two identical digits are not in consecutive positions is $2520 - 720 = 1800$.

(9) Just as in Exercise 8, let us glue the two 1s together. Then we simply have to count permutations of the multiset $\{1, 2, 2, 3, 4, 5\}$. Theorem 3.5 shows that there are $\frac{6!}{2!} = 360$ such permutations.

(10) There are $\binom{7}{5} = \binom{7}{2} = 21$ ways to choose five days of the week. Let us now count the *bad* choices, that is, those that contain both Saturday and Sunday. Clearly, there are $\binom{5}{3} = 10$ of these. Indeed, they contain Saturday, Sunday, and three of the remaining five days. Therefore, the number of good choices is $21 - 10 = 11$.

(11) As we only consider two scenarios different if they result in different bonus payments, we are not interested in the *order* in which the different salespeople sold the seven cars. What matters is how many cars each of them sold. Therefore, we are interested in the number of 7-element multisets whose elements are from the set [5]. By Theorem 3.21, this number is $\binom{5+7-1}{7} = \binom{11}{7} = \binom{11}{4} = 330$.

(12) There are $\frac{20!}{5! \cdot 5! \cdot 5! \cdot 5!}$ ways to visit four cities, each of them five times. Let us determine the number of ways to do this so that we start in city A, and end in city A. In that case, we are free to choose the order in which we make the remaining 18 visits. As three of those visits will

be to city A, and five will be to each of the remaining three cities, this can be done in $\frac{18!}{5!\cdot 5!\cdot 5!\cdot 3!}$ ways. Obviously, the same argument applies for the number of visiting arrangements that start and end in B, that start and end in C, and that start and end in D. So the final answer is

$$\frac{20!}{5!\cdot 5!\cdot 5!\cdot 5!} - 4\cdot \frac{18!}{5!\cdot 5!\cdot 5!\cdot 3!}.$$

(13) No, that is not sure. There are $\binom{15}{2} = \frac{15\cdot 14}{2} = 105$ ways to pick two courses out of 15 courses, and 30 years consist of 60 semesters only.

(14) Joe can make one of six choices on soup as he may decide not to order soup at all. Similarly, he can make one of 11 choices on the main course, and one of seven choices on dessert. So the total number of possibilities is $6\cdot 11\cdot 7 = 462$.

(15) Let us number the days of the semester from 1 to 105, and let us denote the days when the student is in the lab by $a_1, a_2, \cdots, a_5$. Then the conditions imply that $a_5 \leq 95$, and

$$1 \leq a_1 < a_2 - 6 < a_3 - 12 < a_4 - 18 < a_5 - 24 \leq 95 - 24 = 71.$$

Denote $b_1 = a_1$, $b_2 = a_2 - 6$, $b_3 = a_3 - 12$, $b_4 = a_4 - 18$, and $b_5 = a_5 - 24$. Clearly, knowing the numbers b_i is equivalent to knowing the numbers a_i.

Note that $b_5 \leq 105 - 24 = 71$. There is no additional requirement for the numbers b_i besides $1 \leq b_1 < b_2 < b_3 < b_4 < b_5$, so there are $\binom{71}{5}$ possible choices for the set of these numbers. Therefore, our student can make this many choices.

(16)(a) There are $3! = 6$ ways the contest could end, and there are seven days in a week. We know, if from nowhere else, then from the title of Chapter 1, that Seven Is More Than Six. Therefore, the pigeon-hole principle implies that there were two contests with identical results.

(b) If there were no ties at all, the contest could end in $4! = 24$ different ways. If there is a tie, it could only be at the second-third place. The two people who tie can be chosen in $\binom{4}{2} = 6$ ways, then the winner can be either of the remaining two people. So there are $6\cdot 2 = 12$ different outcomes with a tie. Therefore the total number of possible endings for the competition is $24 + 12 = 36$. There are only 35 days in five weeks, so it is possible that there are no two days when the contest ends the same way.

(c) Each tournament consists of six games as we have three choices for the person leading the white pieces, and two choices leading the black pieces. Each of these six games can have three different results: either white wins, or black wins, or it is a draw. So there are $3^6 = 729$ ways the games of a tournament can end. Therefore, the three friends will play for at most 730 days, which is exactly two years as neither 2013, nor 2014 is a leap-year. So D was in fact right, she was not exaggerating.

(17) Let b_{k+1} be a positive integer so that $n = \sum_{i=1}^{k+1} b_i$. Theorem 3.5 then tells us that

$$T = \frac{n!}{b_1!b_2! \cdots b_{k+1}!}$$

is the number of linear orderings of n objects of $k+1$ various kinds, so that b_i objects are of kind i. In particular, $T = \frac{n!}{b_1!b_2!\cdots b_{k+1}!}$ is a positive integer, (as it is the number of elements in a nonempty set), so

$$\frac{n!}{b_1!b_2! \cdots b_k!} = b_{k+1}!T.$$

The right-hand side (and therefore, the left-hand side) is larger than 1 as long as one of T and b_{k+1} is larger than 1. The only way in which $T = 1$ could hold would be if there were no two distinct objects at all, but that is not possible since there is at least one object of type $k+1$, and one other object. So we proved that not only $b_1!b_2! \cdots b_k! < n!$, but also, $b_1 b_2 \cdots b_k$ is a proper divisor of $n!$.

(18) The number of all 6-digit integers is 900000 by Example 3.7. Again, we are going to count those which do *not* satisfy the criteria, that is, those with digit sum of at least 52. There are only four 6-element multisets of digits that sum to at least 52, namely $\{9,9,9,9,9,9\}$, $\{9,9,9,9,9,8\}$, $\{9,9,9,9,9,7\}$, and $\{9,9,9,9,8,8\}$. Theorem 3.5 implies that they have 1,6,6, and 15 multiset permutations (respectively), so altogether there are 28 numbers out of 900000 that violate the criteria. So the number of 6-digit positive integers that satisfy the criteria is 899972.

(19)(a) We have $\binom{30}{11}$ choices for the soccer team. Then we have to choose from the remaining 19 people in $\binom{19}{5}$ ways for the basketball team. Consequently, the final answer is $\binom{30}{11} \cdot \binom{19}{5}$.

(b) If there is no restriction at all, then after choosing the soccer team, we can choose the basketball team in $\binom{30}{5}$ ways, from the set of all students. So the total number of choices is $\binom{30}{11} \cdot \binom{30}{5}$.

(c) All $\binom{30}{11} \cdot \binom{19}{5}$ team compositions (computed in the first part in this exercise) in which no student is on two teams are certainly good. Apart from these, there are those in which there is exactly one student on both teams. We have 30 choices for this person, then there are $\binom{29}{10} \cdot \binom{19}{4}$ ways to choose the remaining players from the rest of the class. Thus the total number of possibilities is

$$\binom{30}{11} \cdot \binom{19}{5} + 30 \cdot \binom{29}{10} \cdot \binom{19}{4}.$$

(20) The digit that occurred three times could be any of ten digits. The positions of its three occurrences could be any of the $\binom{6}{3} = 20$ three-element subsets of $[6]$. The other three digits form a 3-digit word over the remaining 9-letter alphabet without repetition, so we have $9 \cdot 8 \cdot 7 = 504$ choices for them. As all these choices can be made independently from each other, the total number of our choices is $10 \cdot 20 \cdot 504 = 100800$. This is slightly more than ten percent of all license plates, which would be 100000, so the police officer was a little bit too optimistic.

(21) Let A be the country whose players scored, in totality, at most as many points in the international games as players from the other country. Take the n players from A, and let $a_1, a_2, \cdots, a_n$ denote the number of points they accumulated against their countrymen. Let $b_1, b_2, \cdots, b_n$ be the number of points they accumulated against players from country B. Now assume that our claim is false, that is, $a_i < b_i$ for all i. In other words, $a_i \leq b_i - 0.5$ for all i. Summing these inequalities over all $i \in [n]$, we get that

$$\sum_{i=1}^{n} a_i \leq \left(\sum_{i=1}^{n} b_i \right) - n/2. \tag{3.4}$$

On the other hand, note that $\sum_{i=1}^{n} a_i = n(n-1)/2$ as any two players from A played each other once, and in each of those games, one point was up for grabs. Comparing this with (3.4), we get

$$\frac{n(n-1)}{2} + \frac{n}{2} = \frac{n^2}{2} \leq \left(\sum_{i=1}^{n} b_i \right). \tag{3.5}$$

Similarly,

$$\sum_{i=1}^{n} b_i \leq n^2/2 \tag{3.6}$$

as players from A got at most half of all points that were available at the international games.

Comparing (3.5) and (3.6) we see that $\sum_{i=1}^{n} b_i = n^2/2$ must hold. That is, $\sum_{i=1}^{n} b_i$ is *exactly* $n/2$ larger than $\sum_{i=1}^{n} a_i$. Therefore, equality holds in (3.4), and so equality must hold in all equations of the type $a_i \leq b_i - 0.5$. (Recall that (3.4) was obtained by taking the sum of these equations for all i.) Therefore, for all i, we must have $a_i = b_i - 0.5$, meaning that the total score of the ith player from country A was $a_i + b_i = 2a + 0.5$, which is never an integer. Therefore, no player from country A has a final score that is an integer. By the very same argument, no player from country B has a final score that is an integer. Indeed, in totality, players from B scored $n^2/2$ points against players from A, so the same argument works.

This is a contradiction as we know there are no two players with the same final score. The number of possible non-integer final scores is less than $2n$: indeed, they are $0.5, 1.5, 2.5, \cdots (2n-1) - 0.5$, which is only $2n-1$ different scores for the $2n$ players. So there must be a player who did better against his compatriots than against players from the other country.

(22)(a) Let us change the scoring system of chess as follows: a player gets one point for a win, zero points for a draw, and -1 points for a loss. Clearly, this does not change the facts in our problem: people who had different scores in the original scoring system have different scores now, and people who had identical scores in the original scoring system have identical scores now. Indeed, if player x won a_x games, got a draw b_x times, and lost c_x times, then his total score in the old system is $a_x + (b_x/2)$, and his total score in the new system is $a_x - c_x$. Assume player y got the same total score in the old system. That means

$$a_x + \frac{b_x}{2} = a_y + \frac{b_x}{2}.$$

Multiply this equation by 2, and subtract the equation $a_x + b_x + c_x = a_y + b_y + c_y$ from it. (The latter simply shows that both players played the same number of games.) We get

$$a_x - c_x = a_y - c_y,$$

which shows that the two players had the same score in the new system, too.

Let us assume that all n players had different final scores. Let $k = n/2$ if n is even, and let $k = (n-1)/2$ if n is odd. Then we can assume without loss of generality that there are k players with positive final scores. As these scores are all different, their sum is at least $1+2+\cdots+k = k(k+1)/2$. As only wins result in positive scores, there had to be at least $k(k+1)/2$ wins at the tournament. The number of all games is, on the other hand, $\binom{n}{2}$. Therefore, the ratio of wins (games not ended in a draw) and all games is

$$\frac{k(k+1)}{(n-1)n} > \frac{1}{4}. \qquad (3.7)$$

(b) Yes, the same argument will work, except that the total number of games played will be less than $\binom{n}{2}$, therefore the denominator in formula (3.7) will decrease, therefore the ratio of wins will be even larger.

(c) The problem with the previous argument here is that if not all players complete the same number of games, then the new scoring system is *not* the same as the classical one. Indeed, the argument of part (a) would not work here as $a_x + b_x + c_x = a_y + b_y + c_y$ would not hold. The statement is no longer true. A counterexample can then be found for $n = 4$ as follows. Let games $A - B$, $A - C$ end by draws, and let game $B - D$ be won by B. Then B has 1.5 points, A has 1, C has 0.5, and D has 0. (Note that in the $1 - 0 - (-1)$ scoring system, A and C would both have 0 points.)

(d) No. Our counterexample will be a generalization of the preceding example, and also, of Example 1.7 of Chapter 1. Say we have n players, (n is even) $A_1, A_2, \cdots, A_{n-1}$ and B. Let A_{n-1} play with everyone, except for A_1, let A_{n-2} play with everyone except for A_1 and A_2, in general, let A_i play with A_j if $i + j > n$, and let A_i play with B if $i > n/2$. Let all these games end by a draw. Then A_i has $i/2$ points for all i, and B has $\frac{n}{4} - \frac{1}{2}$ points. The only problem now is that B has the same number of points as one of the players A_i. To correct that, let B play with all the A_i he did not (there are $\frac{n}{2}$ of those), and defeat them all. Then B becomes a clear winner of the tournament, and the points of the A_i do not change, so they stay all different. Also note that the number of games played is quadratic in n, whereas that of wins is linear in n, proving that the ratio of draws can be arbitrarily close to 1 if n is large enough.

(23) **First Solution.** We can place the first rook anywhere on the board, that is, we have $8^2 = 64$ choices for its position. The second rook cannot be in the row or column of the first one, leaving $7^2 = 49$ choices for its position. Similarly, we will have $6^2 = 36$ choices for

the position of the third rook, and so on. Therefore, if our rooks were distinguishable, we would have $8^2 \cdot 7^2 \cdots 1^2 = 8!^2$ ways to place them. However, they are indistinguishable, so it does not matter *which rook is in which position* as long as the *set of all rooks* covers the same eight positions. Consequently, we have counted every placement $n!$ times, and the number of all placements is $8!^2/8! = 8! = 40320$.

Second solution. Each $f : [8] \rightarrow [8]$ can be bijectively associated to a non-attacking rook placement as follows. For all $i \in [8]$, put a rook into the square $(i, f(i))$. This ensures that there will be exactly one rook in each row and column. It is also easy to see that this is a bijection, that is, all rook placements define one one-to-one function from [8] onto itself. So the number of rook placements is $n!$ by Exercise 1.

(24) Take any magic square of line sum r and side length 3. It is clear that the four elements shown in the figure determine all the rest of the square.

a	d	
	b	
		c

Indeed, the next table shows our only possible choice for each remaining entry. Thus all we need to do is to compute the number of ways we can choose a, b, c and d so that we indeed have that one choice, i.e., the obtained entries of the magic square are all non-negative.

a	d	$r-a-d$
$r+c-$ $(a+d+b)$	b	$a+d-c$
$b+d-c$	$r-b-d$	c

The previous table shows that the entries of our matrix will be non-negative if and only if the following inequalities hold:

$$a+d \leq r \tag{3.8}$$

$$b+d \leq r \tag{3.9}$$

$$c \leq a+d \tag{3.10}$$

$$c \leq b+d \tag{3.11}$$

$$a+d+b-c \leq r. \tag{3.12}$$

We will consider three different cases, according to the position of the smallest element on the main diagonal. In each of them, at least three of the five conditions above will become redundant, and we will only need to deal with the remaining one or two.

(a) Suppose $0 \leq a \leq b$ and $0 \leq a \leq c$. In this case conditions (3.8), (3.11), and (3.12) are clearly redundant, because they are implied by (3.9) and (3.10).
The crucial observation is that in all the three cases we can collect all our conditions into one single chain of inequalities. In this case we do it as follows:

$$a \leq 2a+d-c \leq a+b+d-c \leq b+d \leq r. \tag{3.13}$$

Indeed, the first inequality is equivalent to (3.10), the second one is equivalent to our assumption $a \leq b$, the third one is equivalent to our assumption $a \leq c$, and the last one is equivalent to (3.9). Moreover, note that once we know the terms of this chain, that is, $a, 2a+d-c, a+b+d-c$ and $b+d$, then we know a, b, c and d, too, thus we have determined the magic square. Thus all we need to do is simply count how many ways there are to choose these four terms. Inequality (3.13) shows that these terms are nondecreasing, therefore the number of ways to choose them is simply the number of 4-combinations of $r+1$ elements with repetitions allowed, which is $\binom{r+4}{4}$. (Recall that 0 is allowed to be an entry.)

(b) Now suppose $a > b$ and $c \geq b$. Then (3.9), (3.11) and (3.12) are redundant. Consider the chain of inequalities

$$b \leq 2b+d-c \leq a+b+d-c-1 \leq a+d-1 \leq r-1. \quad (3.14)$$

We can use the argument of the previous case to prove that (3.14) equivalent to (3.8), (3.12) and our assumptions, as the roles of a and b are completely symmetric. The only change is that here we do not count those magic squares in which $a = b$, and this explains the (-1) in the last three terms. Thus here we have to choose four elements in non-decreasing order out of the set $\{0, 1, \cdots, r-1\}$, which can be done in $\binom{r+3}{4}$ ways.

(c) Finally, suppose that $a > c$ and $b > c$. Then (3.8), (3.9), (3.10) and (3.11) are redundant. Condition (3.12) and our assumptions can be collected into the following chain:

$$c \leq b-1 \leq b+d-1 \leq a+b+d-c-2 \leq r-2 \quad (3.15)$$

Here the first inequality is equivalent to our assumption $c < b$, the second one says that d is non-negative, the third one is equivalent to our assumption $c < a$, and the last one is equivalent to (3.12). The four terms of (3.15) determine a, b, c and d, and they can be chosen in $\binom{r+2}{4}$ ways, which completes the proof.

Thus the number of 3×3 magic squares of line sum r is indeed $\binom{r+4}{4} + \binom{r+3}{4} + \binom{r+2}{4}$. Furthermore, the three terms in this sum count the magic squares in which the (first) minimal element of the main diagonal is the first, second, or third element.

Chapter 4

No Matter How You Slice It. The Binomial Theorem and Related Identities

In the last chapter, we started developing enumerative techniques by finding formulae that covered six basic situations. We will continue in that direction in Chapter 5. Now, however, we take a break and discuss the binomial and the multinomial theorems, as well as several important identities on binomial coefficients. The *proofs* of these identities are probably even more significant than the identities themselves. They will consist of showing that both sides of a given equation count the same kind of objects; they just do it in two different ways. Therefore, the two expressions must be equal to each other. This type of argument is the dream of most combinatorialists when they prove identities.

4.1 The Binomial Theorem

Theorem 4.1 (Binomial Theorem). *For all non-negative integers n,*

$$(x+y)^n = \sum_{k=0}^{n} \binom{n}{k} x^k y^{n-k}. \tag{4.1}$$

Proof. Consider the product of n sums, $(x+y)(x+y)\cdots(x+y)$. When computing this product, we take one summand from each parentheses, multiply them together, then repeat this in all of 2^n possible ways and sum the results. We get a product equal to $x^k y^{n-k}$ each time we take k summands equal to x. There are $\binom{n}{k}$ k-element subsets of the set of all n parentheses, so we will get such a term $\binom{n}{k}$ times, and the proof follows. □

The binomial theorem has a vast array of applications, starting as early as elementary calculus. In this section we will see some of its immediate applications to prove identities on binomial coefficients.

Theorem 4.2. *For all positive integers n, the alternating sum of binomial coefficients $\binom{n}{k}$ is zero. In other words,*

$$\sum_{k=0}^{n}(-1)^k \cdot \binom{n}{k} = 0.$$

Proof. Applying the binomial theorem with $x = -1$ and $y = 1$ we immediately get our claim. □

Theorem 4.3. *For all non-negative integers n and k,*

$$\binom{n}{k} + \binom{n}{k+1} = \binom{n+1}{k+1}. \tag{4.2}$$

Proof. The right-hand side is, by definition, the number of $k+1$-element subsets of $[n+1]$. Such a subset S either contains $n+1$, or it does not. If it does, then the rest of S is a k-element subset of $[n]$, and these are enumerated by the first member of the left-hand side. If it does not, then S is a $k+1$-element subset of $[n]$, and these are enumerated by the second member of the left-hand side. □

Theorem 4.4. *For all non-negative integers n,*

$$2^n = \sum_{k=0}^{n} \binom{n}{k}.$$

Proof. Both sides count the number of all subsets of an n-element set. The left-hand side counts directly, while the right-hand side counts the number of k-element subsets, then sums over k. □

We can get an even shorter proof applying our fresh knowledge.

Proof. (of Theorem 4.4) Apply the binomial theorem with $x = y = 1$. □

The first proof is an example of a classic way of proving combinatorial identities: *by proving that both sides of the identity to be proved count the same objects.* If we count the same objects in two different ways, we should get the same result, so this is a valid reasoning. Such proofs are ubiquitous and well-liked in enumerative combinatorics. This section will contain a handful of them, and many additional examples are listed as exercises.

Now let us write down all binomial coefficients in a triangle as shown in Figure 4.1. That is, the ith element of row n is $\binom{n}{i}$, and the diagram starts with row 0.

$$
\begin{array}{ccccccccccccc}
&&&&&&1&&&&&&\\
&&&&&1&&1&&&&&\\
&&&&1&&2&&1&&&&\\
&&&1&&3&&3&&1&&&\\
&&1&&4&&6&&4&&1&&\\
&1&&5&&10&&10&&5&&1&\\
1&&6&&15&&20&&15&&6&&1
\end{array}
$$

Fig. 4.1 The first few rows of the Pascal triangle.

This diagram is called a *Pascal triangle* and has many beautiful properties. For example, Theorem 4.4 shows that the sum of the nth row is 2^n, when we call the one-element row at the top the zeroth row. Theorem 4.3 shows that each entry of the triangle is the sum of the entries above it. And Theorem 4.2 shows that the alternating sum of the rows is always 0.

Let us prove one more interesting property of the Pascal triangle.

Theorem 4.5. *For all non-negative integers k and n,*

$$\binom{k}{k} + \binom{k+1}{k} + \binom{k+2}{k} + \cdots + \binom{n}{k} = \binom{n+1}{k+1}. \tag{4.3}$$

Proof. The right-hand side clearly counts all $k+1$-element subsets of $[n+1]$. The left-hand side counts the same, separated into cases according to the largest entry. That is, there are $\binom{k}{k}$ subsets of $[n+1]$ that have $k+1$ elements whose largest element is $k+1$; there are $\binom{k+1}{k}$ subsets of $[n+1]$ that have $k+1$ elements whose largest element is $k+2$, and so on. In general, there are $\binom{k+i}{k}$ subsets of $[n+1]$ that have $k+1$ elements whose largest element is $k+i+1$, for all $i \leq n-k$. Indeed, if the largest element of such a subset is $k+i+1$, then its remaining k elements must form a subset of $[k+i]$. □

This means that if we start with the rightmost element of the kth row of the Pascal triangle, and descend diagonally to the southwest for a while, then the sum of all numbers we touch in this procedure is also an entry of the Pascal triangle. The reader should find out where that entry is located.

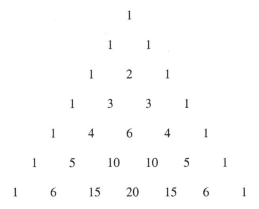

Finally, let us prove some identities about binomial coefficients that do not directly follow from the binomial theorem, but nevertheless are a lot of fun.

Theorem 4.6. *For all non-negative integers n,*

$$\sum_{k=1}^{n} k\binom{n}{k} = n2^{n-1}. \tag{4.4}$$

Before proving the theorem, note that it is not even obvious why

$$\frac{\sum_{k=1}^{n} k\binom{n}{k}}{2^{n-1}}$$

should be *an integer*. Our proof will show that it is not only an integer, it is equal to n. This hopefully convinces the reader that binomial coefficient identities are beautiful.

Proof. (of Theorem 4.6) Both sides count the number of ways to choose a committee among n people, then to choose a president from the committee. On the left-hand side, we first choose a k-member committee in $\binom{n}{k}$ ways, then we choose its president in k ways. On the right-hand side, we first choose the president in n ways, then we choose a subset of the remaining $n-1$-member set of people for the role of non-president committee members in 2^{n-1} ways. □

We provide another proof that uses the binomial theorem. It also gives us an early hint that sometimes very finite-looking problems, such as choice problems, can be solved by using methods from infinite calculus, such as functions and their derivatives.

Proof. (of Theorem 4.6) Apply the binomial theorem with $y = 1$ to get the identity

$$(x+1)^n = \sum_{k=0}^{n} \binom{n}{k} x^k. \tag{4.5}$$

Both sides are differentiable functions of the variable x. So we can take their derivatives with respect to x, and they must be equal. This yields

$$n(x+1)^{n-1} = \sum_{k=1}^{n} k \cdot \binom{n}{k} x^{k-1}.$$

Now substitute $x = 1$ to get (4.4). □

These direct combinatorial arguments are so enjoyable that we cannot refrain from discussing one more of them.

Theorem 4.7. *For all positive integers n, m, and k,*

$$\binom{n+m}{k} = \sum_{i=0}^{k} \binom{n}{i}\binom{m}{k-i}.$$

Proof. The left-hand side counts all k-element subsets of $[n+m]$. The right-hand side counts the same, according to the number of elements *chosen from* $[n]$. Indeed, we can first choose i elements from $[n]$ in $\binom{n}{i}$ ways, then choose the remaining $k-i$ elements from the set $\{n+1, n+2, \cdots, n+m\}$ in $\binom{m}{k-i}$ ways. □

Considering any one row of the Pascal triangle, we note that the binomial coefficients $\binom{n}{0}, \binom{n}{1}, \cdots$ seem to increase as k increases, up to the middle of the row, after which they seem to decrease. As the following theorem shows, this is indeed true for all n.

Theorem 4.8. *For all non-negative integers k and n, such that $k \leq \frac{n-1}{2}$, the inequality*

$$\binom{n}{k} \leq \binom{n}{k+1} \tag{4.6}$$

holds. Furthermore, equality holds if and only if $n = 2k+1$.

Proof. We provide a computational proof here. We need to show that if the conditions hold, then

$$\frac{n!}{k! \cdot (n-k)!} \leq \frac{n!}{(k+1)! \cdot (n-k-1)!}.$$

Let us divide both sides by $n!$, then multiply both sides by $k! \cdot (n-k-1)!$ to get

$$\frac{1}{n-k} \leq \frac{1}{k+1}.$$

Taking reciprocals and rearranging, we get $2k+1 \leq n$, which is equivalent to the condition $k \leq \frac{n-1}{2}$, so the theorem is proved. □

Corollary 4.9. *For all positive integers k and n, such that $k \geq \frac{n-1}{2}$, the inequality*

$$\binom{n}{k} \geq \binom{n}{k+1} \tag{4.7}$$

holds. Furthermore, equality holds if and only if $n = 2k+1$.

Proof. This is immediate from Theorem 4.8, and the fact that $\binom{n}{k} = \binom{n}{n-k}$. □

A sequence of numbers with this property, that is, that it first increases steadily, then it decreases steadily, is called *unimodal.* It can often be quite difficult to prove that a given sequence is unimodal. A more elegant, but less straightforward, non-computational proof of Theorem 4.8 is given in Exercise 19. A stronger statement is proved in Exercise 20.

4.2 The Multinomial Theorem

What if we want to compute the powers of $(x + y + z)$, or $(u + x + y + z)$ instead of just $(x + y)$? The same line of thinking will help, only the result will be a little more complicated to describe.

Example 4.10. We have

$$(x+y+z)^3 = x^3+y^3+z^3+3x^2y+3x^2z+3y^2x+3y^2z+3z^2x+3z^2y+6xyz. \tag{4.8}$$

Solution. We want to compute the product $(x+y+z)\cdot(x+y+z)\cdot(x+y+z)$. To do this, we have to pick one member of each of the three sums, take their product, do this in all $3^3 = 27$ possible ways, then add the obtained 27 products.

All the 27 products we obtain will be terms of degree 3. The only question is what the coefficient of these terms will be. Why is it, for example, that the right-hand side of (4.8) contains $3x^2y$ and $6xyz$?

Let us first examine how can one of our products be equal to x^2y. This happens when two of our three picks is an x, and the third one is a y. There are three ways this can happen as we can pick the single y from any of our three parentheses, then we must pick the two x terms from the remaining three variables. Therefore, the coefficient of $3x^2y$ in $(x + y + z)^3$ is indeed three. Clearly, identical argument applies for all terms of degree three that contain one variable on the second power.

There is only one way for one of our 27 products to be equal to x^3. Indeed, that happens if and only if we choose an x from each of our three parentheses. Therefore, the coefficient of x^3 in $(x + y + z)^3$ is one, and the same is true for y^3 and z^3.

Finally, what about the term xyz? To get such a term, we have to choose an x from one of our three parentheses, which can be done in three ways. Then, we have to choose a y from the remaining two parentheses, which can be done in two ways. At the end, we must pick z from the last parentheses. Therefore, there are six ways we can obtain an xyz-term, completing the proof.

Just as in Theorem 3.21, we need a higher level of abstraction before we can state a general theorem along the lines of Example 4.10. First of all, we want a theorem that works for any number of variables, not just three. Therefore, instead of calling our variables x, y, z, we will call them $x_1, x_2, \cdots, x_k$. The following definition generalizes the notion of binomial coefficients.

Definition 4.11. Let $n = \sum_{i=1}^k a_i$, where n and $a_1, a_2, \cdots, a_k$ are non-negative integers. We define

$$\binom{n}{a_1, a_2, \cdots, a_k} = \frac{n!}{a_1! \cdot a_2! \cdots a_k!}. \tag{4.9}$$

The numbers $\binom{n}{a_1,a_2,\cdots,a_k}$ are called *multinomial coefficients.*

The reader should verify that if $k = 2$, then this definition reduces to that of binomial coefficients.

Now we are in a position to state and prove the general theorem we have been looking for.

Theorem 4.12 (Multinomial theorem). *For all non-negative integers n and k, the equality*

$$(x_1 + x_2 + \cdots + x_k)^n = \sum_{a_1,a_2,\cdots,a_k} \binom{n}{a_1, a_2, \cdots, a_k} x_1^{a_1} x_2^{a_2} \cdots x_k^{a_k} \tag{4.10}$$

holds. Here the sum is taken over all k-tuples of non-negative integers $a_1, a_2, \cdots, a_k$ such that $n = \sum_{i=1}^k a_i$.

Proof. We have to show that the term $x_1^{a_1} x_2^{a_2} \cdots x_k^{a_k}$ can be obtained in exactly $\binom{n}{a_1,a_2,\cdots,a_k}$ ways as a product of k variables, one from each parentheses of $(x_1 + x_2 + \cdots + x_k) \cdots (x_1 + x_2 + \cdots + x_k)$. To obtain such a term, we have to choose x_i from exactly i parentheses, for all $i \in [k]$.

Now let us take a_i copies of x_i, for all $i \in [k]$, and order these n letters linearly. Theorem 3.5 shows that this can be done in exactly $\binom{n}{a_1,a_2,\cdots,a_k}$ ways. On the other hand, each linear ordering p defines a natural way of choosing variables from the parentheses. Indeed, if the jth letter of p is x_i, then from the jth parentheses, we choose x_i. This way our $\binom{n}{a_1,a_2,\cdots,a_k}$

linear orderings will produce exactly $\binom{n}{a_1,a_2,\cdots,a_k}$ terms that are equal to $x_1^{a_1}x_2^{a_2}\cdots x_k^{a_k}$.

It is clear that this procedure establishes a bijection from the set of linear orderings of n letters, a_i of which is equal to x_i for all $i \in [k]$ onto that of terms of $(x_1+x_2+\cdots+x_k)^n$ that are equal to $x_1^{a_1}x_2^{a_2}\cdots x_k^{a_k}$. Therefore, the coefficient of $x_1^{a_1}x_2^{a_2}\cdots x_k^{a_k}$ in $(x_1+x_2+\cdots+x_k)^n$ is precisely $\binom{n}{a_1,a_2,\cdots,a_k}$, and the proof follows. □

There is a close connection between multinomial and binomial coefficients as explained by the following theorem.

Theorem 4.13. *For all non-negative integers n and $a_1, a_2, \cdots, a_k$ such that $n = \sum_{i=1}^{k} a_i$, the equality*

$$\binom{n}{a_1, a_2, \cdots, a_k} = \binom{n}{a_1} \cdots \binom{n - a_1 - \cdots - a_i}{a_{i+1}} \cdots \binom{n - a_1 - \cdots - a_{k-1}}{a_k} \tag{4.11}$$

holds.

Note that $n - a_1 - a_2 - \cdots - a_{k-1} = a_k$, so the last binomial coefficient on the right-hand side of (4.11) is equal to $\binom{a_k}{a_k} = 1$.

Proof. The left-hand side counts all linear orderings of a multiset that consists of a_i copies of the symbol x_i, for all $i \in [k]$. We show that the right-hand side counts the same objects. Indeed, let us first choose the a_1 positions we place all our symbols x_1. This can be done in $\binom{n}{a_1}$ ways. Let us now choose the a_2 positions where we place our symbols x_2. As a_1 positions are already taken, this can be done in $\binom{n-a_1}{a_2}$ ways. Then we can choose the a_3 positions where we place our symbols x_3. As a_1+a_2 positions are already taken, this can be done in $\binom{n-a_1-a_2}{a_3}$ ways. Iterating this procedure, we will choose the positions of all symbols, and we see that the total number of possible outcomes is indeed the right-hand side of (4.11). □

4.3 When the Exponent Is Not a Positive Integer

What can we say about $(1+x)^m$ when m is *not* a positive integer? That is, how can we expand an expression like $(1+x)^{-2/3}$? In order to find a nice, compact answer to this question, first we define the binomial coefficient $\binom{m}{k}$ for all *real numbers* m.

Definition 4.14. Let m be any real number, and let k be a non-negative integer. Then $\binom{m}{0} = 1$, and

$$\binom{m}{k} = \frac{m(m-1)\cdots(m-k+1)}{k!},$$

if $k > 0$.

This definition expands the definition of binomial coefficients for positive integers. Let us consider the Taylor series of $(1+x)^m$ around $x = 0$. Note that the nth derivative of $(1+x)^m$ is $(m)_n(1+x)^{m-n}$, and this expression takes the value $(m)_n = \frac{m!}{(m-n)!}$ when $x = 0$. Therefore, using Taylor's theorem, we get the following identity.

Theorem 4.15 (Binomial Theorem, General Version). *Let m be any real number. Then*

$$(1+x)^m = \sum_{n\geq 0} \binom{m}{n} x^n,$$

where the sum is taken over all non-negative integers n.

Thus $(1+x)^m$ is an *infinite* power series if m is not a positive integer. Note that if m is a positive integer, then $\binom{m}{n} = 0$ if $n > m$, and therefore we only get a sum of $m+1$ elements for $(1+x)^m$.

Example 4.16. Find the power series expansion of $\sqrt{1-4x}$.

Solution. By Theorem 4.15,

$$\sqrt{1-4x} = (1-4x)^{1/2} = \sum_{n\geq 0} \binom{1/2}{n} (-4x)^n. \tag{4.12}$$

To simplify this expression, we have to find a simpler form for $\binom{1/2}{n}$. Note that $\binom{1/2}{0} = 1$, while $\binom{1/2}{1} = 1/2$, and if $n \geq 2$, then

$$\binom{1/2}{n} = \frac{\frac{1}{2} \cdot \frac{-1}{2} \cdot \frac{-3}{2} \cdots \frac{-2n+3}{2}}{n!} = (-1)^{n-1} \frac{(2n-3)!!}{2^n \cdot n!},$$

where $(2n-3)!!$ stands for the product of all *odd integers* from 1 to $2n-3$, and is called $2n-3$ *semifactorial.*

Substituting this into formula (4.12), we get

$$\sqrt{1-4x} = 1 - 2x - \sum_{n\geq 2} \frac{2^n \cdot (2n-3)!!}{n!} x^n.$$

For $n \geq 2$, let us multiply both the numerator and the denominator of $\frac{2^n \cdot (2n-3)!!}{n!}$ by $(n-1)!$, and note that in the numerator, $2^{n-1}(n-1)!$ is equal to the product of all *even* integers from 2 to $2n-2$. Therefore, if $n \geq 2$, then

$$\frac{2^n \cdot (2n-3)!!}{n!} = 2\frac{(2n-2)!}{n!(n-1)!},$$

and so

$$\sqrt{1-4x} = 1 - 2x - \frac{2}{n}\sum_{n\geq 2}\binom{2n-2}{n-1}x^n.$$

Another useful practice problem is to compute $(1-x)^r$, for a *negative* integer r.

Notes

Exercises 19 and 20 concern two interesting areas of Combinatorics. One of them is unimodality and log-concavity, and the other is the combinatorics of lattice paths. Interested readers can consult Chapter 8 of [7] for an introductory text on the topic. Another good starting point is [36], where lattice paths are used to prove unimodality results in a very accessible way. After that, we recommend [13] for unimodality and log-concavity results, and [25] for lattice path enumeration.

Exercises

(1)(a) Is it possible to write a real number into each square of a 5×5 grid so that the sum of the numbers in the entire grid is negative, but the sum of the numbers in any 2×2 square (formed by 4 neighboring boxes) is positive?

(b) What about a 6×6 grid?

(2) (+)

(a) We plant 13 trees at various points in the interior of a garden whose shape is a convex octagon. Then we create some non-intersecting paths joining some of these trees and the eight corners of the garden so that these paths partition the garden into triangles. How many triangles will be created?

(b) What if we also add five trees to the boundary of the garden? (These five trees are not in corners.)

(3) Prove that for all integers $n \geq 2$,

$$2^{n-2} \cdot n \cdot (n-1) = \sum_{k=2}^{n} k(k-1)\binom{n}{k}.$$

How can we generalize this identity?

(4) Let k, m, n be non-negative integers such that $k + m \leq n$. Prove that

$$\binom{n}{m} \cdot \binom{n-m}{k} = \binom{n}{k}\binom{n-k}{m}.$$

(5) Prove that for integers $0 \leq k \leq n-1$,

$$\sum_{j=0}^{k} \binom{n}{j} = \sum_{j=0}^{k} \binom{n-1-j}{k-j} 2^j.$$

(6) A heap consists of n stones. We split the heap into two smaller heaps, neither of which are empty. Denote by p_1 the *product* of the number of stones in each of these two heaps. Now take any of the two small heaps, and do likewise. Let p_2 be the product of the number of stones in each of the two smaller heaps just obtained. Continue this procedure until each heap consists of one stone only. This will clearly take $n-1$ steps, where a step is the splitting of one heap. For what sequence of splits will the sum $p_1 + p_2 + \cdots + p_{n-1}$ be maximal? When is that sum minimal?

(7) Prove that any positive integer n has at least as many divisors of the form $4k+1$ as divisors of the form $4k-1$.

(8) Prove that for all positive integers n, the inequality $\binom{2n}{n} < 4^n$ holds.

(9) How many subsets of $[n]$ are larger than their complements?

(10) Which term of $(x_1 + x_2 + \cdots + x_k)^k$ has the largest coefficient? What is that coefficient?

(11) Let $n < k$. What is the largest coefficient in $(x_1 + x_2 + \cdots + x_k)^n$?

(12) Let $n = rk$, where $r > 1$ is an integer. What is the largest coefficient in $(x_1 + x_2 + \cdots + x_k)^n$?

(13) Let k be a non-negative integer, let m be a positive integer so that $k < 2^m$, and let $n = 2^m - 1$. Prove that $\binom{n}{k}$ is odd.

(14) Let k and m be positive integers so that $k < 2^m$, and let $n = 2^m$. Prove that $\binom{n}{k}$ is even.

(15) Let $p \geq 3$ be a prime number, and let m and $k < p^m$ be positive integers. Show that $\binom{p^m}{k}$ is divisible by p.

(16) Let p be a prime number, and let $x > 1$ be any positive integer. Consider a wheel with p spokes shown in Figure 4.2.

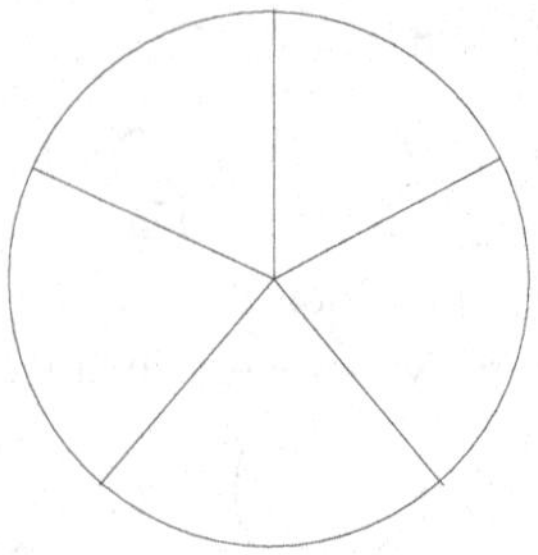

Fig. 4.2 A wheel with five spokes.

(a) We have paints of x different colors. How many ways are there to color the spokes if we want to use at least two colors?
(b) How many ways are there to do the same if we do not consider two paint jobs different if one can be obtained from the other by rotation?
(c) What theorem of number theory does this prove?

(17) Prove that

$$\sum_{a_1+a_2+a_3=n} \binom{n}{a_1, a_2, a_3} = 3^n.$$

(18) Prove that

$$\sum_{a_1+a_2+a_3=n} \binom{n}{a_1, a_2, a_3} (-1)^{a_2} = 1.$$

(19) (+) A walk on the grid of points with integer coordinates that uses the steps $(0,1)$ and $(1,0)$ only is called a *northeastern lattice path.* Let k and n be positive integers so that $k < n/2$. Define an injection from the set of northeastern lattice paths from $(0,0)$ to $(k, n-k)$ into the set of northeastern lattice paths from $(0,0)$ to $(k+1, n-k-1)$. (Recall that the function f is called an injection if $f(x) = f(y)$ implies $x = y$; in other words, different elements have different images.) Why does this prove that the sequence $\binom{n}{0}, \binom{n}{1}, \cdots, \binom{n}{n}$ is unimodal?

(20) Prove that if k and n are positive integers, and $k \leq n-1$, then we have

$$\binom{n}{k-1}\binom{n}{k+1} \leq \binom{n}{k}\binom{n}{k}. \qquad (4.13)$$

We note that the sequence $a_0, a_1, a_2, \cdots, a_n$ of positive real numbers is called *logarithmically concave* or *log-concave* if for $1 \leq i \leq n-1$, the inequality $a_{i-1}a_{i+1} \leq a_i^2$ holds. So the exercise asks us to prove that the sequence $\binom{n}{0}, \binom{n}{1}, \cdots, \binom{n}{n}$ is log-concave.

(21) (+) Give a non-computational proof of the previous exercise, using northeastern lattice paths.

(22) Prove that if the sequence $a_0, a_1, a_2, \cdots, a_n$ of positive real numbers is log-concave, then it is unimodal.

(23) (+) Let C_n be the number of northeastern lattice paths from $(0,0)$ to (n,n) that never go above the diagonal $x = y$. Prove that $C_n = \binom{2n}{n} - \binom{2n}{n-1} = \binom{2n}{n}/(n+1)$.

(24) (+) Let $a \geq b$ be two positive integers. Prove that the number of northeastern lattice paths from $(0,0)$ to (a,b) that never go above the main diagonal is $\binom{a+b}{b} - \binom{a+b}{b-1}$.

(25) Find a closed form for $\sum_{n=1}^{\infty} nx^{n-1}$.

(26) Prove that $\frac{1}{\sqrt{1-4x}} = \sum_{n\geq 0} \binom{2n}{n} x^n$.

(27) Find the power series form of $f(x) = \sqrt{\frac{1+x}{1-x}}$.

Supplementary Exercises

(28) (-) Prove that for all positive integers $n > 1$, the inequality $2^n < \binom{2n}{n}$ holds.

(29) (-) Prove that for all positive integers n, the number $\binom{2n}{n}$ is even.

(30) (-) Prove that for all positive integers $n > k \geq 2$, the inequality $k^n < \binom{kn}{n}$ holds.

(31) (-) The sum of each row of a 10×6 matrix (that means ten rows, six columns) is 36. If each column of the matrix has the same sum r, what is that sum?

(32) (-) How many northeastern lattice paths are there from $(0,0)$ to (n,k)?

(33) Prove, by a combinatorial argument, that for all positive integers n, the number $\binom{3n}{n,n,n}$ is divisible by six.

(34) A computer programmer claims that he generated six real numbers $a_1, a_2, \cdots, a_6$ so that the sum of any four consecutive a_i is positive, but the sum of any three consecutive a_i is negative. Prove that his claim is false.

(35) A school has 105 students, and seven classes. If each student takes three classes, and each class is taken by the same number of students, how many students are taking each class?

(36) How many northeastern lattice paths are there from $(0,0)$ to $(10,10)$

that do not touch the point $(5,5)$, but do touch the point $(3,3)$?

(37) (+) What is the number of northeastern lattice paths from $(0,0)$ to (n,n) that never touch the main diagonal other than in the starting and ending point?

(38) Prove that for all positive integers n,
$$\binom{2n}{n} = \sum_{k=0}^{n} \binom{n}{k}^2.$$

(39) Prove that for all positive integers n,
$$n\binom{2n-1}{n-1} = \sum_{k=1}^{n} k\binom{n}{k}^2.$$

(40) Prove that for all positive integers n,
$$3^n = \sum_{k=0}^{n} 2^k \binom{n}{k}.$$

(41) Prove that for all positive integers $k \leq n$, the equality
$$\sum_{i=0}^{k} \binom{n}{i}(-1)^i = \binom{n-1}{k}(-1)^k$$
holds.

(42) Take the integral of both sides of the equation
$$(1+x)^n = \sum_{k=0}^{n} \binom{n}{k} x^k.$$
Explain what constant C you will need to take on the right-hand side to keep the equation valid.

(43) Prove that for all positive integers $n > 1$,
$$\sum_{k=0}^{n} \frac{1}{k+1}\binom{n}{k}(-1)^{k+1} = \frac{-1}{n+1}.$$

(44) Find a closed formula for the expression
$$\sum_{k=0}^{n} \frac{1}{k+1}\binom{n}{k} t^{k+1},$$
where t is any fixed real number.

(45) Prove that for all positive integers n, the equality
$$\sum_{\substack{k=0 \\ k \text{ even}}}^{n} \binom{n}{k} 2^k = \frac{3^n + (-1)^n}{2}$$
holds.

(46) Prove that for all positive integers n, the equality

$$\sum_{\substack{k=1 \\ k \text{ odd}}}^{n} \binom{n}{k} 5^k = \frac{6^n - (-4)^n}{2}$$

holds.

(47) (+) Let $n = 4k$, with k being a non-negative integer. Prove that

$$\sum_{i=0}^{2k} \binom{n}{2i} (-1)^i = \binom{n}{0} - \binom{n}{2} + \binom{n}{4} - \cdots = 2^{2k}(-1)^k.$$

(48) (++)

(a) Let $n = 3k$. Prove that

$$\lim_{n \to \infty} \frac{\sum_{i=0}^{k} \binom{n}{3i}}{2^n} = \frac{1}{3}.$$

In other words, the sum of every third element of the nth row of the Pascal triangle is roughly one third of the sum of all elements of that row.

(b) Generalize the result of part (a).

(49) What is the coefficient of x^n in the power series form of $\sqrt[3]{1-2x}$?

(50) If we expand the expression

$$(x_1 + x_2 + x_3 + x_4)^6,$$

what will be the largest coefficient that occurs?

(51) Consider the expression

$$(x_1 + x_2 + \cdots + x_k)^n.$$

(a) Let us assume that when we expand this power, there will be an integer that occurs as a coefficient only once. What relation does that imply between k and n?

(b) Can it happen that there will be more than one coefficient that occurs only once in the expansion?

(52) (+) What digit is immediately on the right of the decimal point in $(\sqrt{3} + \sqrt{2})^{2002}$?

(53) (+) What digits are immediately on the left and right of the decimal point in $(\sqrt{11} + \sqrt{10})^{2002}$?

(54) Let $n \geq 2$. We want to select as many subsets of $[n]$ as possible, without selecting two subsets so that one of them contains the other.

(a) Prove that we can always select at least $2^n/n$ subsets.

(b) Can we improve the result of part (a)?

(55) (+) A company specializing in international trade has 70 employees. For any two employees A and B, there is a language that A speaks but B does not, and also a language that B speaks but A does not. At least how many different languages are spoken by the employees of this company?

(56) Find the number of *pairs* of non-intersecting northeastern lattice paths (p, q) so that p goes from $(0,0)$ to $(k, n-k)$ and q goes from $(-1,1)$ to $(k-1, n-k+1)$.

(57) We have written $2n+1$ numbers around a circle. Among these numbers, n are equal to 1, and $n+1$ are equal to -1. Prove that there is exactly one among these $2n+1$ numbers with the following property. If we call this number a_1, and we call the numbers following it in clockwise order $a_2, a_3, \cdots, a_{2n+1}$, then for all $k \in [2n]$, the sums $\sum_{i=1}^{k} a_i$ are non-negative.

(58) Explain the connection between the previous exercise and Exercise 23.

(59) Let $p > 2$ be a prime number. For what values of n will each binomial coefficient $\binom{n}{k}$, with $0 < k < n$, be divisible by p?

(60) Exercise 20 showed that for any fixed n, the sequence $\binom{n}{0}, \binom{n}{1}, \cdots, \binom{n}{k}$ was log-concave. Now let us prove that for any fixed k, the infinite sequence $\binom{k}{k}, \binom{k+1}{k}, \binom{k+2}{k}, \cdots$ is log-concave. That is, show that for any positive integers $n \geq k$, the inequality

$$\binom{n}{k}\binom{n+2}{k} \leq \binom{n+1}{k}^2$$

holds. Try to give a combinatorial proof, similar to the proof of Exercise 20.

Solutions to Exercises

(1)(a) Yes, one example is shown in Figure 4.3.

(b) For 6×6 grids, however, the answer is no. Indeed, if B is a 6×6 grid, then B can be partitioned into nine squares of size 2×2 each, in an obvious way. Then the sum of the elements of B must equal that of the sum of elements of these 2×2 squares.

(2)(a) Let us determine the *angles* of all the k triangles to be created. These angles will be either at one of the vertices of the octogon, and

−1	−1	−1	−1	−1
−1	4	−1	4	−1
−1	−1	−1	−1	−1
−1	4	−1	4	−1
−1	−1	−1	−1	−1

Fig. 4.3 All 2×2 squares have a negative sum.

then their sum is equal to the sum of the vertices of the octogon, which is $6 \cdot 180 = 1080$ degrees, or they are around one of the thirteen trees, and then they clearly sum to $13 \cdot 360 = 4680$ degrees. Thus the total sum of the angles of the k triangles is $1080 + 4680 = 5760$ degrees.

On the other hand, the sum of the degrees of k triangles is $180 \cdot k$ degrees, so we have $5760 = 180k$, and therefore, $k = 32$.

(b) The five trees on the boundary simply add $5 \cdot 180$ degrees to the sum of all angles, so the number of triangles also increases by five, to 37.

(3) Same as the proof of Theorem 4.6, except that now we are choosing a president and vice-president (if we follow the first proof), or we differentiate (4.5) twice (if we follow the second).

To generalize, for any positive integer $m \leq n$, we can differentiate (4.5) m times, or we can choose m committee members for m different offices, to get

$$2^{n-m}(n)_m = \sum_{k=m}^{n} (k)_m \binom{n}{k}.$$

(4) Both sides count the number of ways to choose an m-member soccer team and a k-member basketball team from a group of n people, so that nobody is on two teams. The left-hand side is the result of computing this number by choosing the soccer team first, while the right-hand side is the result of computing this number by choosing the basketball team first.

(5) The left-hand side is the number of 0-1 sequences of length n with at most k ones. The right-hand side is more complicated. Note that if we want to check if a 0-1 sequence S of length n has at most k ones,

and to that end, we test the first, second, third, etc. digits of S in this order, then as soon as we find $n-k$ zeros in S, we can be sure that S has at most k ones. If, on the other hand we do not find $n-k$ zeros in S, then S has more than k ones.
Knowing this, let us count 0-1 sequences with at most k ones according to the position of their $(n-k)$th zeros. The above paragraph shows that such a zero always exists. Let us say that this zero occurs in the $(n-j)$th position. Then $0 \leq j \leq k$ for trivial reasons. There have to be $n-k-1$ zeros on the left of this position- that can happen in $\binom{n-j-1}{n-k-1} = \binom{n-1-j}{k-j}$ ways, and there can be any number of zeros on the right of this position, which can be done in 2^j ways. Summing for j we obtain our claim.

(6) This sum is always the same, namely, it is $\binom{n}{2}$ if $n > 1$. We prove this by strong induction on n. The initial case is trivial. Assume we know the statement for all positive integers less than n, and prove it for n. Let us split our heap of n stones into two small heaps, one of size k, and one of size $n-k$. Then $p_1 = k(n-k)$. Then, by our induction hypothesis, the contribution of the first heap to the sum $p_1 + p_2 + \cdots + p_{n-1}$ is $\binom{k}{2}$, and that of the second heap is $\binom{n-k}{2}$. As

$$k(n-k) + \binom{k}{2} + \binom{n-k}{2} = \binom{n}{2},$$

our claim is proved.

(7) Consider all odd prime divisors of a positive integer n. They are either of the form $4k+1$, or of the form $4k-1$. Denote them by $a_1, a_2, \cdots, a_m$, and $b_1, b_2, \cdots b_p$, respectively. Let

$$n = 2^t \cdot a_1^{x_1} \cdots a_m^{x_m} \cdot b_1^{y_1} \cdots b_p^{y_p}.$$

An odd divisor of n will be of the form $4k-1$ if and only if it contains an odd number of prime factors of the form $4k-1$, multiplicities counted.
Now we construct an *injection* from the set of divisors of n of the form $4k-1$ into the set of divisors of n of the form $4k+1$. Our injection will be very simple as it will only change the exponent of one of the b_i. However, the construction of the injection will depend on n.
Let q be a divisor of n of the form $4k-1$. Then

$$q = a_1^{c_1} \cdots a_m^{c_m} \cdot b_1^{d_1} \cdots b_p^{d_p},$$

with $c_i \leq x_i$, $d_i \leq y_i$, for all i, and the sum of the d_i is odd.

Assume first that n is such that one of the y_i is odd; say y_1. We then define

$$f(q) = a_1^{c_1} \cdots a_m^{c_m} \cdot b_1^{y_1 - d_1} \cdots b_p^{d_p}.$$

Then the parity of the exponent of b_1 changed, all other parities are the same, so the sum of the exponents of the b_i is now even. Therefore, $f(q)$ is of the form $4k + 1$. This is clearly an injection (in fact, a bijection), as $f(f(q)) = q$.

If n is such that all the y_i are even, then we define a different injection g. Let i be minimal so that $d_i < y_i$. (There has to be such an i, otherwise all d_i are even, and q is of the form $4k + 1$.) Then we define

$$a_1^{c_1} \cdots a_m^{c_m} \cdot b_1^{y_1} \cdots b_i^{y_i - 1 - d_i} \cdots b_p^{d_p}.$$

This will again change the parity of the exponent of b_i, therefore $g(q)$ will be of the form $4k + 1$. Also note that i can be read off the image $g(i)$ as it is still the smallest index for which $d_i < y_i$. This function g is an injection. Indeed, to have $g(q) = g(q')$, the integer g' must have the exact same prime decomposition as q, so it must be equal to q'. It is not a bijection though, for $b_1^{y_1} \cdots b_p^{y_p}$ is not in its image.

So for all positive integers, we showed that there is an injection from the set of divisors of the form $4k - 1$ into that of divisors of the form $4k + 1$, and therefore we proved the statement.

(8) The left-hand side is the number of n-element subsets of $[2n]$, while the right-hand side is the number of *all* subsets of $[2n]$.

(9) Arrange all subsets of $[n]$ into pairs, by matching each subset to its complement. If n is odd, then two subsets of the same pair can never be the same size, so exactly one of them has the required property (the larger one). Therefore, half of all subsets, that is, 2^{n-1} subsets are larger than their complements.

If n is even, then there will be $\binom{2n}{n}/2$ pairs, namely those pairs consisting of $n/2$-element subsets and their complements, in which no subset has the required property. So in this case, the answer is $2^{n-1} - \frac{1}{2}\binom{2n}{n}$.

(10) We must find the k-tuple of non-negative integers $a_1, a_2, \cdots, a_k$ for which $\sum_{i=1}^{k} a_i = k$, and $\frac{k!}{a_1! \cdot a_2! \cdots a_k!}$ is maximal. The numerator of this fraction is constant, while its denominator is at least 1 as it is a product of positive integers. (Recall that $0! = 1$.) Therefore, the fraction is largest when its denominator is equal to 1. That happens when $a_1 = a_2 = \cdots = a_k = 1$. In that case, the obtained coefficient is $k!$, and it belongs to $x_1 x_2 \cdots x_k$.

(11) The largest coefficient is $n!$, by the same argument as in the previous exercise.

(12) It is straightforward to verify that if $a+b$ is constant, then $a!b!$ is minimal when $a=b$ (if $a+b$ is even), or when $|a-b|=1$ (when $a+b$ is odd). Now consider $\frac{n!}{a_1!\cdot a_2!\cdots a_k!}$. Again, the numerator is constant, so we need to minimize the denominator. Using the fact we mentioned at the beginning of this solution, one sees that the denominator is minimal when $a_i = r$ for all r. Therefore, the largest coefficient is

$$\binom{rk}{r, r, \cdots, r} = \frac{(rk)!}{r!^k}.$$

(13) Let $i \leq 2^m - 1$ be a positive integer. There is a unique way to write $i = 2^j \cdot p$, where p is an odd integer. Then $2^m - i = 2^m - 2^j \cdot p = 2^j(2^{m-j} - p)$. This shows that the number of times 2 occurs in the prime factorization of i is equal to the number of times 2 occurs in the prime factorization of $2^m - i$. Now note that

$$\binom{n}{k} = \prod_{i=1}^{k} \frac{2^m - i}{i}.$$

Our argument shows that no factor $\frac{2^m-i}{i}$ of the right-hand side is divisible by 2. Therefore, the prime factorization of $\binom{n}{k}$ does not contain 2, and so $\binom{n}{k}$ is odd.

(14) We know from Theorem 4.2 that $\binom{n}{k} = \binom{n-1}{k-1} + \binom{n-1}{k}$. The previous exercise shows that both members of the right-hand side are odd, so the left-hand side is even.

(15) Let j be an integer so that $1 \leq i \leq k$, and let j be the unique integer such that $i = p^j t$, where t is not divisible by j. Then $p^m - i = p^m - p^j t = p^j(p^{m-j} - t)$. So if p occurs j times in the prime factorization of i, then p occurs j times in the prime factorization of $p^m - i$. Now

$$\binom{p^m}{k} = \frac{p^m}{k} \cdot \prod_{i=1}^{k-1} \frac{p^m - i}{i}.$$

Note that the first term of the right-hand side is divisible by p, while in the other terms of the right-hand side, the p-factors cancel out, and the proof is complete.

(16)(a) There are x^p paint jobs, but x of them use only one color, thus the number of good paint jobs is $x^p - x$.

(b) As p is prime, each paint job can be rotated to $p-1$ other paint jobs. Thus the number of different paint jobs is $(x^p - x)/p$.

(c) As the number of different paint jobs must be an integer, this proves that $x^p - x$ is divisible by p. This is called Euler's theorem (or, sometimes, Fermat's theorem).

(17) This follows directly from the multinomial theorem by substituting $x_1 = x_2 = x_3 = 1$.

(18) This follows directly from the multinomial theorem by substituting $x_1 = x_3 = 1$, and $x_2 = -1$.

(19) Let p be a northeastern lattice path from $(0,0)$ to $(k, n-k)$. Let t be the bisector of the segment joining $A = (k, n-k)$ and $B' = (k+1, n-k-1)$. As $k < n/2$, the path p must intersect t at least once. Let L be the intersection point of p and t that is closest to A. Now *reflect* the part of p between L and A through t, to get a path from L to B. Prepending this with the unchanged part of p from $(0,0)$ to L, we get a path p' from $(0,0)$ to B. It is clear that the function f defined this way by $f(p) = p'$ is an injection. Indeed, given a path q from $(0,0)$ to B, either q and t do not intersect, and then q does not have a preimage, or they do intersect, and then L can be found as above, and the preimage of q is obtained by reflecting the part of q between L and B through t.

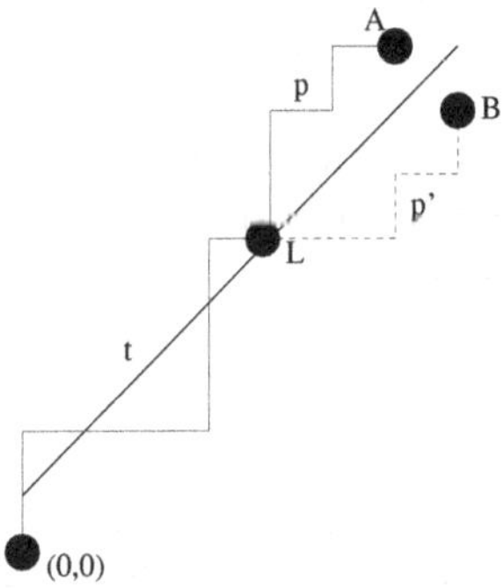

Fig. 4.4 Constructing the injection f.

As we know from Exercise 32, the number of northeastern lattice paths from $(0,0)$ to $(k, n-k)$ is $\binom{n}{k}$. This proves that $\binom{n}{k} \leq \binom{n}{k+1}$ if $k < n/2$. On the other hand, we also know that $\binom{n}{k} = \binom{n}{n-k}$, proving that $\binom{n}{k} > \binom{n}{k+1}$ if $k > n/2$. So the numbers $\binom{n}{k}$ first increase steadily, then decrease steadily, in other words, they form a unimodal sequence.

The technique used in this solution is called the *reflection principle.* See the Notes for references on this subject.

(20) By the definition of the binomial coefficients, (4.13) is equivalent to

$$\frac{n!}{(k-1)!(n-k+1)!} \cdot \frac{n!}{(k+1)!(n-k-1)!} \leq \frac{n!}{k!(n-k)!} \cdot \frac{n!}{k!(n-k)!}.$$

Dividing both sides by $n!^2$ and then multiplying both sides by the product $(k+1)!(k-1)!(n-k+1)!(n-k-1)!$, we get that (4.13) is equivalent to

$$1 \leq \frac{k+1}{k} \cdot \frac{n-k+1}{n-k},$$

which is obviously true as both terms on the right-hand side are larger than one.

(21) Clearly, the binomial coefficient $\binom{n}{k-1}$ enumerates northeastern lattice paths from $A = (1,0)$ to $B = (k, n-k+1)$, whereas the binomial coefficient $\binom{n}{k+1}$ enumerates northeastern lattice paths from $C = (0,1)$ to $D = (k+1, n-k)$. On the other hand, $\binom{n}{k}$ enumerates northeastern lattice paths from A to D and also from C to B.

We are going to define a function g that takes a pair of paths, one from A to B, and one from C to D, and maps them into a pair of paths, one from A to D, and one from B to C. We will then show that g is an injection. That will prove our claim by the easy enumerative considerations of the previous paragraph.

Our map g is simplicity itself. Take a northeastern path p from A to B, and a northeastern path q from C to D. Then p and q must intersect; let X be their first intersection point. *Flip* the parts of paths XB and XD, to get two new paths, one from A to X to D, and one from C to X to B. Call these two paths p' and q', and define $g(p,q) = (p', q')$. To see that the map g is an injection, note that given two paths s and u from A to D, and from B to C, either s and u do not intersect, or they do, but then they have a first intersection point X. In this latter case, their preimage can be obtained by flipping the part XB of s and the part XB of u back.

(22) If the mentioned sequence is log-concave, then

$$\frac{a_1}{a_0} \geq \frac{a_2}{a_1} \geq \frac{a_3}{a_2} \geq \cdots \geq \frac{a_{n-1}}{a_n}.$$

This means that the ratio $\frac{a_i}{a_{i-1}}$ is steadily decreasing, so in particular once it dips below one, it will stay below one. Therefore, once the

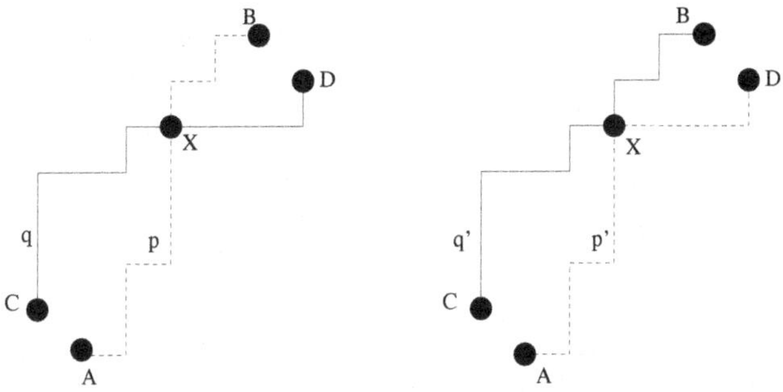

Fig. 4.5 Constructing the injection g.

sequence of the a_i starts decreasing, it will keep decreasing, showing that this is indeed a unimodal sequence.

(23) We know from Exercise 32 that the number of all northeastern lattice paths from $(0,0)$ to (n,n) is $\binom{2n}{n}$. Let us enumerate the bad ones, that is, those that go above the diagonal. In other words, these are the northeastern paths that touch the line $y = x + 1$.

We prove that these paths are in bijection with northeastern paths from $(-1,1)$ to (n,n). Let p be such a path, and let P be the first intersection point of p and the line $y = x+1$. Let us reflect the part p_s of p that is between the origin and P through the line $y = x+1$. This reflection takes $(0,0)$ into $(-1,1)$, and so it take p_s into a northeastern lattice path p'_s from $(-1,1)$ to P. If we append the rest of p to the end of p'_s, we get a path $h(p)$ from $(-1,1)$ to (n,n). To see that h is a bijection, note that every path from $(-1,1)$ to (n,n) must intersect the line $y = x+1$, so P can be recovered, and therefore, by reflection, the preimage of any path can be uniquely recovered.

Thus the number of "bad" paths is $\binom{2n}{n-1}$, therefore the number of good paths is $\binom{2n}{n} - \binom{2n}{n-1} = \binom{2n}{n}/(n+1)$.

(24) Note that the previous problem was a special case of this, i.e., when $a = b$, but we have not used the equality of these two parameters in the proof. Therefore, the same proof will work.

(25) **First solution.** Recall that $1/(1-x) = \sum_{n\geq 0} x^n$. Taking derivatives, we get

$$\frac{1}{(1-x)^2} = \sum_{n\geq 1} nx^{n-1} = \sum_{n\geq 0}(n+1)x^n.$$

Second solution. Apply Theorem 4.15 with $m = -2$, and replace x

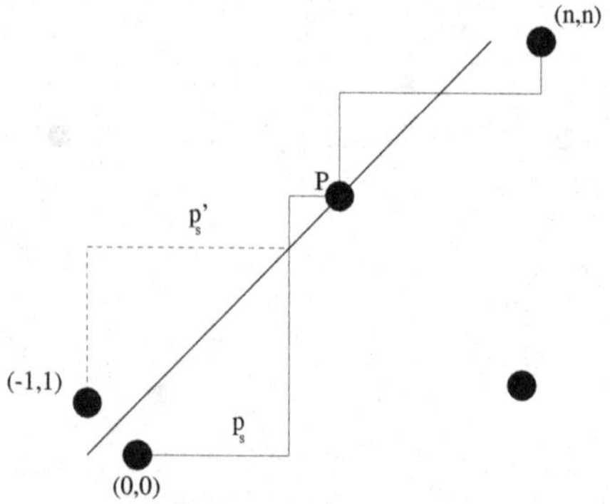

Fig. 4.6 Constructing the bijection h.

by $-x$. Note that

$$\binom{-2}{n} = \frac{(-2)(-3)\cdots(-n-1)}{n!} = (n+1)(-1)^n.$$

Therefore, Theorem 4.15 implies

$$\frac{1}{(1-x)^2} = \sum_{n\geq 0}(n+1)(-1)^n(-x)^n = \sum_{n\geq 0}(n+1)x^n.$$

(26) We know that $\frac{1}{\sqrt{1-4x}} = (1-4x)^{-1/2}$, therefore, the binomial theorem implies

$$\frac{1}{\sqrt{1-4x}} = \sum_{n\geq 0}\binom{\frac{-1}{2}}{n}(-4x)^n$$
$$= \sum_{n\geq 0}\frac{\frac{-1}{2}\cdot\frac{-3}{2}\cdots\frac{-2n+1}{2}}{n!}(-1)^n 2^{2n}x^n$$
$$= \sum_{n\geq 0}2^n\frac{1\cdot 3\cdots(2n-1)}{n!}x^n.$$

So all we have to show is that

$$\binom{2n}{n} = 2^n\frac{1\cdot 3\cdots(2n-1)}{n!},$$

$$\frac{(2n)!}{n!} = 2^n 1\cdot 3\cdots(2n-1),$$

and this is true as on the left-hand side we can simplify all fractions of the form $\frac{2i}{i}$. Then we will be left with 2^n from the n fractions of this form, and all the odd terms $(2i+1)$.

(27) First of all, $f(x) = (1+x)(1-x^2)^{1/2}$. If we replace x by $x^2/4$ in the result of the previous exercise, this implies that

$$\begin{aligned}
f(x) &= (1+x)\sum_{n\geq 0} \frac{\binom{2n}{n}}{4^n} x^{2n} \\
&= (1+x)\sum_{n\geq 0} 2^n \frac{(2n-1)!!}{4^n n!} x^{2n} \\
&= (1+x)\sum_{n\geq 0} \frac{(2n-1)!!}{(2n)!} x^{2n} \\
&= \sum_{n\geq 0} \frac{(2n-1)!!}{(2n)!} x^{2n} + \sum_{n\geq 0} \frac{(2n-1)!!}{(2n)!} x^{2n+1}.
\end{aligned}$$

Chapter 5

Divide and Conquer. Partitions

After the break taken in the last chapter, it is time we returned to our basic enumeration problems. In Chapter 3 we were mainly concerned about lists of objects, distinct or not, with repetitions allowed or not, and with the order of the elements on the list being relevant or not. In this chapter we will go one step further by considering *distribution problems*. We will distribute n objects (balls) into k boxes, and ask in how many ways this can be done.

5.1 Compositions

Let us assume we want to give away twenty identical balls to four children, Alice, Bob, Charlie and Denise. As the balls are identical, what matters is how many balls each child will get. So if we want to know the number of ways we can give away these balls, we simply have to know the number of ways to write 20 as a sum of four non-negative integers. Clearly, the order of the integers will matter, that is, $1+6+8+5$ does not correspond to the same way of distributing the balls as $6+1+5+8$. Indeed, in the first case, Alice gets only one ball, in the second, she gets six.

Definition 5.1. A sequence $(a_1, a_2, \cdots, a_k)$ of integers fulfilling $a_i \geq 0$ for all i, and $(a_1 + a_2 + \cdots + a_k) = n$ is called a *weak composition* of n. If, in addition, the a_i are *positive* for all $i \in [k]$, then the sequence $(a_1, a_2, \cdots, a_k)$ is called a *composition* of n.

Theorem 5.2. *For all positive integers n and k, the number of weak compositions of n into k parts is*

$$\binom{n+k-1}{k-1} = \binom{n+k-1}{n}.$$

Proof. The problem is certainly equivalent to counting the number of ways of putting n identical balls into k different boxes. Place the k boxes in a line, then place the balls in them in some way and align them in the middle of the boxes. This creates a long line consisting of n balls and $k-1$ walls separating the k boxes from each other. Note that simply knowing in which order the n identical balls and $k-1$ separating walls follow each other is the same as knowing the number of balls in each box. So our task is reduced to finding the number of ways to permute the multiset consisting of n balls and $k-1$ walls. Theorem 3.5 tells us that this number is

$$\frac{(n+k-1)!}{n! \cdot (k-1)!}. \qquad \square$$

What if a grandparent insists on giving at least one ball to each child? The problem is not any harder. First we can give one ball to each child, then give away the remaining 16 balls to the four children in any of $\binom{16+4-1}{4-1} = \binom{19}{3}$ ways. The generalization of this argument to n balls and k children is the following statement.

Corollary 5.3. *For all positive integers n and k, the number of compositions of n into k parts is $\binom{n-1}{k-1}$.*

How about the number of *all* compositions, that is, the number of compositions of n into *any* number of parts? Clearly, this question only makes real sense for compositions, not for weak compositions. Indeed, if 0 is allowed to be a part, then any number of zeros can be appended to the end of any composition, therefore any positive integer n has infinitely many weak compositions. For compositions, however, the question has a remarkably compact answer.

Corollary 5.4. *For all positive integers n, the number of all compositions of n is 2^{n-1}.*

Proof. A composition of n will have at least one and at most n parts. So the total number of compositions of n is

$$\sum_{k=1}^{n} \binom{n-1}{k-1} = 2^{n-1}.$$

Indeed, the left-hand side is the number of all subsets of $[n-1]$, first enumerated by their size k, and then summed over $k \in [n]$. □

The reader is hopefully thinking right now that such a nice closed result, 2^{n-1}, must have an alternative explanation, one that really explains why the result is a power of two. Such a proof indeed exists and we provide it below.

Proof. (of Corollary 5.4) We prove the statement by induction on n. For $n = 1$, the statement is true as the integer 1 has one composition. Now assume that the statement is true for n, and take all 2^{n-1} compositions of n. For each such composition C, we will define two different compositions of $n+1$. First, add one to the first element of C. This way we get a composition of $n+1$ with the first element at least 2. Second, take C, and write an additional 1 to its front. This way we get a composition of $n+1$ with first element 1. It is clear that different compositions of n lead to different compositions of $n+1$ this way. Each decomposition of $n+1$ can be obtained in exactly one of these two ways. Therefore, it follows that $n+1$ has twice as many compositions as n, which was to be proved. □

5.2 Set Partitions

Now let us assume that the *balls* are different, but the *boxes* are not. Then we might as well label the balls by numbers 1 through n. In other words, we may simply say that we want to *partition* the set $[n]$ into k *nonempty* subsets.

Definition 5.5. A *partition* of the set $[n]$ is a collection of non-empty blocks so that each element of $[n]$ belongs to exactly one of these blocks.

The number of partitions of $[n]$ into k nonempty blocks is denoted by $S(n,k)$. The numbers $S(n,k)$ are called the *Stirling numbers of the second kind.*

It follows from Definition 5.5 that $S(n,k) = 0$ if $n < k$. We set $S(0,0) = 1$ by convention. In the next chapter, you will see an advantage of this convention. Until then, be comforted in knowing that there is one way to distribute zero objects into zero boxes, namely by not doing anything.

Example 5.6. For all $n \geq 1$, we have $S(n,1) = S(n,n) = 1$. For all $n \geq 2$, the equality $S(n,n-1) = \binom{n}{2}$ holds as a partition of $[n]$ into $n-1$ blocks must consist of one doubleton and $n-2$ singletons.

Example 5.7. The set $[4]$ has seven partitions into two nonempty blocks, namely $\{1,2,3\}\{4\}$; $\{1,2,4\}\{3\}$; $\{1,3,4\}\{2\}$; $\{2,3,4\}\{1\}$, and also $\{1,2\}\{3,4\}$; $\{1,3\}\{2,4\}$; and $\{1,4\}\{2,3\}$. Therefore, $S(4,2) = 7$.

Several questions are in order. The reader may wonder what happened to the Stirling numbers of the first kind. These will be discussed in Chapter 6. The reader may also think that the first thing we will do is to provide a formula for $S(n,k)$, and may in fact wonder why we have not done it yet. However, there exists no closed formula for $S(n,k)$. There is a formula for $S(n,k)$ that contains one summation sign, and we will prove it in Chapter 7 as we need the sieve formula to obtain it.

Nevertheless, we can prove some nice identities about set partitions right now. They will be of recursive nature.

Theorem 5.8. *For all positive integers $k \leq n$,*

$$S(n,k) = S(n-1,k-1) + k \cdot S(n-1,k). \tag{5.1}$$

Proof. As before, we can obtain a combinatorial proof by taking a close look at one particular element, say the maximum element n. If this element forms a singleton block, then the remaining $n-1$ elements have exactly $S(n-1,k-1)$ ways to complete the partition. These partitions are enumerated by the first member of the right-hand side. If, on the other hand, the element n does not form a block by itself, then the remaining $n-1$ elements must form a partition with k blocks in one of $S(n-1,k)$ ways. Then we can add n into any of the k blocks formed by this partition, multiplying the number of all our possibilities by k. These partitions are enumerated by the second member of the right-hand side. As the left-hand side enumerates all partitions of $[n]$ into k blocks, the claim is proved. □

If we have to put n different balls into k different boxes then the number of ways to do this is $k! \cdot S(n,k)$. Indeed, first we can partition $[n]$ into k non-distinguishable blocks in $S(n,k)$ ways, then we can label the k blocks with labels $1, 2, \cdots, k$ in $k!$ different ways.

Corollary 5.9. *The number of all surjective functions $f : [n] \rightarrow [k]$ is $k! \cdot S(n,k)$.*

Proof. Such a function defines a partition of $[n]$. The blocks are the subsets of elements that are mapped into the same element $i \in [k]$. Therefore, the blocks are labeled, and there are exactly k of them, so the proof follows from the previous paragraph. □

An interesting consequence of this is the following unexpected corollary. It is surprising as it shows that x^n, this very compact expression, is in fact a sum of $n+1$ terms involving Stirling numbers.

Corollary 5.10. *For all real numbers x, and all non-negative integers n,*

$$x^n = \sum_{k=0}^{n} S(n,k)(x)_k. \tag{5.2}$$

Proof. Both sides are polynomials of x of degree n. So if we can show that they agree for more than n values of x, we will be done. We will prove an even stronger statement, namely that the two sides agree for *all positive integers x*.

So let x be a positive integer. Then the left-hand side is the number of all functions from $[n]$ to $[x]$. We claim that the right-hand side is the same, enumerated according to the *size of the image*. Indeed, if the image of such a function is of size k, then there are $\binom{x}{k}$ choices for the image I, then, by Corollary 5.9, there are $k! \cdot S(n,k)$ choices for the function itself. As $(x)_k = k! \cdot \binom{x}{k}$, the claim is proved. □

Another way of extending our enumeration of partitions is by enumerating *all* partitions, without restricting the number of parts.

Definition 5.11. The number of *all* set partitions of $[n]$ into nonempty parts is denoted by $B(n)$, and is called the nth *Bell number*. We also set $B(0) = 1$.

So $B(n) = \sum_{i=0}^{n} S(n,i)$. The Bell numbers also satisfy a nice recurrence relation.

Theorem 5.12. *For all non-negative integers n,*

$$B(n+1) = \sum_{i=0}^{n} \binom{n}{i} B(i). \tag{5.3}$$

Proof. We must prove that the right-hand side enumerates all partitions of $[n+1]$. Let us assume the element $n+1$ is in a block of size $n-i+1$. Then there are i elements that are not in the same block as $n+1$. Therefore, there are $\binom{n}{i}$ ways to choose these elements, and then there are $B(i)$ ways to partition them. Summing over all possible values of i, we get the statement of the theorem. □

5.3 Integer Partitions

Now assume that *both* the balls and the boxes are indistinguishable, so when we distribute the balls into the boxes, the only thing that matters is their numbers. In other words, we are interested in finding out the number of ways of writing the positive integer n as a sum of positive integers, where the order of the summands does not matter. That is, $4 = 3+1$ or $4 = 1+3$ will count as only one way of writing four as a sum of positive integers.

As the order of the summands does not matter, we do not lose generality if we assume that they are in weakly decreasing order.

Definition 5.13. Let $a_1 \geq a_2 \geq \cdots \geq a_k \geq 1$ be integers so that $a_1 + a_2 + \cdots + a_k = n$. Then the sequence $(a_1, a_2, \cdots, a_k)$ is called a *partition* of the integer n. The number of all partitions of n is denoted by $p(n)$. The number of partitions of n into exactly k parts is denoted by $p_k(n)$.

We note that the word "partition" is used in a new meaning here. We have used it before, in Definition 5.5, to mean "a way to split the set $[n]$". The new meaning, given in Definition 5.13 is independent of the old one. This double meaning of the same word usually does not result in confusion as the context usually clearly indicates which meaning is relevant. In writing, so too does the notation, that is, we either speak of partitions of $[n]$, or of partitions of n. If there is a danger of confusion after all, it is customary to refer to partitions of $[n]$ as "set-partitions". We also note that some languages, like French, do have two different words for these two notions ("partition" for set-partitions, and "partage" for partitions of the integer n).

Example 5.14. The positive integer 5 has 7 partitions. Indeed, they are (5); (4, 1); (3, 2); (3, 1, 1); (2, 2, 1); (2, 1, 1, 1); (1, 1, 1, 1, 1). Therefore, $p(5) = 7$.

The problem of finding an exact formula for $p(n)$ is even harder than that of finding an exact formula for $S(n, k)$. If we know $p(n-1)$, or, for that matter, $p(i)$ for all $i < n$, we still cannot directly compute $p(n)$ from these data (though some sophisticated recurrence relations do exist, and we will mention them in the Notes section). The approximate size of the number $p(n)$ is provided by the following asymptotic formula.

$$p(n) \sim \frac{1}{4\sqrt{3}} \exp\left(\pi\sqrt{\frac{2n}{3}}\right). \tag{5.4}$$

In other words, $p(n)$ grows faster than any polynomial, but slower than any exponential function $g(n) = c^n$, with $c > 1$.

We will nevertheless find some interesting and useful results concerning $p(n)$ once we will have learned about generating functions. Until then, we will discuss some highly interesting and enjoyable identities. Our main tool in proving them will be the following graphical representation of partitions.

A *Ferrers shape* of a partition $p = (x_1, x_2, \cdots, x_k)$ is a set of n square boxes with horizontal and vertical sides so that in the ith row we have x_i boxes and all rows start at the same vertical line. It is named after the British mathematician Norman MacLeod Ferrers. The Ferrers shape of the partition $p = (4, 2, 1)$ is shown in Figure 5.1. Clearly, there is an obvious bijection between partitions of n and Ferrers shapes of size n.

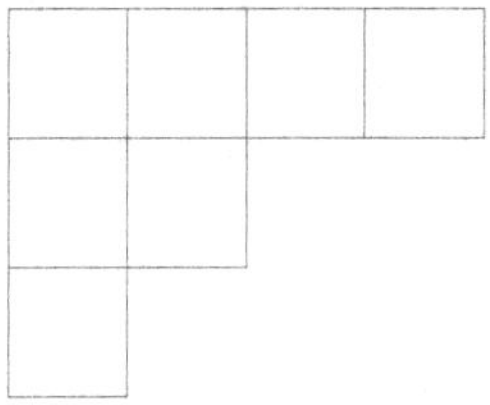

Fig. 5.1 The Ferrers shape of the partition $p = (4, 2, 1)$.

If we reflect a Ferrers shape of a partition p with respect to its main diagonal, we get another shape, representing the *conjugate* partition of p. Thus, in our example, the conjugate of $(4, 2, 1)$ is $(3, 2, 1, 1)$. In particular, the ith row of the Ferrers shape of the conjugate partition of p is as long as the ith column of the Ferrers shape of p.

Definition 5.15. A partition of n is called *self-conjugate* if it is equal to its conjugate.

Example 5.16. Partitions $(4, 3, 2, 1)$, $(5, 1, 1, 1, 1)$, and $(4, 2, 1, 1)$ are all self-conjugate.

Now we are in a position to use Ferrers shapes to prove various partition identities.

Example 5.14 shows that the positive integer 5 has three partitions into at most two parts, 5, $(4, 1)$ and $(3, 2)$, and it also has three partitions into

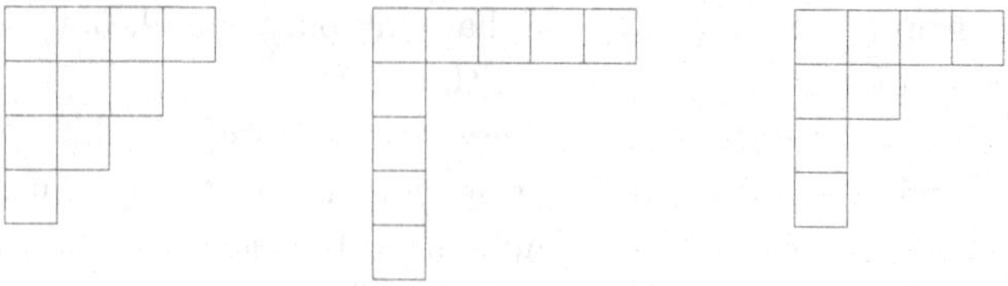

Fig. 5.2 Self-conjugate partitions.

parts that are at most two, namely $(2,2,1),(2,1,1,1)$, and $(1,1,1,1,1)$. This is not by accident.

Theorem 5.17. *The number of partitions of n into at most k parts is equal to that of partitions of n into parts not larger than k.*

Proof. The first number is equal to that of Ferrers shapes of size n with at most k rows. The second number is equal to that of Ferrers shapes with at most k columns. Finally, these two sets of Ferrers shapes are equinumerous as one can see by taking conjugates. □

Theorem 5.18. *The number of partitions of n into distinct odd parts is equal to that of all self-conjugate partitions of n.*

Proof. We define a bijection f from the set of self-conjugate partitions of n onto that of partitions of n into distinct odd parts as follows. Take any self-conjugate partition $\pi = (\pi_1, \pi_2, \cdots, \pi_t)$ of n. Take its Ferrers shape, and remove all the boxes from its first row and first column. As π is self-conjugate, this means removing $2\pi_1 - 1$ boxes. Set $f(\pi)_1 = 2\pi_1 - 1$, that is, make the first part of the image of π of size $2\pi_1 - 1$. Then continue this way. That is, remove the first row and column of the remaining Ferrers shape. This means removing $2\pi_2 - 3$ boxes. So set $f(\pi)_2 = 2\pi_2 - 3$. Continue this way until the entire Ferrers shape is removed. The resulting partition will be of the form $f(\pi) = (2\pi_1 - 1, 2\pi_2 - 3, \cdots, 2\pi_i - (2i-1), \cdots)$. So it will indeed be a partition of n into odd parts, and the parts will all be distinct, as we had $\pi_1 \geq \pi_2 \geq \cdots \geq \pi_t$. We note that the set of all boxes consisting of one fixed box b, all boxes below b, and all boxes on the right of b, is often called a *hook*. Using this terminology, we can say that in each step of our algorithm, we remove the hook of the box that is currently in the top left corner of our Ferrers shape.

To see that f is a bijection, it suffices to prove that for any partition α of n into distinct odd parts, there exists exactly one self-conjugate partition π of n so that $f(\pi) = \alpha$. Indeed, let $\alpha = (2a_1 - 1, 2a_2 - 3, \cdots, 2a_u - (2u-1))$.

Then it follows from the definition of f that if $f(\pi) = \alpha$, then the first row and column of π must each contain a_1 boxes, the second row and second column of π must each contain a_2 additional boxes, and so on. So we can build up the unique Ferrers shape whose partition has image α, proving our claim. □

Example 5.19. If $\pi = (6, 6, 4, 3, 2, 2)$, then $f(\pi) = (11, 9, 3)$. Figure 5.3 shows how $f(\pi)$ is constructed. In step i, the hook consisting of all boxes labeled i in the Ferrers shape of π is removed, and its boxes form row i of the Ferrers shape of $f(\pi)$.

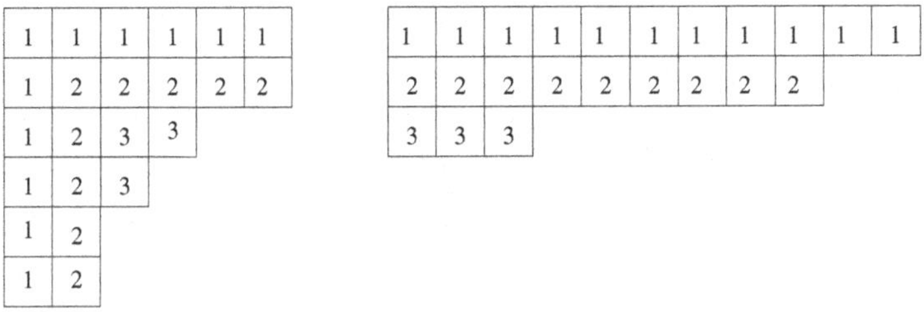

Fig. 5.3 A self-conjugate partition and its image.

Theorem 5.20. *Let $q(n)$ be the number of partitions of n in which each part is at least two. Then $q(n) = p(n) - p(n-1)$, for all positive integers $n \geq 2$.*

Proof. We construct a bijection from the set of all partitions of $n-1$ onto the set of all partitions of n that have at least one part equal to one. The bijection is very simple: just add a part equal to 1 to the end of each partition of $n-1$. The only partitions of $p(n)$ that cannot be obtained this way are those enumerated by $q(n)$, so the claim is proved. □

We will not provide a formula for $p_k(n)$ here (any such formula would be cumbersome anyway). However, we will see in Chapter 8 how to get a good description of these numbers.

Let us try to find some connection between the number of partitions of the integer n, and that of partitions of the set $[n]$. It is clear that the set partitions $\{1, 2, 3\}, \{4\}$ and $\{1, 2, 4\}, \{3\}$ do have something in common. Indeed, they both consist of a block of size three and another block of size one. We are going to generalize this notion as follows. Let $\pi = \{\pi_1, \pi_2, \cdots, \pi_k\}$

be a partition of $[n]$, where the π_i denote the *blocks* of π. Rearrange the sequence of block sizes $|\pi_1|,|\pi_2|,\cdots,|\pi_k|$ in non-increasing order to get the sequence $a_1 \geq a_2 \geq \cdots \geq a_k$. Then $a = (a_1, a_2, \cdots, a_k)$ is a partition of the integer n. We say that a is the *type* of the set partition π.

Example 5.21. The set partition $\{1,5,6\},\{2,7\},\{3,9\},\{4,8\},\{10\}$ is of type $(3,2,2,2,1)$.

We can now state and prove a theorem on the number of set partitions of a given type.

Theorem 5.22. *Let $a = (a_1, a_2, \cdots, a_k)$ be a partition of the integer n, and let m_i be the multiplicity of i as a part of a. Then the number of set partitions of $[n]$ that are of type a is equal to*

$$P_a = \frac{\binom{n}{a_1, a_2, \cdots, a_k}}{\prod_{i\geq 1} m_i!}.$$

Proof. Take a_i balls of color i, for all $i \in [k]$. Order them linearly in $\binom{n}{a_1,a_2,\cdots,a_k}$ ways. Then partition $[n]$ so that i and j are in the same block if and only if the linear order we just created has monochromatic balls in positions i and j. This procedure clearly creates a set partition of type a.

However, the number of different set partitions constructed this way is not necessarily $\binom{n}{a_1,a_2,\cdots,a_k}$. For example, if $a_1 = a_2$, then having the a_1 balls of color 1 in a subset A of positions, and having the a_1 balls of color 2 in a subset B of positions will result in the same partition as having the balls of color 1 in B, and having the balls of color 2 in A. In general, if m_i is the multiplicity of i as a part of a, then there are $m_i!$ ways the m_i color classes having i balls each can be permuted among each other. Therefore, every set partition of type a will be obtained from exactly $\prod_{i\geq 1} m_i!$ linear orders, and the proof follows. □

The following table summarizes our results from this chapter *when no empty boxes are allowed.*

Table 5.1.

	parameters	formula
Surjections	n distinct objects, k distinct boxes	$S(n,k)k!$
	n distinct objects, any number of distinct boxes	$\sum_{i=1}^{n} S(n,i)i!$
Compositions	n identical objects, k distinct boxes	$\binom{n}{k}$
	n identical objects, any number of distinct boxes	2^{n-1}
Set Partitions	n distinct objects, k identical boxes	$S(n,k)$
	n distinct objects, any number of identical boxes	$B(n)$
Integer Partitions	n identical objects, k identical boxes	$p_k(n)$
	n distinct objects, any number of identical boxes	$p(n)$

If empty boxes are allowed, then we have to fix the number of boxes. Indeed, if we do not, then we can add as many empty boxes as we want, yielding infinitely many solutions to all these problems. Therefore, instead of eight different enumeration problems, we only have to treat four. Their results have either been proved in this chapter, or are trivial. Table 5.2 summarizes them.

Table 5.2.

	parameters	formula
Functions	n distinctl objects, k distinct boxes	k^n
Weak Compositions	n identical objects, k distinct boxes	$\binom{n+k-1}{k-1}$
Set Partitions	n distinct objects, k identical boxes	$\sum_{i=1}^{k} S(n,i)$
Integer Partitions	n identical objects, k identical boxes	$\sum_{i=1}^{k} p_i(n)$

Notes

Of the various enumeration problems discussed in this chapter, it is the enumeration of the partitions of the integer n that has been the subject of the most vigorous research. This problem proved to be interesting not only for combinatorialists, but also for number theorists. The interested reader should see [4] for further information on integer partitions. A particularly nice classic result is the following recurrence relation, due to MacMahon. Let us say that a *pentagonal number* is an integer of the form $k(3k-1)/2$, where k is any integer, positive or not. So pentagonal numbers are never negative, and the first few are 0, 1, 2, 5, 7, 12. Then for any positive integer n,

$$p(n) = p(n-1) + p(n-2) - p(n-5) - p(n-7) + \cdots \tag{5.5}$$

where the ith term of the right-hand side has sign $(-1)^{\lfloor i/2 \rfloor}$ and absolute value $p(n-k_i)$, with k_i being the ith largest pentagonal number.

So for instance, for $n = 8$, the above formula shows that

$$p(8) = p(7) + p(6) - p(3) - p(1) = 15 + 11 - 3 - 1 = 22.$$

Pentagonal numbers have other applications besides formula (5.5). The interested reader can consult [8] for an application to permutation enumeration. Euler's famous pentagonal number theorem states that the number of partitions of n into distinct odd parts is equal to the number of partitions of n into distinct even parts as long as n is not pentagonal, and that these two numbers only differ by one if n is pentagonal. A precise statement of this theorem and its detailed proof can be found in the undergraduate textbook [7].

Exercises

(1) Find a formula for $S(n, 3)$.

(2) Prove that if $n \geq 3$, then $B(n) < n!$.

(3) Prove that if $n \geq 2$, then $n! < S(2n, n) < (2n)!$.

(4) (+)

  (a) Let $h(n)$ be the number of ways to place any number (including zero) of non-attacking rooks on the Ferrers shape of the "staircase" partition $(n-1, n-2, \cdots, 1)$. Prove that $h(n) = B(n)$.

  (b) In how many ways can we place k non-attacking rooks on this Ferrers shape?

(5) Let m and n be positive integers so that $m \geq n$. Prove that the Stirling numbers of the second kind satisfy the recurrence relation

$$S(m, n) = \sum_{i=1}^{m} S(m-i, n-1) n^{i-1}.$$

(6) Prove that the number of partitions of n into *exactly* k parts is equal to the number of partitions of n in which the largest part is *exactly* k.

(7) Prove that the number of partitions of n into at most k parts is equal to that of partitions of $n + k$ into exactly k parts.

(8) The *Durfee square* of a partition p is the largest square that fits in the top left corner of the Ferrers shape of p. The Durfee square of $p = (5, 3, 2, 2)$ has side length 2 as shown in the Figure 5.4.

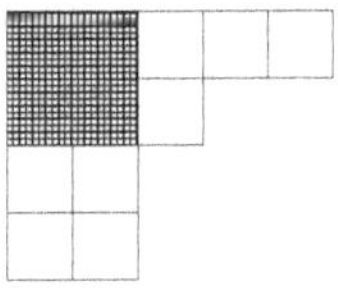

Fig. 5.4 The Durfee square of the partition $(5, 3, 2, 2)$.

If we know the parts of a partition p, how can we figure out the side length of its Durfee square without drawing the Ferrers shape of p?

(9) Let k be a positive integer, and let q be a non-negative integer such that $q < k$. Define $p_{k,q}(r) = p_k(rk+q)$. Prove that $p_{k,q}(r)$ is a polynomial function of r.

(10) Let m be a fixed positive integer. Prove that $S(n, n-m)$ is a polynomial function of n. What is the degree of this polynomial?

(11) Prove that for all integers $n \geq 2$, the number $p(n) - p(n-1)$ is equal to the number of partitions of n in which the two largest parts are equal.

(12) Let $n \geq 4$. Find the number of partitions of n in which the difference of the first two parts is

 (a) at least three,
 (b) exactly three.

(13) Find a formula involving $p(n)$ for the number of partitions of n in which the three largest parts are equal. (You can assume that $n \geq 4$.)

(14) Prove that for all positive integers n,

$$p(1) + p(2) + \cdots + p(n) < p(2n).$$

(15) Our four friends from Exercise 16 of Chapter 3, A, B, C, and D organize a long jump competition every day until the final ranking of the four of them will be the same on two different days. At most how long will they have to wait for that to happen? (Each jump is measured in centimeters, so all kinds of ties, twofold, threefold, fourfold, are possible.)

(16) Prove that for all positive integers n,

$$\sum_{k=1}^{n}(-1)^k k! S(n,k) = (-1)^n.$$

Explain what this identity means without resorting to mathematical formulas.

Supplementary Exercises

(17) (-) How many compositions does the integer 15 have whose first part is not 1?

(18) (-) How many partitions does the set [10] have in which the element 1 does not form a block by itself?

(19) (-) What is the number of partitions of [8] into two blocks in which the two blocks do not have the same size?

(20) (-) What is the number of compositions of 5 with a unique largest part?

(21) (-) Prove that $p(n)$ is equal to the number of partitions of the integer $2n$ with no odd parts.

(22) A student has to take twelve hours of classes a week. Due to her extracurricular activities, she must take at least three hours of classes on Monday, at least two on Thursday, and at least one on Friday.

(a) In how many ways can she do this?

(b) In how many ways can she do this if there is only one class on Tuesday that she may take?

(23) Find the number of compositions of ten into even parts.

(24) Find the number of weak compositions of 25 into five odd parts.

(25) A student has to take eight hours of classes a week. He wants to have fewer hours on Friday than on Thursday. In how many ways can he do this?

(26) Find a closed formula for $S(n, 2)$ if $n \geq 2$.

(27) Find a closed formula for $S(n, n-2)$, for all $n \geq 2$.

(28) Find a closed formula for $S(n, n-3)$, for all $n \geq 3$.

(29) Recall that $p_k(n)$ is the number of partitions of the integer n into exactly k parts.

(a) Prove that for all positive integers $k \leq n$, the inequality $p_k(n) \leq (n-k+1)^{k-1}$ holds.

(b) Is it true that $p_k(n)$ is a polynomial function of n?

(30) Prove that $p(n)$ grows faster than any polynomial function of n. That is, prove that if f is any polynomial function in n, then there exists an integer N so that $f(n) < p(n)$ for all $n > N$. Do not use formula (5.4).

(31) Prove that for all positive integers n, the inequality $p(n)^2 < p(n^2+2n)$ holds.

(32) Let $F(n)$ be the number of all partitions of $[n]$ with no singleton blocks. Prove that $B(n) = F(n) + F(n+1)$. A bijective proof is preferred.

(33) Find a recursive formula for the numbers $F(n)$ in terms of the numbers $F(i)$, with $i \leq n-1$.

(34) Let $B_k(n)$ be the number of partitions of $[n]$ so that if i and j are in the same block, then $|i-j| > k$. Prove that $B_k(n) = B(n-k)$, for all $n \geq k$.

(35) Let a_n be the number of compositions of n into parts that are larger than 1. Express a_n by a_{n-1} and a_{n-2}.

(36) Let b_n be the number of compositions of n into parts that are larger than 2. Find a recurrence relation satisfied by b_n, similar to the one you found for a_n in the previous exercise.

Solutions to Exercises

(1) We can assume that $n \geq 3$. First we determine the number of surjections $f : [n] \to [3]$. The number of all functions $f : [n] \to [3]$ is 3^n. Three of these functions have an image of size one. Moreover, by Exercise 26 and Corollary 5.9, $3 \cdot 2 \cdot (2^{n-1} - 1)$ such functions have an image of size two. Therefore, the number of all surjections $f : [n] \to [3]$ is $3^n - 3 \cdot 2 \cdot (2^{n-1} - 1) - 3$. So Corollary 5.9 shows that

$$S(n,3) = \frac{3^n - 3 \cdot (2^n - 2) - 3}{6} = \frac{3^{n-1} - (2^n - 2) - 1}{2}$$
$$= \frac{3^{n-1} + 1}{2} - 2^{n-1}.$$

(2) We prove the statement by induction on n. For $n = 3$, the statement is true as $3! = 6 > B(3) = 5$. Now assume the statement is true for n and let us prove that it is true for $n + 1$. Equation (5.3) and the induction hypothesis together yield the following upper bound on the left-hand side.

$$B(n+1) = \sum_{i=0}^{n} \binom{n}{i} B(i) < \sum_{i=0}^{n} \binom{n}{i} i!$$
$$= \sum_{i=0}^{n} (n)_i < (n+1)n! = (n+1)!,$$

and the proof follows.

(3) The upper bound follows from the previous exercise. For the lower bound, write the numbers $1, 2, \cdots, n$ in one line in this order, then write the numbers $n+1, n+2, \cdots 2n$ below them in any order. This can be done in $n!$ ways, and each such arrangement defines a partition of $[2n]$ into n blocks of size two each. Strict inequality follows as $n \geq 2$, so partitions with other block sizes are possible.

(4)(a) Number the rows and columns of the staircase Ferrers shape as shown in Figure 5.6. Then each set of non-attacking rooks defines a partition of the set $[n]$ as follows. Let i and j be in the same block if there is a rook in the intersection of row i and column j.

In order to see that this is a bijection, let π be a partition of $[n]$, and let B be a block of π. Let $b_1 < b_2 < \cdots < b_m$ be elements of B. Then it follows from the previous paragraph that B had to be defined by a rook on the box (b_1, b_2), a rook on the box (b_2, b_3), and so on, ending with a rook on the box (b_{k-1}, b_k). Indeed, a block of size k takes $k-1$ rooks to define. No row or column can contain more than one rook. The fact that b_1 and b_2 are in the same block means that there has to be a rook on the box (b_1, b_2), then the fact that b_2 and b_3 are in the same block implies that there has to be a rook on the box (b_2, b_3), and so on.

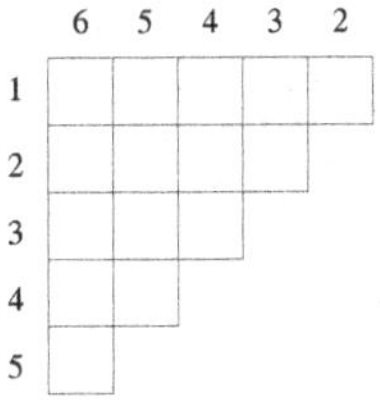

Fig. 5.5 Numbering the rows and columns of the staircase shape.

(b) Continuing the argument from part (a), the placement of no rooks corresponds to the all singleton partition, which has n blocks. The placement of each rook decreases the number of blocks by one (by uniting two blocks), hence the placement of k rooks creates a partition with $n-k$ blocks. So the number of ways of placing k non-attacking rooks on this Ferrers shape is $S(n, n-k)$.

(5) Let π be a partition of $[m]$ into n parts. The left-hand side is the number of such partitions. To see that the right-hand side is the same, let $m-i$ be the largest integer so that the restriction π_i of π into $[i]$ has only $n-1$ blocks. Then we have $S(m-i, n-1)$ possibilities for π_i. It follows from the definition of $m-i$ that $m-i+1$ must be in a new, last block B of π. Then, the numbers $m-i+2, m-i+3, \cdots, m$ can be in any blocks, yielding n^{i-1} choices for the blocks of these elements. Therefore, the total number of possibilities for π is $S(m-i, n-1)n^{i-1}$. Summing over all i, the statement follows.

(6) These are partitions whose Ferrers shape has exactly k rows (resp. columns), so the statement follows by taking conjugates.

(7) Take the Ferrers shape of a partition of n into at most k parts, and add an extra box to the end of each row. If there were less than k

rows, then add additional rows of length one so that the shape has k rows. This way, you get the Ferrers shape of a partition of $n+k$ with exactly k rows.

To see that this is a bijection, it suffices to show that for all Ferrers shapes F with k rows and $n+k$ boxes, one can find a unique Ferrers shape whose image is F. That shape can be obtained by simply deleting the last box of each of the k rows of F.

(8) Let $p = (p_1, p_2, \cdots, p_i)$. Then the side length of the Durfee square is the largest i so that $p_i \geq i$.

(9) We will prove a stronger statement by induction on k, namely that $p_{k,q}(r)$ is a polynomial of degree $k-1$. If $k=1$, then $p_{k,q}(r)=1$, and the statement is true.

It is a well-known fact of calculus (see Exercise 1 of Chapter 2) that a function $f(n)$ is a polynomial of degree d if and only if the function $g(n) = f(n) - f(n-1)$ is a polynomial of degree $d-1$. It is therefore sufficient to prove that $p_{k,q}(r) - p_{k,q}(r-1)$ is a polynomial of degree $n-2$.

Take a partition of $rk+q$ into k parts. Subtract one from each of its parts. We get a partition of $(r-1)k+q$ into *at most k parts*. Indeed, some parts could be equal to 1 in the original partition and now they would disappear.

Therefore,

$$p_{k,q}(r) = p_{k,q}(r-1) + p_{k-1,q}(r-1) + \cdots + p_{0,q}(r-1),$$

where $p_{0,q}(r-1) = 1$ if $q=0$ and $r=1$, and $p_{0,q}(r-1) = 0$ otherwise. After rearrangement, we get

$$p_{k,q}(r) - p_{k,q}(r-1) = \sum_{i=0}^{k-1} p_{i,q}(r-1).$$

By the induction hypothesis, all terms on the right-hand side are polynomials. The last one of them has degree $k-2$, and the rest have smaller degrees. Therefore, the right-hand side is a polynomial of degree $k-2$, and thus so is the left-hand side. Consequently, $p_{k,q}(r)$ is a polynomial of degree $k-1$, by the fact we mentioned in the second paragraph of this solution.

(10) We prove the statement by induction on m, the initial case of $m=0$ being obvious. We will use the same fact of calculus that we used to solve the previous exercise. Applying formula (5.1) with $k = n-m$, we get, after rearrangement

$$S(n, n-m) - S(n-1, n-1-m) = (n-m)S(n-1, n-1-(m-1)).$$

Here, the right-hand side is a polynomial by the induction hypothesis, so the left-hand side must be a polynomial. However, the left-hand side is just the difference of two consecutive values of $S(n, n-m)$, so $S(n, n-m)$ is a polynomial.

The degree of $S(n, n-m)$ as a polynomial is $2m$. This can be proved as a similar statement was proved in the previous exercise.

(11) We know from Theorem 5.20 that $p(n) - p(n-1)$ is equal to the number of partitions of n in which each part is at least two. Taking conjugates, this latter is equal to the number of partitions of n in which the two largest parts are equal.

(12)(a) Take any partition of $n-3$, and add three to its first part. This way we get each partition of the desired property exactly once. Therefore, the answer is $p(n-3)$.

(b) Decreasing the first part of such a partition by three, we get a partition of $n-3$ with the first two parts equal. Exercise 11 shows that the number of these is $p(n-3) - p(n-4)$.

(13) The conjugate of such a partition consists of parts of size at least three. Therefore the number $q(n)$ of such partitions is equal to $p(n) - r(n) - s(n)$, where $r(n)$ is the number of partitions of n with smallest part one, and $s(n)$ is the number of partitions of n with smallest part two.

We know from Theorem 5.20 that $r(n) = p(n-1)$. Let us determine $s(n)$. If π is a partition of n with smallest part two, and we remove the smallest part of π, then we get a partition π' of $n-2$ with smallest part *at least two*. In other words, π' does not contain one as a part, therefore, by Theorem 5.20, we have $p(n-2) - p(n-3) = s(n)$. This shows that

$$q(n) = p(n) - r(n) - s(n) = p(n) - p(n-1) - p(n-2) + p(n-3).$$

(14) If π is a partition of i for $i \leq n$, then its largest part is at most n, so it can be prepended by a new first part $2n - i > n$. The new partition we obtain is a partition of $2n$. This sets up an injection from the set of partitions of all positive integers at most n into the set of partitions of $2n$, and the proof follows.

(15) Each final result of the competition defines an *ordered* partition of $\{A, B, C, D\}$ into k blocks, where k is the number of jumps of different length. In other words, people who tied form the blocks of this partition, and the blocks are ordered according to the sizes of the jumps belonging to people in each block. For example, if B won, A

and D tied for the second place, and C got last, then the ordered partition defined by this result is $\{B\}, \{A, D\}, \{C\}$.
The number of ordered partitions of $[n]$ into k blocks is obviously $S(n,k) \cdot k!$. Therefore, the number of all ordered partitions of $\{A, B, C, D\}$ into at most four blocks is

$$\sum_{k=1}^{4} S(n,k) \cdot k! = 1 \cdot 1 + 7 \cdot 2 + 6 \cdot 6 + 1 \cdot 24 = 75.$$

So the four friends will have their competitions for at most 76 days.

(16) Simply set $x = -1$ in formula (5.2). The meaning of the result is that the number of *surjections* from a given set with an even-sized image and the number of surjections from that same set with an odd-sized image differ by ± 1.

Chapter 6

Not So Vicious Cycles. Cycles in Permutations

We have considered several enumeration problems in the previous three chapters. One of them, that of permutations, stands out by its omnipresence in mathematics. The reason for that is that permutations can be viewed not only as linear orders of different objects, most often elements of $[n]$, but also as *functions* from $[n]$ to $[n]$. In particular, a permutation $p = p_1p_2\cdots p_n$ can be conceived as the unique function $p : [n] \to [n]$ for which $p(i) = p_i$.

Example 6.1. The permutation 312 can be viewed as the (bijective) function $f : [3] \to [3]$ defined by $f(1) = 3$, $f(2) = 1$, and $f(3) = 2$.

The advantage of this approach is that now one can define the *product* of two permutations on $[n]$ by simply taking their composition as a composition of functions.

Example 6.2. Let $f = 312$ and let $g = 213$. Then $(f \cdot g)(1) = g(f(1)) = g(3) = 3$, $(f \cdot g)(2) = g(f(2)) = g(1) = 2$, and $(f \cdot g)(3) = g(f(3)) = g(2) = 1$. Therefore, $fg = 321$.

Example 6.3. Let f and g be defined as in the preceding example. Then $(g \cdot f)(1) = f(g(1)) = f(2) = 1$, $(g \cdot f)(2) = f(g(2)) = f(1) = 3$, and $(g \cdot f)(3) = f(g(3)) = f(3) = 2$. Therefore, $gf = 132$.

As these two examples show, multiplication of permutations is *not* a commutative operation, that is, it is *not* true in general that $fg = gf$. The reader may have seen examples of such operations before, such as matrix multiplication. Exercise 12 explains why multiplication of permutations is a special case of that.

Note that many authors use a slightly different notation for the product of two permutations, in that when they write gf, they mean that we first apply f, then g. In this book, multiplication of permutations will not be as important as *enumeration* of permutations, so even if the reader used that notation before, there will be no danger of confusion.

Operations involving multiplications of permutations are the subject of the theory of *permutation groups.* Our book walks through Combinatorics, and will not contain a digression to that very interesting field. However, some of the exercises at the end of this chapter do relate to the multiplication of permutations.

6.1 Cycles in Permutations

Take the permutation 321564. Again, this permutation can be viewed as a function $g : [6] \rightarrow [6]$. Let us take a closer look at g. First, $g(2) = 2$, in other words, 2 is a *fixed point* of the permutation g. Second, $g(1) = 3$, and $g(3) = 1$. This implies in particular that $g^2(1) = 1$, and $g^2(3) = 3$, moreover, $g^3(1) = 3$, and $g^3(3) = 1$, and so on. In other words, if we repeatedly apply g, the elements 1 and 3 will only be permuted among each other, without any interference from the other entries. Furthermore, g^2 has the effect of the identity permutation $12 \cdots n$ on the entries 1 and 3, but g does not. To describe this phenomenon, we will say that 1 and 3 form a *2-cycle* in g. Similarly, $g(4) = 5, g(5) = 6$, and $g(6) = 4$. Iterating g, we see that $g^2(4) = 6, g^2(5) = 4$, and $g^2(6) = 5$. Finally, $g^3(4) = 4$, $g^3(5) = 5$, and $g^3(6) = 6$. Again, we notice that g permutes elements 4, 5, and 6 among each other so that g^3 has the effect of the identity permutation on the entries 4, 5, and 6, but g and g^2 do not. To describe this phenomenon, we will say that 4, 5 and 6 form a *3-cycle* in g.

Before we can formally define cycles, we need the following lemma.

Lemma 6.4. *Let $p : [n] \rightarrow [n]$ be a permutation, and let $x \in [n]$. Then there exists a positive integer $1 \leq i \leq n$ so that $p^i(x) = x$.*

Proof. Consider the entries $p(x), p^2(x), \cdots, p^n(x)$. If none of them is equal to x, then the Pigeon-hole Principle implies that there are two of them that are equal, say $p^j(x) = p^k(x)$, with $j < k$. Then, applying p^{-1} to both sides of this equation, we get $p^{j-1}(x) = p^{k-1}(x)$. Repeating this step, we get $p^{j-2}(x) = p^{k-2}(x)$, and repeating this step $j - 3$ more times, we get $p(x) = p^{k-j+1}(x)$. □

Time has come for us to make a formal definition of the notion of cycles in permutations.

Definition 6.5. Let $p : [n] \rightarrow [n]$ be a permutation. Let $x \in [n]$, and let i be the smallest positive integer so that $p^i(x) = x$. Then we say that the entries $x, p(x), p^2(x), \cdots, p^{i-1}(x)$ form an *i-cycle* in p.

Corollary 6.6. *All permutations can be decomposed into the disjoint unions of their cycles.*

Proof. Lemma 6.4 shows that each entry is a member of a cycle. By the definition of cycles, distinct cycles are disjoint. □

Example 6.7. The cycles of 321564 are (31), (2), and (564).

Given the cycle decomposition (31)(2)(564) of g, it is easy to reconstruct g as follows: the image $g(i)$ of i is the entry immediately following i in its cycle, or, if i is the last entry in its cycle, then $g(i)$ is the first entry of that same cycle.

While the cycle decomposition of a permutation f is unique, the same cycle decomposition can be written in many different ways. The convention is to write entries that belong to the same cycle in parentheses. The order of the entries in the parentheses is such that j immediately follows i if $f(i) = j$. Furthermore, $f(b) = a$, where b is the last entry and a is the first entry in the parentheses. However, these principles do not preclude multiple notations for the same permutation. For instance, (241)(35) and (53)(412) denote the same permutation. In that permutation, $f(2) = 4$, $f(4) = 1$, $f(1) = 2$, $f(3) = 5$, and $f(5) = 3$.

We would like to avoid the danger of confusion caused by the phenomenon we have just described. Therefore, we will write our permutations in *canonical cycle form*. That is, each cycle will be written with its largest element first, and the cycles will be written in increasing order of their first elements. Thus the permutation f of our previous example has canonical cycle form (412)(53).

Recall that besides using the canonical cycle form, we can also write a permutation $f : [n] \rightarrow [n]$ as a *list*, or *linear order*, by simply writing $f(1)f(2)\cdots f(n)$. This is sometimes called the *one-line notation* of permutations.

Example 6.8. Our running example, (412)(53) would be written as 24513 in the one-line notation.

The next section will show the extreme usefulness of our ability to write permutations in two different notations. Figure 6.1 illustrates the two different ways one can think about the same permutation.

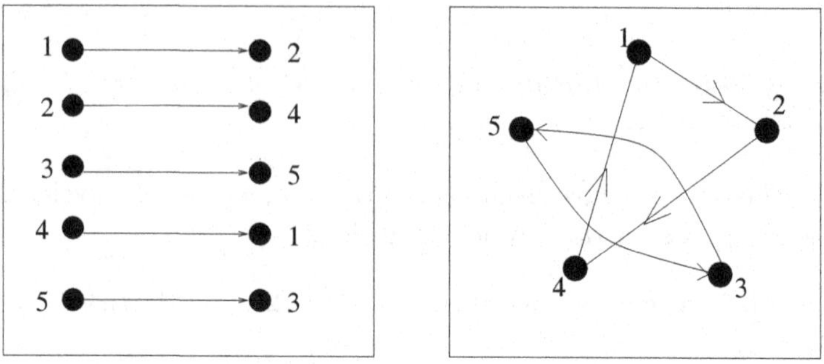

Fig. 6.1 Two ways to look at 24513 = (412)(53).

The cycle decomposition of a permutation contains a lot of information about the permutation. It is therefore important to enumerate permutations according to their cycle decompositions.

In the rest of this section, all permutations will be taken on the set $[n]$, and for shortness we will call them n-permutations. The set of all n-permutations is denoted by S_n. This is because in group theory, this set is called the *symmetric group*.

Theorem 6.9. *Let $a_1, a_2, \cdots, a_n$ be nonnegative integers so that the equality $\sum_{i=1}^{n} i \cdot a_i = n$ holds. Then the number of n-permutations with a_i cycles of length i where $i \in [n]$, is*

$$\frac{n!}{a_1!a_2!\cdots a_n! \cdot 1^{a_1}2^{a_2}\cdots n^{a_n}}. \tag{6.1}$$

Proof. Write down all elements of $[n]$ in a row in some order, then insert parentheses going left to right, according to the required cycle lengths: first a_1 pairs of parentheses creating a_1 1-cycles, then a_2 pairs of parentheses creating a_2 2-cycles, and so on. This way we obtain a permutation in which the cycle lengths are nondecreasing left to right.

There are $n!$ ways to do this- that is the number of ways to write down the elements of $[n]$, and there is only one way to insert the parentheses in the described manner. However, there are several ways of writing down the

n integers that will lead to the same permutation once the parentheses are inserted. We must figure out how many.

The elements within any cycle of length i can be in i different orders and still yield the same cyclic permutation. Therefore, every permutation can be obtained at least $\prod_{i=1}^{n} i^{a_i}$ times as there are a_i cycles of size i. Moreover, if two ways of writing down the elements of $[n]$ result in permutations which have the exact same cycles of length i for all i, just in different order, then again they lead to the same permutation. As a_i cycles can be permuted in $a_i!$ different ways, and permuting the cycles can be done independently from the order of the elements within the cycles, we have shown that each permutation can be obtained $\prod_{i=1}^{n} i^{a_i} a_i!$ ways, and the proof follows. □

If an n-permutation p has a_i cycles of length i, for $i = 1, 2, \cdots, n$, then we say that $(a_1, a_2, \cdots, a_n)$ is the *type* or *cycle type* of p. Thus (6.1) provides a formula for the number of permutations with a given type.

Example 6.10. The number of n-permutations having only one cycle, in other words, the number of n-permutations of type $(0, 0, \cdots, 0, 1)$ is equal to $(n-1)!$.

One combinatorial meaning of Example 6.10 is this. The number of ways n people can sit around a table is $(n-1)!$. (We consider two seating assignments identical if everyone has the same left neighbor in the first seating as in the second.)

Now we are in a position to fulfill an old promise, namely we can define the Stirling numbers of the first kind.

Definition 6.11. The number of n-permutations with k cycles is called a *signless Stirling number* of the first kind, and is denoted by $c(n,k)$. The number $s(n,k) = (-1)^{n-k} c(n,k)$ is called a *Stirling number of the first kind.*

We will explain the reason for including $(-1)^{n-k}$ in the definition of $s(n,k)$ shortly. It will not surprise anyone that $c(n,0) = 0$ if $n > 0$ as nonempty permutations all have cycles. Moreover, we set $c(0,0) = 1$, and $c(n,k) = 0$ if $n < k$, just as was the case with the Stirling numbers of the second kind.

Like the numbers $S(n,k)$, the numbers $c(n,k)$ also satisfy a simple recurrence relation.

Theorem 6.12. *Let n and k be positive integers satisfying $n \geq k$. Then*

$$c(n,k) = c(n-1,k-1) + (n-1)c(n-1,k). \tag{6.2}$$

Proof. We show that the right-hand side counts all n-permutations with k cycles, just as the left-hand side. In such a permutation, there are two possibilities for the position of the entry n.

(1) The entry n can form a cycle by itself, and then the remaining $n-1$ entries have to form $k-1$ cycles. This can happen in $c(n-1, k-1)$ ways, so the first member of the right-hand side of (6.2) enumerates these permutations.
(2) If the entry n does not form a cycle by itself, then the remaining $n-1$ entries must form k cycles, and then the entry n has to be inserted somehow into one of these cycles. The k cycles can be formed in $c(n-1, k)$ ways, then the entry n can be inserted in any of these cycles, after each element. This multiplies the number of possibilities by $n-1$, and explains the second term of the right-hand side of (6.2).

Readers should test their understanding by trying to explain why we did not miss any permutations by inserting n after each entry in each cycle, and not into the front of each cycle. □

The reader is probably wondering whether there is some strong connection between the Stirling numbers of the first kind and the Stirling numbers of the second kind that justifies the similar names. The following Lemma is our main tool in establishing that connection.

Lemma 6.13. *Let n be a fixed positive integer. Then*

$$\sum_{k=0}^{n} c(n,k)x^k = x(x+1)\cdots(x+n-1). \tag{6.3}$$

Proof. We prove that the coefficients of x^k on the right-hand side also satisfy the recursive formula (6.2) that is satisfied by the signless Stirling numbers of the first kind.

Let $G_n(x) = x(x+1)\cdots(x+n-1) = \sum_{k=0}^{n} a_{n,k}x^k$. Then

$$G_n(x) = (x+n-1)G_{n-1}(x) = (x+n-1)\sum_{k=0}^{n-1} a_{n-1,k}x^k$$

$$= \sum_{k=1}^{n} a_{n-1,k-1}x^k + (n-1)\sum_{k=0}^{n-1} a_{n-1,k}x^k.$$

Now we are using a technique that will return in countless applications in Chapter 8. We have just proved that

$$\sum_{k=0}^{n} a_{n,k}x^k = \sum_{k=1}^{n} a_{n-1,k-1}x^k + (n-1)\sum_{k=0}^{n-1} a_{n-1,k}x^k.$$

In other words, we proved that two *polynomials* were identical. The only way that can happen is when the coefficients of the corresponding terms agree in the two polynomials. That is, the equality

$$a_{n,k} = a_{n-1,k-1} + (n-1)a_{n-1,k}$$

must hold for all positive integers n and k such that $n \geq k$. Therefore, the numbers $a_{n,k}$ and $c(n,k)$ do satisfy the same recurrence relation. As their initial terms trivially agree, that is, $c(0,0) = a_{0,0} = 1$, $c(n,0) = a_{n,0} = 0$ if $n > 0$, this implies that $c(n,k) = a_{n,k}$. □

Let us replace x by $-x$ in (6.3), and multiply both sides by $(-1)^n$. We get

$$\sum_{k=0}^{n} s(n,k)x^k = (x)_n. \tag{6.4}$$

Now the reader can see why we included the term $(-1)^{n-k}$ in the definition of $s(n,k)$. Comparing this equation to (5.2), that stated

$$x^n = \sum_{k=0}^{n} S(n,k)(x)_k,$$

we see that the Stirling numbers of the first kind have the "inverse effect" of the Stirling numbers of the second kind. To formulate this observation in a more precise way, we need some notions from linear algebra, and we will assume that the reader has taken a basic course in that field.

It is well-known that the set of all polynomials with real coefficients is a vector space V over the field of real numbers. The most obvious basis of V is $B = \{1, x, x^2, x^3, \cdots\}$, but it is not the only interesting basis. It is easy to show that $B' = \{1, (x)_1, (x)_2, (x)_3, \cdots\}$ is also a basis of V.

Now let S (resp. s) be the infinite matrix whose entry in position (n,k) is $S(n,k)$ (resp. $s(n,k)$). Then (6.4) shows that s is the transition matrix from B to B', while (5.2) shows that S is the transition matrix from B' to B. This proves the promised connection between the two different kinds of Stirling numbers.

Theorem 6.14. *The matrices S and s are inverses of each other, that is, $Ss = sS = I$.*

6.2 Permutations with Restricted Cycle Structure

The following lemma turns the canonical cycle form into a very powerful tool.

Lemma 6.15. *[Transition Lemma] Let $p : [n] \rightarrow [n]$ be a permutation written in canonical cycle notation. Let $g(p)$ be the permutation obtained from p by omitting the parentheses and reading the entries as a permutation in the one-line notation. Then g is a bijection from the set S_n of all permutations on $[n]$ onto S_n.*

Example 6.16. Let p be our running example, that is, $p = (412)(53)$. Then $g(p) = g((412)(53)) = 41253$.

Proof. It suffices to show that for each permutation $q = q_1 q_2 \cdots q_n$ written in the one-line notation, there exists exactly one permutation $p \in S_n$ so that $q = g(p)$. In other words, we have to show that there is exactly one way to insert parentheses into the string $q = q_1 q_2 \cdots q_n$ so that we get a permutation in canonical cycle form.

To see this, note that q_1 certainly starts a new cycle, so the first left parenthesis has to be inserted to the front of the string. Where will this first cycle end? As we are looking for a permutation in canonical cycle form, q_1 has to be the largest of its cycle. Therefore, if i is the smallest index so that $q_1 < q_i$, then the first cycle has to end before q_i. On the other hand, if $j < i$, then the second cycle cannot start with q_j as we know that $q_j < q_1$, and the cycles have to be in increasing order of their first elements. This implies that the second cycle has to start with q_i, and so we have to insert the first right parenthesis and the second left parenthesis between q_{i-1} and q_i.

Then we can continue this deterministic procedure to find all our cycles. By an analogous argument, we have to start a new cycle at q_k if and only if q_k is larger than the leading entries of all previous cycles, which means in particular that q_k is larger than all entries on its left. As these entries are uniquely determined by q, the preimage $g^{-1}(q)$ of q exists and is unique.□

Example 6.17. The preimage of 4356172 under g is (43)(5)(61)(72).

The entries of q that are larger than all entries on their left are called *left-to-right maxima.* Note that if q has t left-to-right maxima, then $g^{-1}(q) = p$ has t cycles. Also note that the leftmost left-to-right maximum of q is always q_1, and the rightmost left-to-right maximum of q is always the entry n. A surprising application of Lemma 6.15 is the following.

Proposition 6.18. *Let i and j be two elements of $[n]$. Then i and j are in the same cycle in exactly half of all n-permutations.*

Proof. As we can relabel our entries by switching n and i, and switching $n-1$ and j, it is sufficient to prove that the entries n and $n-1$ are in the same cycle in exactly half of all n-permutations. Let $q = q_1 q_2 \cdots q_n$ be an n-permutation, and let $g(p) = q$, where g is the bijection of Lemma 6.15. As we said, the entry n of q is always a left-to-right maximum, namely the rightmost left-to-right maximum of q. Therefore, the last cycle of p starts with n, and the entries in that cycle of p are precisely the entries on the right of n in q.

Therefore, p contains n and $n-1$ in the same cycle if and only if $n-1$ is on the right of n in q. As that happens in half of all n-permutations, the proof follows. □

The following surprising result shows that the likelihood that a given entry i is part of a k-cycle is independent of k. In fact, it is $1/n$.

Lemma 6.19. *Let $i \in [n]$. Then for all $k \in [n]$, there are exactly $(n-1)!$ permutations of length n in which the cycle containing i is of length k.*

Proof. Again, it is sufficient to prove the statement for $i = n$, then the general statement follows by relabeling. Let $q = q_1 q_2 \cdots q_n$ be an n-permutation, let $g(p) = q$, where g is the bijection of Lemma 6.15, and let $q_j = n$. Then the cycle C containing n in p is of length $n - j + 1$ as n itself starts the last cycle. So if we want C to have length k, we must have $j = n + 1 - k$. However, there are clearly $(n-1)!$ permutations of length n that contain n in a given position, and the proof follows. □

Theorem 6.9 tells us how to compute the number of permutations of a given type. Sometimes we do not exactly know the type of our permutations, but we at least know something about it. As it turns out, we can still enumerate the relevant permutations in many cases. In what follows, we will show a nice example for that. Other examples can be found in the Exercises.

Let ODD(m), resp. EVEN(m) be the set of m-permutations with all cycle lengths odd, resp. even.

Lemma 6.20. *For all positive integers m, the equality $|\mathrm{ODD}(2m)| = |\mathrm{EVEN}(2m)|$ holds.*

Proof. We construct a bijection Φ from ODD($2m$) onto EVEN($2m$).

Let $\pi \in$ ODD($2m$). Then π consists of an even number $2k$ of odd cycles. Denote by $C_1, C_2, \cdots, C_{2k}$ the cycles in canonical order. For all i, $1 \leq i \leq k$, take the last element of C_{2i-1}, and put it to the end of C_{2i} to get $\Phi(\pi)$, the image of π.

Example 6.21. *If* $p = (4)(513)(726)(8)$, *then* $\Phi(p) = (5134)(72)(86)$.

Note that if C_{2i-1} is a singleton, it disappears. Also note that the canonical form is maintained.

We claim that Φ is a bijection from ODD($2m$) onto EVEN($2m$). Let $\sigma \in$ EVEN($2m$), with cycles $c_1, c_2, \cdots, c_h$. To prove that Φ is a bijection, it suffices to show that we can recover the only permutation $\pi \in$ ODD($2m$) for which $\Phi(\pi) = \sigma$.

While recovering π, we must keep in mind that it might have more than h cycles, because some of its singletons might have been absorbed by the cycles immediately after them. If the last entry in c_h is larger than the first entry in c_{h-1}, then create a singleton cycle with the last entry in c_h, placing it in front of c_h, and repeat the whole procedure using c_{h-2} and c_{h-1}. Otherwise, move the last entry in c_h from c_h to the end of c_{h-1}, and repeat the whole procedure using c_{h-3} and c_{h-2}. If at any point only one cycle remains, create a singleton cycle with the last entry in that cycle. It is then straightforward to check that the permutation π obtained this way fulfills $\Phi(\pi) = \sigma$. It also follows from the simple structure of Φ that at no point of the recovering procedure could we have done anything else. □

Example 6.22. The preimage of $(41)(62)(75)(83)$ under Φ is $(412)(6)(753)(8)$.

Example 6.23. The preimage of $(21)(53)(64)(87)$ under Φ is $(1)(2)(534)(6)(7)(8)$.

Now that we so nicely proved that $|\,\text{ODD}(2m)| = |\,\text{EVEN}(2m)|$, we may well ask if there is a formula describing these numbers. The following Theorem answers that question in the affirmative, and has a touch of surprise in it. Would you have thought that the number of these permutations is always a perfect square?

Theorem 6.24. *For all positive integers* m,

$$|\,\text{ODD}(2m)| = |\,\text{EVEN}(2m)| = 1^2 \cdot 3^2 \cdot 5^2 \cdots (2m-1)^2. \tag{6.5}$$

Solution. Because of Lemma 6.20, it suffices to prove the second equality. Let p be an n-permutation with even cycles only. Clearly, we cannot have $p(1) = 1$, as that would mean that the entry 1 forms a 1-cycle in p. So there are $2m-1$ choices for $p(1)$. Then there are $2m-1$ choices for $p^2(1) = p(p(1))$ as we can choose everything but $p(1)$ itself.

So far we have chosen $p(1)$ and $p(p(1))$. These two elements will either form a 2-cycle (when $p(p(1)) = 1$), or they will not. In either case, we will have $2m - 3$ choices for the image of the next entry. That is, if 1 and $p(1)$ form a 2-cycle, and i is an element outside that cycle, then we have $2m - 3$ choices for $p(i)$. Indeed, we can choose anything except 1, $p(1)$, these have already been chosen, and $p(i)$, as that would mean that i is a 1-cycle. If, on the other hand 1 and $p(1)$ do not form a 2-cycle, then we choose the next element of their cycle, $p^3(1)$ next. The entry $p^3(1)$ cannot be $p(1)$ and $p^2(1)$ as those elements are already chosen, and cannot be 1 either as that would create the 3-cycle $(1, p(1), p^2(1))$. So there are $2m - 3$ choices for the next element in this case too.

Continuing this line of argument, we see that selecting our $(2i - 1)$st entry we always have $2m - 2i + 1$ choices, and selecting our $2i$th entry we always have $2m - 2i + 1$ choices, (as we can close cycles of even length), and the proof follows.

Thus we have a formula for $|\operatorname{ODD}(n)|$ if n is even. If n is odd, then clearly, $|\operatorname{EVEN}(n)| = 0$, but we can still look for a formula for $|\operatorname{ODD}(n)|$.

Theorem 6.25. *For all positive integers m,*

$$|\operatorname{ODD}(2m+1)| = (2m+1)\cdot|\operatorname{ODD}(2m)| = 1^2\cdot 3^2\cdot 5^2\cdots(2m-1)^2(2m+1). \tag{6.6}$$

Proof. We construct a bijection Ψ from $\operatorname{ODD}(2m) \times [2m + 1]$ onto $\operatorname{ODD}(2m + 1)$. In this bijection, we will need the notion of a *gap position*. This notion will be useful to solve some of the exercises, too. An m-permutation has $m + 1$ gap positions, one *after* each element in each cycle, and one at the very beginning of the permutation, before all entries and cycles.

Example 6.26. *The permutation* $(42)(513)$ *has six gap positions, indicated by bars in the following array:* $|(4|2|)(5|1|3|)$.

Let $\pi \in \operatorname{ODD}(2m)$, and let $k \leq 2m + 1$ be a positive integer. We define $\Psi(\pi, k)$ as follows. First, take $\Phi(\pi)$, where Φ is the bijection of Lemma 6.20.

Let us now add 1 to each entry of $\Phi(\pi)$, so it becomes a permutation of the set $\{2, 3, \cdots, 2m+1\}$. Insert the new entry 1 to the kth gap position of $\Phi(\pi)$. That will change one cycle to an odd cycle. Note that the canonical cycle form is preserved. Apply the bijection Φ^{-1} to the permutation given by the remaining cycles to get a permutation that has odd cycles only. This way we obtain a $(2m+1)$-permutation consisting of odd cycles only, and that permutation is our $\Psi(\pi)$.

Lemma 6.27. *The map* Ψ *defined above is a bijection from the set* $\mathrm{ODD}(2m) \times [2m+1]$ *onto the set* $\mathrm{ODD}(2m+1)$.

Proof. To find the reverse of Ψ, take $\pi' \in \mathrm{ODD}(2m+1)$, put the cycle in π' which contains the entry 1 aside, and run the remaining cycles through Φ to get even cycles. Read off k as the gap position in which the entry 1 is. Remove the entry 1 from its odd cycle, and run the obtained permutation, which has all even cycles, through Φ^{-1}, to get $\Psi^{-1}(\pi')$. Note that at every step, we have reversed the corresponding step of Ψ. □

This completes the proof of the theorem. □

Notes

A fair part of the results in Section 6.2 were obtained after Herb Wilf asked some intriguing questions in [49]. Most of the results presented here have been generalized in [11]. For example, it has been proved that if p is prime, then the ratio of n-permutations that have a pth root to all n-permutations is steadily decreasing, and converges to zero. See Exercises 21 and 22 for the relevant definitions.

Exercises

(1) Is it true that $c(n, n-1) = S(n, n-1)$?
(2) Find a formula for $c(n, n-2)$.
(3) Compute the values of $c(5, k)$, for $k = 1, 2, 3, 4, 5$.
(4) Prove that for any fixed k, the function $c(n, n-k)$ is a polynomial function of n. What is the degree of that polynomial?
(5) Let $r(n)$ be the number of n-permutations whose square is the identity permutation. Prove that if $n \geq 1$, then $r(n+1) = r(n) + nr(n-1)$,

where $r(0) = 1$.

(6) Find a recursive formula for the number $t(n)$ of n-permutations whose cube is the identity permutation.

(7) Prove that on average, permutations of length n have H_n cycles, where

$$H_n = \sum_{i=1}^{n} \frac{1}{i}.$$

(8) How many n-permutations contain entries 1, 2 and 3 in the same cycle?

(9) An alpine ski team has n members. They descend a particular slope one by one every day, and no two of them ever record identical times. On an average day, how many times will the best record of that day be broken?

(10) An airplane has n seats, and all of them have been sold for a particular flight, with no overbooking. When the last passenger arrives, he finds that his seat is taken. When he shows his reservation to the passenger at his seat, that passenger stands up, and goes to her own assigned seat. If that seat is empty, she seats down, and the seating procedure is over. If not, she shows her reservation to the person seating at that seat. That person stands up, and goes to his assigned seat, and so on. This procedure continues until someone finds his or her assigned seat empty.

Tom was not the last passenger to board the plane. What is the probability that he has to move during this procedure?

(11) Let p be an n-permutation. We associate a *permutation matrix* A_p to p as follows. Let $A_p(i, j) = 1$ if $p(i) = j$, and let $A_p(i, j) = 0$ otherwise. Here $A_p(i, j)$ denotes the entry of A_p that is in the intersection of the ith row and the jth column. Prove that $|\det A| = 1$.

(12) Prove that if p and q are two n-permutations, then $A_pA_q = A_{pq}$.

(13) The *inverse* of an n-permutation is the permutation q for which $pq = qp = 123\cdots n$. We then write $q = p^{-1}$. Prove that each permutation has a unique inverse.

(14) Prove that permutations f and f^{-1} are of the same type.

(15) What is the combinatorial meaning of A_p^T?

(16) In permutations, 1-cycles are often called fixed points. Prove, using permutation matrices, that permutations pq and qp always have the same number of fixed points.

(17) Let us assume that we know the type $(a_1, a_2, \cdots, a_n)$ of an n-permutation. Determine the smallest positive integer d such that $p^d = 123\cdots n$.

(18) A permutation p is called a nontrivial *involution* if $p^2 = 12\cdots n$, but $p \neq 12\cdots n$. Prove that if $n > 1$, then the number of nontrivial involutions in S_n is odd.

(19) Generalize the previous exercise for all prime numbers t.

(20) Let $n \geq 2$. Prove that $\det A_p = 1$ for exactly one half of all n-permutations p.

(21) We say that a permutation $p \in S_n$ has a square root if there is a permutation $q \in S_n$ so that $q^2 = p$. Find a sufficient and necessary condition for p to have a square root, in terms of its cycle lengths.

(22) We say that a permutation $p \in S_n$ has a kth root if there is a permutation $q \in S_n$ so that $q^k = p$. Is the following statement true?
"A permutation has a kth root if and only if it is of type $(a_1, a_2, \cdots, a_n)$, where whenever i is divisible by k, a_i is divisible by k."

(23) (+) Construct a bijection
$$\tau : \mathrm{ODD}(2m+1) \times [2m+1] \to \mathrm{ODD}(2m+2).$$

(24) (++) Let $SQ(n)$ be the set of n-permutations having at least one square root. Prove that for all positive integers n, we have $|SQ(2n)| \cdot (2n+1) = |SQ(2n+1)|$. Note that this means that $p(2n) = p(2n+1)$, where $p(m)$ denotes the probability that a randomly chosen m-permutation has a square root.

(25) Let k, m, and r be positive integers, and let $kr = m$. Prove that the number of m-permutations all of whose cycle lengths are divisible by k is
$$1 \cdot 2 \cdots (k-1)(k+1)^2(k+2)\cdots(2k-1)(2k+1)^2(2k+2)\cdots(m-1)$$
$$= \frac{m!}{k^r r!} \cdot (k+1)(2k+1) \cdot ((r-1)k+1).$$

Supplementary Exercises

(26) (-) What is the number of n-permutations in which entries 1 and 2 are part of the same 3-cycle?

(27) (-) Find the number of permutations of length six whose square is the identity permutation.

(28) (-) What is the number of permutations of length 20 whose longest cycle is of length 11?

(29) (-) What is the number of n-permutations that have $n-1$ left-to-right maxima?

(30) (-) A group of ten children want to play cards. They split into three groups, one of these groups has four children in it, the other two have three each. Then each group sits around a table. Two seatings are considered the same if everyone's left neighbor is the same.

(a) In how many ways can this be done if the three tables are identical?

(b) In how many ways can this be done if the three tables are distinct?

(31) What is the number of $(2n)$-permutations whose longest cycle is of length n?

(32) Let $p = p_1 p_2 \cdots p_n$ be a permutation. An *inversion* of p is a pair of entries (p_i, p_j) so that $i < j$ but $p_i > p_j$.
Let us call a permutation *even* (resp. *odd*) if it has an even (resp. odd) number of inversions.
Prove that the permutation consisting of the one cycle $(a_1 a_2 \cdots a_k)$ is even if k is odd, and is odd if k is even.

(33) Find a combinatorial proof for the fact that there are $n!/2$ even n-permutations.

(34) What is the relation between the parity of a permutation p and $\det A_p$?

(35) Let us assume that we only know the type of the n-permutation p. How can we decide whether p is odd or even?

(36) Let us assume that we know the length n of a permutation p, and the number k of its cycles. Can we figure out from these data whether p is an odd or an even permutation?

(37) Prove the result of Supplementary Exercise 33 by an appropriate substitution into formula (6.3).

(38) How many permutations $p \in S_6$ satisfy $p^3 = \mathrm{id}$, where $\mathrm{id} = 123456$ (the identity permutation)?

(39) How many even permutations $p \in S_6$ satisfy $p^2 = \mathrm{id}$?

(40) Let n be divisible by 3. Prove that $c(n, n/3) \geq \frac{n!}{3^{n/3}(n/3)!}$.

(41) Prove that for all positive integers n, r and k such that $n = rk$, the inequality

$$(r-1)!^k \leq \frac{c(n,k)}{S(n,k)} \leq (n-k)!$$

holds.

(42)(a) Prove that in the polynomial

$$(1+x)(1+2x)\cdots(1+(n-1)x)$$

the coefficient of x^{n-k} is $c(n,k)$, for all $k \in n$.

(b) State and prove the corresponding fact for the numbers $s(n,k)$.

(43) Let $a(n,k)$ be the number of permutations of length n with k cycles in which the entries 1 and 2 are in the same cycle. Prove that for $n \geq 2$,
$$\sum_{k=1}^{n} a(n,k)x^k = x(x+2)(x+3)\cdots(x+n-1).$$

(44) (+) Let $b_r(n,k)$ be the number of permutations of length n with k cycles in which all entries of $[r]$ are in the same cycle. Prove that for $n \geq r$,
$$\sum_{k=1}^{n} b_r(n,k)x^k = (r-1)!\frac{x(x+1)\cdots(x+n-1)}{(x+1)(x+2)\cdots(x+r-1)}.$$

(45) Let $a(n,k)$ be defined as in Supplementary Exercise 43. Let $t(n,k) = c(n,k) - a(n,k)$ be the number of permutations of length n with k cycles in which the entries 1 and 2 are *not* in the same cycle. Prove that $a(n,k) = t(n,k+1)$, for all $k \leq n-1$.

(46) A group of n tourists arrive at a restaurant. They sit down around an unspecified number of circular tables, leaving no table empty. Then each table orders one of r possible drinks. Prove that the number of ways this can happen is
$$r(r+1)\cdots(n+r-1).$$
Two seating arrangements are considered the same if each person has the same left neighbor in both of them.

(47) We write each element of $[n-1]$ on a separate card, then randomly select any number of cards, and take the product of the numbers of written on them. Then we do this for all 2^{n-1} possible subsets of the set of $n-1$ cards. (The empty product is taken to be 1.) Finally, we take the sum of the 2^{n-1} products we obtained. What is this sum?

(48) Modify the previous exercise so that instead of considering all 2^{n-1} subsets, we only consider all k-element subsets of the $n-1$ cards. What is the sum of all $\binom{n-1}{k}$ products we obtain in that scenario?

(49) Find a recurrence relation satisfied by the numbers $u(n)$ of n-permutations whose fourth power is the identity permutation.

(50) A library has n books. Readers of this library are "almost" careful. That is, after reading a book, they put it back to its shelf, missing its proper place by only one notch. Prove that after a sufficient amount of time, any permutation of the books on the shelves can occur.

(51) Prove that two n-permutations p and q have the same type if and only if there exists an n-permutation g so that $q = gpg^{-1}$ holds.

(52) Inversions of a permutation were defined in Supplementary Exercise 32. Let $I(n,k)$ be the number of n-permutations that have k inversions. Prove that $I(n,k) = I(n, \binom{n}{2} - k)$.

(53) Let $I(n,k)$ be defined as in the previous exercise. Prove that

$$\sum_{k=0}^{\binom{n}{2}} I(n,k)x^k = (1+x)(1+x+x^2)\cdots(1+x+\cdots+x^{n-1}).$$

(54) Deduce from the result of the previous exercise that the number of even n-permutations is the same as the number of odd n-permutations. (See Supplementary Exercise 32 for the definition of even and odd permutations.)

(55) Find an explicit formula for $I(n,3)$.

Solutions to Exercises

(1) Yes, that is true. $S(n,n-1)$ is the number of ways to partition $[n]$ into one doubleton and $n-2$ singletons. To get $c(n,n-1)$, we have to take a permutation consisting of one cycle on each of these $n-1$ subsets. There is only one way to do this, thus $c(n,n-1) = S(n,n-1) = \binom{n}{2}$.

(2) An n-permutation that has $n-2$ cycles can have either two 2-cycles, or one 3-cycle, and the rest must be all 1-cycles. In the first case, we can choose the elements of the first 2-cycle in $\binom{n}{2}$ ways, the elements of the second 2-cycle in $\binom{n-2}{2}$ ways, then take a 2-cycle on each of them in one way. This yields $\binom{n}{2} \cdot \binom{n-2}{2}/2$ permutations as the order of the cycles is irrelevant. In the second case, we have to choose the elements of the 3-cycle in $\binom{n}{3}$ ways, then take a 3-cycle on them in 2 ways. This yields $2\binom{n}{3}$ permutations, and proves that

$$c(n,n-2) = \frac{n(n-1)(n-2)(n-3)}{8} + \frac{n(n-1)(n-2)}{3}.$$

(3) It follows from (6.10) that $c(5,1) = 4! = 24$. Exercise 1 shows that $c(5,4) = \binom{5}{2} = 10$, and Exercise 2 shows that $c(5,3) = 15 + 20 = 35$. It is obvious that $c(5,5) = 1$. As $\sum_{k=1}^{5} c(5,k) = 5! = 120$, the equality $c(5,2) = 50$ follows.

(4) We prove the statement by induction on k. If $k = 1$, then the statement is true by Exercise 1. Now assume we know the statement for $k-1$. This implies

$$c(n,n-k) = c(n-1,n-k-1) + (n-1)c(n-1,n-k),$$

$$c(n, n-k) - c(n-1, n-k-1) = (n-1)c(n-1, n-k).$$

Here the right-hand side is a polynomial by the induction hypothesis, and therefore so is the left-hand side. However, the left-hand side is the difference of two consecutive values of $c(n, n-k)$, therefore $c(n, n-k)$ must be a polynomial by Exercise 1 of Chapter 2. Similarly, the degree of $c(n, n-k)$ is $2n$, by this same inductive setup, and Exercise 1 of Chapter 2.

(5) In such permutations, all cycles must be 1-cycles or 2-cycles. If the entry $n+1$ forms a 1-cycle, then the remaining n entries can form a good permutation in $r(n)$ ways. If the entry $n+1$ is part of a 2-cycle, then there are n choices for the other entry of that 2-cycle, then there are $r(n-1)$ ways for the remaining $n-1$ entries to form a good permutation.

(6) This is similar to the previous exercise. All cycles of such permutations have length one or three. If $n+1$ is in a 3-cycle, then there are $\binom{n}{2}$ choices for the other two elements of the cycle, and there are 2 choices for the cycle itself, once its elements are known. Then the remaining entries can form a good permutation in $t(n-2)$ ways. If the entry $n+1$ forms a 1-cycle, then the remaining n entries can form a good permutation in $t(n)$ ways. Therefore, $t(n+1) = n(n-1)t(n-2) + t(n)$ if $n \geq 3$.

(7) We prove the statement by induction on n. For $n = 1$, the statement is true. Assume it is true for $n-1$. There is $1/n$ chance that entry 1 forms a 1-cycle, and then the remaining $n-1$ elements form H_{n-1} cycles on average. If entry 1 does not form a 1-cycle, then, take any permutation of the elements $\{2, 3, 4, \cdots, n\}$ in the canonical distribution. Insert entry 1 after any of these elements. This will not change the number of cycles as entry 1 will not start a new cycle. Therefore, the number of permutations with k cycles stays the same for all k, so their average stays the same, too, i.e. $H(n-1)$. Therefore, we get

$$H(n) = \frac{1}{n} \cdot (H(n-1) + 1) + \frac{n-1}{n} \cdot H(n-1) = H(n-1) + \frac{1}{n},$$

and the statement follows.

(8) Entries 1, 2, and 3 are together in one cycle exactly as often as elements $n-2, n-1, n$ are. This latter happens exactly when, after omitting all parentheses from the cycle notation, n precedes both $n-2$ and $n-1$. And that clearly happens in $1/3$ of all permutations. So there are $n!/3$ such permutations, for $n \geq 3$.

(9) This is the same as asking that on average, how many left-to-right minima does a random n-permutation have. In accordance with the paragraph following Example 6.17, a left-to-right minimum is an entry of a permutation $p = p_1 \cdots p_n$ that is smaller than all entries on its left. This is the same as the average number of cycles in an n-permutation, and we have computed that in Exercise 7.

(10) If $1, 2, \cdots, n$ denote the passengers, and $f(1), f(2), \cdots, f(n)$ denote their assigned seats, then it is clear that $f(1)f(2)\cdots f(n)$ is a permutation. Tom will have to move if and only if in this permutation, his seat is part of the same cycle as the seat $f(n)$ of passenger n, who arrived last. We know from Proposition 6.18 that the chance of that is one half.

Exercise 7, and the paragraph following Example 6.17 tell us that for left-to-right *maxima*, the answer is $H(n) = \sum_{i=1}^{n} \frac{1}{n}$. To see that this is also the answer for left-to-right minima, note that $p_1p_2\cdots p_n$ has t left-to-right minima if and only if the permutation $q = q_1q_2\cdots q_n$, where $q_i = n + 1 - p_i$, has t left-to-right maxima. By the way, q is called the *complement* of p.

(11) That is true as each row and column will have exactly one nonzero member. Therefore, when expanding the determinant by any row or column, we will only obtain one nonzero product. That product will be the product of many ones, so the only open question is whether that product will occur in the determinant with a positive sign or with a negative sign. That depends on p.

(12) Consider $(A_pA_q)(i, j)$. By the definition of matrix multiplication, this is the inner product of the ith row of A_p and the jth column of A_q. As both of these vectors have exactly one nonzero element in them, their inner product will be 1 if and only if those nonzero elements occur in the (same) kth position in both vectors. That, however, happens if and only if $p(i) = k$, and $q(k) = j$, which is also equivalent to $pq(i) = j$. Therefore, $(A_pA_q)(i, j) = A_{pq}(i, j)$.

(13) The n-permutation q is the inverse of the n-permutation p if and only if $p(i) = j$ implies $q(j) = i$. This relation uniquely defines q.

(14) Reversing each cycle of p results in p^{-1}.

(15) If $p(i) = j$, then $A_p(i, j) = 1$, so $A_p^T(j, i) = 1$. Therefore, A_p^T defines a permutation q in which $q(j) = i$ if and only if $p(i) = j$. This means $pq = qp = 12\cdots n$, so q is the inverse of p. Thus the transpose of a permutation matrix is the permutation matrix of the inverse of the original permutation. This also implies $A_pA_p^T = I$.

(16) The number of fixed points of a permutation can be read off its permutation matrix as the number of ones in the main diagonal. As the remaining entries of the main diagonal are zeros, the number of ones in the main diagonal also equals the sum of diagonal elements, which is called the *trace* of the matrix. It is well known in Linear Algebra that $trace(AB) = trace(BA)$ for all $n \times n$ matrices A and B. Therefore,

$$trace(A_pA_q) = trace(A_qA_p),$$

and the claim is proved.

(17) The smallest positive integer d with that property is the least common multiple of the cycle lengths of the permutation, that is, the indices i so that $a_i > 0$. Indeed, the kth, $2k$th, etc. powers of a k-cycle are equal to the identity permutation.

(18) If $n > 1$, then $n!$ is even. Let us arrange all n-permutations into pairs, by placing p and p^{-1} in the same pair. That will create a t pairs, containing altogether $2t$ permutations, but will not match involutions and $12\cdots n$ to anything. Thus the number of these latter is $n! - 2t$, therefore the number of involutions is $n! - 2t - 1$, and that is an odd number.

(19) If t is prime, and $n \geq t$, then the number of n-permutations p so that $p^t = 12\cdots n$, but $p \neq 12\cdots n$ is congruent to -1 modulo t. The proof is analogous to that of the previous exercise.

(20) Consider $(21) = (21)(3)(4)\cdots(p)$, the permutation that simply swaps the first two entries. For any $p \in S_n$, we define $h(p) = (12)p$. As $\det A_{(12)} = -1$, the matrices A_p and $A_{h(p)}$ have determinants of opposite signs. On the other hand, $h(h(p)) = p$, therefore h creates pairs of permutations $(p, h(p))$. Each pair will contain exactly one permutation whose matrix has determinant 1, and the claim is proved.

(21) Let $r \in S_n$, and consider r^2. It is straightforward to verify that if k is odd, then the k-cycles of r will stay k-cycles in r^2, and if k is even, the k-cycles of r will split into two $\frac{k}{2}$-cycles in r^2. So the only way r^2 can have even cycles is by obtaining them from an even cycle of r, that has split into two cycles of the same size, each of them even. Therefore, r^2 will have an even number of cycles of each even length. On the other hand, we claim that this is sufficient. That is, if p has an even number of cycles of each even length, then p has a square root. Indeed, if p has even cycles $(a_1 \cdots a_t)$ and $(b_1 \cdots b_t)$, then they can be obtained by taking the square of the $(2t)$-cycle $(a_1b_1a_2b_2a_3\cdots a_tb_t)$. Odd cycles of p, such as $(d_1d_3d_5\cdots d_kd_2d_4\cdots d_{k-1})$ can be obtained

as the square of $(d_1 d_2 \cdots d_k)$. After finding square roots for all cycles of p, a good choice for the square root of p is the product of those cycles.

We have proved that p has a square root if and only if p has an even number of cycles of each even length.

(22) No, that is not true in this generality. The claim is true if k is prime, and in that case, it can be proved the same way the previous exercise was proved.

If k is not prime, however, then the statement is not true. For instance, if $k = 4$, then the requirements do not say anything about the number of 2-cycles of p. Thus $p = (21)(3)(4)(5) \cdots (n)$ would have to have a fourth root. That is clearly impossible, however, as this p does not even have a square root. (If there were a q so that $q^4 = p$, then q^2 would be a square root of p.) The reader is invited to construct a similar counterexample for a generic composite number k.

(23) Take a pair $(\pi, k) \in \mathrm{ODD}(2m+1) \times [2m+1]$, and insert the entry 1 to the $(k+1)$st gap position. Note that this implies that the entry 1 cannot create a singleton cycle as it cannot go to the first gap position. Take away the cycle C containing 1, and run Φ (of Lemma 6.20) through the remaining cycles. Then, together with C, we have a permutation in $\mathrm{EVEN}(2m+2)$. Run it through Φ^{-1} to get $\tau(\pi, k) \in \mathrm{ODD}(2m+2)$.

(24) We are going to construct a bijection κ from $SQ(2n) \times [2n+1]$ onto $SQ(2n+1)$. As the growth of $|SQ(n)|$ is equal to that of $|\,\mathrm{ODD}(n)|$ when passing from an even n to an odd $n+1$, we try to integrate Ψ of Lemma 6.27 into κ, by "stretching" the odd cycles part of our permutations. We proceed as follows.

Let $(\pi, k) \in SQ(2n) \times [2n+1]$. Take π, and break it into even cycles part and odd cycles part, or, for short, *odd part* and *even part.* Again, let k mark a gap position in π. If this gap position is in the odd cycles, or at the end of π, then interpret the gap position as a gap position for the odd part only, and simply run the odd part and this gap position through Ψ to get $\kappa(\pi)$, together with the unchanged even parts. Note that $2n+1$ will appear in an odd cycle when we are done.

If the gap position marked by k is in one of the even cycles, say c, we can think of it as marking the member of c immediately following it, say x. Replace x by $2n+1$ in c. To keep the information encoded by x, we interpret x as a gap position in the odd part of π. Indeed, if x is larger than exactly $i-1$ entries in the odd part, then let us mark the

ith gap position in the odd cycles part. So now we are in a situation like in the previous case, that is, the gap position is in the odd part. Run the odd part and this gap position through Ψ. Instead of inserting $2n+1$ to the marked position, however, insert temporarily a symbol B, to denote a number larger than all entries in the odd part. Then decrement all entries in the odd part that are larger than x (including B) by one notch. The obtained odd cycles and the unchanged even cycles (except for the mentioned change in c) give us $\kappa(\pi)$. Note that $2n+1$ will be in an even cycle when we're done.

We claim that the map κ defined above is a bijection from the set $SQ(2n) \times [2n+1]$ onto the set $SQ(2n+1)$. First, let us verify that κ maps into $SQ(2n+1)$. Indeed, (π) and $\kappa(\pi)$ have the same number of cycles of each even length, so by Exercise 21, $\pi \in SQ(2n)$ implies $\kappa(\pi) \in SQ(2n+1)$.

To get the reverse of κ, take a permutation $\pi' \in SQ(2n+1)$, and locate $2n+1$. If it is in an odd cycle, then run the odd cycles through Ψ^{-1}. This will yield an odd part one shorter, and an element of $[2n+1]$. Putting this together with the unchanged even part, we get $\kappa^{-1}(\pi')$. If $2n+1$ is in an even cycle, then run the odd cycles part through Ψ^{-1}. This will specify a gap position in the odd part, and so we recover the entry x. Increment entries larger than x by one notch in the odd part. To get the even part, put x back to the place of $2n+1$. The gap position immediately preceding $2n+1$ is our k in $\kappa^{-1}(\pi')$.

(25) Note that when we proved Theorem 6.24, we proved a special case of this problem, that is, the one when $k=2$. The very same method will prove this general statement.

Chapter 7

You Shall Not Overcount. The Sieve

7.1 Enumerating The Elements of Intersecting Sets

In a high school class there are 14 students who play soccer and there are 17 students who play basketball. How many students play at least one of these two sports?

The above question may sound extremely simple. However, we cannot answer it from the given information. Simply adding the two given numbers could yield an incorrect answer. Indeed, there may be students who play both sports. If we simply added the number of basketball players and the number of soccer players, we would count these students twice. To correct that, we would have to subtract their number once (so that they are counted only once), but we can only do that if we *know* their number.

Example 7.1. There are 14 students in a high school class who play soccer, and there are 17 students who play basketball. Four students play both games. How many students play at least one of the two games?

Solution. By the above argument, the number of students playing at least one of these two games is $14 + 17 - 4 = 27$.

Figure 7.1 illustrates the above situation.

The situation becomes more complicated, but still controllable, if the students are playing up to three different games. This is the content of our next example.

Example 7.2. In a high school class, there are 14 students who play soccer, 17 students who play basketball, and 18 students who play hockey. Four

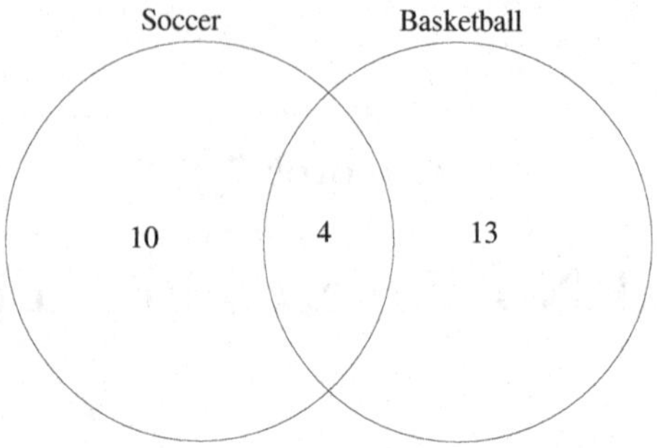

Fig. 7.1 Two intersecting sets.

students play both soccer and basketball, three play both soccer and hockey, and five play both basketball and hockey. There is one student who plays all three games. How many students play at least one of these games?

Solution. We can start our answer as before: adding the numbers of students playing soccer, basketball and hockey $14 + 17 + 18$ results in an overcount because we count students who play two of these games twice, therefore we have to correct this mistake and subtract the number of these students so that they are only counted once. This yields the number $14 + 17 + 18 - 4 - 3 - 5$. This is not a complete answer, however. The only student who plays all three games was counted three times (once for each game), but then she was subtracted three times (once for each pair of games), so right now she is not counted at all. Therefore, we have to correct this mistake by counting her, that is, by adding 1 to our final answer. Thus there are $14 + 17 + 18 - 4 - 3 - 5 + 1 = 38$ students in the class that play at least one of these three games.

We can again represent this situation by a diagram. This diagram is shown in Figure 7.2.

The reader can probably see that as the number of games increases, the same question requires a more and more tedious answer. Therefore, a general theorem is certainly useful to handle situations of this kind.

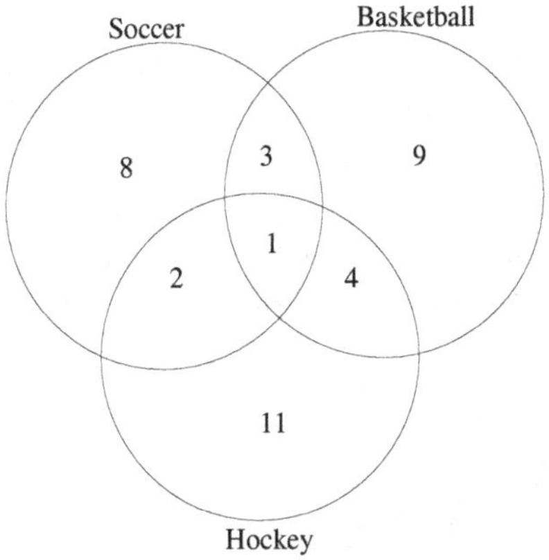

Fig. 7.2 Three intersecting sets.

Theorem 7.3 (Sieve Formula, or Principle of Inclusion-Exclusion). *Let $A_1, A_2, \cdots, A_n$ be finite sets. Then*

$$|A_1 \cup A_2 \cup \cdots \cup A_n| = \sum_{j=1}^{n} (-1)^{j-1} \sum_{i_1, i_2, \cdots, i_j} |A_{i_1} \cap A_{i_2} \cap \cdots \cap A_{i_j}|, \quad (7.1)$$

where $\{i_1, i_2, \cdots, i_j\}$ ranges over all j-element subsets of $[n]$.

Note that Theorem 7.3 is called the Sieve Formula since it counts "good" elements by *sifting* out the "bad" ones. A more formal name for the princple behind that counting idea is the *Inclusion-Exclusion Principle.*

Before proving this quintessential theorem, we would like to stress that the seemingly complicated expression on the right-hand side refers in fact to a conceptually simple sum: the alternating sum of the sizes of all j-fold intersections. The alternating sign is explained by the fact that we have to correct the overcounts. The two examples discussed before the theorem were the special cases when $n = 2$ and $n = 3$. In the first example, the sum on the right-hand side was $|A_1| + |A_2| - |A_1 \cap A_2|$, in the second example, the sum on the right-hand side was

$$|A_1| + |A_2| + |A_3| - |A_1 \cap A_2| - |A_1 \cap A_3| - |A_2 \cap A_3| + |A_1 \cap A_2 \cap A_3|.$$

Proof. [Proof of Theorem 7.3] Notice that an element not in $A_1 \cup A_2 \cup \cdots \cup A_n$ is not counted in any term on the right-hand side of (7.1). Thus we only have to show that each element of $A_1 \cup A_2 \cup \cdots \cup A_n$ is counted exactly once on the right-hand side. To do that, pick any element $x \in A_1 \cup A_2 \cup \cdots \cup A_n$. Let $S \subseteq [n]$ be the set of indices so that $x \in A_i$ if and only if $i \in S$, and let $s = |S|$. Note that $s \geq 1$. As $x \in A_i$ only if $i \in S$, a t-fold intersection $A_{i_1} \cap A_{i_2} \cap \cdots \cap A_{i_t}$ cannot contain x unless $(i_1, i_2, \cdots, i_t) \subseteq S$. So when determining the number of times x is counted by the right-hand side, we only have to consider the intersections involving the A_i which are indexed by S. On the other hand, each of these intersections does contain x. Therefore, the right-hand side counts x once for each of these subsets, with alternating signs. So altogether, the right-hand side counts the element x

$$s - \binom{s}{2} + \binom{s}{3} - \cdots + (-1)^{s-1}\binom{s}{s} = 1 \tag{7.2}$$

times. To see that the left-hand side of (7.2) is indeed 1, subtract 1 from both sides, then multiply both sides by -1, to get

$$1 - s + \binom{s}{2} - \binom{s}{3} - \cdots + (-1)^s\binom{s}{s} = 0 = (1-1)^s,$$

which is true by the Binomial theorem (and is further explained in Theorem 16.21). □

7.2 Applications of the Sieve Formula

Let us discuss some classic applications of the sieve formula. The first is the *problem of derangements.*

Example 7.4. A party was attended by n guests. When the guests arrived, they left their hats in the same coatroom. After the party ended, there was an electrical power failure, so each guest took a hat from the coatroom at random. When the guests were back on the street, they were amused to find out that none of them got his hat back. In how many different ways could that happen?

In a more mathematical formulation: how many permutations of the set $[n]$ have no *fixed points*, that is, have the element i in the ith position for no i? Such permutations are called *derangements.* Indeed, if the hat of the first person is denoted by 1, that of the second person is denoted by 2, and so on, then every way of the n people taking the n hats corresponds to a permutation of the set $[n]$. If the first person takes hat 7, then the first element of this permutation will be 7, if the second person takes hat 3, then the second element of this permutation will be 3, and so on. Now that we showed that the two formulations are in fact equivalent, we will give our answer in the language of permutations.

Solution. (of Example 7.4) It is easy to count permutations in which entry 1 or entry 2 or entry i is not a fixed point, but we want permutations with *no* fixed points. Their number is clearly equal to the number of all permutations minus the number of permutations with *at least one* fixed point. This sounds similar to the two examples we have discussed at the beginning of this section.

Let A_i be the set of all permutations of $[n]$ in which the element i is in the ith position, in other words, in which the element i is fixed. Then Theorem 7.3 will give us the answer to our question if we can compute the sizes of the intersections on the right-hand side of (7.1).

What is the size of A_1? The set A_1 consists of permutations in which the first element is 1. This means that elements $2, 3, \cdots, n$, can be freely permuted among each other, and this can be done in $(n-1)!$ different ways. So $|A_1| = (n-1)!$. Similarly, $|A_2| = (n-1)!$ as in this case element 2 has to be fixed, and all the remaining elements can be freely permuted. An analogous argument shows that $|A_i| = (n-1)!$ for all n values $i \in [n]$. Therefore, the total contribution of the first term of the right-hand side to the total value of the right-hand side is $(n-1)! \cdot n = n!$.

Now we move up to the next member of the right-hand side of (7.1), that is, to intersections of the type $|A_i \cap A_j|$. The set $A_i \cap A_j$ consists of permutations in which elements i and j are fixed, and the remaining $n-2$ entries can be permuted freely, in $(n-2)!$ ways. As there are $\binom{n}{2}$ choices for i and j, the total contribution of the second term is

$$-\binom{n}{2}(n-2)! = -\frac{n!}{2! \cdot (n-2)!} \cdot (n-2)! = -\frac{n!}{2!}.$$

In general, a similar argument shows that the contribution of the ith term is

$$(-1)^{i-1}\binom{n}{i} \cdot (n-i)! = (-1)^{i-1}\frac{n!}{i! \cdot (n-i)!} \cdot (n-i)! = (-1)^{i-1}\frac{n!}{i!}.$$

Indeed, if i given elements are fixed, the remaining $n-i$ elements can be permuted in $(n-i)!$ ways. On the other hand, there are $\binom{n}{i}$ possibilities for the set of i given elements.

Therefore, Theorem 7.3 yields

$$|A_1 \cup A_2 \cdots \cup A_n| = \sum_{j=1}^{n}(-1)^{j-1} \sum_{i_1, i_2, \cdots, i_j} |A_{i_1} \cap A_{i_2} \cap \cdots \cap A_{i_j}|$$

$$= n! - \frac{n!}{2!} + \frac{n!}{3!} - \cdots + (-1)^{n-1}\frac{n!}{n!} = \sum_{i=1}^{n}(-1)^{i-1}\frac{n!}{i!}.$$

So we have computed the number of permutations of $[n]$ with at least one fixed point. Consequently, the number $D(n)$ of permutations of $[n]$ with no fixed points, or the number of *derangements*, is $n!$ minus this number, that is,

$$D(n) = \sum_{i=0}^{n}(-1)^i \frac{n!}{i!}. \tag{7.3}$$

This completes the proof.

The right-hand side of formula (7.3) strongly depends on n. Still, the reader may want to get a feeling about roughly how likely it is that a random permutation has no fixed point. One can get such an intuition by dividing the number of favorable outcomes by that of all outcomes, that is, dividing the number of all derangements of $[n]$ by that of all permutations of $[n]$. This yields

$$\frac{D(n)}{n!} = \sum_{i=0}^{n}(-1)^i \frac{1}{i!} = \sum_{i=0}^{n} \frac{(-1)^i}{i!}.$$

This shows that if n diverges to infinity, then $D(n)/n!$ converges to e^{-1}, so for large values of n, roughly $1/e$ (so more than one third) of all permutations are derangements. So there is a fairly high chance all people will be looking for their hats.

The reader is invited to prove that for all positive integers n, the number $D(n)$ is the closest integer to $n!/e$.

We have promised in Section 5.2 that we will obtain a formula for the Stirling numbers of the second kind. Time has come to fulfill that promise.

Theorem 7.5. *For all positive integers n and k, the equality*

$$S(n,k) = \frac{1}{k!}\sum_{i=0}^{k}(-1)^i \binom{k}{i}(k-i)^n = \sum_{i=0}^{k}(-1)^i \frac{1}{i!(k-i)!}(k-i)^n$$

holds.

Proof. Instead of finding a formula for $S(n,k)$, we will find a formula for $k! \cdot S(n,k)$. We know from Corollary 5.9 that the latter is the number of all surjections from $[n]$ to $[k]$.

It is clear that the number of all *functions* from $[n]$ to $[k]$ is k^n as any element of the domain can be mapped into one of k elements. However, not all these functions will be surjections; many will miss one, two, or more elements of $[k]$ in their image. We have to enumerate those that do not miss *any* element of k. This sounds a little bit similar to the previous problem (there we were also interested in the number of certain objects *no part of which* had a certain property). It is therefore hopeful that the same approach will work here.

Let $i \in [k]$ and let A_i denote the set of all functions from $[n]$ to $[k]$ whose image does not contain i. It is then clear that $|A_i| = (k-1)^n$ as such functions can map any element of $[n]$ into any one of $k-1$ elements. Similarly,

$$|A_{i_1} \cap A_{i_2} \cap \cdots \cap A_{i_j}| = (k-j)^n,$$

for all $j \leq k$. Therefore, the sieve formula yields:

$$|A_1 \cup A_2 \cdots \cup A_n| = \sum_{j=1}^{n} (-1)^{j-1} \sum_{i_1, i_2, \cdots, i_j} |A_{i_1} \cap A_{i_2} \cap \cdots \cap A_{i_j}|$$

$$= \sum_{i=1}^{k} (-1)^{i-1} \binom{k}{i} (k-i)^n.$$

This is the number of functions from $[n]$ to $[k]$ whose range is *not* the entire set $[k]$. So the number of those with range $[k]$, in other words, the number of surjections, can be obtained by subtracting this number from that of all functions from $[n]$ to $[k]$, and our claim follows. □

The following Theorem is just a version of the sieve formula. We state it separately as its formulation goes in a direction we will continue in Chapter 16.

Theorem 7.6. *Let f and g be functions that are defined on the subsets of $[n]$, and whose range is the set of real numbers. Let us assume that f and g are connected by*

$$g(S) = \sum_{T \subseteq S} f(T).$$

Then

$$f(S) = \sum_{T \subseteq S} g(T)(-1)^{|S-T|}. \tag{7.4}$$

Proof. If we express $g(T)$ by values of f on the right-hand side of (7.4), we see that for all $U \subseteq S$, the value $f(U)$ will appear once for each set T satisfying $U \subseteq T \subseteq S$. Each such appearance of $f(U)$ will be counted with a sign given by $(-1)^{|S-T|}$. The number of such subsets T for which $|S-T| = i$ is equal to $\binom{|S-U|}{i}$, since T is determined by the elements of S that are not in T, and T contains U.

Therefore, $f(U)$ will appear on the right-hand side of (7.4) exactly $\sum_{i=0}^{|S-U|}(-1)^i\binom{|S-U|}{i} = (1-1)^{|S-U|}$ times. This number is always zero, except when $|S-U| = 0$, that is, when $S = U$. So the only term on the right-hand side that does not cancel out will be $f(S)$, and the claim is proved. □

Notes

Chapter 2 of "Enumerative Combinatorics", (Volume 1) by Richard Stanley [41] provides a higher-level review of the applications of the Sieve Formula.

Exercises

(1) A grade school class has two sports teams. For any two students in the class, there is at least one team so that the two students are members of that team. Prove that there is a team that contains all students of that class.

(2) A grade school class has three sports teams. For any two students in the class, there is at least one team so that the two students are members of that team. Prove that there is a team that contains at least 2/3 of the students of the class.

(3) How many positive integers $k \leq 210$ are relatively prime to 210?

(4) Let m be a positive integer. Denote by $\phi(m)$ the number of integers in $[m]$ that are relatively prime to m. Let p, q, and r be distinct prime numbers. Compute $\phi(pqr)$.

(5) Let $p_1, \cdots, p_k$ be distinct prime numbers. Find a formula for $\phi(p_1 \cdots p_k)$.

(6) Is it true that $\phi(mn) = \phi(n)\phi(m)$, for all positive integers m and n?

(7) Find a formula for $\phi(p^k)$, where p is a prime number.

(8) Find a formula for $\phi(n)$ if the prime factorization of n is known.

(9) Let $p = p_1p_2\cdots p_n$ be an n-permutation. We say that i is a *descent* of

p if $p_i > p_{i+1}$. The *descent set* of p is the set of all of its descents. How many 8-permutations have descent set T that is a subset of $\{1, 4, 6\}$?

(10) How many 8-permutations have descent set $\{1, 4, 6\}$?

(11) How many 8-permutations have descent set $\{1, 2, 4, 5, 7\}$?

(12) (This is a dual version of Theorem 7.6.) Let h and r be functions that are defined on the subsets of $[n]$, and whose range is the set of real numbers. Assume that h and r are connected by

$$r(S) = \sum_{S \subseteq T} h(T).$$

Prove that then

$$h(S) = \sum_{S \subseteq T} r(T)(-1)^{|T-S|}.$$

(13) Let $p = p_1 p_2 \cdots p_n$ be an n-permutation, and assume $n \geq 3$. We say that i is an *excedance* of p if $p_i > i$. Compute the number of n-permutations whose excedance set contains at least one of $n-2$ and $n-1$.

(14) (+) Let $f(n, k)$ be the number of ways to select a subset of $[n]$, and then select an involution on that subset that has k fixed points. (The empty set has one involution, and that involution has no fixed points.) Let $g(n) = \sum_{k=0}^{n} f(n, k)(-1)^k$. Prove that $g(n)$ is equal to the number of fixed point-free involutions on $[n]$.

Supplementary Exercises

(15) (-) How many partitions of $[n]$ contain at least one of the singleton blocks $\{1\}$ and $\{n\}$?

(16) (-) How many n-permutations contain at least one of the 1-cycles (1), (2), and (3)?

(17) (-) How many compositions does the integer n have in which neither the first nor the last entry is 1?

(18) (-) How many permutations of length n contain at least one of the 2-cycles (12) and (34)?

(19) (-) How many n-permutations $p = p_1 p_2 \cdots p_n$ are there in which at least one of p_1 and p_n is even?

(20) Give a combinatorial proof of the identity $D(n+1) = n(D(n) + D(n-1))$ for $n \geq 1$. Do not use the formula for the numbers $D(n)$ that we proved in the text. Set $D(0) = 1$ and $D(1) = 0$.

(21) How many n-permutations are there that contain exactly one cycle of length one?
(22) Let $d(n, k)$ be the number of derangements of length n that consist of k cycles. Find a formula for $d(n, k)$ in terms of signless Stirling numbers of the first kind.
(23) Let $F_k(n)$ be the number of partitions of $[n]$ into k blocks, each block consisting of more than one element. Express the numbers $F_k(n)$ in terms of Stirling numbers of the second kind.
(24) How many three-digit positive integers are divisible by at least one of six and seven?
(25) How many two-digit positive integers are relatively prime to both two and three?
(26) In how many ways can we list the digits $\{1, 1, 2, 2, 3, 4, 5\}$ so that two identical digits are not in consecutive positions?
(27) How many positive integers are there that are not larger than 1000 and are neither perfect squares nor perfect cubes?
(28) Show an example of four infinite subsets of the set of all positive integers so that the intersection of any three of them is an infinite set, but the intersection of all four of them is empty.
(29) How many n-permutations are there with exactly one descent?
(30) (+) How many n-permutations are there with exactly two descents?
(31) How many 2×2 matrices are there with entries from the set $\{0, 1, \cdots, k\}$ in which there are no zero rows and no zero columns?
(32) Let $F(n)$ denote the number of partitions of $[n]$ which contain no singleton blocks. Find a formula for the numbers $F(n)$ in terms of the Bell numbers $B(n)$.
(33) (+) Prove that $\lim_{n\to\infty} \frac{F(n)}{B(n)} = 0$.
(34) How many 2×2 matrices are there with entries from the set $\{0, 1, 2, \cdots, k\}$ that contain neither 0 rows nor 0 columns?
(35) (+) Find a *combinatorial* proof for the result of Exercise 16 of Chapter 5. Do not use equation (5.2).
(36) Find a closed formula (no summation signs) for $\sum_{i=0}^{n} \binom{n}{i} D(i)$.

Solutions to Exercises

(1) Consider the diagram of this situation. It will look similar to Figure 7.1. We note, however, that in this case, it cannot happen that the

leftmost and the rightmost domains of that diagram both contain a nonzero number. Indeed, that would mean that there is one student who is only a member of team A, and there is another one who is only a member of team B, so they are not on any common teams. Therefore, all positive integers of the diagram are contained in one of the two circles. Thus one team must contain all students.

(2) Again, consider the diagram of this situation. It will be similar to Figure 7.2. However, there cannot be any positive numbers in the domains that belong to one circle only. (Unless, that is, all students are on that team.) Denote A, B, C, D the numbers in the remaining domains as shown in Figure 7.3. Assume without loss of generality that $C \leq A$, and $C \leq B$. Then

$$\frac{A+B+D}{A+B+C+D} \geq \frac{A+B}{A+B+C} \geq \frac{A+B}{A+B+(A+B)/2} = \frac{2}{3}.$$

We used the fact that a fraction that is less than one increases if we increase its numerator and denominator by the same positive number.

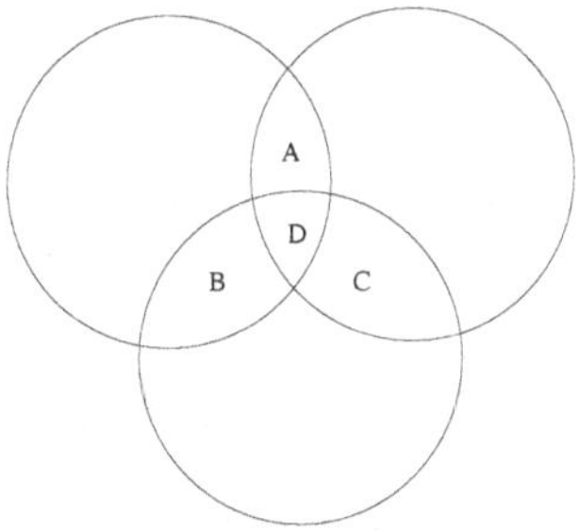

Fig. 7.3 The situation of Exercise (2).

(3) Let A_i be the set of those positive integers from [210] that are divisible by p_i, where $p_1 = 2$, $p_2 = 3$, $p_3 = 5$, and $p_5 = 7$. Then $|A_i| = \frac{210}{p_i}$, $|A_i \cap A_j| = \frac{210}{p_i p_j}$, and $|A_i \cap A_j \cap A_k| = \frac{210}{p_i p_j p_k}$. Therefore, we have all the ingredients for the application of the sieve formula. We get, after routine computation,

$$210 - |\bigcup_{i=1}^{4} A_i| = 48.$$

This method takes a long time, even for small numbers like 210. The following exercises will show a much faster method.

(4) We count those positive integers that are *not relatively prime* to pqr instead. Clearly, there are pq integers in $[pqr]$ that are divisible by r, there are pr that are divisible by q, and there are qr that are divisible by p. On the other hand, there are p integers in $[pqr]$ that are divisible by qr, there are q that are divisible by pr, and there are pq that are divisible by r. Finally, pqr is the only integer in this interval that is divisible by pqr. Therefore, the sieve implies

$$\phi(pqr) = pqr - pq - pr - qr + p + q + r - 1 = (p-1)(q-1)(r-1).$$

Note that the function ϕ is called *Euler's totient function.*

(5) Let $m = p_1 \cdots p_k$, and let $m' = p_1 \cdots p_{k-1}$. We claim that $\phi(m) = \Pi_{i=1}^{k}(p_i - 1)$. We are going to prove this claim by induction on k. The initial case of $k = 1$ is obviously true. Assume the statement is true for $k - 1$, and prove it for k.
By the induction hypothesis, there are $\phi(m') = \Pi_{i=1}^{k-1}(p_i - 1)$ integers in $[m']$ that are relatively prime to $[m']$. Moreover, if $q \leq p_k - 1$, then $n = m'q + r$ is relatively prime to m' if and only if r is relatively prime to m'. Therefore, there are $p_k \cdot \phi(m')$ integers in $[m]$ that are relatively prime to $[m']$. As divisibility by p_i for $i < k$ does not influence divisibility by p_k, exactly $\frac{1}{p_k}$ of these numbers is divisible by p_k, and exactly $\frac{p_k-1}{p_k}$ of them is relatively prime to p_k. Therefore, we get

$$\phi(m) = \frac{p_k - 1}{p_k} \cdot p_k \cdot \phi(m') = \Pi_{i=1}^{k}(p_i - 1).$$

(6) No, that is not true. For instance, $\phi(2) = 1$, $\phi(4) = 2$, however $\phi(8) = 4$.

(7) In this case, m is relatively prime to p^k if and only if m is not divisible by p. As exactly one integer in $[p]$ is divisible by p, we have $\phi(p^k) = \frac{p-1}{p} \cdot p^k = p^{k-1}(p-1)$.

(8) Let $n = p_1^{a_1} \cdots p_k^{a_k}$, where the p_i are the prime divisors of n. We claim that $\phi(n) = \Pi_{i=1}^{k} p_i^{a_i - 1}(p_i - 1)$. We prove this claim by induction on k. The initial case of $k = 1$ is true as it was proved in the previous exercise. The induction step is analogue to that of Exercise 5.

(9) In 8-permutations with a descent set contained in $\{1, 4, 6\}$, we know that $p_2 < p_3 < p_4$, moreover $p_5 < p_6$, and $p_7 < p_8$. There is no requirement on the relations not listed here. Therefore, we can get such a permutation if we split the set $[8]$ into four subsets, of sizes 1, 3, 2, and 2, arrange each of these subsets in increasing order, then

concatenate the four increasing strings in this order. The number of ways to do this is

$$\binom{8}{1}\binom{7}{3}\binom{4}{2}\binom{2}{2} = 8 \cdot 35 \cdot 6 \cdot 1 = 1680,$$

so this is the number of permutations with the required property.

(10) We are going to use Theorem 7.6. Denote by $g(S)$ the number of permutations with descent set contained in S, and denote by $f(S)$ the number of permutations with descent set equal to S. In order to be able to use Theorem 7.6, we have to compute the values of $g(T)$, for all $T \subseteq \{1,4,6\}$. This has been done for $T = \{1,4,6\}$ in the previous exercise. It is also obvious that $g(\emptyset) = 1$. For the other subsets, we proceed as in the previous exercise, and get

- $g(\{1\}) = \binom{8}{1} = 8$,
- $g(\{4\}) = \binom{8}{4} = 70$,
- $g(\{6\}) = \binom{8}{6} = 28$,
- $g(\{1,4\}) = \binom{8}{1}\binom{7}{3} = 280$,
- $g(\{1,6\}) = \binom{8}{1}\binom{7}{5} = 168$,
- $g(\{4,6\}) = \binom{8}{4}\binom{4}{2} = 420$.

Therefore, Theorem 7.6 shows

$$f(\{1,4,6\}) = 1680 - 280 - 168 - 420 + 8 + 70 + 28 - 1 = 917.$$

(11) It would take a long time to proceed as in the previous exercise, so we apply the following trick. Instead of counting these permutations $p = p_1p_2\cdots p_8$, count their *reverses* $p' = p_8p_7\cdots p_1$. As p had descent set $\{1,2,4,5,7\}$, its reverse will have descent set $\{2,5\}$. Indeed, if i was not a descent in p, then $p_i < p_{i+1}$. So in the reverse permutation p', the entry p_{i+1}, that is in position $8-i$, will be larger than the entry immediately following it. Therefore, $8-i$ is a descent of p'. Consequently, we only have to compute $f(\{2,5\})$, and that is relatively simple. We have

- $g(\emptyset) = 1$,
- $g(\{2\}) = \binom{8}{2} = 28$,
- $g(\{5\}) = \binom{8}{5} = 56$,
- $g(\{2,5\}) = \binom{8}{2}\binom{6}{3} = 560$.

Therefore, Theorem 7.3 implies

$$f(\{2,5\}) = 560 - 28 - 56 + 1 = 477 = f(\{1,2,4,5,7\}).$$

(12) Define new functions f and g by $f(A) = h(A^c)$, and $g(A) = r(A^c)$, where A^c denotes the complement of A in $[n]$. Then the condition of this exercise translates into the condition of Theorem 7.6, and the result of Theorem 7.3 translates back to the result of this exercise.

(13) Let $f(S)$ be the number of n-permutations whose excedance set *contains* S. Then $f(n-1) = (n-1)!$ as in such permutations, the entry n must be in position $n-1$. Similarly, $f(n-2) = 2(n-1)!$ as in permutations enumerated by $f(n-2)$, either the entry $n-1$ or the entry n has to be in position $n-2$. Finally, $f(n-2, n-1) = (n-2)!$ as in such permutations, the entry n must be in position $n-1$, and the entry $n-1$ must be in position $n-2$. Therefore, by the sieve formula, there are

$$f(n-1) + f(n-2) - f(n-2, n-1) = 3(n-1)! - (n-2)!$$

permutations with the required property.

(14) The statement to prove is equivalent to

$$\sum_{k \ even} f(n,k) - \sum_{k \ odd} f(n,k) = g(n).$$

Let $S \subseteq [n]$, and let p be an involution on S. Now consider the following map. Find the largest element M of $[n]$ that is either a fixed point of p or is not in S. If $M \in S$, remove M from S, and remove M from p. This results in a new, shorter involution $F(p)$ that has one less fixed points than p. If $M \notin S$, then add M to S, and add M to p as a fixed point. This results in a new, longer involution $F(p)$ that has one more fixed points than p.

So the parity of the number of fixed points of p and $F(p)$ is always different. In other words, the involution F matches involutions with an even number of fixed points with involutions with an odd number of fixed points. The only time F is not defined is when M is not defined, that is, when the set whose maximum was defined to be M is empty. That happens exactly when $S = [n]$ and p has no fixed points.

Chapter 8

A Function Is Worth Many Numbers. Generating Functions

As Herb Wilf said, a generating function of a sequence is a clothesline on which you hang all elements of the sequence. That single clothesline contains all elements of the sequence, and all information about them. This great idea, that is, to comprise data given by infinitely many numbers into a single function, leads to what is arguably the most powerful tool in Enumerative Combinatorics, namely to the technique of generating functions.

8.1 Ordinary Generating Functions

8.1.1 *Recurrence Relations and Generating Functions*

The frog population of an infinitely large lake grows fourfold each year. On the first day of each year, 100 frogs are taken out of the lake and shipped into another lake. Assuming that there were 50 frogs in the lake originally, how many frogs will be in the lake in 20 years? In 30 years? In 100 years? In n years?

The difficulty here does *not* lie in finding some kind of an answer. It is very easy to find a *recursive* answer. Indeed, if a_i denotes the number of frogs at the end of the ith year, so that $a_0 = 50$, $a_1 = 4 \cdot 50 - 100 = 100$, $a_2 = 4 \cdot 100 - 100 = 300$, and so on, then it is not difficult to prove that $a_{n+1} = 4a_n - 100$ if $n \geq 0$. In the computer age, such an answer is very useful, as we can go ahead and compute the values of a_n for all n as long as the memory of our computer lasts. There is, however, a tremendous waste in this method. Let us assume that we are only interested in the number of frogs after 87 years. Then, using the formula $a_{n+1} = 4a_n - 100$, we would have to compute the values of $a_1, a_2, \cdots, a_{86}$ in order to be able to compute

a_{87} at the end. So we would have to compute 86 values in which we were not interested.

To avoid such a waste of time and energy, it is best to find an *explicit formula* for a_n. That is, we would like to deduce a formula for a_n that does not contain a_{n-1}, or any other elements of the sequence; a formula that depends only on n, and is therefore directly computable.

All we have to work with is the equation

$$a_{n+1} = 4a_n - 100, \tag{8.1}$$

for all integers $n \geq 0$, and the initial condition $a_0 = 50$. This seems to be precious little at first sight. However, (8.1) holds for *all non-negative integer* values of n. So we in fact have infinitely many equations, in infinitely many variables. To collect all the information scattered in these infinitely many equations into just one equation, we will introduce the technique of *generating functions.*

Definition 8.1. Let $\{f_n\}_{n\geq 0}$ be a sequence of real numbers. Then the formal power series $F(x) = \sum_{n\geq 0} f_n x^n$ is called the *ordinary generating function* of the sequence $\{f_n\}_{n\geq 0}$.

As this section discusses ordinary generating functions only, we will sometimes omit the word "ordinary" for shortness. In what follows, we will manipulate (8.1) so that the ordinary generating function of the sequence $\{a_n\}$ appears. To that end, let us multiply both sides of (8.1) by x^{n+1}, then sum over all $n \geq 0$. This may well be a new operation for the reader, and it is crucial for the rest of this chapter, so we repeat it one more time. *Take a copy of (8.1) for each non-negative integer n, multiply both sides by x^{n+1}, and then take the sum of the infinitely many equations obtained.* We get

$$\sum_{n\geq 0} a_{n+1}x^{n+1} = \sum_{n\geq 0} 4a_n x^{n+1} - \sum_{n\geq 0} 100x^{n+1}. \tag{8.2}$$

The left-hand side is almost the generating function $G(x)$ of the sequence $\{a_n\}_{n\geq 0}$. Indeed, after replacing $n+1$ by n, the only missing term is a_0. So the left-hand side of (8.2) is $G(x) - a_0$. The first term of the right-hand side is $4xG(x)$, while the second term of the right-hand side is $\frac{100x}{1-x}$, by elementary calculus. So (8.2) is equivalent to

$$G(x) - a_0 = 4xG(x) - \frac{100x}{1-x}. \tag{8.3}$$

We have completed our first task: we compressed the information given by the infinitely many equations of the type $a_{n+1} = 4a_n - 100$ into just one equation.

The reader may think something along these lines "Big deal. True, the number of equations is only one, but that one equation contains the function $G(x)$, which has infinitely many terms, and is a weird thing anyway. So where is the great progress?" We cannot blame the reader for such thoughts at this point; they are quite natural. However, equation (8.3) is very useful, mainly because $G(x)$ is not just any function, it is a (formal) power series. Though the reader has probably met power series while studying calculus, the word *formal* is important there. By definition, a *formal* power series is an expression of the form $\sum_{n\geq 0} b_n x^n$, where the b_i are real numbers. Thus formal power series are *defined* by their coefficients, and are not necessarily equal to the Taylor series of some function. For example, the formal power series $\sum_{n\geq 0} n!x^n$ is not equal to the Taylor series of any function as it is not convergent for any $x \neq 0$.

Rearranging (8.3) we get

$$G(x) = \frac{a_0}{1-4x} - \frac{100x}{(1-x)(1-4x)}. \tag{8.4}$$

Remember that $a_0 = 50$, so the right-hand side does not contain any unknowns, in other words, it is a formal power series in x. Therefore, we have obtained an *explicit formula* for $G(x)$, the generating function of the sequence $\{a_n\}$.

Finally, we want to obtain an explicit formula for the numbers a_n themselves. Note that (8.4) is an equation on formal power series, and two formal power series are equal if and only if for all n, the coefficient of x^n is the same in both of them. The coefficient of x^n in $G(x)$ (so on the left-hand side of (8.4)) is a_n by definition. Therefore, in the formal power series on the right-hand side of (8.4), the coefficient of x^n is also a_n. On the other hand, we can also compute this coefficient as the sum of the coefficients of x^n in the two members of the right-hand side. The first term is easier. Indeed,

$$\frac{a_0}{1-4x} = 50\sum_{n\geq 0}(4x)^n = 50\sum_{n\geq 0} 4^n x^n,$$

so in the first term of the right-hand side, the coefficient of x^n is $50 \cdot 4^n$.

The second term is a little bit more complicated. That term is

$$\frac{100x}{(1-x)(1-4x)} = 100x \cdot \left(\sum_{n\geq 0} x^n\right) \cdot \left(\sum_{n\geq 0} 4^n x^n\right).$$

So the constant (the coefficient of x^0) is 0 in this term, and if $n \geq 1$, then we have to find the coefficient of x^{n-1} in the product $\left(\sum_{n\geq 0} x^n\right) \cdot \left(\sum_{n\geq 0} 4^n x^n\right)$. This, that is, finding the coefficient of a_n in a product, is something we will have to do very often while using generating functions. There are two ways to do this; we will show one now, and the other one after the completion of this solution.

The method we show now is that of *partial fractions*, which the reader may have well seen before in a Calculus or Differential Equations class.

Let us try to find constants A and B so that

$$\frac{A}{1-x} + \frac{B}{1-4x} = \frac{100x}{(1-x)(1-4x)}.$$

Multiplying both sides by $(1-x)(1-4x)$ yields

$$A(1-4x) + B(1-x) = 100x,$$

$$(-B-4A)x + A + B = 100x.$$

The polynomial on the left-hand side will be equal to the polynomial on the right-hand side if the coefficients of the two linear terms are the same and the two constants are the same. That is, $-B-4A = 100$, and $A+B=0$. Solving this system, we get that $A = -100/3$ and $B = 100/3$. Therefore,

$$\begin{aligned}\frac{100x}{(1-x)(1-4x)} &= \frac{100}{3}\cdot\frac{1}{1-4x} - \frac{100}{3}\cdot\frac{1}{1-x}\\ &= \frac{100}{3}\left(\sum_{n\geq 0} 4^n x^n - \sum_{n\geq 0} x^n\right)\\ &= \sum_{n\geq 0}(4^n-1)x^n\frac{100}{3}.\end{aligned}$$

Now that we have computed both terms on the right-hand side of (8.3), we can conclude that the coefficient of x^n there (and thus, the left-hand side of (8.3)) is

$$a_n = 50\cdot 4^n - 100\cdot\frac{4^n-1}{3}. \tag{8.5}$$

We have completed our task, that is, we have found an explicit formula for a_n. It is easy to check that (8.5) is indeed the correct formula. Substituting $n=0$ we indeed get $a_0 = 50$. Moreover,

$$4a_n - 100 = 4(50\cdot 4^n - 100\cdot\frac{4^n-1}{3}) - 100 = 50\cdot 4^{n+1} - 100\cdot\frac{4^{n+1}-4}{3} - 100$$

$$= 50 \cdot 4^{n+1} - 100 \cdot \frac{4^{n+1} - 1}{3},$$

so the sequence of numbers given by our explicit formula (8.5) satisfies the recurrence relation (8.1).

Let us summarize the technique we have just learned to turn recursive formulae into explicit ones.

(1) Define the ordinary generating function $G(x)$ of the sequence $\{a_n\}_{n\geq 0}$.
(2) Transform the recursive formula into an equation in $G(x)$. This can usually be done by multiplying both sides of the recursion by x^n, or x^{n+1}, sometimes x^{n+k}, and summing for all non-negative n.
(3) Solve for $G(x)$.
(4) Find the coefficient of x^n in $G(x)$. As this coefficient is a_n, this will provide an explicit formula for a_n.

Remarks.

(1) Here is an alternative way of handling the expression $\frac{100x}{(1-x)(1-4x)} = 100x \cdot (\sum_{n\geq 0} x^n) \cdot (\sum_{n\geq 0} 4^n x^n)$. There are many ways we can get a term in our product $(\sum_{n\geq 0} x^n) \cdot (\sum_{n\geq 0} 4^n x^n)$ in which the exponent of x is $n-1$. (We are interested in that coefficient because when we multiply $\sum_{n\geq 0} 4^n x^n$ by $100x$, this coefficient will turn into the coefficient of x^n.) We can take 1 from the first sum, and multiply it by $4^{n-1}x^{n-1}$ from the second sum. Or we could take x from the first sum, and multiply it by $4^{n-2}x^{n-2}$ from the second sum. In general, if i is such that $0 \leq i \leq n-1$, we can take x^i from the first sum, and multiply it by $4^{n-1-i}x^{n-1-i}$ from the second sum, getting the term $4^{n-1-i}x^{n-1}$. There are no other ways to get x^{n-1} in our product as the coefficients of x are non-negative in both sums. So the coefficient of x^{n-1} in $(\sum_{n\geq 0} x^n) \cdot (\sum_{n\geq 0} 4^n x^n)$ is

$$4^{n-1} + 4^{n-2} + \cdots + 4 + 1 = \frac{4^n - 1}{4 - 1} = \frac{4^n - 1}{3}.$$

Therefore, the coefficient of x^n in $\frac{100x}{(1-x)(1-4x)}$ is $100 \cdot \frac{4^n-1}{3}$, agreeing with our previous computation.

(2) There are several software packages that can compute the partial fraction decomposition of $\frac{100x}{(1-x)(1-4x)}$. For instance, in Maple, we can simply type

```
convert(100*x/((1-x)*(1-4*x),parfrac,x));
```

to obtain the desired decomposition.

Let us practice the technique of generating functions with another example.

Example 8.2. We have invested 1000 dollars into a savings account that pays five percent interest at the end of each year. At the beginning of each year, we deposit another 500 dollars into this account. How much money will be in this account after n years?

Solution. It is again very easy to find a recurrence relation. Let a_n be the account balance after n years. Then $a_0 = 1000$, and $a_{n+1} = 1.05 \cdot a_n + 500$. Let us go through the steps of our strategy one by one.

(1) Let $G(x) = \sum_{n\geq 0} a_n x^n$ be the generating function of the sequence $\{a_n\}_{n\geq 0}$.
(2) Multiplying both sides of the recurrence relation by x^{n+1} and summing over all non-negative integers n, we get

$$\sum_{n\geq 0} a_{n+1}x^{n+1} = \sum_{n\geq 0} 1.05 a_n x^{n+1} + \sum_{n\geq 0} 500x^{n+1}. \tag{8.6}$$

Here the left-hand side is clearly $G(x) - a_0$, while the first term of the right-hand side is $1.05xG(x)$, and the second term of the right-hand side is simply $\frac{500x}{1-x}$. So (8.6) is equivalent to

$$G(x) - a_0 = 1.05xG(x) + \frac{500x}{1-x}.$$

(3) Therefore,

$$G(x) = \frac{1000}{1-1.05x} + \frac{500x}{(1-x)(1-1.05x)}. \tag{8.7}$$

(4) To find a_n, it suffices to find the coefficient of x^n on the right-hand side, which is the sum of the coefficient of x^n in the first term, and the coefficient of x^n in the second term. Note that

$$\frac{1000}{1-1.05x} = 1000 \cdot \sum_{n\geq 0} 1.05^n x^n,$$

so the coefficient of x^n in the first term is $1000 \cdot 1.05^n$. For the second term, note that

$$\frac{500x}{(1-x)(1-1.05x)} = 500x \cdot \left(\sum_{n\geq 0} x^n\right)\left(\sum_{n\geq 0} 1.05^n x^n\right). \tag{8.8}$$

In order to find the coefficient of x^n in this expression, we will now use the alternative method shown in the Remarks after the previous example. If the reader is less than certain that he could apply the method of partial fractions here, we encourage the reader to try that method and compare his result to ours.

Note that to find the coefficient of x^n in (8.8), it suffices to find the coefficient of x^{n-1} in $\left(\sum_{n\geq 0} x^n\right)\left(\sum_{n\geq 0} 1.05^n x^n\right)$. In this product, we will get a term with exponent $n-1$ if and only if we take x^i from the first sum, and $1.05^{n-1-i}x^{n-1-i}$ from the second sum, for some i so that $0 \leq i \leq n-1$. (Because then the coefficients of x will add up to $n-1$, as needed.) Therefore, the coefficient of x^n in the second term of the right-hand side of (8.7) is

$$500\sum_{i=0}^{n-1} 1.05^i = 500\frac{1.05^n - 1}{1.05 - 1} = 10000 \cdot (1.05^n - 1).$$

Therefore, the coefficient of x^n on the right-hand side, and therefore, the left-hand side of (8.7) is

$$a_n = 1000 \cdot 1.05^n + 10000 \cdot (1.05^n - 1) = 1.05^n \cdot 11000 - 10000.$$

The following example shows how we could use the technique of generating functions to turn a recurrence relation to an explicit formula if the recurrence relation has more terms.

Example 8.3. Let $a_{n+2} = 3a_{n+1} - 2a_n$ if $n \geq 0$, and let $a_0 = 0$, and let $a_1 = 1$. Find an explicit formula for a_n.

Solution. Let $G(x) = \sum_{n\geq 0} a_n x^n$. Multiply both sides of the recurrence relation by x^{n+2}, and sum over all natural numbers n, to get

$$\sum_{n\geq 0} a_{n+2}x^{n+2} = 3\sum_{n\geq 0} a_{n+1}x^{n+2} - 2\sum_{n\geq 0} a_n x^{n+2},$$

which is equivalent to

$$G(x) - x = 3xG(x) - 2x^2G(x).$$

Expressing $G(x)$, we get

$$G(x) = \frac{x}{1 - 3x + 2x^2}.$$

The denominator of the right-hand side is again a quadratic polynomial. Note that $1 - 3x + 2x^2 = (x-1)(2x-1)$. Therefore, we are going to find real numbers A and B so that

$$G(x) = \frac{x}{1 - 3x + 2x^2} = \frac{A}{x-1} + \frac{B}{2x-1}. \tag{8.9}$$

After rearranging (8.9), we get

$$x = (2A + B)x - (A + B).$$

Two polynomials are the same if and only if their corresponding coefficients are the same. Therefore, it follows that $2A + B = 1$, and $A + B = 0$. So $A = 1$, and $B = -1$. Consequently, (8.9) yields

$$G(x) = \frac{x}{1 - 3x + 2x^2} = \frac{-1}{1 - x} + \frac{1}{1 - 2x}. \tag{8.10}$$

Both terms on the right-hand side are very easy to expand now. So

$$G(x) = -\sum_{n \geq 0} x^n + \sum_{n \geq 0} 2^n x^n = \sum_{n \geq 0} (2^n - 1)x^n$$

and therefore, $a_n = 2^n - 1$.

8.1.2 *Products of Generating Functions*

Our examples in the previous subsection showed how to use generating functions to turn a recurrence relation into an explicit formula. However, they only contained *one* generating function. Time has come for us to learn about the combinatorial use of the *product* of several generating functions.

Lemma 8.4. *Let $\{a_n\}_{n\geq 0}$ and $\{b_n\}_{n \geq 0}$ be two sequences, and let $A(x) = \sum_{n\geq 0} a_n x^n$, and $B(x) = \sum_{n \geq 0} b_n x^n$ be their respective generating functions. Define $c_n = \sum_{i=0}^{n} a_i b_{n-i}$, and let $C(x) = \sum_{n \geq 0} c_n x^n$. Then*

$$A(x)B(x) = C(x).$$

In other words, the coefficient of x^n in $A(x)B(x)$ is $c_n = \sum_{i=0}^{n} a_i b_{n-i}$.

Proof. When we multiply the infinite sum $A(x) = a_0 + a_1 x + a_2 x^2 + \cdots$ and the sum $B(x) = b_0 + b_1 x + b_2 x^2 + \cdots$, we multiply each term of the first sum by each term of the second sum, then add all these products. So a typical product is of the form $a_i x^i \cdot b_j x^j$. The exponent of x in this product will be n if and only if $j = n - i$, and the claim follows. □

The combinatorial consequence of Lemma 8.4 is the following theorem.

Theorem 8.5 (The Product formula). *Let a_n be the number of ways to build a certain structure on an n-element set, and let b_n be the number of way to build another structure on an n-element set. Let c_n be the number of ways to separate the set $[n]$ into the intervals $S = \{1, 2, \cdots, i\}$ and*

$T = \{i+1, i+2, \cdots, n\}$, (the intervals S and T are allowed to be empty), and then to build a structure of the first kind on S, and a structure of the second kind on T. Let $A(x)$, $B(x)$, and $C(x)$ be the respective generating functions of the sequences $\{a_n\}$, $\{b_n\}$, and $\{c_n\}$. Then

$$A(x)B(x) = C(x).$$

Proof. There are a_i ways to build a structure of the first kind on S, and b_{n-i} ways to build a structure of the second kind on T. This is true for all i, as long as $0 \leq i \leq n$. Therefore, $c_n = \sum_{i=0}^{n} a_i b_{n-i}$, and our claim follows from Lemma 8.4. □

Example 8.6. A semester at a Technical University consists of n days. At the beginning of each semester, the Dean of Engineering designs the term in the following way. She splits the term into two parts. The first k days of the term will form the theoretical part of the semester, and the second $n-k$ days will form the laboratory part (here $1 \leq k \leq n-2$). Then she chooses one holiday in the first part, and two holidays in the second part. In how many different ways can she design the term with these constraints?

Solution. Let f_n be the number of ways the Dean can plan the semester. It is straightforward to see that $f_n = \sum_{k=1}^{n-2} k\binom{n-k}{2}$. Looking at this expression, however, it is not so easy to see if it has a closed form (that is, a form without a summation sign), and if it does, what it is.

Let us separate problems of finding holidays in the two parts of the semester. There are k ways to do it in the first part, and $\binom{m}{2}$ ways to do it in the second part, where $m = n-k$.

The generating functions of these two sequences are $A(x) = \sum_{k\geq 1} kx^k$, and $B(x) = \sum_{m\geq 2}\binom{m}{2}x^m$. Recall from Calculus that $\sum_{i\geq 0} x^i = \frac{1}{1-x}$. Taking derivatives, (see Exercise 25 of Chapter 4 for another argument) this implies

$$A(x) = \frac{x}{(1-x)^2},$$

$$B(x) = \frac{x^2}{(1-x)^3}.$$

Now let $F(x)$ be the generating function of the sequence $\{f_n\}$. Then $A(x)B(x) = F(x)$. Therefore,

$$F(x) = A(x)B(x) = \frac{x^3}{(1-x)^5} = x^3 \sum_{n\geq 0}\binom{n+4}{4}x^n = \sum_{n\geq 3}\binom{n+1}{4}x^n.$$

This shows that $f_n = \binom{n+1}{4}$.

Note that the solution of the previous example used the fact that $(1-x)^{-5} = \sum_{n\geq 0}\binom{n+4}{4}x^n$. The reader is encouraged to find two proofs of this fact.

Example 8.7. Now let us assume that instead of holidays, the Dean chooses some days for independent study in both parts of the semester. In how many different ways can she plan the semester with these constraints?

Solution. Let g_n be the number of ways the dean can complete this task. Again, let us split the problem into two parts. Let $C(x)$ be the generating function for the number of ways to pick a set of days for independent study in the first part. As a k-element set has 2^k subsets, we have $C(x) = \sum_{k\geq 0} 2^k x^k = \frac{1}{1-2x}$. Note that the summation starts at $k=0$ here, since it is possible that the dean will not choose any days for independent study in one or both parts. This was different in the preceding example. Clearly, the second part has the same generating function, as the task at hand is the same. Therefore, we get

$$F(x) = C(x)C(x) = \frac{1}{(1-2x)^2}.$$

This shows that $F(x) = \frac{1}{2}C'(x)$. Therefore,

$$F(x) = \frac{1}{2}\cdot\sum_{n\geq 1} n\cdot 2^n x^{n-1} = \sum_{n\geq 0}(n+1)2^n x^n,$$

showing that $g_n = (n+1)\cdot 2^n$.

A little thought shows that Theorem 8.5 can easily be generalized from two generating functions into any fixed number of generating functions. The following example is an application of this generalized Product formula.

Example 8.8. Find the number of ways to split an n-day semester into three parts, choose any number of holidays in the first part, an odd number of holidays in the second part, and an even number of holidays in the third part.

Solution. Let g_n be the number of ways the one can plan such a semester. Let $A(x)$, $B(x)$, and $C(x)$ be the generating functions for the sequences for the three individual tasks. That is $A(x) = \sum_{n\geq 0} 2^n x^n = \frac{1}{1-2x}$ since there are 2^n ways to choose an unspecified number of holidays from a set of n days. As we have seen in Exercise 2 of Chapter 3, the number of subsets of $[n]$ that are of odd size is 2^{n-1} if $n \geq 1$, and 0 if $n=0$. Therefore, $B(x) = \sum_{n\geq 1} 2^{n-1}x^n = \frac{x}{1-2x}$. Finally, the reader is asked to prove that the number of subsets of $[n]$ that are of even size is 2^{n-1} if $n \geq 1$, and 1 if $n = 0$. Therefore, $C(x) = 1 + \frac{x}{1-2x} = \frac{1-x}{1-2x}$.

Now let $G(x)$ be the generating function of the sequence $\{g_n\}$. Then $G(x) = A(x)B(x)C(x)$.

Therefore,

$$\begin{aligned} G(x) = A(x)B(x)C(x) &= \frac{1}{1-2x} \cdot \frac{x}{1-2x} \cdot \frac{1-x}{1-2x} \\ &= \frac{x(1-x)}{(1-2x)^3}. \end{aligned}$$

The partial fraction decomposition leads to the equation

$$G(x) = -\frac{1}{4} \cdot \frac{1}{1-2x} + \frac{1}{4} \cdot \frac{1}{(1-2x)^3}.$$

Finally, using the binomial theorem, we get that

$$(1-2x)^{-3} = \sum_{n \geq 0} \binom{-3}{n} (-2x)^n = \sum_{n \geq 0} \binom{n+2}{2} 2^n x^n.$$

Therefore,

$$G(x) = -\frac{1}{4} \left(\sum_{n \geq 0} 2^n x^n \right) + \frac{1}{4} \left(\sum_{n \geq 0} \binom{n+2}{2} 2^n x^n \right).$$

So $g_n = (\binom{n+2}{2} 2^n - 2^n)/4 = 2^{n-3} n(n+3)$, for $n \geq 0$.

Example 8.9. If $p_{\leq k}(n)$ denotes the number of partitions of the integer n into parts of size at most k, then

$$\sum_{n \geq 0}^{\infty} p_{\leq k}(n) x^n = \prod_{i=1}^{k} \frac{1}{1-x^i} \tag{8.11}$$

$$= (1 + x + x^2 + x^3 + \cdots)(1 + x^2 + x^4 + x^6 + \cdots) \cdots (1 + x^k + x^{2k} + x^{3k} + \cdots).$$

Solution. Let us determine the coefficient of x^n on the right-hand side. The right-hand side is a sum of k-term products, such that each member comes from a different parentheses. The member from the ith parentheses is of the form x^{ij_i}, and the sum of the exponents of the k terms is n. In other words, $1j_1 + 2j_2 + \cdots kj_k = n$. If we write $1 + 1 + \cdots + 1$ (j_1 copies of 1) instead of $1j_1$, and in general, $i + i + \cdots + i$ (j_i copies of i) instead of ij_i in the previous equation, we obtain a partition of n into the sum of parts that are at most k.

Using this procedure, each time a product on the right-hand side is equal to x^n, we obtain a partition of n into the sum of parts that are at most k. Conversely, each partition of n into parts at most k can be associated to a product on the right-hand side, and the statement follows.

In Chapter 5, we proved that $p_{\leq k}(n)$ is also the number of partitions of n into at most k parts. Thus $\prod_{i=1}^{k} \frac{1}{1-x^i}$ is also the generating function of those partitions.

You could ask what the use of all this is if the above generating function does not yield a particularly nice closed formula for the numbers $p_{\leq k}(n)$. A quick answer is that any mathematics software can provide the expansion of (8.11) up to several dozen terms, so (8.11) provides a painless way to obtain a lot of numerical data.

A much deeper answer, and we will see examples of that soon, is that the generating function of a sequence contains a lot of information about the sequence, sometimes even more than an exact formula.

Example 8.10. If $p(n)$ denotes the number of partitions of the integer n, then

$$\sum_{n\geq 0}^{\infty} p(n)x^n = \prod_{k=1}^{\infty} \frac{1}{1-x^k} \tag{8.12}$$

$$= (1+x+x^2+x^3+\cdots)(1+x^2+x^4+x^6+\cdots)(1+x^3+x^6+x^9+\cdots)\cdots .$$

Solution. Same as the proof of the previous example, just here there is no limit on the size of the parts, and therefore, there are infinitely many parentheses on the right-hand side.

The reader may think that such a generating function, that is, the infinite product of sums, is not very useful. Indeed, a computer would have a hard time to handle an infinite formula. The following example refutes that belief. It is a stunning example of a problem that is much easier to handle with generating functions than without them.

Example 8.11. The number $p_{odd}(n)$ of partitions of n into odd parts is equal to the number $p_d(n)$ of partitions of n into all distinct parts.

Solution. The crucial idea is this. *It suffices to show that the generating functions of the two sequences are equal.* It is clear that

$$F(x) = \sum_{n \geq 0} p_{odd}(n)x^n = \prod_{\substack{i \geq 1 \\ i \text{ odd}}} \frac{1}{1-x^i}$$

and

$$G(x) = \sum_{n \geq 0} p_d(n)x^n = \prod_{i \geq 1}(1+x^i) = \prod_{i \geq 1} \frac{1-x^{2i}}{1-x^i}.$$

Note that after cancellations, the denominator of $G(x)$ will contain $(1-x^i)$ if and only if i is odd, and will therefore be the same as the denominator of $F(x)$. As both numerators are equal to 1, the proof follows.

8.1.2.1 *The Catalan Numbers*

A student moves into a new room, and upon his arrival, he puts an empty jar on his kitchen counter. From that on, every day he either puts a dollar coin in the jar, or takes a dollar coin out of the jar. After $2n$ days, the jar is empty again. In how many different ways could this happen?

This easily defined problem leads to a very famous, and exceptionally well-studied sequence of positive integers. Let c_n be the number of ways in which the events described in the preceding paragraph could take place, with $c_0 = 1$. Formally speaking, c_n is the number of sequences $b_1, b_2, \cdots, b_{2n}$ so that $b_j = \pm 1$ for all j, with $\sum_{j=1}^{2n} b_j = 0$, and, crucially, for all $k \in [2n]$, the inequality $\sum_{j=1}^{k} b_j \geq 0$ holds. Indeed, the jar never holds a *negative* number of coins. Let us call such sequences *good sequences* of length $2n$.

Let $C(x) = \sum_{n \geq 0} c_n x^n$ be the ordinary generating function of the sequence of the numbers c_n. The beauty of this example lies in the fact that we can use the Product formula to get a functional equation for $C(x)$. Though it is not immediately obvious on the outset, good sequences have a natural way of decomposing into an ordered pair of two structures, and therefore, the Product formula is relevant here. Of further interest is the fact that one of these structures will also have $C(x)$ for its generating function, while the other structure will have $xC(x)$ for its generating function.

In order to find this decomposition, note that if $n > 0$, and the jar is empty after $2n$ days, then there had to be a *first* day other than the starting day when the jar was empty. Let us say that the first such day was day $2i$, for some $i \in [n]$. (The reader is asked to prove that the jar can only be empty after an even number of days.) It is then clear that what happened from day $2i$ to day $2n$ is equivalent to a good sequence of length $2(n-i)$.

However, we must be a little bit more careful when we describe what happened during the first $2i$ days. There, we do not simply have a good sequence of length $2i$, but a good sequence of length $2i$ in which all the partial sums $\sum_{j=1}^{k} b_j$ are *positive* if $0 < k < 2i$. Let us call such sequences *very good.* Indeed, this positivity is equivalent to the fact that day $2i$ is the *first* day when the jar is empty. Note that there is a bijective correspondence between the very good sequences $B = (b_1, b_2, \cdots, b_{2i})$ of length $2i$, and the good sequences $B' = (b_2, b_3, \cdots, b_{2i-1})$ of length $2(i-1)$. Indeed, a very good sequence must start with a 1 and end in a -1; removing these two entries we get a good sequence that is two bits shorter. By the removal of these two entries, the sums of initial segments got only one smaller, so they are still non-negative.

In other words, if $n \geq 1$, then the number of very good sequences of length $2i$ is c_{i-1}. Their number is 0 if $i = 0$ since there will be no initial segments with positive sums then. So the generating function for the numbers of very good sequences is $\sum_{n \geq 1} c_{n-1} x^n = xC(x)$.

Now we are ready to use the Product formula. The above discussion shows that each good sequence of length $2n > 0$ decomposes in a natural and unique way into a very good sequence of length $2i$ and a good sequence of length $2(n-i)$ for some $i \in [n]$. Therefore, the Product formula implies that

$$C(x) - 1 = xC(x) \cdot C(x).$$

Note that we wrote $C(x) - 1$, and not $C(x)$, on the left-hand side, since for $n = 0$, the above decomposition does not exist.

The last displayed equation can be rearranged as

$$xC(x)^2 - C(x) + 1 = 0, \tag{8.13}$$

which is a quadratic equation for $C(x)$. We can solve this equation using the well-known formula for solving quadratic equations. However, there is a last hurdle to clear. The quadratic formula implies that (8.13) has two solutions, namely $\frac{1+\sqrt{1-4x}}{2x}$ and $\frac{1-\sqrt{1-4x}}{2x}$. How do we know which one to choose for $C(x)$? In order to answer this question, note that $C(x)$ has constant term 1, so we have to choose the solution which also has constant term 1. Substituting $x = 0$, we see that the second solution has this property, therefore,

$$C(x) = \frac{1 - \sqrt{1-4x}}{2x}. \tag{8.14}$$

Recall that we computed in Example 4.16 that $\sqrt{1-4x} = 1 - 2x - 2\sum_{n\geq 2} \frac{\binom{2n-2}{n-1}}{n} x^n$. Comparing this with (8.14), we get

$$C(x) = \sum_{n\geq 0} \frac{\binom{2n}{n}}{n+1} x^n,$$

so $c_n = \frac{\binom{2n}{n}}{n+1}$.

The numbers c_n are called the *Catalan numbers*, named after the French mathematician Eugene Catalan. They count at least 150 different kinds of combinatorial objects, and we will mention a comprehensive reference for these objects in the Notes section of this chapter. We have already seen some of them, in Exercises 23 and 57 of Chapter 4. We will see some more in the exercises of this chapter, and some in Chapters 14 and 16. Starting with $n = 0$, the first few values of the sequence of the Catalan numbers c_n are 1, 1, 2, 5, 14, 42.

8.1.3 *Compositions of Generating Functions*

How could we possibly define the composition of two generating functions? Let us, for simplicity, that $F(x) = 1/(1-x) = 1 + x + x^2 + x^3 + \cdots$, and let $G(x)$ be any generating function. Our knowledge of the composition of functions suggests that $F(G(x))$ should be defined as $1/(1 - G(x)) = 1 + G(x) + G(x)^2 + G(x)^3 + \cdots$. It is here that the problems could start. The sum of infinitely many power series is defined only if for each n, the coefficient of x^n is zero in all but a finite number of summands. In our case, this will happen if and only if the constant term of $G(x)$ is 0. Indeed, in that case $G(x)^n$ is divisible by x^n, thus there are at most $n-1$ summands that contain x^{n-1}, and this holds for all $n > 0$. (If $n = 0$, then there is one such summand.) Therefore, $F(G(x))$ is defined in this case. If F is a formal power series other than $1/(1-x)$, the same argument holds. This is the basis of the following definition.

Definition 8.12. Let $F(x) = \sum_{n\geq 0} f_n x^n$ be a formal power series, and let G be a formal power series with constant term 0. Then we define

$$F(G(x)) = \sum_{n\geq 0} f_n (G(x))^n = f_0 + f_1 G(x) + f_2 (G(x))^2 + \cdots.$$

The following theorem is a major application of compositions of generating functions.

Theorem 8.13. *Let a_n be the number of ways to build a certain structure on an n-element set, and let us assume that $a_0 = 0$. Let h_n be the number of ways to split the set $[n]$ into an unspecified number of disjoint non-empty intervals, then build a structure of the given kind on each of these intervals. Set $h_0 = 1$. Denote $A(x) = \sum_{n\geq 0} a_n x^n$, and $H(x) = \sum_{n\geq 0} h_n x^n$. Then*

$$H(x) = \frac{1}{1 - A(x)}.$$

Note that unlike in Theorem 8.5, here we do *not* allow empty intervals. The reason for this is that if we did, we would have infinitely many ways to split up $[n]$ as we could insert as many empty intervals as we like. This problem did not arise in Theorem 8.5, because we only had a specified number (two) of intervals there.

Proof. (of Theorem 8.13) It follows from Theorem 8.5 that $A(x)^k$ is the generating function for the number of ways to split $[n]$ into exactly k intervals, then to build a structure of the given kind on each interval. Summing over all k, we get $\sum_{k\geq 1} A(x)^k$. As $a_0 = 0$, none of the power series $A(x)^k$ has a nonzero constant term. On the other hand, $H(x)$ has constant term 1 by definition. This shows

$$H(x) = 1 + \sum_{k\geq 1} A(x)^k = \sum_{k\geq 0} A(x)^k = \frac{1}{1 - A(x)}.$$

$\square$

Example 8.14. All n soldiers of a military squadron stand in a line. The officer in charge splits the line at several places, forming smaller (non-empty) units. Then he names one person in each unit to be the commander of that unit. Let h_n be the number of ways he can do this. Find a closed formula for h_n.

Solution. Denote by h_k the number of ways the officer in charge can proceed. Let a_k be the number of ways to choose a commander from a unit of k people. Then clearly $a_k = k$, and therefore $A(x) = \sum_{k\geq 1} kx^k = \frac{x}{(1-x)^2}$, as we have computed in Example 8.6. Then Theorem 8.13 applies, and we get that

$$H(x) = \frac{1}{1 - A(x)} = \frac{1}{1 - \frac{x}{(1-x)^2}} = 1 + \frac{x}{1 - 3x + x^2},$$

where $H(x)$ is the generating function of the sequence $\{h_n\}_{n\geq 0}$.

The evaluation of the fraction $1/(1 - 3x + x^2)$ is somewhat more complicated than in the earlier examples. We will use the method of partial fractions. The roots of $x^2 - 3x + 1$ are $\alpha = (3 + \sqrt{5})/2$, and $\beta = (3 - \sqrt{5})/2$. Therefore, we want to obtain $1/(1 - 3x + x^2)$ in the following form.

$$\frac{1}{1 - 3x + x^2} = \frac{1}{(x - \alpha)(x - \beta)} = \frac{A}{x - \alpha} - \frac{B}{x - \beta}.$$

After cross-multiplying, we get

$$1 = (A - B)x - A\beta + B\alpha.$$

Therefore, we must have $A = B$, and $B(\alpha - \beta) = B\sqrt{5} = 1$. So $A = B = 1/\sqrt{5}$. This yields

$$\frac{1}{1 - 3x + x^2} = \frac{1}{\sqrt{5}}\left(\frac{1}{x - \alpha} - \frac{1}{x - \beta}\right).$$

Now note that $\alpha \cdot \beta = 1$. Therefore, we can multiply both the numerator and the denominator of the first (respectively, second) term in the parentheses by α (respectively, β). After routine steps, we get

$$\frac{1}{1 - 3x + x^2} = \frac{1}{\sqrt{5}}\left(\frac{\alpha}{1 - \alpha x} - \frac{\beta}{1 - \beta x}\right) = \frac{1}{\sqrt{5}}\left(\alpha \sum_{k=0}^{\infty} \alpha^n x^n - \beta \sum_{k=0}^{\infty} \beta^n x^n\right).$$

Therefore, the coefficient of x^n in $\frac{1}{1-3x+x^2}$ is $\frac{1}{\sqrt{5}}(\alpha^{n+1} - \beta^{n+1})$. Thus the coefficient of x^n in $H(x)$ is 1 if $n = 0$, and

$$h_n = \frac{1}{\sqrt{5}}(\alpha^n - \beta^n)$$

if $n > 0$.

Would you have guessed that our answer to this problem, that was defined totally within the kingdom of integers, will involve powers of $\alpha = (3 + \sqrt{5})/2$, and $\beta = (3 - \sqrt{5})/2$? The first few values of the sequence h_n are, (starting at h_1), 1, 3, 8, 21, 55. These numerical data may be helpful in some of the exercises.

In Theorem 8.13, we first split $[n]$ into non-empty intervals, then we take a structure of the same kind on each of these intervals. However, we do not take a structure *on the set of the intervals.* Translating this to our example, the officer in charge did not ask the units to choose a unit on duty, or to form a new line. The following theorem generalizes Theorem 8.13 in that direction.

Theorem 8.15 (The Compositional formula). *Let a_n be the number of ways to build a certain structure on an n-element set, and assume $a_0 = 0$. Let b_n be the number of ways to build a second structure on an n-element* *set, and let* $b_0 = 1$. *Let* g_n *be the number of ways to split the set* $[n]$ *into an unspecified number of non-empty intervals, build a structure of the first kind on each of these intervals, and then build a structure of the second kind on the set of the intervals. Set* $g_0 = 1$. *Denote by* $A(x)$, $B(x)$, *and* $G(x)$ *the generating functions of the sequences* $\{a_n\}$, $\{b_n\}$, *and* $\{g_n\}$. *Then*

$$G(x) = B(A(x)).$$

Proof. Let us assume that we split $[n]$ into k intervals. Then there are b_k ways to take a structure of the second kind on the k-element set of these intervals. The product formula shows that the generating function for the number of ways to take a structure of the first kind on each interval is $A(x)^k$. Therefore, the contribution of this case to $G(x)$ is $b_k A(x)^k$. Summing over all k, we get that $G(x) = \sum_{k\geq 0} b_k A(x)^k$, which was to be proved. □

Example 8.16. All n soldiers of a military squadron stand in a line. The officer in charge splits the line at several places, forming smaller (non-empty) units. Then he chooses a (possibly empty) subset of the newly formed units for night duty. In how many different ways can he do this?

Solution. Let us keep the notation of Theorem 8.15. Then $a_k = 1$ for all $k \geq 1$, as there is one way to put the trivial structure (that is to say, no structure at all) on the individual units. Furthermore, $b_m = 2^m$ as we simply choose a subset of the set of all intervals. Therefore, $A(x) = x/(1-x)$, and $B(x) = 1/(1-2x)$. So

$$G(x) = B(A(x)) = \frac{1}{1-\frac{2x}{1-x}} = \frac{1-x}{1-3x} = \frac{1}{1-3x} - \frac{x}{1-3x},$$

$$G(x) = \sum_{n\geq 0} 3^n x^n - \sum_{n\geq 1} 3^{n-1}x^n = 1 + \sum_{n\geq 1} 2\cdot 3^{n-1}x^n.$$

Consequently, if $n \geq 1$, the officer in charge has $2\cdot 3^{n-1}$ options.

8.2 Exponential Generating Functions

8.2.1 *Recurrence Relations and Exponential Generating Functions*

Not all recurrence relations can be turned into a closed formula by using an ordinary generating function. Sometimes, a closed formula may not exist. Some other times, it could be that we have to use a different kind of generating function.

Example 8.17. Let $a_0 = 1$, and let $a_{n+1} = (n+1)(a_n - n + 1)$, if $n \geq 0$. Find a closed formula for a_n.

If we try to solve this recurrence relation by ordinary generating functions, we run into trouble. The reason for this is that this sequence grows too fast, and its ordinary generating function will therefore not have a closed form. Let us instead make the following definition.

Definition 8.18. Let $\{f_n\}_{n\geq 0}$ be a sequence of real numbers. Then the formal power series $F(x) = \sum_{n\geq 0} f_n \frac{x^n}{n!}$ is called the *exponential generating function* of the sequence $\{f_n\}_{n\geq 0}$.

The word "exponential" is due to the fact that the exponential generating function of the constant sequence $f_n = 1$ is e^x. Let us use this new kind of generating function to solve the example at hand.

Solution. (of Example 8.17.) Let $A(x) = \sum_{n=0}^{\infty} a_n \frac{x^n}{n!}$ be the exponential generating function of the sequence $\{a_n\}_{n\geq 0}$. From this point on, we proceed in a way that is very similar to the method of the previous section. Let us multiply both sides of our recursive formula by $x^{n+1}/(n+1)!$, and sum over all $n \geq 0$ to get

$$\sum_{n=0}^{\infty} a_{n+1} \frac{x^{n+1}}{(n+1)!} = \sum_{n=0}^{\infty} a_n \frac{x^{n+1}}{n!} - \sum_{n=0}^{\infty} (n-1) \frac{x^{n+1}}{n!}. \qquad (8.15)$$

Note that the left-hand side is $A(x)-1$, while the first term of the right-hand side is $xA(x)$. This leads to

$$A(x) - 1 = xA(x) - x^2e^x + xe^x,$$

$$A(x) = \frac{1}{1-x} + xe^x = \sum_{n\geq 0} x^n + \sum_{n\geq 0} \frac{x^{n+1}}{n!}.$$

The coefficient of $x^n/n!$ in $\sum_{n\geq 0} x^n$ is $n!$, while the coefficient of $x^n/n!$ in $\sum_{n\geq 0} \frac{x^{n+1}}{n!}$ is n. Indeed, this second term has summand $x^n/(n-1)!$. Therefore, the coefficient of $x^n/n!$ in $A(n)$ is $a_n = n! + n$.

Example 8.19. Let $f_0 = 0$, and let $f_{n+1} = 2(n+1)f_n + (n+1)!$ if $n \geq 0$. Find an explicit formula for f_n.

Solution. Let $F(x) = \sum_{n\geq 0} f_n \frac{x^n}{n!}$ be the exponential generating function of the sequence f_n. Let us multiply both sides of our recursive formula by $x^{n+1}/(n+1)!$, then sum over all $n \geq 0$. We get

$$\sum_{n\geq 0} f_{n+1}\frac{x^{n+1}}{(n+1)!} = 2x\sum_{n\geq 0} f_n\frac{x^n}{n!} + \sum_{n\geq 0} x^{n+1}. \tag{8.16}$$

As $f_0 = 0$, the left-hand side of (8.16) is equal to $F(x)$, while the first term of the right-hand side is $2xF(x)$, and the second term of the right-hand side is $x/(1-x)$. Therefore, we get

$$F(x) = 2xF(x) + \frac{x}{1-x},$$

$$F(x) = \frac{x}{(1-x)(1-2x)}.$$

Therefore,

$$F(x) = \sum_{n\geq 0}(2^n - 1)x^n,$$

and so the coefficient of $x_n/n!$ in $F(x)$ is $f_n = (2^n - 1)n!$.

8.2.2 *Products of Exponential Generating Functions*

Just as we have seen for ordinary generating functions, the product of two exponential generating functions has a very natural combinatorial meaning.

Lemma 8.20. *Let $\{a_i\}$ and $\{b_k\}$ be two sequences, and let $A(x) = \sum_{i\geq 0} a_i \frac{x^i}{i!}$ and $B(x) = \sum_{k\geq 0} b_k \frac{x^k}{k!}$ be their exponential generating functions. Define $c_n = \sum_{i=0}^{n} \binom{n}{i} a_i b_{n-i}$, and let $C(x)$ be the exponential generating function of the sequence $\{c_n\}$. Then*

$$A(x)B(x) = C(x).$$

In other words, the coefficient of $x^n/n!$ in $A(x)B(x)$ is $c_n = \sum_{i=0}^{n} \binom{n}{i} a_i b_{n-i}$.

Proof. Just as in the proof of Lemma 8.4, multiplying $A(x)$ by $B(x)$ involves multiplying each term of $A(x)$ by each term of $B(x)$. A general term in this product is of the form

$$a_i\frac{x^i}{i!}\cdot b_j\frac{x^j}{j!} = a_ib_j\cdot\frac{x^{i+j}}{i!j!}\cdot\frac{(i+j)!}{(i+j)!} = a_ib_j\cdot\frac{x^{i+j}}{(i+j)!}\cdot\binom{i+j}{i}.$$

Such a product is of degree n if and only if $i + j = n$, and the statement follows. □

Theorem 8.21 (Product formula, exponential version). *Let a_n be the number of ways to build a certain structure on an n-element set, and let b_n be the number of way to build another structure on an n-element set. Let c_n be the number of ways to separate $[n]$ into the disjoint subsets S and T, ($S \cup T = [n]$), and then to build a structure of the first kind on S, and a structure of the second kind on T. Let $A(x)$, $B(x)$, and $C(x)$ be the respective exponential generating functions of the sequences $\{a_n\}$, $\{b_n\}$, and $\{c_n\}$. Then*

$$A(x)B(x) = C(x).$$

Note that while Theorems 8.5 and 8.21 sound very similar, they apply in different circumstances. Theorem 8.5 applies when $[n]$ is split into two parts so that one part is $[i]$. That is, $[n]$ is split into *intervals.* Theorem 8.21 applies when $[n]$ is split into two parts with no restrictions. In other words, the first theorem applies when our objects are linearly ordered (like days in a calendar, or people in a line), and we cut that linear order somewhere to get two subsets. The second theorem applies when we are free to choose our two subsets, that is, they do not have to be consecutive objects in a previously ordered line.

Proof. (of Theorem 8.21) If S has i elements, then there are $\binom{n}{i}$ ways to choose the elements of S. Then there are a_i ways to build a structure of the first kind on S, and b_{n-i} ways to build a structure of the second kind on T, and this is true for all i, as long as $0 \leq i \leq n$. Therefore, $c_n = \sum_{i=0}^{n} \binom{n}{i} a_i b_{n-i}$, and our claim follows from Lemma 8.20. □

Example 8.22. A football coach has n players to work with at today's practice. First he splits them into two groups, and asks the members of each group to form a line. Then he asks each member of the first group to take on an orange shirt, or a white shirt, or a blue shirt. Members of the other group keep their red shirt. In how many different ways can all this happen?

Solution. Let us assume that the coach selects k people to form the first group. Let a_k be the number of ways these k people can take on an orange or white or blue shirt, and then form a line. Then $a_k = k!3^k$, so the exponential generating function of the sequence $\{a_k\}$ is

$$A(x) = \sum_{k \geq 0} k!3^k \frac{x^k}{k!} = \frac{1}{1-3x}.$$

Similarly, assume there are m people in the second group. Let b_m be the number of ways these m people can form a line. Then $b_m = m!$, and the exponential generating function of the sequence $\{b_m\}$ is

$$B(x) = \sum_{m \geq 0} m! \frac{x^m}{m!} = \frac{1}{1-x}.$$

Let c_n be the number of ways the players can follow the instructions of the coach, and let $C(x)$ be the exponential generating function of the sequence $\{c_n\}$. Then the Product formula implies

$$C(x) = A(x)B(x) = \frac{1}{1-3x} \cdot \frac{1}{1-x},$$

as $\frac{1}{1-3x} = \sum_{k=0} 3^k x^k$, and $\frac{1}{1-x} = \sum_{m=0} x^m$, it follows that the coefficient of $x^n/n!$ in $C(x)$ is $c_n = n!(3^{n+1} - 1)/2$.

A particularly useful property of exponential generating functions is that their derivatives are very easy to describe. Indeed, $\left(\frac{x^{n+1}}{(n+1)!}\right)' = \frac{x^n}{n!}$, and therefore

$$\left(\sum_{n \geq 0} a_n \frac{x^n}{n!}\right)' = \sum_{n \geq 0} a_{n+1} \frac{x^n}{n!}.$$

The following example makes good use of this observation. Recall that the Bell numbers and the recurrence relation used in the example were proved in Chapter 5.

Example 8.23. Let $B(x)$ be the exponential generating function of the Bell numbers $B(n)$. Prove that $B(x) = e^{e^x - 1}$.

Solution. We know that $B(n+1) = \sum_{i=0}^{n} B(i)\binom{n}{i}$ if $n \geq 0$, and $B(0) = 1$. Multiply both sides by $x^n/n!$ and sum over all $n \geq 0$ to get

$$\sum_{n \geq 0} B(n+1) \frac{x^n}{n!} = \sum_{n \geq 0} \sum_{i=0}^{n} B(i) \binom{n}{i} \frac{x^n}{n!}.$$

Now note that the left-hand side is $B'(x)$, while the right-hand side is $B(x)e^x$ by Lemma 8.20. Therefore, we get

$$B'(x) = B(x)e^x,$$

$$\frac{B'(x)}{B(x)} = e^x,$$

and, taking integrals,

$$\ln B(x) = e^x + C.$$

Setting $x = 0$, the left-hand side is $\ln 1 = 0$, therefore we must choose $C = -1$ on the right-hand side. Therefore, $\ln B(x) = e^x - 1$, and $B(x) = e^{e^x - 1}$ as claimed.

8.2.3 Compositions of Exponential Generating Functions

The compositions of exponential generating functions can be defined in the same circumstances, and in the same way, as those of ordinary generating functions. In this subsection we will see that the corresponding versions of Theorems 8.13 and 8.15 also hold.

Theorem 8.24 (The Exponential formula). *Let a_n be the number of ways to build a certain structure on an n-element set, and assume $a_0 = 0$. Let h_n be the number of ways to partition the set $[n]$ into an unspecified number of non-empty subsets, then build a structure of the given kind on each of these subsets. Set $h_0 = 1$. Denote by $A(x)$ and $H(x)$ the exponential generating functions of these sequences. Then*

$$H(x) = e^{A(x)}.$$

Proof. Since in a set partition the order of blocks is irrelevant, it follows from Theorem 8.21 that $A(x)^k/k!$ is the exponential generating function for the number of ways to partition $[n]$ into exactly k subsets, then build a structure of the given kind on each subset. Summing over all k, we get $\sum_{k \geq 1} A(x)^k/k!$. As $a_0 = 0$, none of the power series $A(x)^k/k!$ has a constant term. On the other hand, $H(x)$ has constant term 1 by definition. This shows

$$H(x) = 1 + \sum_{k \geq 1} \frac{A(x)^k}{k!} = \sum_{k \geq 0} \frac{A(x)^k}{k!} = e^{A(x)}.$$

$\square$

Example 8.25. In how many different ways can we arrange n people into groups, and then have each group sit at a circular table?

Solution. There are $(k-1)!$ ways for a k-member group to sit at a circular table. Therefore, keeping the notation of Theorem 8.24, $a_k = (k-1)!$. This yields

$$A(x) = \sum_{k \geq 1} (k-1)! \cdot \frac{x^k}{k!} = \sum_{k \geq 1} \frac{x^k}{k} = \ln \left(\frac{1}{1-x} \right).$$

Therefore, the Exponential formula implies that

$$H(x) = e^{\ln\left(\frac{1}{1-x}\right)} = \frac{1}{1-x} = \sum_{n \geq 0} x^n = \sum_{n \geq 0} n! \cdot \frac{x^n}{n!}.$$

This shows that there are $h_n = n!$ ways to arrange our n people around circular tables.

The reader should try to find an immediate combinatorial proof of this result.

The following is an example of combined applications of the Product formula and the Exponential formula.

Example 8.26. Find the exponential generating function $F(x)$ for the sequence $\{f_n\}$ that denotes the number of partitions of $[n]$ into blocks of size three, four, and nine.

Solution. Let a_n, b_n, and c_n denote the number of partitions of $[n]$ into blocks of size three only, size four only, and size nine only, and let $A(x)$, $B(x)$, and $C(x)$ denote the respective exponential generating functions. We will determine these exponential generating functions by the Exponential formula. To that end, consider the following very simple sequence. Let t_n be the number of ways an n-element set can form a block of size three. Obviously, $t_3 = 1$, and $t_n = 0$ if $n \neq 3$. Thus the exponential generating function of this sequence is $T(x) = x^3/3!$. It then follows by the Exponential formula that

$$A(x) = e^{T(x)} = e^{x^3/3!}.$$

An analogous argument shows that $B(x) = e^{x^4/4!}$, and $C(x) = e^{x^9/9!}$.

Now let us split n into three (possibly empty) subsets, and take a partition with blocks of size three on the first subset, a partition with blocks of size four on the second subset, and a partition with blocks of size nine on the third subset. Then the Product formula shows that

$$F(x) = A(x)B(x)C(x) = e^{\frac{x^3}{3!}+\frac{x^4}{4!}+\frac{x^9}{9!}}.$$

Theorem 8.27 (Compositional formula, Exponential version). *Let a_n be the number of ways to build a certain structure on an n-element set, and assume $a_0 = 0$. Let b_n be the number of ways to build a second structure on an n-element set, and let $b_0 = 1$. Let g_n be the number of ways to partition the set $[n]$ into an unspecified number of non-empty subsets, then build a structure of the first given kind on each of these subsets, then build a structure of the second kind on the set of the subsets. Denote by $A(x)$, $B(x)$, and $G(x)$ the generating functions of the sequences $\{a_n\}$, $\{b_n\}$, and $\{g_n\}$.*

Then

$$G(x) = B(A(x)).$$

Proof. Let us assume that we partition $[n]$ into k subsets. Then there are b_k ways to take a structure of the second kind on the k-element set of these subsets. Therefore, it follows from Theorem 8.21 that $b_k A(x)^k/k!$ is the exponential generating function for the number of ways to partition $[n]$ into exactly k subsets, then build a structure of the given kind on each subset, and then take a structure of the second kind on the k-element set of these subsets. As $a_0 = 0$, none of the power series $b_k A(x)^k/k!$ has a constant term. On the other hand, $G(x)$ has constant term 1 by definition. This shows

$$G(x) = 1 + \sum_{k \geq 1} b_k \frac{A(x)^k}{k!} = \sum_{k \geq 0} b_k \frac{A(x)^k}{k!} = B(A(x)).$$

$\square$

Example 8.28. We have n distinct cards. We want to split their set into non-empty subsets so that each of them contains an even number of cards. Then we want to order the cards within each subgroup. Finally, we want to order these subgroups into a line. Find an explicit formula for the number of ways g_n we can do this.

Solution. Keeping the notation of Theorem 8.27, we see that $a_n = n!$ if $n \geq 2$ is even, and $a_n = 0$ if n is odd, or $n = 0$. Moreover, $b_n = n!$ for all $n \geq 0$. Therefore,

$$A(x) = \sum_{n \geq 0} a_n \frac{x^n}{n!} = \sum_{\substack{n \geq 2 \\ n \ even}} x^n = \frac{x^2}{1 - x^2},$$

and

$$B(x) = \sum_{n \geq 0} b_n \frac{x^n}{n!} = \sum_{n \geq 0} x^n = \frac{1}{1 - x}.$$

Therefore, by the Compositional formula,

$$G(x) = B(A(x)) = \frac{1}{1 - \frac{x^2}{1 - x^2}} = \frac{1 - x^2}{1 - 2x^2}$$

$$= 1 + \frac{x^2}{1 - 2x^2} = 1 + x^2 \sum_{m \geq 0} (2x^2)^m = 1 + \sum_{m \geq 0} 2^m x^{2m+2}.$$

So the coefficient g_n of $x^n/n!$ in $G(x)$ is 0 if n is odd, and $2^{m-1}(2m)!$ if $n = 2m$ and $m > 0$. Consequently, for even n, there are $g_n = 2^{\frac{n}{2}-1} \cdot n!$ ways to proceed.

Notes

The theory of generating functions is certainly rich enough to be the subject of several books. A classic in that area is "Generatingfunctionology" by Herb Wilf [49]. For a far-reaching analysis of exponential generating functions, we recommend "Enumerative Combinatorics" Volume 2, by Richard Stanley [42]. Chapter 6 of that book contains a very extensive list of objects that are counted by the Catalan numbers.

Exercises

(1) Find an explicit formula for a_k if $a_0 = 0$ and $a_{k+1} = a_k + 2^k$ for $k \geq 0$.

(2) Let $\{a_n\}_{n\geq 0}$ and $\{b_n\}_{n\geq 0}$ be two sequences, and let $b_n = \sum_{i=0}^{n} a_i$. What is the relationship between the ordinary generating functions of these sequences?

(3) Let $\{a_n\}_{n\geq 0}$ and $\{b_n\}_{n\geq 0}$ be two sequences, and let $A(x)$ and $B(x)$ be their respective exponential generating functions. Let us assume we know that $B(x) = A(x)/(1-x)$. What is the relationship between the two sequences?

(4) A child wants to walk up a stairway. At each step, she moves up either one or two stairs. Let $f(n)$ be the number of ways she can reach the nth stair. Find a closed explicit formula for $f(n)$.

(5) Let h_n be defined as in Example 8.14. Prove that if $n \geq 1$, then $h_{n+2} = 3h_{n+1} - h_n$.

(6) If we consider the sequence of the numbers h_n defined in Example 8.14, and that of the numbers $f(n)$ defined in Exercise 4, we note that the equality $f(2n-1) = h_n$ seems to hold, for all $n \geq 1$.

    (a) Prove this fact (by any method).
    (b) (+) Give a *direct bijective proof* of this fact. Do not use generating functions, or recursive formulae.

(7) Let a_n be the number of ways to pay n dollars using ten-dollar bills, five-dollar bills, and one-dollar bills only. Find the ordinary generating function $A(x) = \sum_{n\geq 0} a_n x^n$.

(8) Find a simple, closed form for the generating function of the sequence defined by $a_n = n^2$.

(9) Let $f(n)$ be the number of subsets of $[n]$ in which the distance of any two elements is at least three. Find the generating function of $f(n)$.

(10) Find the ordinary generating function of the sequence $p_k(n)$. Recall

that $p_k(n)$ is the number of all partitions of n into exactly k parts.

(11) **[C]** Use your favorite software package to find the numbers $p_4(n)$ for $n \leq 20$.

(12) Find a combinatorial proof for the result of Example 8.6.

(13) Find a combinatorial proof for the result of Example 8.7.

(14) (+) Find a combinatorial proof for the result of Example 8.16.

(15) Let a_n be the number of monotonic functions f from $[n]$ to $[n]$ such that $f(i) \leq i$ for every $i \in [n]$. Find a closed formula for a_n.

(16) (+) Let M_n denote the number of lattice paths from $(0,0)$ to $(n,0)$ which never dip below $y = 0$ and are made up only of the steps $(1,0)$, $(1,1)$, and $(1,-1)$. Find the ordinary generating function $\sum_{n\geq 0} M_n x^n$. The numbers M_n are called the *Motzkin numbers.*

(17) (+) Let f_n be the number of paths with steps $(1,0)$, $(1,1)$ and $(0,1)$ from $(0,0)$ to (n,n) that never run above the diagonal $x = y$. Find the ordinary generating function $F(x) = \sum_{n\geq 0} f_n x^n$. The numbers f_n are called the *Schröder numbers.*

(18)(a) **[C]** Use your favorite software package to find the Motzkin numbers of Exercise 16, for $n \leq 10$.

(b) **[C]** Use your favorite software package to find the Schröder numbers of Exercise 17, for $n \leq 10$.

(19) (+) Let $r(n)$ be the number of n-permutations whose square is the identity permutation. We proved in Exercise 5 of Chapter 6 that

$$r(n+2) = r(n+1) + (n+1)r(n), \tag{8.17}$$

if $n \geq 0$, while $r(0) = r(1) = 1$. Use this recurrence relation to find an explicit formula for the generating function $R(x) = \sum_{n\geq 0} r(n)\frac{x^n}{n!}$.

(20) Find the exponential generating function $F(x)$ for the number of n-permutations having cycles of length $a_1, a_2, \cdots, a_k$ only.

(21) Let $H_{2,3}(n)$ be the number of n-permutations in which all cycles are of length two or three. Use the result of the previous exercise to find a recurrence relation for $H_{2,3}(n)$.

(22) Let $b(n)$ be the number of compositions of n in which each part is an odd integer. Find a closed formula for $\sum_{n\geq 0} b(n)x^n$. Express $b(n)$ by the numbers $f(n)$ defined in Exercise 4.

Supplementary Exercises

(23) (-) Find an explicit formula for a_n if $a_0 = 1$ and $a_{n+1} = 3a_n + 2^n$ if $n \geq 0$.

(24) (-) Find an explicit formula for a_n if $a_0 = 1$, $a_1 = 4$, and $a_{n+2} = 8a_{n+1} - 16a_n$ for $n \geq 0$.

(25) (-) A certain kind of insect population multiplies so that at the end of each year, its size is the double of its size a year before, plus 1000 more insects. Assuming that originally we released 50 insects, how many of them will we have at the end of the nth year?

(26) (-) A permutation is called *indecomposable* if it cannot be cut into two parts so that everything before the cut is smaller than everything after the cut. For example, 3142 is indecomposable, but 2143 is not as you can cut it after the first two elements.
Let $f(n)$ be the number of indecomposable permutations of length n, and set $f(0) = 0$. Find the generating function $F(x) = \sum_{n\geq 0} f(n)x^n$. Note: you can give your result in terms of $G(x) = \sum_{n\geq 0} n!x^n$, the generating function of all permutations.

(27) (-) Find an explicit formula for the numbers a_n if $a_{n+1} = (n+1)a_n + 2(n+1)!$ for $n \geq 0$, and $a_0 = 0$.

(28) Let $a_0 = a_1 = 1$, and let $a_n = na_{n-1} + n(n-1)a_{n-2}$ for $n \geq 2$. Find the exponential generating function of the numbers a_n. Compare your result to the result of Exercise 4.

(29) Let $a_0 = 0$, and let $a_{n+1} = (n+1)a_n + n!$ for $n \geq 0$. Find an explicit formula for a_n. In what earlier chapter did you see your answer as the answer to a combinatorial enumeration problem? Explain the connection.

(30) *Exponential formula, permutation version.* Let $C = \{c_1, c_2, \cdots\}$ be a set of positive integers. Let $g_C(n)$ be the number of n-permutations in which each cycle length belongs to C. Set $g_\emptyset(n) = 0$. Prove that

$$G_C(x) = \sum_n g_C(n)\frac{x^n}{n!} = \exp\left(\sum_{i\geq 1} \frac{x^{c_i}}{c_i}\right).$$

(31)(a) Explain how the result of the previous exercise is a generalization of Example 8.25.

(b) Use the result of the previous exercise to find the exponential generating function for the number of n-permutations whose square is the identity permutation.

(c) Use the result of the previous exercise to provide generating function proofs of the two formulae given in Theorem 6.5.

(32) Find a closed form (no summation signs) for the generating function $G(x) = \sum_{n \geq 0} c(n,k) \frac{x^n}{n!}$.

(33) Find a closed form (no summation signs) for the generating function $G(x) = \sum_{n \geq 0} S(n,k) \frac{x^n}{n!}$.

(34) Let $h_0 = 1$, and let h_n be the number of compositions of n into parts equal to 2 or 3. Find a closed formula for $H(x) = \sum_{n \geq 0} h_n x^n$.

(35) Let h_n be the number of ways to tile a $1 \times n$ rectangle with 1×1 tiles that are red or blue and 1×2 tiles that are green, yellow, or white. Find a closed formula for $H(x) = \sum_{n \geq 0} h_n x^n$.

(36) Let h_n be the number of sequences of length n consisting of letters A and B in which there is no subsequence of two letters A in consecutive positions. Find a closed formula for $H(x) = \sum_{n \geq 0} h_n x^n$.

(37) (+) Let $p_{odd}(n)$ denote the number of partitions of n into an odd number of parts, and let $p_{even}(n)$ denote the number of partitions of n into an even number of parts. Prove that $|p_{even}(n) - p_{odd}(n)|$ is equal to the number of partitions of n into distinct odd parts.

(38) Let g_n be the number of ways of selecting a permutation of length n, and then selecting a cycle of that permutation. Use the Compositional formula to find the exponential generating function $G(x) = \sum_{n \geq 0} g_n \frac{x^n}{n!}$, then deduce an explicit formula for g_n. What earlier result does your formula confirm?

(39) We have two bookshelves. Let t_n be the number of ways to first partition a set of n distinct books into two non-empty blocks, and then to line up one block on the top bookshelf and to line up the other block on the bottom bookshelf. Find a closed formula for t_n.

(40) Find the exponential generating function $D(x)$ for the number of derangements, defined in Example 7.4. Look for several different ways to obtain $D(x)$.

(41) Let $D(n)$ be the number of derangements of length n. Prove that for $n \geq 1$, the equality $D(n) - nD(n-1) = (-1)^n$ holds. Recall that $D(0) = 1$ and $D(1) = 0$.

(42) Let $D_e(n)$ (resp. $D_o(n)$) denote the number of derangements of length n that are even (resp. odd) permutations. Prove that $D_e(n) - D_o(n) = (-1)^{n-1}(n-1)$.

(43) We divide a group of people into subgroups A, B, and C, and ask each subgroup to form a line. We also require that A have an odd number of people, and that B have an even number of people. How

many ways are there to do this?

(44) We select an odd number of people from a group of n people, to serve on a committee. Then we select an even number from this committee to serve on a subcommittee. (Zero is an even number, too.) In how many different ways can we do this?

(45) We have n cards. We want to split them into an even number of non-empty subsets, form a line within each subset, then arrange the subsets in a line. In how many different ways can we do this?

(46) Find a direct combinatorial proof for the result of the previous exercise.

(47) Let $f(n)$ be defined as in Exercise 4. Prove that for all positive integers n,

$$f(n) = \sum_{k=0}^{[n/2]} \binom{n-k}{k}.$$

Do not use the closed formula proved in Exercise 4.

(48) (+) Generalize the result of Example 8.11.

(49)(a) Let $a_1, a_2, \cdots, a_k$ be non-negative integers, and let $a(n)$ be the number of compositions of n into k parts so that ith part is not larger than a_i. Find the ordinary generating function $A(x) = \sum_{n\geq 0} a(n)x^n$.

(b) Let $b(n)$ be the number of compositions of n into $k+1$ parts so that the ith part is not larger than a_i, and there is no constraint on the last part. Find the ordinary generating function $B(x) = \sum_{n\geq 0} b(n)x^n$.

(50) (+) We say that a permutation $p = p_1p_2\cdots p_n$ is a has an ascent in position i if $p_i < p_{i+1}$. How many permutations of length n are there in which the first ascent occurs in an even position? For the sake of this problem only, let us say that p always has an ascent in position n. (We can justify this convention by saying that p_n is followed by an infinitely large last symbol.) What other class of n-permutations has the same number of elements?

Solutions to Exercises

(1) Let $A(x) = \sum_{n\geq 0} a_k x^k$. Multiplying the recurrence relation by x^{k+1} and summing over all $k \geq 0$ we get

$$\sum_{k\geq 0} a_{k+1}x^{k+1} = \sum_{k\geq 0} a_k x^{k+1} + x\sum_{k\geq 0}(2x)^k.$$

This means, in the language of generating functions,

$$A(x) = xA(x) + \frac{x}{1-2x},$$

$$A(x) = \frac{x}{(1-x)(1-2x)} = x(1+x+x^2+\cdots)(1+2x+4x^2+\cdots),$$

so $a_k = \sum_{i=0}^{k-1} 2^i = 2^k - 1$.

(2) If $A(x)$ and $B(x)$ are the two generating functions, then we have

$$\begin{aligned} B(x) &= \frac{A(x)}{1-x} = A(x)(1+x+x^2+\cdots) \\ &= (a_0 + a_1 x + a_2 x^2 + \cdots)(1+x+x^2+\cdots). \end{aligned}$$

Indeed, let us take a look at the exponent of x^n in $A(x)(1+x+x^2+\cdots)$. To get x^n, we have to choose $a_i x^i$ from $(a_0+a_1x+a_2x^2+\cdots)$, then we must choose x^{n-i} from $(1+x+x^2+\cdots)$. This results in the product $a_i x^n$. We can do this for all i such that $0 \leq i \leq n$, and, on the other hand, this is the only way we can obtain a constant multiple of x^n in our product $A(x)(1+x+x^2+\cdots)$. Therefore, the coefficient of x^n in $(a_0+a_1x+a_2x^2+\cdots)(1+x+x^2+\cdots)$ is $\sum_{i=0}^{n} a_i$, and the statement follows.

(3) If you look at $A(x)$ and $B(x)$ as the *ordinary generating functions* of sequences $\{a_n/n!\}_{n\geq 0}$ and $\{b_n/n!\}_{n\geq 0}$, then the previous exercise shows that

$$\frac{b_n}{n!} = \sum_{i=0}^{n} \frac{a_i}{i!},$$

$$b_n = \sum_{i=0}^{n} a_i (n)_i.$$

(4) As the child can move at most two stairs at a time, she can get to the nth stair either from the $(n-1)$st, or from the $(n-2)$nd stair. Therefore, $f(n) = f(n-1) + f(n-2)$, for $n \geq 2$. In other words, $f(n+2) = f(n+1) + f(n)$ for all $n \geq 0$, and $f(0) = f(1) = 1$.

Let $F(x) = \sum_{n\geq 0} f(n)x^n$ be the ordinary generating function of the numbers $f(n)$. Multiplying both sides of the (last version of the) recursive formula by x^{n+2}, and summing for all $n \geq 0$, we get

$$\sum_{n\geq 0} f(n+2)x^{n+2} = \sum_{n\geq 0} f(n+1)x^{n+2} + \sum_{n\geq 0} f(n)x^{n+2},$$

which is equivalent to

$$F(x) - x - 1 = x(F(x) - 1) + x^2F(x),$$

$$F(x) = \frac{1}{1-x-x^2}.$$

The two roots of $1-x-x^2$ are $\alpha = -\frac{1+\sqrt{5}}{2}$ and $\beta = -\frac{1-\sqrt{5}}{2}$. Therefore, we look for the partial fraction decomposition

$$\frac{1}{1-x-x^2} = \frac{A}{x-\alpha} + \frac{B}{x-\beta}.$$

After rearranging, this yields

$$1 = (A+B)x + \alpha B + \beta A,$$

therefore, we must have $-B = A$, and thus $A(\alpha - \beta) = -1$, which implies $A = \frac{-1}{\sqrt{5}}$ and $B = \frac{1}{\sqrt{5}}$. So we have shown that

$$F(x) = \frac{-1}{\sqrt{5}} \cdot \frac{1}{x-\alpha} + \frac{1}{\sqrt{5}} \cdot \frac{1}{x-\beta}.$$

A computation similar to that of Example 8.14 then implies

$$f_n = \frac{1}{\sqrt{5}}\left(\frac{1+\sqrt{5}}{2}\right)^{n+1} - \frac{1}{\sqrt{5}}\left(\frac{1-\sqrt{5}}{2}\right)^{n+1}.$$

The first few values of this sequence, starting at $f(0) = 1$, are 1, 1, 2, 3, 5, 8, 13, 21, 34, 55. This sequence is called the *Fibonacci sequence.* Often, the shifted indexing is used. In that indexing, $F_i = f(i-1)$, leading to $F_0 = 0$, $F_1 = F_2 = 1$, $F_3 = 2$, etc. Then F_n is called the nth *Fibonacci number.*

(5) Let us distinguish three different cases according to the situation of the last soldier in the line of $n+2$ soldiers. She can form a unit herself, (and of course, be the commander of it), which happens in h_{n+1} cases. She can be part of the last unit as a non-commander, which happens again in h_{n+1} cases. Finally, she can be the commander of a unit that has more than one person in it. If the first soldier in the line who is in her unit is in position $i+1$, then there are h_i ways to arrange the

first i soldiers. Summing for i, we see that in this last case, there are $\sum_{i=0}^{n} h_i$ possibilities. This proves that

$$h(n+2) = 2h(n+1) + \sum_{i=0}^{n} h_i, \tag{8.18}$$

$$h(n+2) - h(n+1) = \sum_{i=0}^{n+1} h_i.$$

If we replace n by $n-1$ in this last equation, we get

$$\sum_{i=0}^{n} h_i = h(n+1) - h(n),$$

adding this to (8.18) the proof follows.

(6)(a) Induction on n. If $n = 1$, then $h_n = h_1 = 1$, and $f_{2n-1} = f_1 = 1$, and the initial condition holds. Let us assume that the statement is true for all positive integers smaller than $n+1$. Then, using the induction hypothesis, and the fact that $f_m = f_{m-1} + f_{m-2}$,

$$h_{n+1} = 3h_n - h_{n-1} = 3f_{2n-1} - f_{2n-3} = 2f_{2n-1} + f_{2n-2}$$

$$= f_{2n-1} + f_{2n} = f_{2n+1},$$

and the statement is proved.

(b) Note that f_{2n-1} is in fact the number of all compositions of $2n-1$ into parts that are equal to 1 or 2. We are going to define a bijection from the set of all such compositions onto that of all arrangements the officer in charge of Example 8.14 can make. Let α be such a composition, and say that α consists of $2k-1$ parts equal to 1, and $n-k$ parts equal to 2. Now we start reading the string of 1s and 2s in α, from left to right. Every time we read a 2, we will declare the corresponding soldier in the line to be a non-commander. Therefore, we will get $n-k$ non-commanders. The first time we read a 1, we declare the corresponding soldier in the line to be a commander. The second time we read a 1, we make the corresponding soldier (that is, the soldier who has just been named a commander or non-commander) in the line the last soldier of his unit by starting a new unit right after him. Then we continue this alternating procedure, that is, when we read the third, fifth, seventh, etc. 1, we declare the corresponding soldier in the line a commander, and when we read the fourth, sixth, eighth, etc. 1, we make the corresponding soldier in the line the last soldier of his unit. This way, we create k units, and name k commanders, each unit having a commander.

For example, if $n = 8$, and we have the composition $\alpha = 2+2+1+1+2+1+2+2+1+1 = 15$, then we get a line of soldiers $b(\alpha) = NNC|NCNN|C$, where N denotes a non-commander, C denotes a commander, and the bars denote the end of each unit.

To see that this is a bijection, it suffices to show that for each arrangement β of the officer, there exists a unique composition α so that $b(\alpha) = \beta$. This unique preimage can be constructed easily, by replacing all the N symbols in β by 2s, and all the C symbols and bars by 1s. This completes the proof.

(7) It would be troublesome to find a nice recurrence relation here as it is clear that the number a_n of ways to pay n dollars with these bills will strongly depend on the divisibility of n by five and ten. We will instead obtain the ordinary generating function $A(x) = \sum_{n\geq 0} a_n x^n$ in a different way.

Let $f(n)$ be the number of ways to pay n dollars with *ten-dollar bills only*. Then $f(n) = 1$ if n is divisible by 10, and $f(n) = 0$ otherwise. Then $F(x) = \sum_{n\geq 0} f(n)x^n = 1 + x^{10} + x^{20} + \cdots = \frac{1}{1-x^{10}}$. Similarly, let $g(n)$ be the number of ways to pay n dollars with *five-dollar bills only*. Then $g(n) = 1$ if n is divisible by 5, and $g(n) = 0$ otherwise. Then $G(x) = \sum_{n\geq 0} g(n)x^n = 1 + x^5 + x^{10} + \cdots = \frac{1}{1-x^5}$. Finally, if $h(n)$ is the number of ways to pay n dollars with *one-dollar bills only*, then clearly $h(n) = 1$ for all $n \geq 0$, and $H(x) = \sum_{n\geq 0} h(n)x^n = 1 + x + x^2 + \cdots = \frac{1}{1-x}$.

It is high time we explained why we are interested in these seemingly bland generating functions. Consider the product

$$F(x)G(x)H(x) = \frac{1}{(1-x^{10})(1-x^5)(1-x)}$$

$$= (1 + x^{10} + x^{20} + \cdots)(1 + x^5 + x^{10} + \cdots)(1 + x + x^2 + \cdots).$$

Let us try to find the coefficient of, say x^{53} on the right-hand side. To get a term whose coefficient is 53, we must choose a member of each of the three sums so that their exponents sum to 53. That means, one exponent that is divisible by ten, one that is divisible by 5, and one last exponent, say 30+20+3. However, this provides a way to pay 53 dollars with our bills: three ten-dollar bills (to pay 30 dollars), four five-dollar bills (to pay 20 dollars), and three one-dollar bills (to pay

3 dollars). This way we can set up an obvious bijection between ways to pay n dollars, and ways to choose one term from each of the three parentheses so their product is x^n. So the coefficient of x^n on the right-hand side (which is precisely the number of ways we can pick three such terms) is exactly a_n. So we have proved that

$$A(x) = F(x)G(x)H(x) = \frac{1}{(1-x^{10})(1-x^5)(1-x)}.$$

(8) Recall that $\frac{1}{(1-x)^3} = \sum_{n\geq 0}\binom{n+2}{2}x^n$. In other words,

$$\frac{x^2}{(1-x)^3} = \sum_{n\geq 0}\binom{n+2}{2}x^{n+2} = \sum_{n\geq 2}\binom{n}{2}x^n.$$

Also recall that $\frac{1}{(1-x)^2} = \sum_{n\geq 1} nx^{n-1}$, in other words, $\frac{x}{(1-x)^2} = \sum_{n\geq 1} nx^n$. (If you need a reminder: these can be proved by either taking the derivative of $1/(1-x)$, or by considering the powers of $(1+x+x^2+\cdots)$, and the coefficient of x^n there.)
Finally, note that $n^2 = 2\binom{n}{2} + n$, so

$$\begin{aligned}\sum_{n\geq 0} n^2x^n &= 2\sum_{n\geq 2}\binom{n}{2}x^n + \sum_{n\geq 1} nx^n \\ &= 2\frac{x^2}{(1-x)^3} + \frac{x}{(1-x)^2} = \frac{x(x+1)}{(1-x)^3}.\end{aligned}$$

(9) Try to construct such a subset. If n is part of the subset, then we cannot have $n-1$ or $n-2$ in the subset, so we have $f(n-3)$ ways to choose such a subset. Indeed, we can append n to the end of any good subset of $[n-3]$. If n is not part of our subset, then we obviously have $f(n-1)$ choices. So $f(n) = f(n-1) + f(n-3)$, for all integers $n \geq 3$. Moreover, $f(0) = 1$, $f(1) = 2$, and $f(2) = 3$.
Let $F(x) = \sum_{n\geq 0} f(n)x^n$. Multiplying the recurrence relation by x^n and summing over $n \geq 3$, we get

$$\sum_{n\geq 3} f(n)x^n = x\sum_{n\geq 3} f(n-1)x^{n-1} + x^3\sum_{n\geq 3} f(n-3)x^{n-3}.$$

In other words,

$$F(x) - 3x^2 - 2x - 1 = x(F(x) - 2x - 1) + x^3F(x),$$

from where we get

$$F(x) = \frac{1+x+x^2}{1-x-x^3}.$$

(10) It follows from Exercises 6 and 7 of Chapter 5 that we always have $p_k(n) = p_{\leq k}(n-k)$. Therefore,

$$\sum_{n\geq 0}^{\infty} p_k(n)x^n = x^k \sum_{n\geq 0}^{\infty} p_{\leq k}(n)x^n = \frac{x^k}{(1-x)(1-x^2)\cdots(1-x^k)}.$$

(11) The previous exercise shows that

$$\sum_{n\geq 0}^{\infty} p_k(n)x^n = \frac{x^k}{(1-x)(1-x^2)\cdots(1-x^k)}.$$

Therefore, the numbers $p_4(n)$ are the coefficients of the above power series, with $k = 4$. To get the first 20 coefficients, type the following in the software package Mathematica.

```
Series[x^4/((1-x)(1-x^2)(1-x^3)(1-x^4)),{x,0,20}],
```

then press Shift Return. (Do not type the last comma.) You will see that the numbers $p_4(n)$ are, starting with $p_4(4)$, 1, 1, 2, 3, 5, 6, 9, 11, 15, 18, 23, 27, 34, 39, 47, 54, 64.

(12) We need to choose three holidays, and the last day of the first part of the semester. These four days will completely determine the structure of the term. Out of these four days, the first holiday may be the same as the last day of the first part of the semester, but there cannot be any other coincidences. Thus we have to choose positive integers a, b, c, d so that $1 \leq a \leq b < c < d \leq n$. This is equivalent to choosing non-negative integers $0 \leq a-1 < b < c < d \leq n$, and that can be done in $\binom{n+1}{4}$ ways.

(13) We have to choose the set of all holidays, which can be done in 2^n ways, then the last day of the first part of the semester, which can be done in $n+1$ ways as 0 is a choice, too. Thus the total number of choices is $(n+1) \cdot 2^n$.

(14) Each soldier can be either the first soldier of a unit chosen for night duty, the first soldier of a unit not chosen for night duty, or not the first soldier of any unit. The only exception is the soldier who is at the top of the line as he only has the first two possibilities. This proves that the number of all arrangements is $2 \cdot 3^{n-1}$.

(15) Let f be such a function, and let i be the largest number in $[n]$ so that $f(i) = i$. There is always such a number, as $f(1) = 1$. Then we have, of course, a_{i-1} possibilities for the restriction of f to $[i]$. The restriction of f to $\{i+1, i+2, \cdots, n\}$ is a slightly different function as $f(j) = j$ is

not allowed there. In particular, we must have $f(i+1) = i$. In general, f satisfies the criteria on this interval if and only if $f(i+1) = i$, and

$$i+1 \leq f(i+2)+1 \leq f(i+3)+1 \cdots \leq f(n)+1 \leq n,$$

or, in other words, $f(i+1) - (i-1) = 1$, and

$$1 \leq f(i+2)-(i-1) \leq f(i+3)-(i-1) \leq \cdots \leq f(n)-(i-1) \leq n-i.$$

If we set $g(j) = f(j+i)-(i-1)$, we see that the latter clearly happens in a_{n-i} cases. Therefore, we proved that if $n \geq 1$, then

$$a_n = \sum_{i=1}^{n} a_{i-1}a_{n-i}, \tag{8.19}$$

with $a_0 = 1$. Now let $C(x) = \sum_{n\geq 0} a_n x^n$, and note that (8.19) is equivalent to (8.13). Therefore $a_n = c_n = \binom{2n}{n}/(n+1)$, and we have found another occurrence of the Catalan numbers.

(16) If your first step is horizontal, then you clearly have M_{n-1} ways to complete your path. If not, then let us say that you will first touch the line $y = 0$ at $(k, 0)$. Then, to go from $(k, 0)$ to $(n, 0)$, you have M_{n-k} ways to go. How many ways do you have to go from $(0, 0)$ to $(k, 0)$ without touching the $y = 0$ line? Clearly, your first step will be to $(1, 1)$, and the last one will be from $(k-1, 1)$ to $(k, 0)$. So the question is the number of ways to get from $(1, 1)$ to $(k-1, 1)$ without dipping below the $y = 1$ line, and that is clearly M_{k-2}. So $M_0 = M_1 = 1$ and for $n \geq 2$,

$$M_n = M_{n-1} + \sum_{k=2}^{n} M_{k-2}M_{n-k}.$$

Now let $M(x) = \sum_{n\geq 0} M_n x^n$. Multiply both sides of the previous equation by x^n, and sum for all non-negative n to get

$$M(x) = xM(x) + 1 + x^2M^2(x).$$

Therefore,

$$M(x) = \frac{1 - x - \sqrt{1-2x-3x^2}}{2x^2}.$$

(17) Let us first find a recursive formula for f_n. If our first step is $(1, 1)$, then we clearly have f_{n-1} ways to complete our path, from $(1, 1)$ to (n, n). Otherwise, let (i, i) be the first point (other than the origin) on the diagonal (x, x) that our path touches. Then there are f_{n-i} ways to complete this path, from (i, i) to (n, n). Moreover, the number of

ways we could go from $(0,0)$ to (i,i) without touching the diagonal is f_{i-1}. Indeed, we had to start with a $(1,0)$ step, and with a $(0,1)$ step, and never go above the diagonal $(x, x-1)$ that is spanned by the points $(1,0)$ and $(i, i-1)$.
Therefore, we proved that $f_n = f_{n-1} + \sum_{i=1}^{n} f_{i-1}f_{n-i}$ if $n \geq 1$, while $f_0 = 1$. Multiplying both sides by x^n and summing over $n \geq 1$, we get

$$F(x) - 1 - xF(x) = xF(x)^2, \tag{8.20}$$

which yields

$$F(x) = \frac{1 - x - \sqrt{x^2 - 6x + 1}}{2x}.$$

Again, (8.20) has two solutions, so we had to choose the one in which the constant term is 1.

(18)(a) We computed the generating function $M(x)$ of the Motzkin numbers in Exercise 16. The numbers M_n are the coefficients of $M(x)$. We can expand this $M(x)$ by typing

```
Series[(1-x-Sqrt[1-2x-3x^2])/(2x^2),{x,0,10}]
```

in Mathematica, and then hitting Shift Return. We get that the Motzkin numbers are, starting at M_0, 1, 1, 2, 4, 9, 21, 51, 127, 323, 835, 2188.

(b) Using the result of Exercise 17, type

```
Series[(1-x-Sqrt[1-6x+x^2])/(2x),{x,0,10}]
```

in Mathematica. You get that the numbers we were looking for are, starting with f_0, 1, 2, 6, 22, 90, 394, 1806, 8558, 41586, 206098, 1037718.

(19) We define $R(x) = \sum_{n\geq 0} r(n)\frac{x^n}{n!}$, the exponential generating function of the numbers $r(n)$. Let us multiply both sides of equation (8.17) by $x^n/n!$, then sum over all positive integers n, to get

$$\sum_{n\geq 0} r(n+2)\frac{x^n}{n!} = \sum_{n\geq 0} r(n+1)\frac{x^n}{n!} + \sum_{n\geq 0}(n+1)r(n)\frac{x^n}{n!}.$$

Now note that the left-hand side is $R''(x)$, and the first member of the right-hand side is $R'(x)$. The second member of the right-hand side is somewhat harder to recognize, but with a little practice, one can see that it is in fact $(xR(x))'$. Therefore, we get

$$R''(x) = R'(x) + (xR(x))' = R'(x) + xR'(x) + R(x).$$

Solving this, we get $R(x) = e^{x+x^2/2}$.

(20) This is very similar to Example 8.26. The only difference is in the definition of t_n. Let t_n be the number of ways an n-element set can be arranged in an a_1-cycle. Then $t_{a_1} = (a_1 - 1)!$, and $t_n = 0$ if $n \neq a_1$. Therefore, the exponential generating function of that sequence is $T(x) = x^{a_1}/a_1$. Then the same application of the Exponential formula, and then the Product formula shows that

$$F(x) = \exp\left(\sum_{i=1}^{k} \frac{x^{a_i}}{a_i}\right).$$

(21) The previous exercise shows that the exponential generating function of the sequence $\{H_{2,3}(n)\}_{n\geq 0}$ is $H(x) = \exp(\frac{x^2}{2} + \frac{x^3}{3})$. Therefore, $H'(x) = (x + x^2)H(x)$. This implies that the coefficient of $x^n/n!$ on the left-hand side is $H_{2,3}(n+1)$, while the coefficient of $x^n/n!$ on the right-hand side is $nH_{2,3}(n-1) + n(n-1)H_{2,3}(n-2)$. Therefore, $H_{2,3}(n+1) = nH_{2,3}(n-1) + n(n-1)H_{2,3}(n-2)$, when $n \geq 4$, and $H_{2,3}(n) = 0$ if $n = 0$, or $n = 1$, $H_{2,3}(2) = 1$, and $H_{2,3}(3) = 2$.

(22) A composition of n into odd parts is equivalent to splitting up $[n]$ into an unspecified number of nonempty intervals, and covering each interval by a single tile of odd length. The number of ways of covering an interval that way is one if the interval has odd length and 0 otherwise. The generating function of these numbers is $A(x) = x + x^3 + x^5 + \cdots = x/(1-x^2)$, and hence the generating function of the combined task is

$$B(x) = 1/(1 - A(x)) = \frac{1 - x^2}{1 - x - x^2}$$

by Theorem 8.13. Comparing this to the result of Exercise 4, we see that $b(n) = f(n) - f(n-2) = f(n-1)$. In other words, $b(n)$ is the difference of two Fibonacci numbers, and so $b(n)$ is a Fibonacci number itself.

Chapter 9

Dots and Lines. The Origins of Graph Theory

In the eighteenth century, the city of Königsberg consisted of islands where two branches of the river Pregel joined. (Today the city is called Kaliningrad, and is in Russia, on the Baltic Sea.) Seven bridges connected various islands as shown in Figure 9.1. Mathematics for centuries to come was greatly enhanced by this innocent fact. In 1736, the most prolific mathematician of all time, Leonhard Euler, became interested in the following question. Is it possible to walk through town, starting and ending at the same place, so that we use each bridge exactly once?

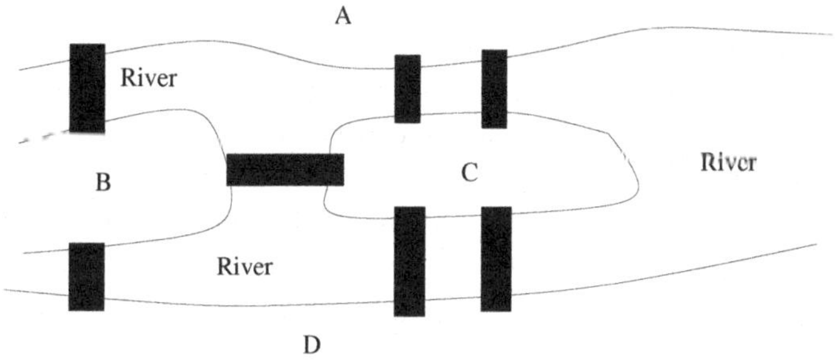

Fig. 9.1 A map of Königsberg.

9.1 The Notion of Graphs. Eulerian Trails

Euler understood that the shape of the islands and the river does not influence the answer to this question. He recognized that the only relevant pieces of information here are those of connectivity, that is, the number of bridges between any two islands. Therefore, instead of using the map of Königsberg, he used the simple diagram shown in Figure 9.2.

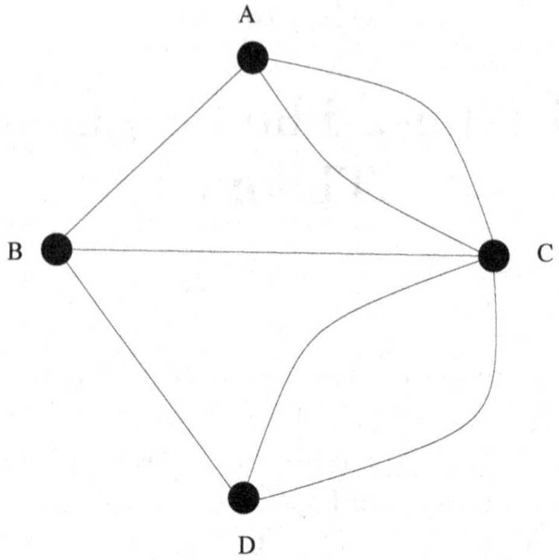

Fig. 9.2 The graph of the Königsberg bridges.

Here the dots represent the land masses, and the lines represent the bridges between them. It is clear that a trail Euler was looking for exists if and only if you can draw the diagram of Figure 9.2 so that you never lift your pencil, you go through each line exactly once, and you start and end at the same point.

Such a diagram, made up from points, and lines connecting some pairs of those points, is called a *graph*. The dots are called the *vertices* of the graph, and the lines are called the *edges* of the graph. In this book, we will only discuss graphs with a *finite* number of vertices, and a *finite* number of edges. The number of edges connected to vertex A is called the *degree* of A.

This simple model proves to be incredibly useful. The theory of graphs is a very extensive part of combinatorics as there are plenty of problems of various nature that can be solved by this simple model. (Recall that we have in fact used graphs in a surprisingly powerful way to solve the problem of Example 1.7.) In our Walk through Combinatorics, we would like to emphasize the diversity of these problems. First, however, we need to introduce some basic terminology.

It is possible that in a graph, there are multiple edges joining the same pair of points, or there are edges that start and end in the same vertex (such edges are called *loops*). If a graph G has no loops, and has no multiple edges, then we will say that G is a *simple* graph.

A sequence of *distinct* edges $e_1e_2\cdots e_k$ is called a *trail* if we can take a continuous walk in our graph, first walking through the edge e_1, then the edge e_2, and so on. In other words, the endpoint of e_i is the starting point of e_{i+1}. Note that this happens if and only if we can draw the set of edges $e_1e_2\cdots e_k$ so that we never lift our pencil from the paper, and we first draw e_1, then e_2, and so on. A *walk* is like a trail, except that all edges do not need to be distinct.

If, in addition, we start the drawing at the same vertex where we end it, then we say that $e_1e_2\cdots e_k$ is a *closed trail.* If a trail uses all edges of G, then we call it an *Eulerian trail.* If a trail does not touch any vertex twice, then we call it a *path.*

If we put two or more graphs next to each other, we can certainly call the union obtained this way a graph. Still, it is natural to think that this new graph is *not quite as good* as the original graphs. For instance, there are pairs of vertices so that you cannot get from one vertex to another through a path. This is a very important difference, and motivates the following definition.

Definition 9.1. If the graph G has the property that for any two vertices x and y, one can find a path from x to y, then we say that G is a *connected* graph.

If G is not connected, then let k be the smallest integer so that G can be obtained as the union of k connected graphs. Then we say that G has k *connected components.* We also say that vertices u and v are in the same connected component if there is a path from u to v. In other words, the connected components are the maximal connected subgraphs of G, that is, connected subgraphs to which we cannot add any new vertex of G without forcing them to lose the connected property. Now we are in a position to state and prove Euler's theorem.

Theorem 9.2. *A connected graph G has a closed Eulerian trail if and only if all vertices of G have even degree.*

Proof. First we prove the "only if" part, that is, we show that if G has a closed Eulerian trail, then all vertices of G must have even degree. Indeed, when we take the closed Eulerian trail W, we visit each vertex a certain number of times. Let A be a vertex that was not where W started, and assume we visited A exactly a times. This means we entered A exactly a times, and we left A exactly a times. As we assumed W was a trail, we had to do this using different edges, so we used $2a$ edges. On the other hand, W contains all edges of G, so A cannot have any additional edges, therefore the degree of A is $2a$. This shows that the degree of any vertex other than the starting point S of W is even. Finally, note that S is not only the starting point of W, but also the endpoint, so if we visit S exactly t times between the start and the end of W, then we use $1+2t+1=2(t+1)$ edges. Therefore, the degree of S is $2(t+1)$, and our claim is proved.

Now assume all vertices of G have even degree and prove that G has a closed Eulerian trail. Take any vertex S, and start walking along an edge e_1, to the other endpoint A_1 of that edge, then walk along any new edge e_2 that starts in A_1. Continue this way, using new (previously unused) edges at each step, until a closed trail C_1 is formed. As G is finite, such a closed trail will always be formed. The first closed trail will be formed when we first revisit a vertex already visited. We cannot get stuck at some vertex before completing a closed trail as each vertex has even degree, so each time we enter a vertex, we can also leave it, except possibly the initial vertex. If $C_1 = G$, then we are done. If not, then choose a vertex V in C_1 so that C_1 does not contain all edges adjacent to V.

The alert reader can ask now how do we know that there *is* such a vertex V. Let us assume that there is not. As C_1 contains less edges than G, and supposedly C_1 contains all edges adjacent to all vertices it contains, there must be a vertex A that is not in C_1. However, G is a connected graph, so there must be a path connecting A to any vertex in C_1. Start walking on this path from A to any given vertex of C_1. When you reach C_1 the first time, you will reach it in a vertex V that is in C_1, but not all the edges adjacent to it are in C_1. Indeed, the one that has just ended in V is not. This proves by contradiction that such a vertex V always exists. Figure 9.3 illustrates this situation.

Let us now omit all edges of C_1 from G. We get a graph in which again all vertices have even degree. Starting at V, let us take another closed trail C_2 in the remaining graph. We can then unite C_1 and C_2 into one closed trail in G. Indeed, if we start walking by C_1, we can stop at V, walk through C_2, then complete our trail by using the remaining part of C_1. If the new trail $C_1 \cup C_2$ contains all edges of G, we are done. If not, then let us omit $C_1 \cup C_2$ from G, and find a new closed trail C_3 in the remaining graph.

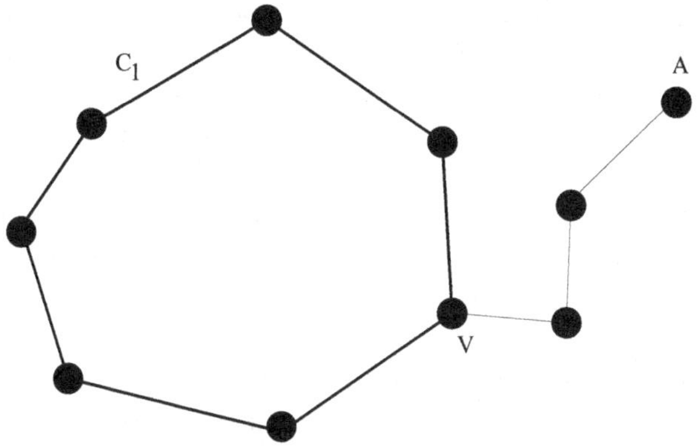

Fig. 9.3 The cycle C_1 does not contain all edges adjacent to V.

As G has a finite number of edges, this procedure has to stop after a finite number of steps. Therefore, after a finite number of steps, $C_1 \cup C_2 \cup \cdots \cup C_k$ will be a closed trail containing all edges of G. □

This proves that we *cannot* walk through all bridges of Königsberg so that we end where we started, and use each bridge exactly once. Indeed, the graph shown in Figure 9.2 has four vertices of odd degree.

What happens if we relinquish the requirement that our trail start and end at the same place? The answer to this question is a direct consequence of Theorem 9.2.

Corollary 9.3. *Let G be a connected graph. Then G has an Eulerian trail starting at vertex S and ending at a different vertex T if and only if S and T have odd degree, and all other vertices of G have even degree.*

Proof. Add a new edge joining S and T, and call the new graph obtained H. Then H has a *closed* Eulerian trail if and only if G has an Eulerian trail from S to T, so the claim follows from Theorem 9.2. □

We have seen that the parity of the degrees is an important property of a graph. The following theorem shows a basic fact about these parities.

Theorem 9.4. *In a graph G without loops, the number of vertices of odd degree is even.*

Proof. Take such a graph with e edges. Let $d_1, d_2, \cdots, d_n$ be the degrees of the n vertices of G. We claim that

$$d_1 + d_2 + \cdots + d_n = 2e.$$

Indeed, each edge contributes one to the degree of exactly two vertices, namely its two endpoints. So a total of e edges will result in a total of $2e$ in the sum of degrees. Therefore, the sum of degrees is $2e$, which is an even number. This implies that there has to be an even number of odd summands in $d_1 + d_2 + \cdots + d_n$. □

9.2 Hamiltonian Cycles

A *cycle* in a graph is a closed trail that does not touch any vertex twice, except, of course the initial vertex, that must also be the ending vertex. This implies that if a cycle has k vertices, then it has k edges. A cycle that includes all vertices of a graph is called a *Hamiltonian cycle*, whereas a path that includes all vertices of a graph is called a *Hamiltonian path.*

A real-life scenario in which Hamiltonian cycles are relevant is the following. Suppose many people are invited to a party, and they will all be seated around a circular table. Is it possible to find seating arrangements so that each guest knows both people seated next to him?

In this scenario, we can define a graph in which people are represented by vertices, and two vertices are connected by an edge if the corresponding people know each other. Then a Hamiltonian cycle in this graph, if it exists, provides an appropriate seating.

Whether a Hamiltonian cycle exists in this graph depends, of course, on the graph itself. For example, if there is a person who does not know anyone, then it is clear that there is no Hamiltonian cycle. If there is no such person, but the graph is not connected, there will not be a Hamiltonian cycle either. If everyone knows everyone, then of course, there will be a Hamiltonian cycle.

These were all very special situations. What can be said about the general case, though? That is, given a simple graph G, how can we quickly decide whether it has a Hamiltonian cycle or not?

The answer to this question is that we *cannot.* It is easy to prove that an appropriate seating exists (when it exists). Indeed, you can prove that by simply exhibiting one. There is, however, no quick way known to prove that *no appropriate seating* exists (when it does not). By "quick way" we mean an algorithm that uses only $f(n)$ steps, where n is the number of guests, and $f(n)$ is a *polynomial function of* n, such as n^3, or $n^7 + 3n^5 + 6n + 3$. We can certainly prove that no good seating exists by verifying all $(n-1)!$ possible seating arrangements, and concluding that none of them are good, but that takes too long. The function $g(n) = (n-1)!$ is not a polynomial function of n.

This problem is interesting on its own, but it is also related to a vast array of very important problems of an exciting area in Theoretical Computer Science, called Complexity Theory, which is the topic of Chapter 20 of this book. (So it is well worth reading the book till the very end!) It can be proved that the problem of deciding whether a given simple graph has a Hamiltonian cycle is equivalent to about 5000 other problems, which are all very different at first sight. By "equivalent", we mean that if a polynomial-time algorithm were found for the Hamiltonian cycle problem, then that would provide a polynomial-time algorithm for any of those 5000 problems, and vice versa. The set of all these equivalent problems is called NP-complete problems. It is believed by most, but not all, researchers, that such polynomial-time algorithm *does not exist.* You can try to find one, but do not try too hard...

There are nevertheless some nontrivial theorems about the existence of Hamiltonian cycles.

Theorem 9.5. *Let $n \geq 3$, let G be a simple graph on n vertices, and let us assume that all vertices in G are of degree at least $n/2$. Then G has a Hamiltonian cycle.*

Proof. Let us assume that G does not have a Hamiltonian cycle. Let us add new edges to G as long as we can without creating a Hamiltonian cycle. When we stop, we have a graph G' in which all vertices have degree at least $n/2$, there is no Hamiltonian cycle, but adding any new edge would create a Hamiltonian cycle.

Let x and y be two vertices in G' that are not connected by an edge. As adding the edge xy would create a Hamiltonian cycle, it follows that G' has a Hamiltonian path P that starts at x and ends in y. Let $x = z_1, z_2, z_3, \cdots, z_{n-1}, z_n = y$ be the vertices of this path, from x to y. Vertices x and y together have at least n neighbors. Therefore, the pigeon-hole principle implies that there must be an index i so that $2 \leq i \leq n-1$, while xz_i is an edge, and also, $z_{i-1}y$ is an edge. (Otherwise the set of neighbors of y and the set of vertices that immediately precede a neighbor of x on the xy-path would be disjoint, which is impossible since these sets are too large.) This is a contradiction, however, for this would mean that $xz_2 \cdots z_{i-1}yz_{n-1} \cdots z_i$ is a Hamiltonian cycle as shown in Figure 9.4. □

There are several additional results proving that a simple graph in which

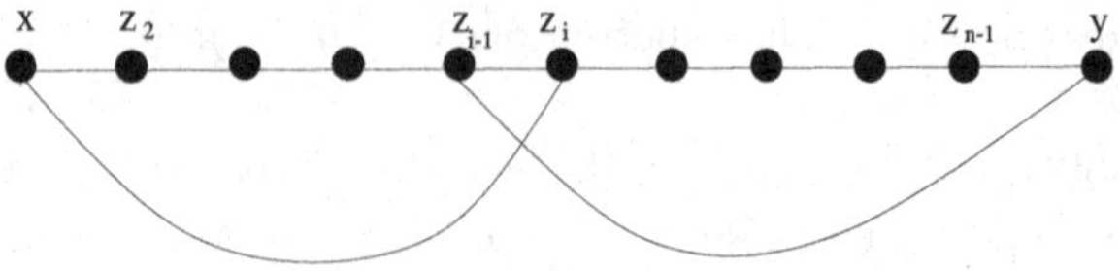

Fig. 9.4 The cycle $xz_2 \cdots z_{i-1}yz_{n-1} \cdots z_i$ is a Hamiltonian cycle.

the degrees are, in some sense, large, has a Hamiltonian cycle. We will see some of these results in the Exercises.

9.3 Directed Graphs

In the previous section, the edges of a graph were not assigned a direction. We could walk through them in both ways. As anyone with big city driving experience knows, this is not always the case in real life, that is, there are one way streets, too. A graph in which each edge is assigned a direction, such as in Figure 9.5, is called a *directed graph.*

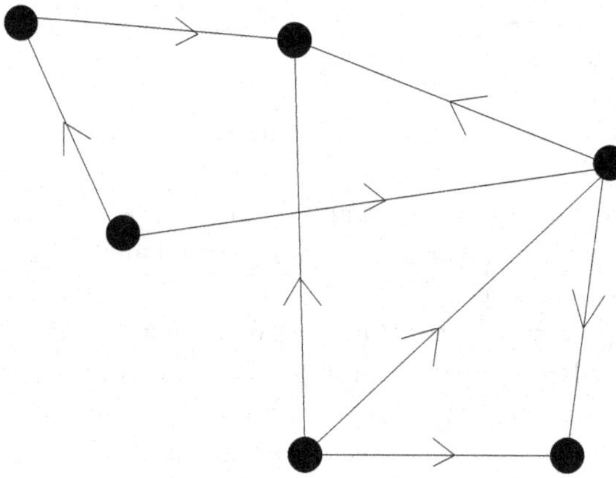

Fig. 9.5 A directed graph.

It is natural to wonder under what conditions does a directed graph have a closed Eulerian trail. Of course, a trail in a directed graph must contain all edges in the right direction, that is, we can only walk through an edge from its "tail" to its "head". Paths and closed trails are defined in an analogous way.

Clearly, in this case it is not enough to require that all vertices have an even number of edges adjacent to them. For example, if no edge starts in a given vertex, then there will be no Eulerian trail in that graph.

In order to answer this question, we introduce some new definitions. We say that a directed graph G is *strongly connected* if for all vertices a and b of G, there is a directed path from a to b. The *in-degree* of a vertex of a directed graph is the number of edges that end at that vertex. The *out-degree* of a vertex is the number of edges that start at that vertex. A directed graph H is called *balanced* if for each vertex V of H, the equality $indegree(V) = outdegree(V)$ holds.

Theorem 9.6. *A directed graph G has a closed Eulerian trail if and only if it is balanced and strongly connected.*

Proof. First we prove that these conditions are necessary. As a closed Eulerian trail W leaves each vertex as many times as it enters that vertex, G must be balanced. Similarly, W provides a trail from any vertex to any vertex, so G is strongly connected.

These two conditions are sufficient. To see this, copy the proof of Theorem 9.2, replacing edges by directed edges. □

A simple undirected graph is called *complete* if there is an edge between every pair of distinct vertices. Thus a complete graph on n vertices has $\binom{n}{2}$ edges. If we direct each edge of a complete graph, then the resulting directed graph is called a *tournament*. The reason for this name is the following. If n players participate at a round robin tennis tournament, and we define a directed graph in which the vertices represent the players, and ij is an edge if i has beaten j, then we get a tournament. We have met tournaments before, in Exercises 2 and 3 of Chapter 2.

Hamiltonian paths and cycles can be defined in directed graphs, too, in the obvious way. While it is trivial that all complete (undirected) graphs have Hamiltonian paths, the corresponding statement for directed graphs is not that obvious. This is not surprising; while there is only one complete undirected graph on n vertices, there are many, (in some sense, $2^{\binom{n}{2}}$) tournaments. Nevertheless, they all have Hamiltonian paths. This is the content of the next theorem.

Theorem 9.7. *All tournaments have a Hamiltonian path.*

Proof. We prove the claim by induction on n, the number of vertices of our tournament T. If T has one, or two vertices, then the statement is clearly true. Now assume that we know the statement for all tournaments having $n-1$ vertices. Let T be any tournament on n vertices. Separate any vertex V, and call the remaining graph on $n-1$ vertices T'. By the induction hypothesis, T' has a Hamiltonian path $h = h_1h_2\cdots h_{n-1}$. The question is how we can insert V into h. If there is an index i so that h_iV is an edge and Vh_{i+1} is an edge, then we can insert V between h_i and h_{i+1}.

If no such i exists, then there must exist an index k so that $0 \leq k \leq n-1$, and for all $j \leq k$, Vh_j is an edge, and for all $j > k$, h_jV is an edge. Therefore, either Vh_1 is an edge, or $h_{n-1}V$ is an edge. So we can affix V either to the front, or to the end of h. □

What can we say about the existence of Hamiltonian cycles in tournaments? Clearly, not all tournaments will contain them. For example, if T has a vertex that has in-degree 0, or out-degree 0, then T does not have a Hamiltonian cycle. It turns out that it is fairly easy to describe the tournaments that do have Hamiltonian cycles.

Theorem 9.8. *A tournament T has a Hamiltonian cycle if and only if it is strongly connected.*

Proof. If T has a Hamiltonian cycle, then that cycle provides a directed path from any vertex to any vertex, so G is strongly connected.

Now assume that T is strongly connected, and let $E(T)$ denote the set of edges of T. First we prove that T does contain a cycle. Indeed, if it did not, then $xy \in E(T)$ and $yz \in E(T)$ would imply $xz \in E(T)$, so T would be a *transitive tournament.* In such a tournament, the vertices can be listed from left-to-right so that $ij \in E(T)$ if and only if j is on the right of i. However, such a tournament is not strongly connected as no paths go to the right. So T does have a cycle.

Let $C = y_1y_2\cdots y_k$ be a cycle of maximal length in T, and assume C is not a Hamiltonian cycle. As T is strongly connected, it contains an edge from C to some vertex x that is not in C. We can assume without loss of generality that this edge is y_1x. If xy_2 were an edge, then $y_1xy_2y_3\cdots y_k$ would be a cycle having more vertices than C. Therefore, y_2x has to be an edge, and then similarly, $y_3x, y_4x, \cdots, y_kx$ must all be edges.

Let Z be the set of all vertices z so that $y_1z \in E(T)$. Then $y_iz \in E(T)$ for all $z \in Z$ and all $i \in [k]$ by the same argument as the one we applied for y_ix in the previous paragraph. Let zt be an edge, with $z \in Z$, and $t \notin Z$. Such an edge exists as T is strongly connected. Then $t \notin C$, and therefore $t \notin Z$ implies that $ty_1 \in E(T)$. Then, however, $zty_1y_2y_3\cdots y_k$ is a longer cycle than C. Figure 9.6 shows our construction. This contradiction completes the proof.

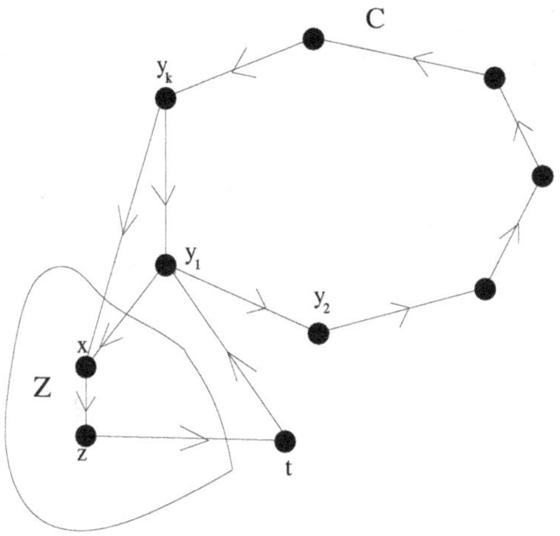

Fig. 9.6   Constructing a cycle that is larger than C.

$\square$

9.4   The Notion of Isomorphisms

When are two graphs considered the same? This question can be answered in several different ways. For the time being, we will only discuss two of them.

We will say that the two graphs shown in Figure 9.7 are identical because for any pair of vertices X and Y, the number of edges between X and Y is the same in both graphs.

The fact that the two graphs are not drawn the same way does not matter here. What matters is that exactly the same pairs of vertices have edges between them.

Now consider Figure 9.8. The two graphs shown there are certainly not identical. Indeed, the first one contains the edge AB, and the second one does not.

However, we certainly get the impression that these two graphs are not completely unrelated either. For instance, if we omit all labels from the vertices, then we get the two graphs shown in Figure 9.9, that surely look the same.

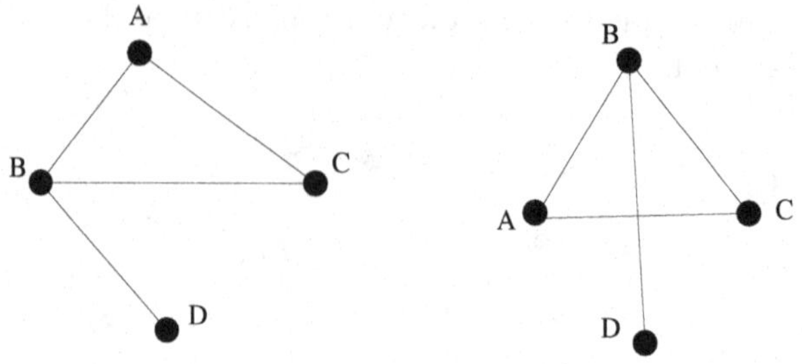

Fig. 9.7 Two identical graphs with labeled vertices.

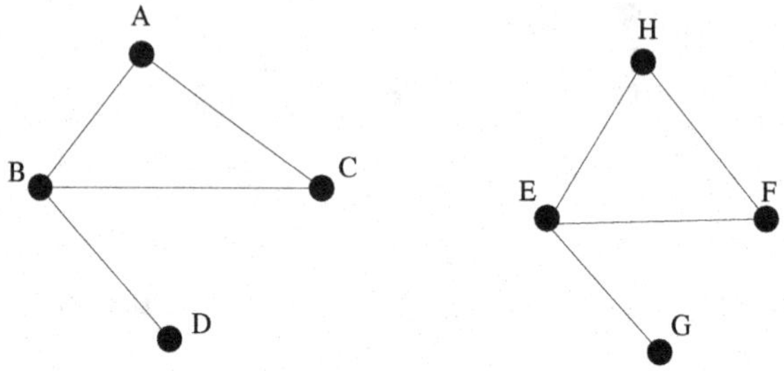

Fig. 9.8 Two isomorphic graphs with labeled vertices.

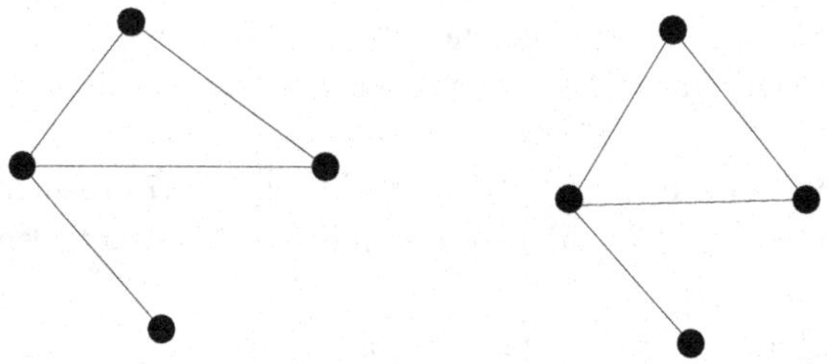

Fig. 9.9 These two unlabeled graphs are identical.

We will express this by saying that the two graphs shown in Figure 9.8 are *identical as unlabeled graphs*, or, in one word, *isomorphic.* Let us make this definition more precise.

Definition 9.9. We say that graphs G and H are isomorphic if there is a bijection f from the vertex set of G onto that of H so that the number of edges between any pair of vertices X and Y of G is equal to the number of edges between vertices $f(X)$ and $f(Y)$ of H. The bijection f is called an isomorphism.

Example 9.10. Let G and H be the graphs shown in Figure 9.8. Then the map f defined by $f(A) = H$, $f(B) = E$, $f(C) = F$, $f(D) = G$ is an isomorphism, and therefore, the two graphs are isomorphic.

Note that an isomorphism maps a pair of connected vertices into a pair of connected vertices. In particular, if the degree of A is d, then the degree of $f(A)$ is d, for all isomorphisms f. Therefore, two graphs can be isomorphic only if the multisets of their degrees are the same. Exercise 21 shows that this condition is *not* sufficient for isomorphism, indeed, there are graphs with the same multiset of degrees that are not isomorphic.

In order to prove that two graphs are isomorphic, we have to exhibit an isomorphism between them. To prove that two graphs are *not* isomorphic is a more difficult issue. In certain cases we get lucky. If the two graphs do not have the same number of vertices, or the same multiset of degrees, or they do not have the same number of cycles, or the same number of paths of length k, and so on, then it is clear that they are not isomorphic. Indeed, isomorphisms preserve all these parameters. (You should think about this for a while.)

There is no general way, however, to test whether two graphs are isomorphic. Unless, that is, you verify all $n!$ bijections from G to H, where n is the number of vertices of each graph. It is not known whether this problem belongs to the class of NP-complete problems, the class of problems that we mentioned when we discussed the problem of deciding whether a graph has a Hamiltonian cycle.

To summarize, we have seen two different answers to the question of when two graphs are different. In one of them, the vertices were distinguishable (labeled), in the other one, they were indistinguishable (unlabeled). The way the graph was drawn did not matter in either case. We will see situations, in Chapters 12 and 14, when that will matter.

Notes

Graph Theory is the subject of Chapters 9–12 of this book. If the reader wants a book-length treatment of the topic, an obvious place to start is "Introduction to Graph Theory" by Douglas West [47].

We will return to graphs in several later chapters as well, essentially in all chapters following the Graph Theory part of the book. This shows how omnipresent graphs are in combinatorics.

Exercise 18 contains the definition of graphical partitions. Let $g(n)$ be the number of all graphical partitions of the even integer n. Paul Erdős conjectured that $\lim_{n\to\infty} \frac{g(n)}{p(n)} = 0$. This conjecture was open for twenty years, and has only been recently proved by Boris Pittel [33], who used sophisticated techniques from Probability Theory in proving it.

Exercises

(1) Let G be a loopless undirected graph. Prove that the edges of G can be directed so that no directed cycle is formed. (To put this into a real-life context, it is possible to make all the streets of a city one-way so that you can never return to a point you have left. This seems rather likely, by experience....)

(2) Is it true that if a graph has a closed Eulerian trail, then it has an even number of edges?

(3) Let G be a simple graph on 10 vertices and 28 edges. Prove that G contains a cycle of length 4.

(4) Let G be a simple graph on 9 vertices, and assume we know that the sum of all degrees in G is at least 27. Is it true that G has a vertex of degree at least four?

(5) Let G be a graph. We say that H is an *induced subgraph* of G if the vertex set of H is a subset of that of G, and if x and y are two vertices of H, then xy is an edge in H if and only if xy is an edge in G.
Let G be a simple graph that has 10 vertices and 38 edges. Prove that G contains K_4 (the complete graph on four vertices) as an induced subgraph.
Remark. The word *induced* in the name of induced subgraph is important. The notion of *subgraphs* is different from that of induced subgraphs. If G is a graph, we say that J is a *subgraph of* G if the vertex set of J is a subset of that of G, and if x and y are two vertices of J, then xy is an edge in J **only if** xy is an edge in G. In other words, a subgraph of G does not necessarily contain all the edges of G that connect two of its vertices, while an induced subgraph of G does.

(6) Let G be a simple graph in which all vertices have degree four. Prove

that it is possible to color the edges of G orange or blue so that each vertex is adjacent to two orange edges and two blue edges.

(7) How many different simple graphs are there on the vertex set $[n]$?

(8) An *automorphism* of a graph G is an isomorphism between G and G itself. That is, the permutation f of the vertex set of G is an automorphism of G if for any two vertices x and y of G, the number of edges between x and y is equal to the number of edges between $f(x)$ and $f(y)$. How many automorphisms do the following (labeled) graphs have?

(a) The complete graph K_n on n vertices.
(b) The cycle C_n on n vertices.
(c) The path P_n on n vertices.
(d) The star S_n on n vertices. (This graph has one vertex of degree $n-1$, and $n-1$ vertices of degree 1.)

(9) Prove that there are more than 6600 pairwise non-isomorphic graphs on eight labeled vertices.

(10) Is it true that the number of people currently living on our planet and having an odd number of siblings is even?

(11) Is it true that

(a) if a simple graph has a closed Eulerian trail, then it has a Hamiltonian cycle?
(b) if a simple graph has a Hamiltonian cycle, then it has a closed Eulerian trail?

(12) A simple graph is called *regular* if all its vertices have the same degree. Let G be a connected regular graph with 22 edges. How many vertices can G have?

(13) The previous exercise defines a regular graph as a simple graph in which each vertex has the same number of neighbors. Is it true that in such a graph, each vertex will have the same number of second neighbors? (The vertex X is a second neighbor of a vertex Y if XY is not an edge, and there is a path of length 2 joining X and Y.)

(14) The graph shown in Figure 9.10 is called the *Petersen graph.* Does this graph have a Hamiltonian cycle?

(15) Find all ways to omit edges from the Petersen graph shown in Figure 9.10 so that the remaining graph, that still has ten vertices, has a closed Eulerian trail.

(16) The *ordered degree sequence* of a graph is the list of the degrees of its vertices in non-increasing order. So if a graph G has e edges, then

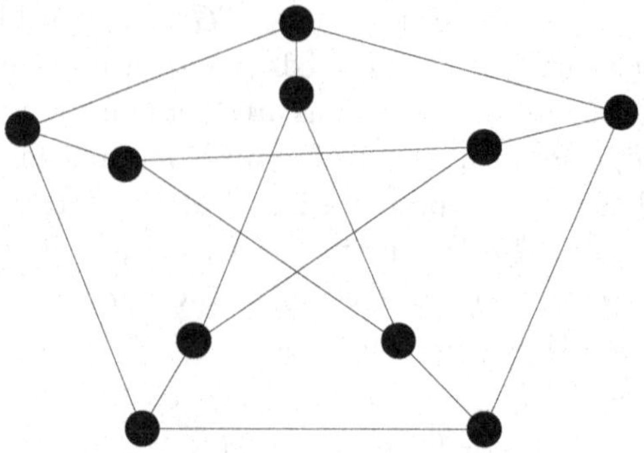

Fig. 9.10 The Petersen graph.

the positive members of its degree sequence form a partition $\Pi(G)$ of the integer $2n$. Prove that if G is a simple graph, then $\Pi(G)$ is never self-conjugate.

(17) Is there a simple graph on 6 vertices with ordered degree sequence 4, 4, 4, 2, 1, 1?

(18) Let p be a partition of the integer $2n$. We say that p is *graphical* if there exists a simple graph G (necessarily with n edges) that has ordered degree sequence p. Prove that $p = (4, 4, 3, 2, 1)$ is not graphical.

(19) (+) How many automorphisms does the graph shown in Figure 9.11 have?

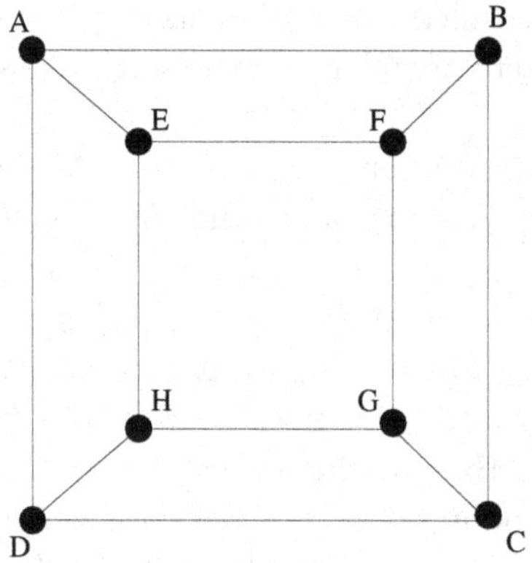

Fig. 9.11 Find the number of automorphisms of this graph.

(20) (+) How many automorphisms does the graph shown in Figure 9.12 have?

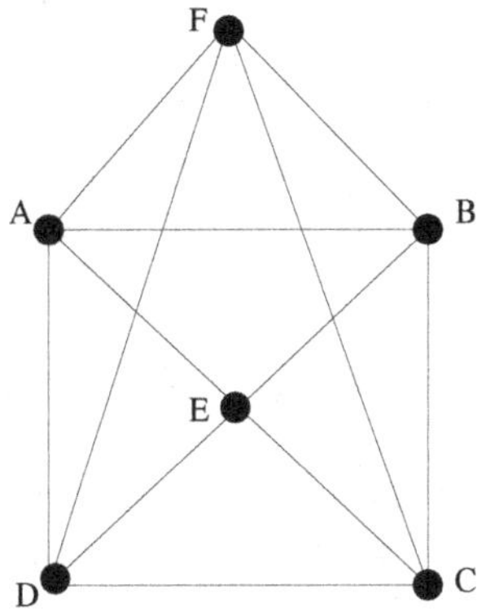

Fig. 9.12 Find the number of automorphisms of this graph.

(21) Two graphs have the same ordered degree sequence. Show that they are not necessarily isomorphic.

(22) Let $c(n)$ be the number of connected graphs on the vertex set $[n]$, and let $C(x)$ be the exponential generating function of the sequence $\{c(n)\}$. Find $C(x)$. Do not look for a closed form. Look for a functional equation that enables us to compute the values $c(n)$.

Supplementary Exercises

(23) (-)

(a) How many simple *directed* graphs are there on vertex set $[n]$?

(b) How many tournaments are there on vertex set $[n]$?

(24) (-) A tournament is called *transitive* if the fact that there is an edge from i to j and an edge from j to k implies the fact that there is an edge from i to k. How many transitive tournaments are there on vertex set $[n]$?

(25) (-) Prove that a tournament is transitive if and only if it has only one Hamiltonian path.

(26) (-) Is it true that a directed graph with a finite number of vertices and with no directed cycles has at least one vertex whose outdegree

is zero?

(27) (-) Prove that for any integers $n \geq 1$, there exists a set S of $\binom{n}{2} + 1$ graphs on vertex set $[n]$ so that no two elements of S are isomorphic.

(28) Prove that if in a simple graph G, there is a trail or walk from vertex A to vertex B, then there is also a path from A to B.

(29) (+) A high school has 90 alumni, each of whom has ten friends among the other alumni. Prove that each alumni can invite three people for lunch so that each of the four people at the lunch table will know at least two of the other three.

(30) Prove that in any simple graph, there are two vertices with the same degree.

(31) There are several people in a classroom; some of them know each other. It is true that if two people know the same number of people in the classroom, then there is nobody in the classroom both of these people know. Prove that there in someone in the classroom who knows exactly one other person in the classroom.

(32) Prove that the number of people who have shaken hands at an odd number of times (in their life so far) is even.

(33) Ten players participate at a chess tournament. Eleven games have already been played. Prove that there is a player who has played at least three games.

(34) Find all non-isomorphic simple graphs on four vertices.

(35) Find a simple graph G on n vertices so that G has no non-trivial automorphisms, $n > 1$, but otherwise n is minimal under these conditions. Explain how your answer changes if we drop the requirement that G be simple.

(36) Let G be the union of k disjoint cycles of length r. How many automorphisms does G have?

(37) (+) At most how many edges can a simple graph G on n vertices have if G is not to have a Hamiltonian cycle?

(38) For what values of n can K_n be decomposed into a union of edge-disjoint Hamiltonian cycles?

Note: In the following several exercises, we will ask how many Hamiltonian cycles various graphs have. All these graphs have labeled vertices, and two Hamiltonian cycles are considered distinct if their set of (undirected) edges are different.

(39) How many Hamiltonian cycles does K_n have?

(40) Let $K_{m,n}$ be the simple graph whose vertex set consists of the m-element vertex set A, and the n-element vertex set B, and which has

a total of mn edges, each between a vertex in A and a vertex in B. Find the number of Hamiltonian cycles of $K_{m,n}$. Note that in the special case of $m = n$, the answer will differ from the other cases. We point out that $K_{m,n}$ is called a *complete bipartite graph.*

(41) For graph theoretical purposes, the *n-dimensional hypercube* Q_n is a simple graph whose vertices are the 2^n points $(x_1, x_2, \cdots, x_n) \in \mathbf{R^n}$ so that for each $i \in [n]$, either $x_i = 0$ or $x_i = 1$, and in which two vertices are adjacent if they agree in exactly $n - 1$ coordinates. Prove that if $n \geq 2$, then Q_n has a Hamiltonian cycle.

(42) Prove that if $n \geq 2$, then Q_n has at least $n!/2$ Hamiltonian cycles.

(43) Find the number of Hamiltonian cycles of Q_3 (the regular, three-dimensional cube).

(44) Is there a simple graph G on seven vertices such that it is not connected, and each vertex of G has degree at least three?

(45) Each vertex of a simple graph G has degree k. Prove that G contains a cycle of length at least $k + 1$.

(46) Prove that if G is a simple graph on n vertices, and for any two vertices X and Y of G, it is true that $d_x + d_z \geq n$, then G has a Hamiltonian cycle. (Here d_z denotes the degree of the vertex z.)

(47) Prove that the statement of the previous exercise is not true if we only assume that $d_x + d_z \geq n - 1$.

(48) Let G be a simple graph on vertex set $[n]$ in which each vertex has degree two.

   (a) Prove that G is a union of disjoint cycles.

   (b) Let $g(n)$ be the number of graphs described above, and set $g(0) = 1$. Prove that

$$\sum_{n\geq 0} g(n)\frac{x^n}{n!} = \frac{e^{-\frac{x}{2}-\frac{x^2}{4}}}{\sqrt{1-x}}.$$

   (c) Explain why the generating function computed in part (b) is different from the exponential generating function $\sum_{n\geq 0} n!\frac{x^n}{n!} - \frac{1}{1-x}$ of the numbers of n-permutations, when permutations are in fact also unions of disjoint cycles on the set $[n]$.

(49) Let $h(n)$ be the number of simple graphs G on vertex set $[n]$ in which no vertex has degree more than two. Find the exponential generating function $\sum_{n\geq 0} h(n)\frac{x^n}{n!}$.

(50) Let $z(n)$ be the number of simple graphs G on vertex set $[n]$ in which no connected component has more than three vertices. Find the exponential generating function $\sum_{n\geq 0} z(n)\frac{x^n}{n!}$.

Solutions to Exercises

(1) Label the vertices of G by the integers $1, 2, \cdots, |G|$ using each integer once. Then orient the edges so that the arrow on the edge ij points to j if and only if $i < j$. This way, the labels *increase* along any directed path, so no directed cycle can exist.

(2) No, that is not true. A triangle is a counterexample.

(3) The sum of all degrees of G is 56. Therefore, G has two vertices so that the sum of their degrees is at least 12, by the pigeon-hole principle. Let X and Y be these two vertices. They may be connected by an edge, but even then, they are connected to ten other vertices. However, G has only eight other vertices, so there must be at least two vertices, C and D, that are connected to both A and B. Therefore, $ACBD$ is a cycle of length four.

(4) Yes, that is true. The sum of all degrees of a graph is always an even number. Therefore, if this sum is at least 27, then it is at least 28, and the statement follows by the pigeon-hole principle.

(5) There are $\binom{10}{4} = 210$ four-element vertex sets in G. Denote by $a_1, a_2, \cdots, a_{210}$ the number of edges in the subgraphs induced by each of them. Then we have

$$\frac{a_1 + a_2 + \cdots + a_{210}}{28} = 38$$

as the numerator of the left-hand side counts each edge 28 times. Indeed, the edge xy is counted 28 times there, as there are $\binom{8}{2} = 28$ ways to add two vertices to xy, and obtain a four-element vertex set. So $a_1 + a_2 + \cdots + a_{210} = 28 \cdot 38 = 1064$. This implies, by the Pigeon-hole Principle, that the largest of the a_i must be at least $1064/210 = 5.07$. As the a_i are all integers, this means that the largest a_i is in fact at least 6, which means that the corresponding induced subgraph is K_4.

(6) Theorem 9.2 shows that G has a closed Eulerian trail W. Go through C edge by edge, and color its first edge orange, the second one blue, the third one orange again, the fourth one blue again, and so on. As W leaves a vertex right after entering it, the statement follows. Indeed, each time W passes through a vertex, it contributes one orange edge and one blue edge to that vertex. As W passes through each vertex twice, the statement follows.
The only exception to this is the starting (and ending) vertex V. The trail W passes through V only once, but it starts and ends in V, too. To see that the starting and ending edges of W have different colors,

we must prove that W, and therefore, G, has an even number of edges. We know from the proof of Theorem 9.4 that in any loop-less graph, $e = \frac{1}{2}\sum_{i=1}^{n} d_i$. In our case, $d_i = 4$ for all i, therefore $e = \frac{1}{2}4n = 2n$, which is indeed an even number. This completes the proof.

(7) There are $\binom{n}{2}$ pairs of vertices in such a graph, and each of them is connected by either 0 edges, or by 1 edge. Thus for each pair of vertices, we have to make a choice of two possibilities. Therefore, the total number of simple graphs on $[n]$ is $2^{\binom{n}{2}}$.

(8)(a) As any bijection from the vertex set of G onto itself is an automorphism, the answer is $n!$.

(b) Let us first assume that $n \geq 3$. Let A and B be two adjacent vertices of G, and let f be an automorphism of G. Then $f(A)$ and $f(B)$ have to be adjacent vertices, and they completely determine f. Indeed, if C is the other neighbor of B in G, then $f(C)$ must be the other neighbor of $f(B)$ in G, and so on. If we choose $f(A)$ first, then $f(B)$, then we have n choices for $f(A)$, and then 2 choices for $f(B)$. Therefore, we have $2n$ possibilities for f.

If $n = 1$, or $n = 2$, then there are only n automorphisms. (In these cases, we are only free to choose the image of one vertex.)

(c) If E and F are the two endpoints of P_n, then an automorphism either leaves them fixed, or interchanges them. Indeed, these are the only vertices of degree one in P_n, and any automorphism preserves degree. Once we know $f(E)$ and $f(F)$, the rest of f is determined. Therefore, P_n has two automorphisms.

(d) If C is the center (the only vertex of degree $n-1$) of S_n, then it is clear that in any automorphism f of S_n, we must have $f(A) = A$. There is no restriction on the other vertices; f can permute them in any way. Thus S_n has $(n-1)!$ automorphisms.

(9) As we saw in Exercise 7, the number of all simple graphs on $[8]$ is $2^{\binom{8}{2}} - 2^{28} = 268435456$. On the other hand, the number of bijections from $[8]$ onto $[8]$ is $8!$. Therefore, any labeled graph on eight vertices can be isomorphic to at most $8! = 40320$ other graphs. It then follows from the pigeon-hole principle that the number of isomorphism classes must be at least $268435456/40320 = 6657.625$.

(10) Yes, consider the graph whose vertices are all people currently living on our planet, and two vertices are joined by an edge if and only if the corresponding people are siblings.

(11)(a) No. A counterexample is shown in Figure 9.13.

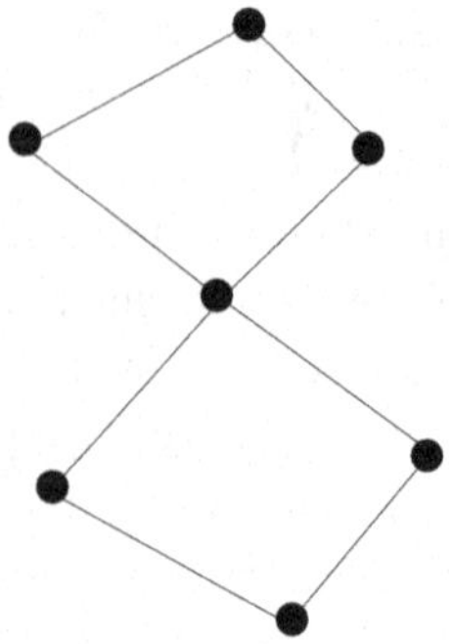

Fig. 9.13 A graph with no Hamiltonian cycle.

(b) No. A counterexample is a complete graph on $2n$ vertices, for $n \geq 2$.

(12) Let d be the common degree of the vertices of G, and let v be the number of vertices of G. Then we have $44 = v \cdot d$. So v must be a divisor of 44, that is, it cannot be anything other than 1, 2, 4, 11, 22 or 44. As G is simple, it cannot have more edges than K_n, which excludes the three smallest divisors of 44. If $v = 22$, then $d = 2$, and this is indeed possible if G is a cycle of 22 vertices. If $v = 11$, then we must have $d = 4$, and this is indeed possible. Simply take a cycle on 11 vertices, then join each vertex to both of its second neighbors by an edge. Finally, $v = 44$ is not possible, because that would mean $d = 1$, so G would consist of vertex-disjoint edges, and thus it would not be connected.

(13) No, that is not necessarily true. Figure 9.14 shows a regular graph in which each vertex has three neighbors. However, vertices B, D, F and H have four second neighbors, while vertices A, C, E, and G have three.

(14) No, it does not. Call the five edges joining an outer vertex to an inner vertex *sticks*. Then any Hamiltonian cycle would have to contain a positive even number of sticks, that is, two or four of them. Two sticks are impossible as then the Hamiltonian cycle would have to contain four outer edges and four inner edges, that is, there would be a path of length four between the two outer endpoints of the two sticks, and a path of length four between their two inner endpoints. That is clearly impossible. Four sticks are also impossible. Indeed, if AB is the only stick that is not in our purported Hamiltonian cycle h, then both the

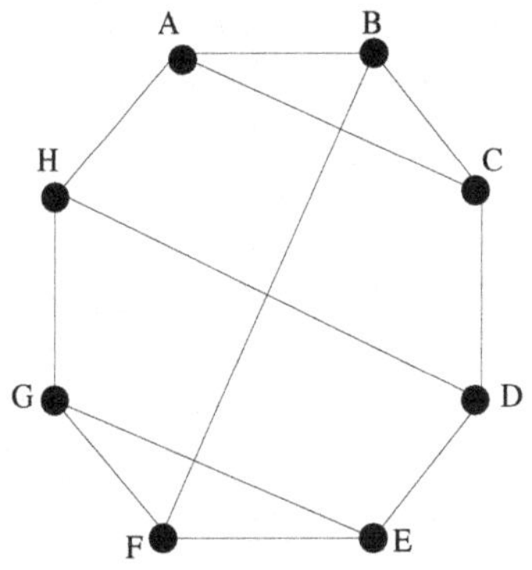

Fig. 9.14 A regular graph.

four non-stick edges adjacent to A and B must all be part of h. Indeed, all vertices have degree two in h. If we continue reconstructing h using this observation, we quickly run into a contradiction by obtaining a cycle with less than ten vertices.

(15) It is not possible to omit edges so that the obtained graph has a closed Eulerian trail. Indeed, for that, we would have to make all the degrees even, which means two in this case. That, however, would mean that our closed Eulerian trail is a Hamiltonian cycle, and the previous exercise shows that the Petersen graph has no Hamiltonian cycle.

(16) If G has n vertices, then $G(\Pi)$ will have $k \leq n$ parts, where k is the number of vertices attached to at least one edge. On the other hand, the largest part of $G(\Pi)$ is at most $k-1$ as G is simple, so each vertex can be connected to each other vertex at most once. So the number of parts and the size of the largest part is not the same, therefore $G(\Pi)$ fails the first test of self-conjugacy.

(17) No, there is not. Let us assume that G is such a graph, and let S be the set of vertices of G that have degree 4. Then S has three elements, so there can be at most three edges that join two vertices of S. This forces each vertex of S to be connected to at least two vertices of $G - S$. That, however, would mean that there are at least six edges between S and $G - S$, which is more than the sum of all degrees in $G - S$. We reached a contradiction, so no such graph G can exist.

(18) Let us assume that G has degree sequence $p = (4, 4, 3, 2, 1)$. Then G has five vertices and seven edges. In particular, there are two vertices,

say A and B, that are connected to all other vertices. That, however, would mean that all vertices have degree at least two as they are connected to both A and B.

(19) Note that this graph is in fact the graph of a cube. Therefore, we will talk about it as such. A cube has six faces. Once we know the image of the vertices of one face by an automorphism f, we know the entire automorphism. Indeed, assume that we know what $f(A)$, $f(B)$, $f(C)$, and $f(D)$ are. Then these four vertices must form a face. Moreover, $f(E)$ must be the only unused vertex adjacent to $f(A)$, $f(F)$ must be the only unused vertex adjacent to $f(B)$, and so on. The question is, therefore, how many different possibilities are there for the images $f(A)$, $f(B)$, $f(C)$, and $f(D)$. First count those automorphisms in which the orientation of the cube does not change. In this case, there are six faces into which the face $ABCD$ can be mapped, and then there are four ways the images $f(A)$, $f(B)$, $f(C)$, and $f(D)$ can be rotated on each face. So there are 24 automorphisms that preserve the orientation of the cube. After each of these, we can perform a reflection through a plane that bisects the cube. This provides 24 automorphisms that reverse the orientation of the cube. Therefore, the graph of the cube has altogether 48 automorphisms.

(20) Note that this graph is in fact the graph of an octahedron. We can get an octahedron by taking the center of each face of a cube (these will be the vertices), and adding an edge between two vertices if the corresponding cube-faces are adjacent. We can get a cube from an octahedron the very same way.
This implies that there is a bijection between the automorphisms of a cube and the automorphisms of an octahedron. The previous exercise shows that the cube has 48 automorphisms, therefore the octahedron also has 48 automorphisms.

(21) The ordered degree sequences of both graphs shown in Figure 9.15 are $(3, 3, 2, 2, 2)$. However, they are not isomorphic. Indeed, if they were, then any isomorphism f would have to map the set $\{A, B\}$ of vertices of the first graph onto the set $\{A, B\}$ of vertices of the second graph. (Isomorphisms preserve degree, and these are the only vertices of degree three in our graphs.) However, there is an edge between A and B in the first graph, but not in the second one, contradicting the definition of isomorphism.

(22) Let $g(n)$ be the number of all simple graphs on $[n]$, and let $G(x)$ be the exponential generating function of the sequence $\{g(n)\}$. Then

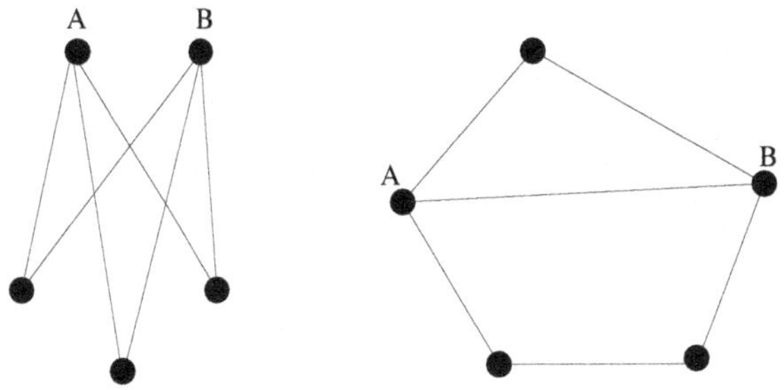

Fig. 9.15 Two non-isomorphic graphs with the same degree sequence.

$g(n) = 2^{\binom{n}{2}}$, and therefore,

$$G(x) = \sum_{n \geq 0} 2^{\binom{n}{2}} \frac{x^n}{n!}.$$

Thus $G(x) = 1 + x + 2x^2 + 8x^3 + 64x^4 + 1024x^5 + \cdots$. The exponential formula (Chapter 8, Section 2) implies that

$$G(x) = e^{C(x)},$$

$$C(x) = \ln G(x).$$

Note that the power series $\ln G(x)$ is defined using the identity

$$\ln(1 + x) = \sum_{n \geq 1} (-1)^n \frac{x^n}{n}$$

Chapter 10

Staying Connected. Trees

Being or not being connected is a crucial property of graphs, as any telecommunications company, airline, or railroad will tell you. It is certainly desirable to be able to create a connected network with relatively few edges, but one can intuitively feel that he will not be able to decrease the number of edges too much. For example, one edge will certainly not do if the graph has more than two vertices. This chapter is devoted to the study of minimally connected graphs, which we will call *trees*.

10.1 Minimally Connected Graphs

Theorem 10.1. *Let G be a connected simple graph on n vertices. Then the following are equivalent.*

(1) G is minimally connected, that is, if we remove any edge of G, then the obtained graph G' will not be connected.
(2) G does not contain a cycle.

Before proving the theorem, let us give a name to this extremely useful class of graphs.

Definition 10.2. A connected simple graph G satisfying either, and therefore, both, criteria of Theorem 10.1 is called a *tree*.

Proof. (of Theorem 10.1)

(1)$\Rightarrow$ (2) Let us assume that G is minimally connected, but it contains a cycle C. Remove the edge ab of C. We claim that G is still connected. Indeed, let x and y be two vertices in G. As G was connected, G

contained a path p from x to y. If p did not contain the edge ab, then it still connects x and y. If p did contain ab, then let us replace ab by the other (longer) arc ab, to get a new walk from x to y. As there is a walk from x to y in G', there must also be a path, as we saw in Exercise 28 of Chapter 9. Therefore, G' is connected, which is a contradiction.

(2)$\Rightarrow$(1) We prove that the *opposite* of (1) implies the *opposite* of (2). That will suffice, because it will imply that if (2) holds, the opposite of (1) cannot hold as that would imply the opposite of (2), therefore (1) has to hold. So "(2) implies (1)" will follow.

Let us assume that G is not minimally connected. That means that there is an edge in G, say AB, so that $G' = G - \{AB\}$ is still connected. Then there is a path P from B to A in G'. However, $AB \cup P$ must then be a cycle in G as it defines a path that starts in A and ends in A. So G contains a cycle. $\square$

Corollary 10.3. *A connected graph H is a tree if and only if for each pair of vertices (x, y), there is* exactly one *path joining x and y.*

Proof. If for each pair of vertices (x, y), there is *exactly one* path joining x and y, then H is minimally connected. Indeed, suppose you can omit edge rs from H and get a connected graph. Then in the original graph H, there were at least two paths from r to s, namely the edge rs, and the path that joins them in the new graph.

Conversely, suppose H is a tree, but there are two paths P and Q joining vertices x and y. Now take the *symmetric difference* of P and Q, that is, the edges that are part of exactly one of P and Q. It is straightforward to see (why?) that this symmetric difference will be a union of cycles, which is impossible in a tree. $\square$

So trees are connected graphs that do not contain a cycle. An easy way to obtain a tree on n vertices is to take a full (n-vertex) cycle on it, then to delete one edge. This will be a tree with $n - 1$ edges. We can experiment for a while and draw trees of very different structures, Some of these are shown in Figure 10.1. After some time, we start suspecting that *all* trees on n vertices have $n - 1$ edges. The following theorem shows that even more is true.

Theorem 10.4. *All trees on n vertices have $n - 1$ edges. Conversely, all connected graphs on n vertices with exactly $n - 1$ edges are trees.*

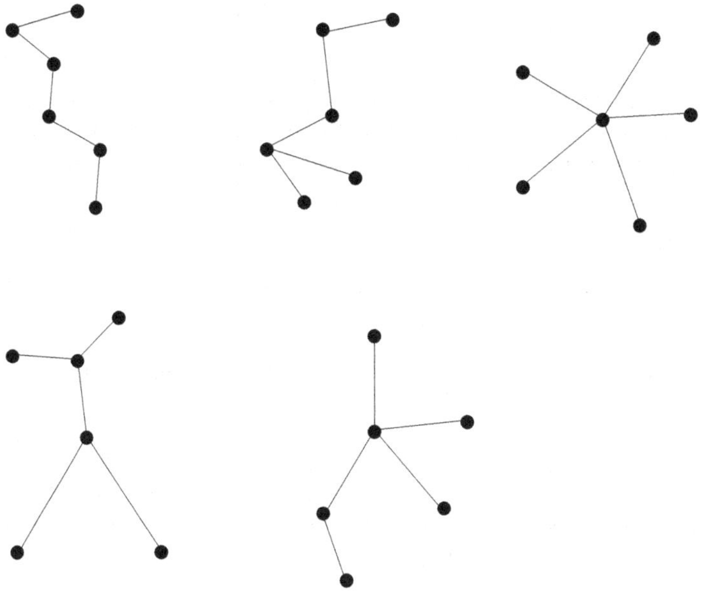

Fig. 10.1 Trees on six vertices.

In the proof of Theorem 10.4, we will need the following lemma.

Lemma 10.5. *Let T be a tree on n vertices, where $n \geq 2$. Then T has at least two vertices whose degree is 1.*

Proof. Take any path p of maximum length in T. The endpoints of p must then be leaves. Indeed, if one of them, say a, were not a leaf, then p could be extended by one of the edges that are adjacent to a but not currently part of p. □

Vertices of trees that have degree one are called *leaves*. Now we are ready to prove Theorem 10.4.

Proof. (of Theorem 10.4) We use induction on n. If $n = 1$, the statement is trivially true as a 1-vertex cycle-free graph has no edges. Let us assume that the statement is true for trees on n vertices. Let T be a tree on $n+1$ vertices. Find a leaf l in T (the previous lemma ensures the existence of two leaves), then delete l and the only edge e adjacent to it from T, to get a new tree T'. (Note that T' is always a tree as it is connected and cycle-free.) This new tree T' has n vertices, so by the induction hypothesis, it has $n-1$ edges. But then $T = T' \cup e$ has n edges, and the Theorem is proved. □

Just as in nature, a set of trees is called a *forest*. So a forest is a graph in which each connected component is a tree. This hopefully explains the cover page illustration of this book. Some of the following theorems might explain the wondering/lost facial expression of the person shown in that picture as he is walking through the woods.

Proposition 10.6. *Let F be a forest on n vertices with k connected components. Then F has $n-k$ edges.*

Proof. By Theorem 10.4, the number of vertices exceeds that of edges by one in each connected component, and the proof follows. □

How many trees are there on n vertices? After reading Section 9.3, we know that there are at least two ways to interpret this question. One is when the vertices are indistinguishable, and then the two trees shown in Figure 10.2 are considered the same (we will return to this question in Chapter 18), and the other is when the vertices are distinguishable. In this case we can say that we are counting all trees with vertex set $[n]$. In this case, the two trees in Figure 10.2 are considered different.

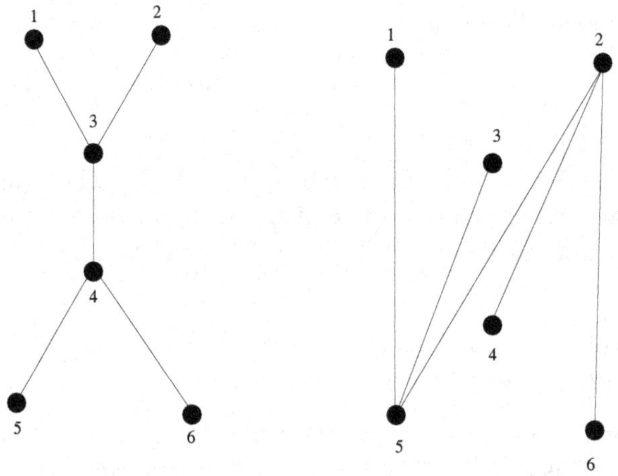

Fig. 10.2 Two isomorphic trees.

Trying the first few values of n, one sees that there is one tree on $[1]$, one tree on $[2]$, there are three trees on $[3]$, and 16 trees on $[4]$. After this, we might find enumerating all trees on $[n]$ by hand cumbersome. These scarce data suggest that there are n^{n-2} trees on $[n]$, but the reader may think that this was far too little data, and that it is very unlikely that such an incredibly nice closed formula would exist for the number of things as diverse as all trees on $[n]$. In most cases, the reader would be right to make such an argument. The dreaded "Law of Small Numbers" says that if you know just a few elements of a sequence, and those elements are small numbers, then you can always find a nice formula that is verified by those first few elements, but is incorrect in general. This case, however, is the exception.

Theorem 10.7 (Cayley's formula). *For any positive integer n, the number of all trees with vertex set $[n]$ is $A_n = n^{n-2}$.*

This beautiful result has received its fair share of attention and has at least 16 known proofs. Many of them require additional knowledge. Here we cover what may be the shortest proof on the books, and is due to André Joyal. Several other proofs will be included in the Exercises.

While reading the proof, the reader is encouraged to study the example immediately following it.

Proof. (of Theorem 10.7) Take all A_n trees on $[n]$, and in each of them, choose two vertices, which do not have to be different, and call one of them Start, and the other one End. Do this in all possible n^2 ways for each tree. Call the $n^2 A_n$ objects obtained this way *doubly rooted trees.*

We are going to show that the number of doubly rooted trees on $[n]$ is n^n by constructing a bijection from the set of all functions from $[n]$ to $[n]$ to that of doubly rooted trees on $[n]$. This will prove our Theorem.

Let f be a function from $[n]$ to $[n]$. Let $C \subseteq [n]$ be the subset of elements $x \in [n]$ which are part of a cycle under the action of f, that is, for which there is a positive integer i so that $f^i(x) = x$. Let $C = \{c_1 < c_2 < \cdots < c_k\}$. Now let $d_i = f(c_i)$, and write the integers $d_1, d_2, \cdots, d_k$ in this order to the nodes of a tree consisting of one line of k vertices. In other words, we write down the elements of C in the order given by the permutation that is the product of the cycles on C. Also, we mark d_1 by Start, and d_k by End.

Finally, if $j \in [n]$, but $j \notin C$, then join the vertex j to the vertex $f(j)$ by an edge. This way we always get a tree. Indeed, we get a connected graph as the Start-End line is connected to all vertices, and we get a cycle-free graph as the only cycles created by f involved vertices from C, and C corresponds to a single line. The tree is doubly rooted, as the vertices

Start and End are marked.

To see that this is a bijection, take a doubly rooted tree on $[n]$. For vertices j not on the Start-End line, define $f(j)$ to be the first neighbor of j on the unique path from j to the Start-End line. For the vertices on the Start-End line, define f so that the image of the ith smallest of them is the one that is in the ith position from Start.

This shows that there is exactly one function $f : [n] \to [n]$ corresponding to each doubly rooted tree, and our theorem is proved. □

Example 10.8. Let $n = 8$, and let $f : [8] \to [8]$ be the function defined by $f(1) = 3$, $f(2) = 4$, $f(3) = 1$, $f(4) = 5$, $f(5) = 5$, $f(6) = 7$, $f(7) = 8$, $f(8) = 6$. Then the action of f is shown in Figure 10.3.

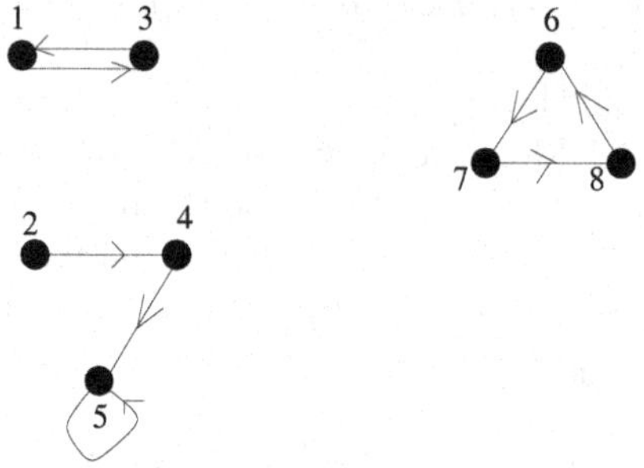

Fig. 10.3 The action of f.

The function f creates the cycles (13), (5), and (678). Therefore, $C = \{1, 3, 5, 6, 7, 8\}$, and $d_1 = 3$, $d_2 = 1$, $d_3 = 5$, $d_4 = 7$, $d_5 = 8$, and $d_6 = 6$. Therefore, our Start-End line will contain the integers 3, 1, 5, 7, 8, and 6, in this order. As $f(2) = 4$, and $f(4) = 5$, we connect the vertex 2 to 4, and the vertex 4 to 5. The obtained doubly rooted tree is shown in Figure 10.4.

To the analogy of doubly rooted trees, we can define *rooted trees*, which are trees with one vertex called the root. So the number of rooted trees on $[n]$ is n^{n-1}. A *rooted forest* is a forest in which each component is a rooted tree.

Corollary 10.9. *For all positive integers n, the number of rooted forests on $[n]$ is $(n+1)^{n-1}$.*

Proof. Take a rooted forest on $[n]$, and join all roots to the new vertex $n+1$ by an edge. This transforms the original rooted forest to an unrooted

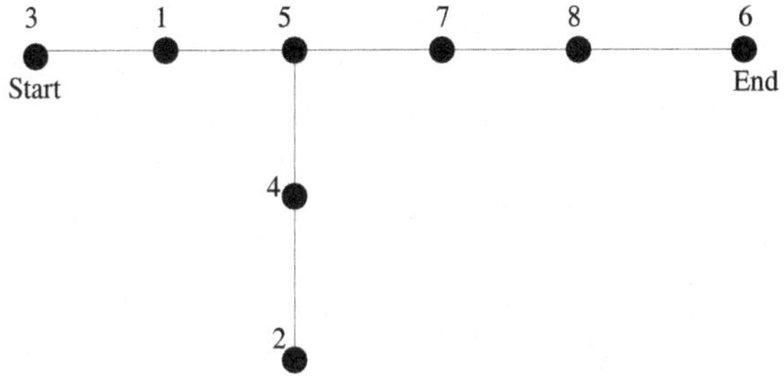

Fig. 10.4 The doubly rooted tree of f.

tree on $[n+1]$. This map is a bijection: given a tree on $[n+1]$, we can simply mark all neighbors of $n+1$ as roots, then delete all edges adjacent to $n+1$, to get the original rooted forest on $[n]$ back.

So the set of all rooted forests on $[n]$ is in bijection with that of all trees on $[n+1]$, therefore they are equinumerous. Theorem 10.7 then shows that each of them has $(n+1)^{n-1}$ elements. □

There are several other structures that are in bijection with trees or forests. See Exercises 6 and 7 for some examples.

10.2 Minimum-weight Spanning Trees. Kruskal's Greedy Algorithm

Let us return to the applications of trees. If G is a connected graph, we say that T is a *spanning tree* of G if G and T have the same vertex set, and each edge of T is also an edge of G.

Clearly, any connected graph G will have at least one spanning tree. Indeed, if G is a tree, then G is its own spanning tree; if not, then G is not a minimally connected graph, so we can omit an edge from G so that we get a connected graph G'. If G' is still not a tree, then we can continue this same procedure. We will only have to stop when we get a minimally connected graph, that is, a tree.

In general, a connected graph will have many spanning trees. Theorem 10.7 shows for example that K_n has n^{n-2} spanning trees. Sometimes it can be quite difficult to find the number of all spanning trees of a connected graph.

Spanning trees have a plethora of practical applications, especially in graphs with weighted edges. A classic example is the following.

A railroad wants to expand into a 20-city area where presently they have no lines. They thoroughly analyzed the relevant data, and for each of the $\binom{20}{2} = 190$ pairs of cities they know the exact amount they would have to spend to build a direct link between those two cities. The railroad wants to build a connected network, that is every city should be reachable from every city, but they want no redundant lines. How can they find the cheapest possible network?

A graph theoretical description of this problem is the following.

Example 10.10. Let G be a connected simple graph. Let $w : E(G) \rightarrow \mathbf{R}^+$ be a function. Find the spanning tree T of G so that $\sum_{e \in T} w(e)$ is minimal.

The function w is usually called the *weight function* or *cost function* of G, and $w(e)$ is called the *weight* or *cost* of e, while $\sum_{e \in T} w(e)$ is called the weight of T. It is common practice to write the weights of the edges on the edges, as shown in Figure 10.5. If G has only a few edges, then we might try to find its minimum-weight spanning tree by examining all spanning trees. For only slightly larger graphs, however, this approach would take too long. Indeed, if $n = 20$ and $G = K_n$ as in the railroad example, then we would have to compute the total weight of $20^{18} > 2.5 \cdot 10^{23}$ spanning trees. If our computer could handle one billion spanning trees per second, it would still need $2.5 \cdot 10^{14}$ seconds, or more than 91 years to do it!

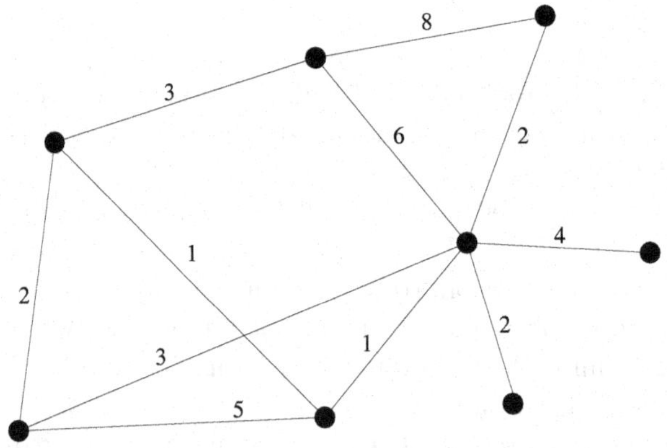

Fig. 10.5 A graph and its weight function.

Therefore, the quest for a general method to find the minimum weight spanning tree is undoubtedly well motivated. How would we start building up such a tree T? One can try the *greedy* way. That is, take the edge with the smallest weight (or one of the edges with the smallest weight, if there are several), and put it in T. Second, look for the edge that has the smallest weight among those not in T, and add it to T. In the third step, and all subsequent steps, we must be a little more careful. We have to make sure that by adding the new edge, we will not create a cycle.

In general, in the ith step of this *greedy algorithm* we look for the edge e_i that has the following properties.

(i) The edge e_i is not yet in T, and
(ii) if we add e_i to T, the obtained graph does not contain a cycle, and
(iii) the weight of e_i is minimal among all edges that have properties (i) and (ii).

When we found this edge e_i, we add it to T. It is clear that we can continue this procedure until T has $n-1$ edges, as a graph on n vertices and less than $n-1$ edges cannot be connected. However, G is connected, so if T has less than $n-2$ edges, we can find an edge of G that lies between two connected components of T, and can therefore be added to T.

The alert reader could note that the graph T that we obtain this way after i steps is not necessarily connected; all we know is that T is a cycle-free graph, that is, a forest. However, if we continue this algorithm up to step $n-1$, then T will be a forest with $n-1$ edges, that is, a tree.

Will the greedy algorithm give us the minimum weight spanning tree? The answer to this question is *not* obvious. There are problems for which the greedy algorithm does give the good answer, such as finding the three-element subset with the largest sum in any finite set of integers. There are also problems, however, for which the greedy algorithm does *not* give the correct answer, because greedy steps at the beginning adversely influences our choices later. An example for this is finding *two vertex-disjoint edges with minimum total weight* in the graph shown in Figure 10.6.

Here the greedy algorithm results in a pair of disjoint edges with total weight 11, though the correct answer is clearly 4. This problem, called the *minimum-weight matching* problem is another very important problem. We will learn about matchings in the next chapter.

For the task at hand, however, that is, for finding a minimum-weight spanning tree, the greedy algorithm works. To prove this, we will need the following interesting property of forests.

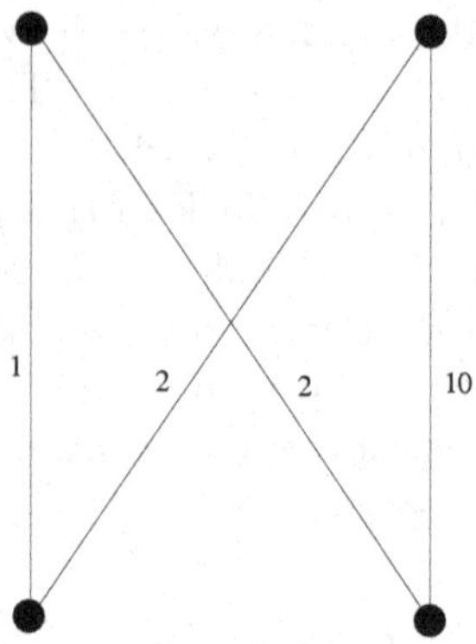

Fig. 10.6 $2+2<1+10$.

Lemma 10.11. *Let F and F' be two forests on the same vertex set V, and let F have less edges than F'. Then F' has an edge e that can be added to F so that the obtained graph $F \cup e$ is still a forest.*

Proof. Assume that there is no such edge $e \in F'$. Then adding any edge of F' to F would create a cycle in F. So all edges of F' are between two vertices of *the same* component of F. Therefore, F' has at least as many components as F. This is a contradiction, however, as we know that a forest on n vertices and with k components has $n-k$ edges, so if F' has more edges than F, it must have less components. □

Now we are in a position to prove the main result of this section.

Theorem 10.12. *The greedy algorithm always finds the minimum-weight spanning tree.*

Proof. Again, we use an indirect argument. Assume the greedy algorithm gives us the spanning tree T, whereas our graph G has a spanning tree H whose total weight is less than that of T. Let $h_1, h_2, \cdots, h_{n-1}$ be the edges of H so that $w(h_1) \leq w(h_2) \leq \cdots \leq w(h_{n-1})$ holds. Similarly, let $t_1, t_2, \cdots, t_{n-1}$ be the edges of T so that $w(t_1) \leq w(t_2) \leq \cdots \leq w(t_{n-1})$ holds.

Let i be the step at which H first "beats" T. That is, let i be the smallest integer so that $\sum_{j=1}^{i} w(h_j) < \sum_{j=1}^{i} w(t_j)$. Such an index i exist as at the end of the entire selection procedure H beats T, so there has to be a time H takes the lead. It is also clear that $i > 1$ as $w(t_1)$ is minimal among all the edge-weights of G.

As i is the first index at which H took the lead, the inequality $w(h_i) < w(t_i)$ must hold. Indeed, this is the only way

$$\sum_{j=1}^{i} w(h_j) < \sum_{j=1}^{i} w(t_j)$$

and

$$\sum_{j=1}^{i-1} w(h_j) \geq \sum_{j=1}^{i-1} w(t_j)$$

can both hold.

We will deduce a contradiction from this, that is, we will prove that with $w(h_i) < w(t_i)$ holding, the greedy algorithm could not possibly choose t_i at step i. Let T_{i-1} be the forest the greedy algorithm produced in $i-1$ steps, that is, the union of the edges $t_1, t_2, \cdots, t_{i-1}$, and let H_i be the forest formed by the edges $h_1, h_2, \cdots, h_i$. Applying Lemma 10.11 to T_{i-1} and H_i, we see that there is an edge h_j (for some $j \leq i$) that can be added to T_{i-1} without forming a cycle. However, our definitions show that $w(h_j) \leq w(h_i) < w(t_i)$, so at step i, the greedy algorithm could not add t_i to T_{i-1} as t_i did not have minimum weight among the edges that could be added to T_{i-1} without forming a cycle.

This proves by contradiction that no spanning tree H can have a smaller total weight than T, the tree obtained by the greedy algorithm. □

We would like to point out that there are several ways to attack the problem of finding a minimum-weight spanning tree with a greedy algorithm. We could for instance insist on keeping the graph we are building *connected* in each step. The particular algorithm we covered in this section is called *Kruskal's algorithm*, or *the Kruskal algorithm*, named after his inventor, the American mathematician Joseph Kruskal.

10.3 Graphs and Matrices

There are several ways to associate a matrix to a graph. These matrices are often useful for enumerating graphs. Perhaps the most widely used such matrix is the *adjacency matrix* of a graph.

10.3.1 *Adjacency Matrices of Graphs*

Definition 10.13. Let G be an undirected graph on n labeled vertices, and define an $n \times n$ matrix $A = A_G$ by setting $A_{i,j}$ equal to the number of edges between vertices i and j. Then A is called the *adjacency matrix* of G.

Example 10.14. If G is the graph shown in Figure 10.7, then

$$A_G = \begin{pmatrix} 0\,1\,1\,1 \\ 1\,0\,0\,1 \\ 1\,0\,0\,0 \\ 1\,1\,0\,0 \end{pmatrix}.$$

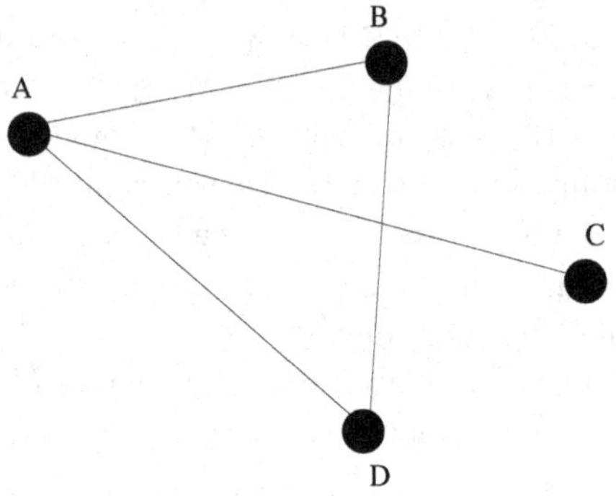

Fig. 10.7 The graph whose adjacency matrix is A_G.

If G is directed, then we can define its adjacency matrix by setting $A_{i,j}$ equal to the number of edges *from i to j*. Thus the adjacency matrix of a directed graph is not necessarily symmetric, while that of an undirected graph is.

Example 10.15. If H is the directed graph shown in Figure 10.8, then

$$A_H = \begin{pmatrix} 0\,1\,1\,0 \\ 0\,0\,0\,1 \\ 0\,0\,0\,0 \\ 1\,0\,0\,0 \end{pmatrix}.$$

The adjacency matrix of a graph comprises almost all properties of that graph. There are several situations when it is actually easier to solve an

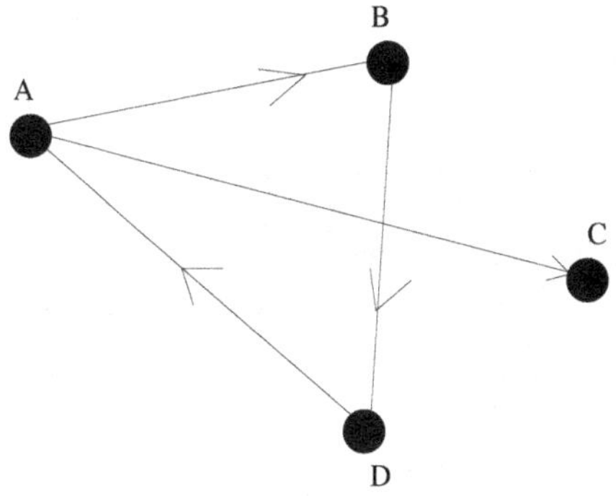

Fig. 10.8 The directed graph whose adjacency matrix is A_H.

enumeration problem working with A_G than working with G. A basic result in that direction is the following.

Theorem 10.16. *Let G be a graph on labeled vertices, let A be its adjacency matrix, and let k be a positive integer. Then $A^k_{i,j}$ is equal to the number of walks from i to j that are of length k.*

Proof. By induction on k. For $k = 1$, the statement is true as a walk of length one is an edge. Now assume that the statement is true for k, and prove it for $k+1$. Let z be any vertex of G. If there are $b_{i,z}$ walks of length k from i to z, and there are $a_{z,j}$ walks of length one (in other words, edges) from z to j, then there are $b_{i,z}a_{z,j}$ walks of length $k+1$ from i to j whose next-to-last vertex is z. Therefore, the number of all walks of length $k+1$ from i to j is

$$c(i,j) = \sum_{z \in G} b_{i,z}a_{z,j}.$$

It follows from the induction hypothesis that the matrix B defined by $B_{i,j} = b_{i,j}$ fulfills $B = A^k$. It is immediate from the definition of the adjacency matrix A of G that $A_{i,j} = a_{i,j}$.

Therefore, it follows from the definition of matrix multiplication that $c(i,j) = \sum_{z \in G} b_{i,z}a_{z,j}$ is in fact the (i,j)-entry of $BA = A^{k+1}$, (indeed, it is the scalar product of the ith row of B and the jth column of A), and our claim is proved. □

The adjacency matrix of a graph provides a quick way of testing whether the matrix has certain properties. We will discuss testing of connectivity here.

Theorem 10.17. *Let G be a simple graph on n vertices, and let A be the adjacency matrix of G. Then G is connected if and only if $(I + A)^{n-1}$ consists of strictly positive entries.*

Proof. We know from Exercise 28 of Chapter 9 that if there is a walk from i to j in G, then there is a path, too. The length of a path in G is at most $n-1$. Therefore, G is connected if and only if, for any pair of distinct vertices i and j, there is a positive integer $k \leq n-1$ so that $A^k_{i,j} > 0$. As

$$(I + A)^{n-1} = \sum_{k=0}^{n-1} \binom{n-1}{k} A^k,$$

the statement follows. □

10.4 The Number of Spanning Trees of a Graph

The adjacency matrix of a graph, surprisingly, can be used to compute the number of all spanning trees of that graph. To see this, we first need to extend our investigation to directed graphs. If G is a directed graph, then we say that H is a spanning tree of G if H is a subgraph of G, and if we remove the orientations of all edges, obtaining the undirected graphs G_1 and H_1, then H_1 is a spanning tree of G_1. We need one additional definition before we can enumerate spanning trees.

Definition 10.18. Let G be a *directed* graph without loops. Let $\{v_1, v_2, \cdots, v_n\}$ denote the vertices of G, and let $\{e_1, e_2, \cdots, e_m\}$ denote the edges of G. Then the *incidence matrix* of G is the $n \times m$ matrix A defined by

- $a_{i,j} = 1$ if v_i is the head of e_j,
- $a_{i,j} = -1$ if v_i is the tail of e_j, and
- $a_{i,j} = 0$ otherwise.

Theorem 10.19. *Let G be a* directed *graph without loops, and let A be the incidence matrix of G. Remove any row from A, and let A_0 be the remaining matrix. Then the number of spanning trees of G is* $\det A_0 A_0^T$.

This is very surprising. At first sight, it is not even obvious why $\det A_0 A_0^T$ will always be the same, no matter which row we remove, let alone have such a nice combinatorial meaning.

Proof. Let us assume, without loss of generality, that the last row of A was omitted. Let B be an $(n-1) \times (n-1)$ submatrix of A_0. (If $m < n-1$, then G cannot be connected, and it has no spanning trees.) We claim that $|\det B| = 1$ if and only if the subgraph G' corresponding to the columns of B is a spanning tree, and $\det B = 0$ otherwise.

We prove this claim by induction on n.

(a) Let us first assume that there is a vertex v_i $(i \neq n)$ of degree one in G'. (The degree of a vertex in an undirected graph is the number of all edges adjacent to that vertex.) Then the ith row of B contains exactly one nonzero element, and that element is 1 or -1. Expanding $\det B$ by this row, and using the induction hypothesis, the claim follows. Indeed, G' is a spanning tree of G if and only if $G' - v_i$ is a spanning tree of $G - v_i$.

(b) Now let us assume that G' has no vertices of degree one (except possibly v_n, the vertex associated to the deleted last row). Then G' is not a spanning tree. Moreover, as G' has $n-1$ edges, and is not a spanning tree, there must be a vertex in G' that has degree zero. If this vertex is not v_n, then B has a zero row, and $\det B = 0$. If this vertex is v_n, then each column of B contains one 1, and one -1 as each edge has a head and a tail. Therefore, the sum of all rows of B is 0, so the rows of B are linearly dependent, and $\det B = 0$.

So we have proved that indeed, $|\det B| = 1$ exactly if the subgraph G' corresponding to the columns of B is a spanning tree, and $\det B = 0$ otherwise.

Now we can finish the proof of Theorem 10.19. The Binet–Cauchy formula, that can be found in most Linear Algebra textbooks, says that

$$\det A_0 A_0^T = \sum (\det B)^2,$$

where the sum ranges over all $(n-1) \times (n-1)$ submatrices B of A_0. However, we have just seen that $(\det B)^2 = 1$ if and only if B corresponds to a spanning tree of A, and $(\det B)^2 = 0$ otherwise. Therefore, we have proved Theorem 10.19. □

You could have several remarks at this point. First, you could say, "fair enough, but it could take a long time to compute $\det A_0 A_0^T$, or even $A_0 A_0^T$ for a given graph". More generally, you could say, "what about undirected graphs?" These concerns will be simultaneously alleviated by the following theorem.

Theorem 10.20 (Matrix-Tree theorem). *Let U be a simple undirected graph. Let $\{v_1, v_2, \cdots, v_n\}$ be the vertices of U. Define the $(n-1)\times(n-1)$ matrix L_0 by*

$$l_{i,j} = \begin{cases} \text{the degree of } v_i \text{ if } i = j, \\ -1 \text{ if } i \neq j, \text{ and } v_i \text{ and } v_j \text{ are connected, and} \\ 0 \text{ otherwise,} \end{cases}$$

where $1 \leq i, j \leq n-1$. Then U has exactly $\det L_0$ *spanning trees.*

Note that while U has n vertices, L_0 is an $(n-1)\times(n-1)$ matrix, since it does not contain the row and column that would belong to v_n. This does not mean that L_0 does not contain information about v_n. Indeed, if a vertex v_i is connected to v_n, then the edge between v_i and v_n counts towards the degree of v_i, hence towards the entry $l_{i,i}$. Therefore, if we are given L_0, we can find out which vertices of U are adjacent to v_n.

Proof. (of Theorem 10.20) First we turn U into a directed graph G by replacing each edge of U by a *pair* of directed edges, one edge going in each direction.

Let A_0 be the incidence matrix of G. We claim that $A_0A_0^T = 2L_0$. The entry of $A_0A_0^T$ in position (i,j) is the scalar product of the ith and jth row of A_0. If $i = j$, then every edge that starts or ends at v_i contributes 1 to this inner product. Therefore, the entry of $A_0A_0^T$ in position (i,i) is the degree of v_i in G, or, in other words, twice the degree of v_i in U.

If $i \neq j$, then every edge that starts at v_i and ends at v_j, and every edge that starts at v_j and ends at v_i contributes -1 to this inner product. Recall that U was simple, so there is either 0 or 1 edge from v_i to v_j in G. Thus the entry of $A_0A_0^T$ in position (i,j) is -2 if v_iv_j is an edge of U, and 0 otherwise. This proves that indeed, $A_0A_0^T = 2L_0$.

This implies that $2^{n-1}\det L_0 = \det(A_0A_0^T)$. Note that each spanning tree of U can be turned into 2^{n-1} different spanning trees of G by orienting its $n-1$ edges. Therefore, our statement immediately follows from Theorem 10.19. □

Let us use our fresh knowledge for our classic example, the number of all trees on $[n]$.

Example 10.21. The number of spanning trees of K_n is n^{n-2}.

Solution. The matrix L_0 associated to K_n will have the following simple structure

$$\begin{pmatrix} n-1 & -1 & \cdots & -1 \\ -1 & n-1 & \cdots & -1 \\ \cdots \\ -1 & -1 & \cdots & n-1 \end{pmatrix}.$$

To compute this determinant, add all rows to the first, to get

$$\begin{pmatrix} 1 & 1 & \cdots & 1 \\ -1 & n-1 & \cdots & -1 \\ \cdots \\ -1 & -1 & \cdots & n-1 \end{pmatrix}.$$

Now add the first row to all other rows to get the triangular matrix

$$\begin{pmatrix} 1 & 1 & \cdots & 1 \\ 0 & n & \cdots & 0 \\ \cdots \\ 0 & 0 & \cdots & n \end{pmatrix}.$$

This shows that $\det L_0 = n^{n-2}$ as claimed.

Theorem 10.20 is a powerful tool. Let us use it to compute the number of spanning trees of some interesting graphs.

Example 10.22. Let A be a set of m vertices, and let B be a set of n vertices. Connect each vertex of A to each vertex of B by an edge. Denote this graph by $K_{m,n}$. Find the number of spanning trees of $K_{m,n}$.

The graph $K_{m,n}$ is called a *complete bipartite graph.* We will learn more about these graphs in the next chapter. For now, note that there is no edge *within* A or *within* B in $K_{m,n}$.

Solution. (of Example 10.22) The matrix L_0 associated to $K_{m,n}$ has the following block structure

$$\begin{pmatrix} n & \cdots & 0 & -1 & \cdots & -1 \\ \cdots & & & & & \\ 0 & \cdots & n & -1 & \cdots & -1 \\ -1 & \cdots & -1 & m & \cdots & 0 \\ \cdots & & & & & \\ -1 & \cdots & -1 & 0 & \cdots & m \end{pmatrix},$$

that is, the first m rows look "similar", then the last $n-1$ rows look "similar". The same is true for columns.

To compute this determinant, use the same trick as in the proof of Theorem 10.20. That is, add all rows to the first one to get a row of the form $(1, 1, \cdots 1, 0, \cdots, 0)$, then add this row to each of the last $n-1$ rows, to get

$$\begin{pmatrix} 1 & \cdots & 1 & 0 & \cdots & 0 \\ \cdots & & & & & \\ 0 & \cdots & n & -1 & \cdots & -1 \\ 0 & \cdots & 0 & m & \cdots & 0 \\ \cdots & & & & & \\ 0 & \cdots & 0 & 0 & \cdots & m \end{pmatrix}.$$

This shows that $\det L_0 = n^{m-1}m^{n-1}$.

Your sense of symmetry might be slightly disturbed by our disregarding the vertex v_n. You may be thinking that in situations when our graph has many vertices of different degree, it may not be obvious which vertex should be chosen for the role of v_n. Of course, Theorem 10.20 is true with any choice of v_n, but the computation of $\det L_0$ may become more complex if we do not make the right choice.

One way of getting around this is to use the following alternative form of the Matrix-tree theorem.

Theorem 10.23 (Matrix-Tree theorem, eigenvalue version). *Let U be a graph as in Theorem 10.20, and let L be defined the same way as L_0 in*

Theorem 10.20, except that let L be an $n \times n$ matrix. Denote $\lambda_1, \lambda_2, \cdots, \lambda_n$ the eigenvalues of L, with $\lambda_n = 0$. Then the number of spanning trees of U is

$$\frac{1}{n}\lambda_1 \cdot \lambda_2 \cdot \cdots \cdot \lambda_{n-1}.$$

Remarks. By now, you should be asking "how do we know that 0 is always an eigenvalue of L?" The answer is that the rows of L sum to a zero row, and therefore, they are linearly dependent. So $\det L = 0$, which implies that 0 is an eigenvalue of L. The matrix L is called the *Laplacian* of U.

We do not prove Theorem 10.23 here. It can be proved from Theorem 10.20 by algebraic manipulations that do not involve additional combinatorics.

In order to be able to use Theorem 10.23, we have to be able to find the eigenvalues of L. You may remember from your studies in Linear Algebra that there is no universal method for this if L is larger than 4×4. For *nice* graphs, however, that is, for graphs that have a lot of automorphisms, we can find these eigenvalues by some clever tricks, and then use Theorem 10.23 to compute the number of spanning trees of U. We will see examples for this in the Exercises.

For now, let us discuss one particular situation. If U is a *regular* graph, that is, all vertices of U have degree d, then we see that $dI - A = L$, where A is the adjacency matrix of U. Therefore, if $\alpha_1, \alpha_2, \cdots, \alpha_n$ are the eigenvalues of A, then $d - \alpha_1, d - \alpha_2, \cdots, d - \alpha_n$ are the eigenvalues of A. This means that to find the eigenvalues of L, it suffices to find the eigenvalues of A.

Example 10.24. Let $U = K_n$. Then the eigenvalues of the adjacency matrix A of U are $n-1, -1, -1, \cdots -1$, therefore the eigenvalues of L are $n, n, \cdots, n, 0$, showing again that K_n has n^{n-2} spanning trees.

Solution. Note that $A + I = J$, the matrix whose entries are all equal to 1. This matrix is obviously of rank 1, therefore $n-1$ of its eigenvalues are equal to 0. As the trace of J is n, and we know that the trace of any matrix is equal to the sum of its eigenvalues, the remaining eigenvalue must be n. However, $A = J - I$, so the eigenvalues of A are the eigenvalues of J decreased by 1, and the statement is proved.

Notes

A more general discussion of the Matrix-Tree theorem, as well as a survey of results connecting the number of spanning trees of a graph to the number of certain Eulerian cycles can be found in *Enumerative Combinatorics*, Volume 2, by Richard Stanley [42]. Additional proofs of Cayley's formula can be found in *Combinatorial Problems and Exercises* by László Lovász [27], which is a comprehensive source of difficult exercises in graph theory anyway.

An introductory text about graphical enumeration is Chapter 5 of *Introduction to Enumerative Combinatorics* [7]. A book-length treatment is "Graphical Enumeration", by F. Harary and E. M. Palmer [23].

Structures for which the greedy algorithm works are so important in Combinatorics (and other fields) that they have their own name, and are the subject of several books on their own. They are called *matroids*. The reason for this name is that in some sense, matroids are generalizations of matrices. The interested reader is encouraged to consult [34].

Exercises

(1) Let $n \geq 2$ be an integer, and let $a_1 \geq a_2 \geq \cdots \geq a_n$ be a sequence of positive integers satisfying $a_1 + a_2 + \cdots + a_n = 2n - 2$. Prove that there exists a tree T on n vertices so that the ordered degree sequence of T is $a_1, a_2, \cdots, a_n$.

(2) A *complete k-ary tree* is a rooted tree in which every vertex has either k or 0 descendants. Let T be such a tree with m non-leaf vertices. How many leaves does T have?

(3) Prove that for all $n \geq 3$, the number t_n of non-isomorphic trees on n vertices is at least $p(n-2)$.

(4) Prove that if n is sufficiently large, then there exists a lower bound for t_n that is better than that of the previous exercise. Find such a lower bound.

(5) Let T be a tree on $[n]$, with $n \geq 3$. Cut off the leaf of T that has the smallest label, and write down its single neighbor. Then continue this same procedure on the remaining tree until there are only two vertices (and one edge) left. This procedure results in a sequence of elements of $[n]$ that has length $n-2$, called the *Prüfer sequence*, or *Prüfer code* of T.

Prove that this algorithm defines a bijection from the set of all trees on $[n]$ onto that of sequences of length $n-2$ with elements from $[n]$. Deduce Theorem 10.7.

(6) A function $f : [n] \to [n]$ is called *acyclic* if there are no cycles longer than one under its action on $[n]$. Prove that the number of acyclic functions on n is $(n+1)^{n-1}$.

(7) There are n parking spots $1, 2, \cdots, n$ on a one-way street. Cars $1, 2, \cdots, n$ arrive in this order. Each car i has a favorite parking spot $f(i)$. When a car arrives, it first goes to its favorite spot. If the spot is free, the car will take it, if not, it goes to the next spot. Again, if that spot is free, the car will take it, if not, the car goes to the next spot. If a car had to leave even the last spot and did not find the space, then its parking attempt has been unsuccessful.

If, at the end of this procedure, all cars have a parking spot, we say that f is a *parking function* on $[n]$. Prove that the number of parking functions on $[n]$ is $(n+1)^{n-1}$.

(8) How many parking functions are there on $[n]$ without like consecutive elements? That is, we want to enumerate all parking functions on $[n]$ in which there is no $i \in [n]$ so that $f(i) \neq f(i+1)$.

(9) Prove that if G is a simple graph on $[n]$, then at least one of G and its complement is connected. Show an example when they are both connected. The complement $\bar{G}$ of G has the same vertex set as G and xy is an edge in $\bar{G}$ if and only if it is not an edge in G.

(10) How many edges can a simple graph G on $[n]$ have if it is not connected?

(11) Let H be a simple graph on n vertices that has m edges. Prove that H contains at least $m-n+1$ cycles.

(12) Let F be a rooted forest on n vertices, and view F as a directed graph, in which all edges are directed away from the root. If F' is another rooted forest, then we say that F contains F' if F contains F' as a directed graph. Clearly, in that case F has less components than F'.

We say that $F_1, F_2, \cdots, F_k$ is a *refining sequence* if, for all $i \in [k]$, F_i is a rooted forest on $[n]$ having i components, and F_i contains F_{i+1}. Now fix F_k.

(a) Find the number $N^*(F_k)$ of refining sequences ending in F_k.
(b) Find the number $N(F_k)$ of rooted trees containing F_k.
(c) Deduce Cayley's formula.

This proof of Cayley's formula is due to James Pitman.

(13) Find a formula for the number of rooted forests on $[n]$ having k com-

ponents.

(14) Let G be a simple graph, and let A be the adjacency matrix of G. Decide whether the following statements are true or false.

(a) A has only real eigenvalues.
(b) The sum of the eigenvalues of A is 0.
(c) The determinant of A is always positive.

(15) Let G be a graph on $n > 1$ vertices having no isolated vertices, and let A be the incidence matrix of G. Prove the following statements.

(a) For all i, we have $(A^4)_{i,i} > 0$.
(b) If both $(A^5)_{i,j}$ and $(A^6)_{i,j}$ are positive for some fixed indices $i < j$, then G contains a cycle of odd length.
(c) Let $i < j$ be two fixed indices. If $(A^k)_{i,j} = 0$ for all $k \leq n-1$, then $(A^k)_{i,j} = 0$ for all k.

(16) Let G be the complete bipartite graph of Example 10.22, and let A be the adjacency matrix of G. For any positive integer m, explain which entries of A^m have to be equal to 0.

(17) A *complete tripartite graph* is a simple graph defined as follows. The vertices are split into three subsets, A, B, and C, and there is an edge between two vertices if and only if they belong to different subsets. This graph is denoted $K_{|A|,|B|,|C|}$. Find a formula for the number of spanning trees of the complete tripartite graph $K_{m,m,n}$.

(18)(a) Find the eigenvalues of the adjacency matrix A_1 of the two-vertex tree.
(b) Find the eigenvalues of the adjacency matrix A_2 of the square (cycle of four edges).
(c) Find the eigenvalues of the adjacency matrix A_3 of the cube.
(d) Find the eigenvalues of the adjacency matrix A_n of the n-dimensional cube. (The n-dimensional cube is obtained by taking two copies of the $(n-1)$-dimensional cube, and then joining the corresponding vertices.)

(19) Find the exponential generating function $F(x)$ for the numbers f_n of forests on vertex set $[n]$ having components of size at most three.

(20) Let $G(x)$ be the exponential generating function for the numbers g_n of all rooted trees on vertex set $[n]$. Prove that $G(x) = xe^{G(x)}$.

Supplementary Exercises

(21) (-) Find a simple combinatorial proof showing that the number of forests on vertex set $[n]$ is at least the Bell number $B(n)$.

(22) (-) Let us call a vertex v of the graph G a *cut vertex* if the removal of v and the edges adjacent to it from G increases the number of components of G. Prove that any graph with at least two vertices has at least two vertices that are *not* cut vertices.

(23) (-) How many different labeled trees are there on $[n]$ that have no vertices with degree more than 2?

(24) (-) How many non-isomorphic forests are there on vertex set $[5]$?

(25) (-) Prove that the number of non-isomorphic labeled forests on vertex set $[n]$ is at least $p(n)$ (the number of partitions of the integer n).

(26) Prove that in any tree T, any two longest paths cross each other.

(27) Prove that in any tree T, all longest paths cross one another in one vertex.

(28) A *unicycle* is a simple graph that contains exactly one cycle. Let u_n be the number of unicycles on vertex set $[n]$. Find a formula for u_n. Your formula may contain one summation sign.

(29) The *distance* $d(x, y)$ between two vertices x and y of the graph G is defined as the number of edges in the shortest path from x to y. For every vertex $v \in G$, let us define

$$td(v) = \sum_{w \in G} d(v, w).$$

In other words, $td(v)$ measures the total distance of v from all vertices of G. Now define the *center* of G as the set of vertices v for which $td(v)$ is minimal. Prove that if G is a tree, then the center of G consists of either a vertex, or two adjacent vertices.

(30) Show an example for a tree on vertex set $[n]$ that has more than 2^{n-1} induced subgraphs that are trees. Try to find an example that works for all $n \geq 1$.

(31) Find the smallest tree that has at least one edge and has no non-trivial automorphisms.

(32) Let a be any positive real number so that $a < e$. Prove that there exists a natural number N so that if $n > N$, then there exist at least a^n non-isomorphic trees on n vertices.

(33) How many non-isomorphic trees are there on seven vertices?

(34) Let T be a tree on 101 vertices so that the largest degree in T is ten. Is it true that T contains a path of length five?

(35) Prove that a tree always has more leaves than vertices of degree at least three.

(36) Find two non-isomorphic trees with the same ordered degree sequence.

(37) At most how many automorphisms can a tree with n vertices have?

(38) Prove that if n is large enough, then the following statement is true. For all graphs G on n vertices, at least one of G and $\bar{G}$ contains a cycle. How large must n be for this to hold?

(39) Decide whether the following statements are true or false.

(a) If G is a connected simple graph and e is an edge of G, then there is a spanning tree of G that contains e.

(b) If G is a connected simple graph and e and f are edges of G, then there is a spanning tree of G that contains e and f.

(c) If G is a connected simple graph and e, f and g are edges of G, then there is a spanning tree of G that contains e, f and g.

(d) If G is a connected simple graph and F is a cycle-free set of edges in G, then there is a spanning tree of G that contains F.

(40) Let G be a connected graph, and let T_1 and T_2 be two of its spanning trees. Prove that T_1 can be transformed into T_2 through a sequence of intermediate trees, each arising from the previous one and adding another.

(41) (+) (Knowledge of linear algebra required.) Let T be a tournament on n vertices. Prove that the adjacency matrix of T is either of rank n or of rank $n-1$. Give an example for both.

(42) (++) (Knowledge of linear algebra required.) Prove that for any undirected graph G, the number of different eigenvalues of $A(G)$ is larger than the diameter of G. The *diameter* of G is given by

$$\max_{x,y \in G} d(x,y),$$

where $d(x,y)$ is the distance between x and y as defined in Exercise 29.

(43) Let A be the graph obtained from K_n by deleting an edge. Find a formula for the number of spanning trees of A.

(44) Let G be a regular graph, that is, let all vertices of G have degree d. Express the eigenvalues of $L(G)$ by the eigenvalues of $A(G)$.

(45) Use the result of the previous exercise to find the number of all spanning trees for each graph of Exercise 18.

Solutions to Exercises

(1) We use induction on n. For $n = 2$, the statement is trivially true. Now let us assume the statement is true for n. Take the sequence $a_1 \geq a_2 \geq \cdots \geq a_{n+1}$ satisfying $a_1 + a_2 + \cdots + a_{n+1} = 2n$. The last two elements, a_n and a_{n+1} must be equal to one, otherwise the sum of all the a_i would be at least $2n+1$. So we have $a_{n+1} = 1$. Delete a_{n+1}. Let j be the largest index so that $a_j > 1$. (There must be such an index as long as $n > 2$, otherwise the sum of the a_i is only $n+1 < 2n$.) Decrease a_j by one. This way we obtain a new sequence S which has only n elements, and sums to $2n-2$.
Therefore, the induction hypothesis applies, so there is a tree T whose ordered degree sequence is S. Now add a new leaf to T by joining it to the vertex corresponding to a_j. This new tree T' will have the desired ordered degree sequence.

(2) After trying a few specific trees, one can easily conjecture that T will have $(k-1)m+1$ leaves. This can be proved by induction on m as follows. If $m = 1$, then T has k leaves, and the claim is true. Now let us assume that the claim is true for m. Let T have $m+1$ non-leaf vertices. Pick a non-leaf vertex V that has k successors, and all of them are leaves. (As T is finite, there is always such a vertex.) Omit all the k successors of V, to get a new tree T'. This new tree T' has m non-leaf vertices (as V has just become a leaf), so by the induction hypothesis, it must have $(k-1)m+1$ leaves. Since T had k leaves more than T', it is indeed true that T had $km+1$ leaves, and the proof is complete.

(3) By Exercise 1, it suffices to show that there are $p(n-2)$ ordered degree sequences $d_1 \geq d_2 \geq \cdots \geq d_n$ so that $\sum_{i=1}^{n} d_i = 2n-2$, and $d_{n-1} = d_n = 1$. The number of these sequences is clearly the same as that of the number of ordered sequences $d_1 \geq d_2 \geq \cdots \geq d_{n-2}$ whose sum is $2n-4$. Now let $c_i = d_i - 1$, then the *positive* numbers c_i form a partition of $(2n-4)-(n-2) = n-2$. Conversely, if $(c_1, c_2, \cdots, c_k)$ is a partition of $n-2$, then $k \leq n-2$. Add zeros to the end of $(c_1, c_2, \cdots, c_k)$ if necessary to have $n-2$ entries, then add 1 to each of them to get $d_1 \geq d_2 \geq \cdots \geq d_{n-2}$ back. This shows that the number of valid ordered degree sequences is exactly $p(n-2)$. As trees with different ordered degree sequences are non-isomorphic, the statement follows.

(4) There are n^{n-2} labeled trees, and no isomorphism class can contain

more than $n!$ of them. Therefore, the number of non-isomorphic trees is at least $n^{n-2}/n!$, which is larger than $\frac{e^n}{n^2}$, if n is large enough. Formula (3.1) shows that this is a much larger number than the $p(n-2)$ we got in the previous exercise.

(5) We show that for each such sequence $S = \{s_1, s_2, \cdots, s_{n-2}\}$, there exists a unique tree T whose Prüfer code is s. Take S, and note that the elements of $[n]$ that do not occur in S must precisely be the leaves of the purported tree T. Indeed, if $j \in S$, then there was a leaf that was cut off from j, so j is not a leaf. If j is not a leaf, then there are two possibilities. Either j is cut off from the tree at some point, but then at some point of time before that j had to be made a leaf, and that was made by cutting off one of the neighbors of j, and therefore, by putting j into S. Or, j is one of the two vertices that are never cut off. However, in this case, the degree of j in the final, 2-element tree is one, while its degree in the original tree T was at least two as j was not a leaf. So again, at some point a vertex was cut off from j, putting j into S.

So S tells us what the leaves of the original tree were; denote them by $b_1, b_2, \cdots, b_k$ in increasing order. We know that first we have cut off the leaf with the smallest label (in what follows, the smallest leaf). Therefore, we must start reconstructing the tree by joining b_1 to a_1, as a_1, by definition, is the single neighbor of the smallest leaf. We must continue this way, but carefully. It could be that after cutting off b_1 from T, the smallest leaf of the new tree T' was not b_2 but a_1, that might have become a leaf after b_1 was cut off. How do we know whether a_1 became a leaf after that first step? If and only if it did, it does not occur in S any additional times, as in that case nothing else can be cut off from it. So if the integer a_1 occurs in S after the first position, then in the second step of our reconstruction, we join $\min(a_1, b_2)$ to a_2 by an edge. Otherwise, we simply join b_2 to a_2 by an edge.

In general, in the ith step of recovering T, we have to find the minimal element b_j that has not yet been assigned to any edge (then necessarily $j \leq i$), and the minimal element a_k that has not been assigned to any edge yet, and does not occur in S anywhere after position a_{i-1}. Then we join $\min(a_k, b_j)$ to a_i by an edge. This is the only thing we can do as the ith step of the Prüfer coding algorithm has cut off the smallest leaf of the tree that remained after $i-1$ steps, and this is precisely what we are reversing here.

So we have shown that for any S, the set of leaves of any tree with Prüfer sequence S is unique. Then we showed that there was a unique sequence of edges that could lead to S, so there was a unique tree with Prüfer sequence S. So these two sets are in bijection. As the number of Prüfer sequences is clearly n^{n-2}, we have reproved Cayley's theorem.

(6) Take any acyclic function on $[n]$, and for all $i \in [n]$, draw an arrow from i to $f(i)$. This way we get a graph G whose edges are directed. As f is acyclic, the connected components of G will be tree-like graphs except that each of them will have a one-element cycle (loop) at one of its vertices. Mark these vertices as roots, and delete all the loops, and delete the arrows from the edges. Then G will become a rooted forest on $[n]$.
To see that this is a bijection, take a rooted forest on $[n]$ and define f by $f(i) = i$ if i is a root and $f(i) = j$ if j is the parent of i (the first vertex on the unique path from i to the root of its component).
So there are as many acyclic functions as rooted forests, and the statement follows from Corollary 10.9.

(7) Let us assume that instead of a linear street, the cars arrive at a circular street with $n + 1$ parking spots. The parking procedure is the same, except that if a car leaves spot $n + 1$, it does not give up, but goes to spot 1, and keeps trying. There are still n cars, but their favorite spots can be anything from 1 to $n + 1$.
At the end of this procedure, all cars will always have a spot (as nobody is ever forced to give up), and one spot will be left empty. The crucial observation is that if that one spot is spot $n+1$, then that spot has never been used in the procedure, (indeed, cars do not leave a spot that they have already taken), so the procedure would have worked without spot $n + 1$, that is, in the original linear street. So f is a parking function on $[n]$ if and only if $n + 1$ is the empty spot at the end.
On the other hand, all spots have the same chance to remain empty for symmetry reasons. Indeed, adding 1 to the parking preference of each car shifts the empty spot by one. Therefore, spot $n+1$ will be left empty in exactly $1/(n + 1)$ of all cases, that is in $\frac{(n+1)^n}{n+1} = (n + 1)^{n-1}$ cases.

(8) Same argument as in Exercise 7, except that only the first car can have $n + 1$ parking preferences. The other cars can have only n, as they cannot have the same one as the previous car. Therefore, the number of parking functions without like consecutive elements is

$\frac{(n+1)n^{n-1}}{n+1} = n^{n-1}$.

(9) Let us assume G is not connected. Let $G_1, G_2, \cdots, G_k$ be its connected components. Then in the complement of G, all vertices of G_i are connected to all vertices of G_j (if $i \neq j$) by an edge. So in the complement of G, any vertex is reachable from any vertex, either by a path of length one (if the two vertices are in two different components of G), or by a path of length two (if not, the path can go through any vertex of a different component).
For an example when both G and its complement are connected, take a pentagon and its complement (which is another pentagon). For an example on four vertices, take a tree that consists of a single path and its complement.

(10) If a vertex of G has degree 0, then G is certainly not connected, even if the remaining $n-1$ vertices form a complete subgraph. So G can certainly have $\binom{n-1}{2}$ edges without being connected.
We are going to show that this is the maximum number of edges that will not cause G to be connected. In other words, we prove that if G has $\binom{n-1}{2}+1$ edges, then G must be connected.
We proceed by induction on n. For $n = 2$, the statement is true. Now let us assume that the statement is true for n, and prove it for $n+1$. Take a graph G on $[n+1]$ with $\binom{n}{2}+1$ edges. It is clear that G has no isolated vertices, otherwise it could not have that many edges. If G has a vertex of degree n, then we are done, since there is a path of length at most two from every vertex to every vertex, through G. Otherwise, take any vertex V, and remove it from G, together will all edges adjacent to V. This leaves a graph G' that has n vertices, and at least $\binom{n-1}{2}+1$ edges, since at most $n-1$ edges were removed with V. So the induction hypothesis applies to G', and G' is connected. Therefore, G is connected since V is connected to at least one vertex in G'.

(11) We prove the statement by induction on m, the number of edges. If $m \leq n-1$, then the statement is trivial. Therefore, we can restrict our attention to the case when $m \geq n$. Now assume that we know the statement for m, and prove it for $m+1$. Let H have $m+1$ edges. As $m \geq n$, there is at least one cycle C in H. Let e be an edge of this cycle. Remove e, then the remaining graph H' has m edges, and, by the induction hypothesis, it contains at least $m-n-1$ cycles. However, H contains C as well, therefore, H contains at least $m-n$

cycles.

(12)(a) Let us build the refining sequence from F_k up. First, we need to choose F_{k-1} by adding one edge e to F_k. The starting vertex of e can be any of our n vertices. The ending vertex of e, however, must be the root of one of the components of F_k not containing e. Therefore, we have $n(k-1)$ choices for e, and thus we have $n(k-1)$ choices for F_{k-1}. Repeating this argument, we have $n(k-2)$ choices for F_{k-2} (for each choice of F_{k-1}), and so on. Therefore, repeating this argument $k-1$ times, we get $N^*(F_k) = n^{k-1}(k-1)!$.

(b) If F_1 is a rooted tree containing F_k, then F_1 has $k-1$ more edges than F_k. We can remove these $k-1$ edges in $(k-1)!$ different ways, showing

$$N^*(F_k) = (k-1)!N(F_k), \tag{10.1}$$

and comparing this to the result of part a, we see that $N(F_k) = n^{k-1}$.

(c) Choose $k = n$, then F_k is the empty forest (n isolated vertices), and all rooted trees contain F_k. Then (10.1) shows that $N(F_k) = n^{n-1}$, so this is the number of all rooted trees on $[n]$. The number of unrooted trees on $[n]$ is therefore n^{n-2}.

(13) Keeping the notation of the previous exercise, $N^*(F_n) = n^{n-1}(n-1)!$ as a special case of (10.1). Now let $N^{**}(F_k)$ be the number of those refining sequences $F_1, F_2, \cdots, F_n$ whose kth term is F_k. There are $N^*(F_k)$ choices for the part $F_1, F_2, \cdots, F_k$ of such a sequence, then there are $(n-k)!$ different orders to remove the remaining $n-k$ edges. This shows that

$$N^{**}(F_k) = N^*(F_k)(n-k)! = n^{k-1}(k-1)!(n-k)!,$$

using (10.1). This number does not depend on the choice of F_k. On the other hand, each refining sequence $F_1, F_2, \cdots, F_n$ contains exactly one rooted forest of k components. Therefore, the number of rooted forests on $[n]$ with k components is the number of all refining sequences divided by the number of refining sequences each of these rooted forests with k components occur, that is,

$$\frac{n^{n-1}(n-1)!}{n^{k-1}(k-1)!(n-k)!} = \binom{n}{k} kn^{n-1-k}.$$

(14)(a) True as A is always symmetric.

(b) True that the trace of A is always zero as G has no loops.

(c) False. For example, if G is the only tree on two vertices, then A has determinant -1.

(15)(a) As there are no isolated vertices, each vertex is adjacent to at least one edge. Therefore, there is a walk of length four from i to i as we can walk back and forth twice on any edge adjacent to i.

(b) Let W and W' be two walks from i to j that are of length 5, resp. 6. Then the symmetric difference of W and W' (that is, the edges that are contained in exactly one of W and W') is a set of cycles, that have altogether an odd number of edges. Indeed, they altogether have $11-2e$ edges, where $e = |W \cap W'|$. Therefore, one of these cycles must consist of an odd number of edges.

(c) The claim says that if there is a walk from i to j, then there is also a walk from i to j that is of length at most $n-1$. This is true as we know from Exercise 28 of Chapter 9 that if there is a walk from i to j, then there is a path from i to j, and that has at most $n-1$ edges in it.

(16) The answer depends on the parity of m. Note that there is no walk of even length from A to B or vice versa, and there is no walk of odd length that starts in A and ends in A, or starts in B and ends in B.

(17) We will use the eigenvalue version of the Matrix-Tree theorem. The Laplacian L of this graph has an obvious block structure, the diagonal blocks being $(m+n)I_m$, $(m+n)I_m$, and $2mI_n$, and the other blocks consisting of -1s only. This means that $L-(m+n)I_{2m+n}$ has a set of m rows that are equal, and another set of m rows that are equal. Therefore, its rank is at most $2m+n-(2m-2)$, and so it has at least $2m-2$ eigenvalues equal to zero. Similarly, $L-2mI_n$ has n equal rows, and therefore, $n-1$ eigenvalues equal to zero. Thus L has $2m-2$ eigenvalues equal to $m+n$, and $n-1$ eigenvalues equal to $2m$. One eigenvalue of L is certainly 0, so we are still missing two eigenvalues. Note that the vector $(1,1,\cdots,1,-1,-1,\cdots,-1,0,0,\cdots,0)$, consisting of m entries equal to 1, then m entries equal to -1, then n entries equal to 0, is an eigenvector of $L-(m+n)I_{2m+n}$ with eigenvalue m. Thus $2m+n$ is an eigenvalue of L. Therefore, the last eigenvalue of L must also be $2m+n$, to fulfill the trace condition.
This yields that the number of all spanning trees is

$$\frac{(m+n)^{2m-2}\cdot(2m)^{n-1}\cdot(2m+n)^2}{2m+n}$$

$$=(m+n)^{2m-2}\cdot(2m)^{n-1}\cdot(2m+n).$$

(18)(a) We know that $A_1 = \begin{pmatrix} 0 & 1 \\ 1 & 0 \end{pmatrix}$. Therefore, it follows from elementary linear algebra that the eigenvalues are $\lambda_1 = 1$, and $\lambda_2 = -1$. The corresponding eigenvectors are $\mathbf{v_1} = \begin{pmatrix} 1 \\ 1 \end{pmatrix}$, and $\mathbf{v_2} = \begin{pmatrix} 1 \\ -1 \end{pmatrix}$. Note that multiplying a vector $\mathbf{x}$ by A simply interchanges the coordinates of $\mathbf{x}$.

(b) We know that $A_2 = \begin{pmatrix} 0 & 1 & 0 & 1 \\ 1 & 0 & 1 & 0 \\ 0 & 1 & 0 & 1 \\ 1 & 0 & 1 & 0 \end{pmatrix}$. Note that this is in fact four copies of A_1, arranged in a block. As the square is a regular graph in which each vertex is of degree 2, we have $\lambda_1 = 2$. As A_2 has only two linearly independent rows, the rank of A_2 is 2, and therefore we have $\lambda_2 = 0$, and $\lambda_3 = 0$. As the trace of A_2 is 0, it follows that $\lambda_4 = -2$. Knowing all this, it is a routine linear algebra exercise to find the eigenvectors. They are $\mathbf{v_1} = \begin{pmatrix} 1 \\ 1 \\ 1 \\ 1 \end{pmatrix}$, $\mathbf{v_2} = \begin{pmatrix} 1 \\ 0 \\ -1 \\ 0 \end{pmatrix}$, $\mathbf{v_3} = \begin{pmatrix} 0 \\ 1 \\ 0 \\ -1 \end{pmatrix}$, and $\mathbf{v_4} = \begin{pmatrix} 1 \\ -1 \\ 1 \\ -1 \end{pmatrix}$. Note the similarities between this answer and that of part (a)

(c) and (d) We answer part (d)first. Let Q_n be the n-dimensional cube. The adjacency matrix A_n of Q_n can be obtained by putting two copies of $A_{2^{n-1}}$ in the diagonal, and two identity matrices of size 2^{n-1} in the remaining two corners. In other words, A_n is obtained from A_1 by replacing the diagonal elements by A_{n-1}, and replacing the 1's by copies of $I_{2^{n-1}}$. Thus the characteristic polynomial of A_n arises from that of A_1 by replacing the -1s by $-I_{2^{n-1}}$, and replacing the λ's by $\lambda I - A_{n-1}$.

A little computation then yields that the characteristic polynomial of A_n is

$$\prod_{i=1}^{2}\prod_{j=1}^{2^{n-1}}(\lambda - \lambda_i - \mu_j),$$

where the λ's are the eigenvalues of A_1 and the μ's are those of $A_{2^{n-1}}$. So the eigenvalues of A_n are the *sums* of these.

The eigenvalues of A_1 are $+1$ and -1. To get those of A_n, you can use induction, or you can note that Q_n can be obtained by multiplying Q_1 by itself n times. So the eigenvalues of A_n are all the numbers that can be obtained by choosing one of $+1$ and -1 from each component, then adding them. Therefore, the eigenvalues are $n, n-2, n-4, \cdots, -n$, and the multiplicity of $n-2k$ is $\binom{n}{k}$. So for A_3, we get $3, 1, 1, 1, -1, -1, -1, -3$.

(19) There is one tree on one labeled vertex. There is one tree on two labeled vertices. There are three trees on three labeled vertices. Now any forest in which the connected components have size at most three partitions our set $[n]$ into three subsets in a natural way: for each $i \in [3]$, vertices that are part of a component of size i will be in the same block. Let $E_i(x)$ be the exponential generating function for the number of graphs that are possible on the ith block, that is, that have components of size exactly i only.

It is easy to find the generating function $E_i(x)$ by the exponential formula. Indeed, $E_i(x)$ counts forests in which each component has size exactly i. That is, we first partition our set into blocks, then put one of $f_i(j)$ different structures on each block, where $f_i(j) = 0$ if $j \neq i$. If $i = j$, then $f_i(j)$ equals the number of trees on i vertices, that is, 1,1, and 3, respectively. Then it follows by the exponential formula that $E_i(x) = \exp E_{f_i}(x)$. Finally, by the product formula,

$$F(x) = \prod_{i=1}^{3} E_i(x) = \exp\left(x + \frac{x^2}{2} + \frac{x^3}{2}\right).$$

(20) If we cut off the root r of a rooted tree, we get two different structures, one of which is the root itself, and the other is a rooted forest, in which the vertices that were adjacent to r became the roots of their respective components.

Therefore, to get a rooted tree on n, we first split n into two parts. One part will have only one vertex r, and that will be the root of the tree; the other part will have $n-1$ vertices, and will be the vertex set of a rooted forest. These two parts completely determine a rooted tree as the root of each tree in the forest is to be connected to r. The exponential generating function of the first part is obviously x, and that of the second part is $e^{G(x)}$ by the exponential formula. Therefore, the product formula implies $G(x) = xe^{G(x)}$.

Chapter 11

Finding A Good Match. Coloring and Matching

11.1 Introduction

A cellular phone company provides service on three different frequencies. They expand into a new area, and they plan to build ten communication towers there, at locations already selected. Each tower will broadcast signals on one frequency only. The company has to make sure that the distance between any two towers broadcasting on the same frequency is more than 50 miles. Let us decide (knowing the exact locations of the towers) if this is possible, in other words, whether a *proper assignment of frequencies* exist.

How can we translate this problem into the language of combinatorics? The reader probably conjectures that we will somehow find a graph-theoretical representation for this problem, otherwise we would not have brought it up in this chapter. The natural candidates for the vertices of the graph G representing a given set of towers are the towers themselves. And when should two vertices be connected by an edge? There are only two kinds of pairs of towers for the purposes of this problem: those whose distance from each other is at most 50 miles (such pairs of towers cannot broadcast on the same frequency), and those whose distance from each other is more than 50 miles (such pairs of towers can do so). Therefore, it is plausible to define the edge set of G by requiring that there be an edge between A and B if and only if the distance between the corresponding two towers is at most 50 miles.

Fine, you could say, we figured out how to express all relevant information about the distances between our towers by a graph G. However, does this help us decide whether the frequencies can be assigned to the towers in a proper way? After all, G does not contain any information about different frequencies, not even their number.

This is a valid concern. So that we could incorporate more information into G, we will *color* its vertices. If frequency 1 gets assigned to a tower, we will color the corresponding vertex red, if frequency 2 gets assigned to a tower, we will color the corresponding vertex blue, if frequency 3 gets assigned to a tower, we will color the corresponding vertex green, and finally if frequency 4 gets assigned to a tower, we will color the corresponding vertex yellow.

Now the following Proposition is a direct consequence of the definition of our graph G.

Proposition 11.1. *Let C be any set of ten towers, and let G be the graph defined by C as described above. Then one can assign the four frequencies to the ten towers of C if and only if it is possible to color the vertices of G with four colors so that there are no two monochromatic vertices that are adjacent.*

The following definition provides a simple way to describe how difficult it is to color the vertices of a given graph without creating adjacent monochromatic pairs.

Definition 11.2. The *chromatic number* of a graph H, denoted by $\chi(H)$, is the smallest integer k for which the vertices of H can be colored by k colors so that adjacent vertices are colored by different colors.

If the vertices of a graph can be colored by k colors so that there are no adjacent monochromatic vertices, then that graph is called k-colorable.

Example 11.3. The chromatic number of the pentagon is three. Indeed, two colors do not suffice, while three colors do as shown in Figure 11.1.

All graphs in the remainder of this chapter are assumed to be connected. This will not result in any loss of generality as colorings of different connected components of an unconnected graph are certainly independent from each other. Similarly, we can assume that our graphs are simple, as adding one or more new edges between the same pair of adjacent vertices does not impose any new restriction on those two vertices (they could not be the same color anyway).

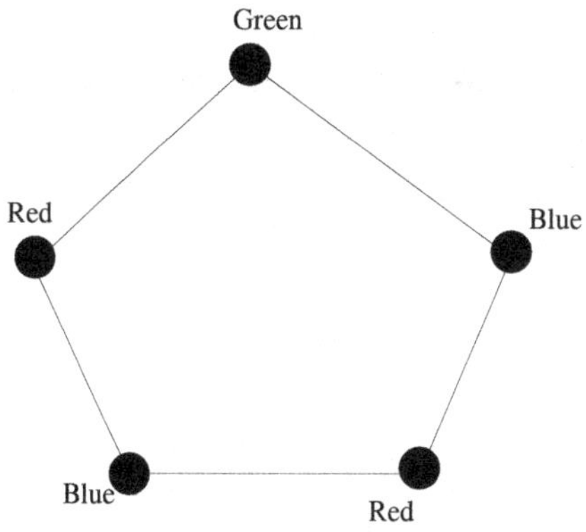

Fig. 11.1 The pentagon is 3-colorable.

11.2 Bipartite Graphs

The most important special case of k-colorable graphs is when $k = 2$. This case is so omnipresent in combinatorics that it has its own name.

Definition 11.4. A 2-colorable graph is called *bipartite*. Equivalently, G is bipartite if the vertex set of G can be split into the disjoint sets A and B (the color classes) so that each edge of G is adjacent to one vertex of A and one vertex of B.

A generic example of a bipartite graph is shown in Figure 11.2. Note that there are no edges within either color class.

For example, all trees are bipartite as one can start at any given vertex, color it red, then color all its neighbors blue, then color all the second neighbors red, and so on. This coloring algorithm works as there is no cycle in the tree, so we will never get back to a vertex that has already been colored. Another example of a bipartite graph is, say, a square, hexagon, or octogon, where we can color vertices alternatingly.

An easy example of a graph that is *not* bipartite is a triangle. Indeed, if a triangle has a red vertex A, then one of the two neighbors of A can be colored blue, but the third vertex of the triangle cannot be properly colored. It is also clear that no graph that contains a triangle can be bipartite (as not even that triangle could be 2-colored, let alone the whole graph).

Is it true that if a graph does not contain a triangle, then it is bipartite? As we can see in Figure 11.1, this is not true as the pentagon provides a counterexample. There is nothing magic about the pentagon here, the

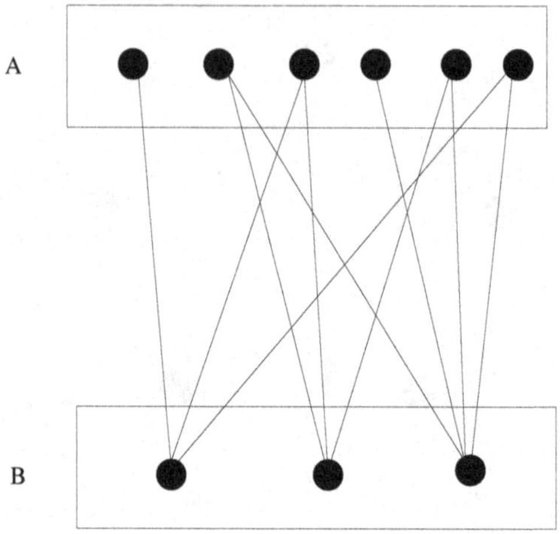

Fig. 11.2 A generic bipartite graph.

reader can easily see that no cycle of odd length can be 2-colored, and therefore no graph containing an odd cycle can be 2-colored.

The following Theorem shows that with this we have completely characterized bipartite graphs.

Theorem 11.5. *A graph G is bipartite if and only if it does not contain a cycle of an odd length.*

Proof. As we have mentioned, the "only if" part is easy. Suppose G contains the odd cycle $A_1A_2\cdots A_{2m+1}$. Let us assume without loss of generality that A_1 is red. Then A_2 must be blue, therefore A_3 must be red, A_4 must be blue, and so on, and at the end, A_{2m+1} must be red, too. This is not allowed, however, as A_1A_{2m+1} is an edge.

To prove the "if" part, let G be a graph with no odd cycles. Let V be a vertex of G, and color V red. Define the color of any other vertex W as follows. If the shortest path from V to W has even length, then let W be red, and if the shortest path from V to W has odd length, then let W be blue. We show that this is a good coloring, that is, there are no two adjacent vertices that are the same color.

Let us assume the contrary, by first assuming that P and Q are two red vertices that are joined by an edge. Let the shortest path from V to P be p, and let the shortest path from V to Q be q. Then p and q both have an even number of edges, so walking from V through p to P, then through

PQ, then back from Q through q to V, we get a closed walk C with an odd number of edges. Taking away edges that were used both by p and q, this walk C splits into the union of edge-disjoint cycles. As the total number of edges in these cycles is still odd, there has to be at least one cycle with an odd number of edges, which is a contradiction.

If we assumed instead that both P and Q were blue, the same proof would work as the sum of two odd numbers is still even, so C would still have an odd number of edges. □

How many edges can a simple bipartite graph G on n vertices have? The alert reader should have an intuition at this point that the answer to this question will be some kind of an upper bound. Indeed, it is not difficult to create bipartite graphs with few edges. For example, forests have no cycles at all, so they cannot have odd cycles either. Thus all forests are bipartite. We could think, however, that if we keep adding new edges, without adding new vertices, then sooner or later an odd cycle will be formed. In fact, the complete graph K_n certainly has an odd cycle if $n \geq 3$.

So where is the threshold? How many edges can we have in G without having an odd cycle? The forests only allow us to go to $n-1$ edges. Will the final answer be some linear function of n, or maybe around n^{α}, where $1 \leq \alpha < 2$, or will it be just a constant factor below $\binom{n}{2}$, the number of edges in the complete graph? The following theorem shows that the answer to this question is closer to the maximum.

Theorem 11.6. *Let G be a simple bipartite graph on n vertices. Then G has at most $n^2/4$ edges if n is even, and at most $(n^2-1)/4$ edges, if n is odd.*

Proof. Choose G so that no other simple bipartite graph on n vertices has more edges than G. Denote by a and b the sizes of the two color classes of G. It is clear that each vertex of one color class is connected to each vertex of the other color class in G. Indeed, if there was a missing edge between the two color classes, we could add it to G, contradicting to our assumption. So G has $ab = a(n-a)$ edges, and the proof follows from elementary calculus. (One simply has to find the integer $a \in [1, n]$ for which the number $f(a) = a(n-a)$ is maximal.) □

The class of bipartite graphs we used in this proof, that is, bipartite graphs in which each vertex of one color class is connected to each vertex of the other color class, is an important one, therefore, such graphs have a name. They will be called *complete bipartite graphs.* These graphs played a role in several exercises of earlier chapters. If a complete bipartite graph has color classes of size a and b, then we will denote that graph by $K_{a,b}$.

So bipartite graphs can have a lot more edges than trees. We will see that accordingly, they have a much richer structure, too. To start, let us take a closer look at the consequences of Theorem 11.6. Let H be a simple graph on $2m$ vertices. If H has only m^2 edges, then H can be bipartite; indeed, H can be $K_{m,m}$. If H has more than m^2 edges, then Theorem 11.6 implies that H is not bipartite, in other words, H has an odd cycle. The following Lemma shows that more is true.

Lemma 11.7. *Let H be a simple graph on $2m$ vertices ($m \geq 2$) and at least $m^2 + 1$ edges. Then H contains a triangle.*

Proof. We prove our statement by induction on m. If $m = 2$, then H is a subgraph of K_4 with at least five edges. Theorem 11.6 shows that H is not bipartite, so it must have an odd cycle. This odd cycle must be a triangle as H has only four vertices.

Now assume we know that the statement is true for all integers that are smaller than m, and are at least 2. Let H be as in the statement of the Theorem, and let F and G be two adjacent vertices in H. If the sum of the degrees of F and G is more than $2m$, then they have a common neighbor T, and so FGT is a triangle. If, on the other hand, the sum of the degrees of F and G is at most $2m$, then deleting F, G, and all the edges adjacent to them from H will decrease the number of edges in our graph by at most $2m - 1$. (Note that the edge FG is contained twice in the sum of the two degrees.) Therefore, after the deletion of these vertices and edges, we are left with a graph of $2m - 2$ vertices, and at least $m^2 + 1 - (2m - 1) = m^2 - 2m + 2 = (m-1)^2 + 1$ edges. Such a graph contains a triangle by the induction hypothesis, so our claim is proved. □

Thus we know that graph on $2m$ vertices with just one more edge than what is possible in bipartite graphs does not simply have an odd cycle, but also has a triangle, the shortest odd cycle possible. The real surprise, however, comes now.

Theorem 11.8. *Let H be a simple graph on $2m$ vertices ($m \geq 2$) and at least $m^2 + 1$ edges. Then H contains at least m triangles.*

So if H has m^2 edges, it may not have any odd cycles at all, but with only one more edge, H must have at least m triangles! Note that there is nothing similar that would be true for trees on m vertices. A connected graph on m vertices and $m-1$ edges is a tree. Adding an extra edge we get a cycle (in fact, exactly one cycle), but that cycle can be of many different lengths depending on the tree.

Proof. Clearly, we can assume that H has *exactly* m^2+1 edges as additional edges will not destroy any triangles.

We prove our statement by induction on m. If $m = 2$, then our graph has four vertices and five edges, so it is K_4 with an edge missing, and therefore does contain two triangles.

Now assume the statement is true for all positive integers smaller than m, but at least 2. Let H be as in the statement of the theorem. Lemma 11.7 shows that H contains at least one triangle ABC. We have to find $m-1$ other triangles.

We will distinguish three cases based on the number of edges connecting outside vertices to the vertices of the triangle ABC. We claim that if the number of all these edges is $2m-3+x$ for some $x \geq 1$, then there are x triangles formed by two vertices of the triangle ABC, and a third vertex that comes from outside that triangle. Indeed, if such an *outside vertex* is connected to two vertices of ABC, then it forms a triangle with them. As there are $2m-3$ outside vertices, our claim follows by pigeon-hole principle.

The outline of the proof will be this. If there are many edges between ABC and the outside vertices, then there are many triangles spanned by two vertices of ABC and an outside vertex. If, on the other hand, there are only a few such edges, then there have to be so many edges among outside vertices that we can apply the induction hypothesis for their subgraph (and an extra vertex).

(1) If $x \geq m-1$, then we are done as we found our missing $m-1$ triangles.
(2) If $1 \leq x < m-1$, then the total number of edges between ABC and the outside vertices is at most $(2m-3)+(m-2) = 3m-5$. As ABC itself contains three edges, it follows that there are at least $m^2+1-(3m-5)-3 = m^2-3m+3 = (m-1)(m-2)+1$ edges within the subgraph R spanned by all $2m-3$ outside vertices. If we omit the vertex of R which has the smallest degree in R, it follows by the Pigeon-hole Principle that we get a graph R' on $2m-4$ vertices that still has more than

$$(m-1)(m-2) \cdot \frac{2m-4}{2m-3} = (m-2)^2 \cdot \frac{2m-2}{2m-3} > (m-2)^2$$

edges. So R' has strictly more than $(m-2)^2$ edges, that is, it has at least $(m-2)^2+1$ of them. Therefore, by the induction hypothesis, there are at least $m - 2$ triangles within R'. As we said in the previous paragraph, there are x triangles spanned by two vertices of ABC and an outside vertex. In our case, $x \geq 1$, so we have again found the $m - 1$ needed triangles.

(3) Finally, consider the case when the number of edges connecting outside vertices to ABC is not more than $2m - 3$. Note that we can assume that there is at least one such edge, otherwise R has $m^2 - 2$ edges, so adding any vertex of ABC to R creates a graph on $2m - 2$ vertices and $m^2 - 2 \geq (m - 1)^2 + 1$ vertices, and the proof follows by the induction hypothesis. That said, the number of edges within R is at least $m^2 + 1 - (2m - 3) - 3 = (m - 1)^2$. Adding a vertex of ABC that is adjacent to at least one outside vertex to R creates a graph with $2m - 2$ vertices and at least $(m - 1)^2 + 1$ edges, and again, the induction hypothesis shows that such a graph must contain at least $m - 1$ triangles.

$\square$

In other words, if we start with the empty graph on $2m$ vertices, and keep adding edges to it at random, then as soon as we can be sure (without looking) that our graph has one triangle, we can also be sure that it has m triangles!

11.3 Matchings in Bipartite Graphs

Bipartite graphs abound in real life. Consider for example m job openings and n applicants for these jobs. Define the graph G on $m + n$ vertices as follows. The first m vertices correspond to the jobs, and the second n vertices correspond to the applicants, and two vertices are connected by an edge if and only if the corresponding applicant is qualified for the corresponding job. Then G is certainly bipartite as edges are only possible *between* the sets of the first m and last n vertices, not *within* these sets. Figure 11.3 shows an example for such a graph.

We have to fill each job opening by hiring exactly one qualified person for that opening. How can we translate this problem to the language of graph theory? Just as in Section 11.1, we will refine our existing model so that it can encode more information. If we fill a given opening A by hiring applicant a, then we will represent this by changing the edge aA of G to a bold edge. Then, if we fill another opening B by hiring the qualified candidate B, then we will represent this by changing the edge bB to a bold

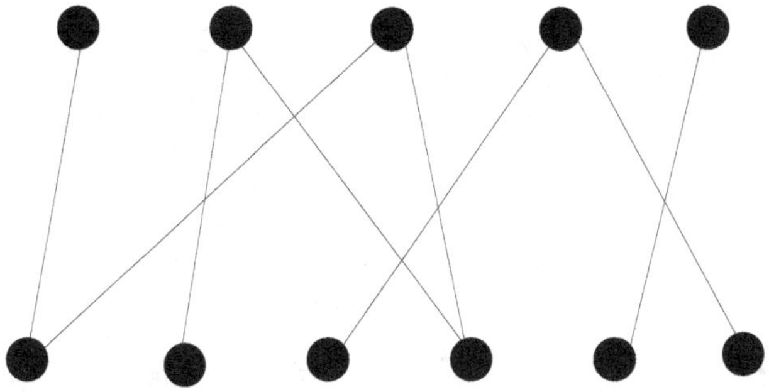

Fig. 11.3 Try to fill all jobs with qualified applicants.

edge, and so on. As the hiring procedure goes ahead, we will have more and more bold edges. The crucial property of the set of bold edges is that at any point of time throughout the hiring process *it will always consist of vertex-disjoint edges.* Indeed, no job opening can be filled by more than one person, and no person can accept more than one job offer.

If the hiring process is complete, and we filled all m positions, there will be m bold edges. If we filled less than m positions, but cannot find any qualified candidates for any of the remaining openings, then that means that we cannot change any non-bold edge to a bold edge so that all bold edges are still pairwise vertex disjoint. Therefore, the following Proposition is immediate.

Proposition 11.9. *Let S be an instance of the hiring problem, that is, a set of m job openings and n applicants, and all the relevant information about the qualifications of each applicant. We can simultaneously fill all m job openings in S if and only if, in the graph G defined above, we can find m vertex-disjoint edges.*

In our running example, the graph shown in Figure 11.3, we can fill all openings, as shown in Figure 11.4.

We see that for a set of edges in a graph, it can be an important question whether they are pairwise vertex-disjoint or not. This warrants the following definition.

Definition 11.10. Let G be any graph, and let S be a set of edges in G so that no two edges in G have a vertex in common. Then we say that S

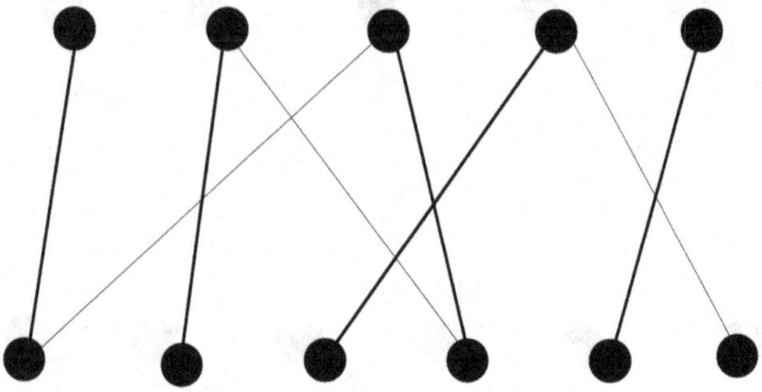

Fig. 11.4 A maximal set of vertex-disjoint edges.

is a *matching* in G. If each vertex in G is covered by an edge in S, then we call S a *perfect matching*.

A matching is also called an *independent set of edges* in certain contexts. Note that the above definition does not require that G be a bipartite graph. For the time being, however, we will restrict our discussion to matchings in bipartite graphs, which are very useful in the practice.

Definition 11.11. Let $G = (X, Y)$ be a bipartite graph. If S is a matching in G that covers all vertices of X, then we say that S is a perfect matching of X into Y.

If we are not particularly interested in the matching S, just the fact that there is a perfect matching of X into Y, then we will say that X *has a matching into* Y or X *can be matched into* Y.

Let $G = (X, Y)$ be a bipartite graph. At least two questions are in order. Does X have a perfect matching into Y? (In the language of the previous discussion this is the question whether all job openings can be filled at the same time.) How do we find the largest matching of G?

First let us try to decide if X has a perfect matching into Y. Let us look for necessary conditions first; for properties G certainly must have if it is to have a perfect matching. For one thing, $|X| \leq |Y|$ is a necessary condition as all edges in our purported matching S would have to have a vertex in X, and one in Y, setting up an injection from X to Y. For another trivial observation, if there are two vertices a and b of X that are both of degree

1, and they are both connected to the same vertex $y \in Y$, then we are in trouble. Indeed, X cannot have a perfect matching into Y as we cannot even match the two vertices a and b into Y. One of them can be matched into y, by an edge e, but that would leave no possibility to find any edge starting at the other one that is vertex-disjoint from e.

It is not hard to generalize these easy necessary conditions. If $T \subseteq X$ is a subset of vertices in X, then let $N(T)$ denote the set of all neighbors of the vertices in T. In other words, $y \in Y$ is an element of $N(T)$ if and only if there is a vertex $x \in T$ so that xy is an edge. The neighbor set $N(T)$ is relevant to matchings because if we just want to match T into Y, then we can certainly restrict our attention to the bipartite graph $(T, N(T))$. Indeed, $N(T)$ contains all possible Y-endpoints of the edges of a matching of T into Y.

If there is a danger of confusion as to in which graph we count the neighbors of a vertex set, we use the notation $N_G(T)$ to identify the graph.

Proposition 11.12. *Let $G = (X, Y)$ be a bipartite graph. Then X has a perfect matching into Y only if for all $T \subseteq X$, the inequality $|T| \leq |N(T)|$ holds.*

Proof. Let us assume that there is a $T \subseteq X$ so that $|T| > |N(T)|$. Then T certainly cannot be matched to $N(T)$ as T has more vertices than $N(T)$. However, this means that T cannot be matched into Y either as any such matching would match T into $N(T)$. Finally, this means that X cannot be matched into Y as any such matching would obviously contain a matching of T into Y. □

This Proposition was, after all, not too surprising. It basically said that if, among our job openings, there are k for which we only have $k-1$ qualified applicants, then we cannot fill all positions. This is pretty clear. What is much more interesting is that *the converse* of Proposition 11.12 is also true. This remarkable result is known as Philip Hall's theorem.

Theorem 11.13 (Philip Hall's theorem). *Let $G = (X, Y)$ be a bipartite graph. Then X has a perfect matching into Y* if and only if *for all $T \subseteq X$, the inequality $|T| \leq |N(T)|$ holds.*

Proof. As we provided a proof of the "only if" part when we proved Proposition 11.12, we only have to prove the "if" part. The proof we present is due to Halmos and Vaughn, dated 1950.

We prove the statement by induction on $|X|$, the initial case being trivial. Now assume we know the statement for all nonnegative integers less than $|X|$, and prove it for $|X|$. Let us assume that for all $T \subseteq X$, the inequality $|T| \leq |N_G(T)|$ holds. We distinguish two cases.

(1) First let us assume that for each subset $T \subset X$, even the *strict* inequality $|T| < |N_G(T)|$ holds. Let x and y be adjacent vertices, with $x \in X$. Let $G' = G - x - y$, and let A be any nonempty subset of $X - x$. Our assumption then shows that $|A| < |N_G(A)|$, therefore $|N_{G'}(A)| \geq |N_G(A)| - 1 \geq |A|$. Consequently, the induction hypothesis implies that $X - x$ can be matched into $Y - y$ in G'. Adding the edge xy to this matching, we get a perfect matching of X into Y.

(2) Now assume there is a subset $B \subset X$ so that $|B| = |N_G(B)|$ holds. We split G into two smaller subgraphs G_1 and G_2, and then show that each of these subgraphs satisfies the induction hypothesis separately. Let G_1 be the subgraph induced by $B \cup N(B)$, and let G_2 be the graph obtained from G by deleting all vertices that belong to $B \cup N(B)$.

To see that G_1 satisfies the induction hypothesis, choose any subset $T \subseteq B$. Then $N_G(T) \subseteq N_G(B)$, and therefore, $N_{G_1}(T) = N_G(T)$, (all neighbors of T are within G_1), and therefore, $|N_{G_1}(T)| = |N_G(T)| \geq |T|$.

To see that G_2 satisfies the induction hypothesis, choose any subset $U \subseteq X - B$. Then $N_G(U \cup B) = N_{G_2}(U) \cup N_G(B)$, and because this is a union of disjoint sets, $|N_{G_2}(U)| = |N_G(U \cup B)| - |N_G(B)| \geq |U \cup B| - |B| = |U|$.

If we apply the induction hypothesis to both G_1 and G_2, we see that B can be matched into (and therefore, onto), $N_G(B)$, and $X - B$ can be matched into $Y - N_G(B)$. Therefore, X can be matched into Y as claimed.

$\square$

This theorem has many interesting applications to problems that look unrelated at first. Exercise 9 is one of them.

While Theorem 11.13 is undoubtedly useful, it does not answer all our questions. It does not tell us *how* to find a perfect matching if there is one, or how to find *a maximum matching* in any given graph.

The last sentence brings up an important issue in our terminology. Henceforth, the words *maximal* and *maximum* will have *different meanings*. In a graph G, a matching M is called **maximal** if we cannot extend M by adding a new edge to it. A matching N is called **maximum** if no matchings of G contain more edges than N.

At this point, the reader should test her understanding of this subtle difference by showing that a maximum matching is always maximal, but a maximal matching is not always maximum. After doing that, the reader can find one example for the latter in Figure 11.5.

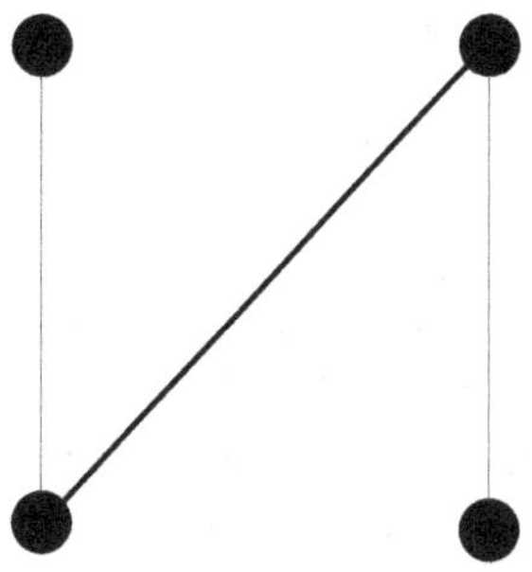

Fig. 11.5 A maximal, but not maximum, matching.

Let G be a bipartite graph, and let M be a matching in G. A path $P = v_1 v_2 \cdots v_r$ is called an M-alternating path if $v_i v_{i+1}$ is in M if and only if $v_{i+1} v_{i+2}$ is not in M. In other words, every other edge of P belongs to M. If, in addition, P starts and ends at vertices that are not adjacent to any edge of M, then M is clearly not a maximum matching. Indeed, we get a larger matching if we discard the edges in $P \cap M$ and replace them by the edges $P - M$. Therefore, if this happens, we call P an M-augmenting path. See Figure 11.6 for an example. The bold lines are the edges of M, and the dotted lines are the edges of $P - M$.

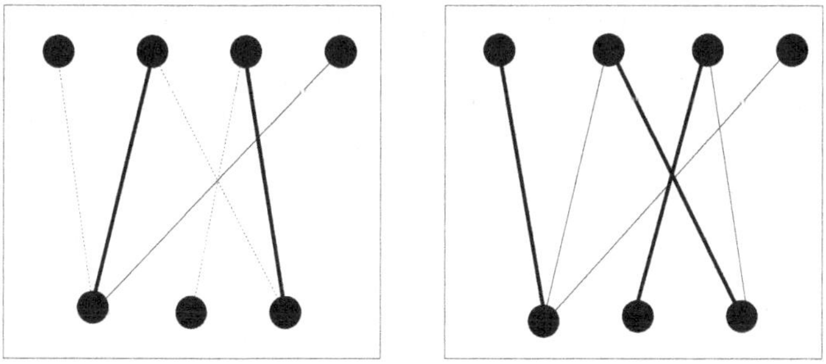

Fig. 11.6 Extending a matching by an augmenting path.

Note that we have not used the fact that G is bipartite, so all we said about alternating and augmenting paths holds for all simple graphs.

The non-existence of augmenting paths actually characterizes maximum matchings.

Theorem 11.14. *Let G be any simple graph, and let M be a matching in G. Then M is maximum if and only if G has no M-augmenting paths.*

Proof. We have already shown the "only if" part in our discussion preceding Figure 11.6.

To prove the "if" part, assume there is no M-augmenting path in G, and let $M' \neq M$ be any maximum matching in G. Consider $M \oplus M'$, the set of edges that are part of exactly one of M and M'. As M and M' are both matchings, the connected components of $M \oplus M'$ can only be even cycles or alternating paths. However, M' is maximum, and there is no M-augmenting path, therefore all these alternating paths are of even length. This implies $|M| = |M'|$, and our claim is proved. □

11.4 More Than Two Colors

We have seen in Theorem 11.6 that a bipartite graph cannot have too many edges. We have also seen that if we want the bipartite graph on n vertices that has the largest number of edges, we have to take the bipartite graph in which the numbers of vertices in the two color classes are equal (if n is even), or differ by 1 (if n is odd).

Let us generalize this question into k-colorable graphs instead of bipartite (2-colorable) graphs. Is it still true that the best strategy to maximize the number of edges is to split the vertices among the color classes as equally as possible? The following famous theorem of Pál Turán shows that this is indeed the case.

To prepare the statement and proof of Turán's theorem, let $n = tk + r$, with $0 \leq r \leq k-1$, and divide the n vertices into k subsets, r of them of size $t+1$, and the rest of size t. In other words, we divided the n vertices into k blocks, whose sizes are "as equal as possible". Let two vertices be joined by an edge if and only if they are in different subsets. The graph H obtained is a *complete k-partite* graph. The number of its edges (see Exercise 1) is

$$T(n,k) = \frac{k-1}{2k} \cdot n^2 - \frac{r(k-r)}{2k}. \tag{11.1}$$

It goes without saying that H is k-colorable as we can assign color 1 to the vertices of the first subset, color 2 to the vertices of the second subset, and so on. Now we are going to show that no k-colorable simple graph on n vertices can have more edges than H.

Theorem 11.15. *Let G be a simple graph on n vertices that contains more than $T(n,k)$ edges. Then G contains a K_{k+1} subgraph. In particular, G is not k-colorable.*

Proof. Let G have n vertices, let G contain no K_{k+1}, and let it have the maximum number of edges possible with these conditions. We will prove that G can contain at most $T(n,k)$ edges.

We will proceed by induction on t, where t has been defined in the paragraph preceding the Theorem. If $t = 0$, then the statement is obvious. Now assume we know that the statement is true for $t - 1$. Our conditions imply that adding any edge to G would create a K_{k+1} subgraph. Therefore, G must contain a K_k subgraph, say S.

Now we will count how many edges G can have. The edges of G can be

- within S, or
- between a vertex of S and a vertex of $G - S$, or
- within $G - S$.

There are $\binom{k}{2}$ edges within S. Each of the vertices of $G - S$ can be connected to at most $k - 1$ of the vertices of S. Finally, $G - S$ has $n - k = (t-1)k + r$ vertices, so the induction hypothesis implies that there are at most $T(n-k,k)$ edges within $G - S$. Therefore, the number of edges in G is at most

$$\binom{k}{2} + (n-k)(k-1) + T(n-k,k) = T(n,k). \tag{11.2}$$

This shows that if G has more than $T(n,k)$ edges, it must contain a K_{k+1}, and therefore, it cannot be k-colorable. □

We admit that we swept two technicalities under the rug here. One was the computation of the number $T(n,k)$ of edges in our complete k-partite graph H. The other is the proof of equality (11.2). See Exercises 1 and 2 for these details.

Theorem 11.15 proved in a rather strong way that certain graphs are not k-colorable. Indeed, it proved that graphs containing too many edges will always contain a K_{k+1}-subgraph, and *therefore* are not k-colorable.

We certainly know that a graph does not have to contain K_{k+1} in order to have chromatic number at least $k+1$. Indeed, if G is an odd cycle of length more than three, then it does not contain K_3 and still has chromatic number three. What is interesting is that in some sense, odd cycles and complete graphs are alone in forcing high chromatic numbers. This is the content of the following theorem of Brooks that we state without proof.

Theorem 11.16. *Let G be a graph which is not an odd cycle, and not a complete graph, and let $d \geq 3$ be a positive integer, so that each vertex of G has degree at most d. Then $\chi(G) \leq d$.*

On the other hand, note that for all n, there exists a graph that contains no triangles, but has chromatic number n. This is the content of Exercise 6. In other words, if a graph has a high chromatic number, then it has a vertex with a high degree, but it may or may not have a large complete subgraph.

11.5 Matchings in Graphs That Are Not Bipartite

There are many real-life situations when finding a matching (a set of vertex-disjoint edges) in a non-bipartite graph is needed. Let us assume for example that a big company wants to form pairs of employees for certain assignments, and wants to do it in a way that the two employees within each pair know each other. Or take a set of football teams, and find pairings for this week-end so that teams that have played each other within the last two years do not play each other again.

In these examples, we have a graph that is not necessarily bipartite, but we still want to find a set of vertex-disjoint edges in it. Fortunately, there is a sufficient and necessary condition for a perfect matching to exist. If G is a graph, and S is a subset of the vertex-set of G, then let $G-S$ be the graph obtained from G by deleting the elements of S, and all the edges that are adjacent to them. Let $c_o(G-S)$ be the number of components of $G-S$ that have an odd number of vertices.

Theorem 11.17 (Tutte's theorem). *A graph G has a perfect matching if and only if, for all subsets S of the vertex set of G, the inequality $c_o(G-S) \leq |S|$ holds.*

There are several proofs of this theorem. We will present one that is due to Gábor Hetyei Sr. (1972) and László Lovász (1975). We will need some tools for the proof of the "if" part. The "only if" part, however, is trivial. Indeed, if there is an S violating the conditions, then no perfect matching could exist. In order to see this, let us assume that M is a perfect matching, then each odd component of $G - S$ must contain at least one vertex M matches with a vertex of S. This would imply $c_o(G - S) \leq |S|$, contradicting our assumption.

Let us call the graph G *saturated non-factorizable* if G has no perfect matching, but added any new edge, the resulting graph does. To prove the "if" part of Tutte's theorem, we need the following, somewhat technical, Lemma.

Lemma 11.18. *If the graph G is saturated non-factorizable, and if S is the set of vertices of G that are joined to every other point of G, (that is, the set of vertices of degree $|G| - 1$), then the components of $G - S$ are complete graphs.*

Proof. Let ab and bc be two adjacent edges in $G - S$. To prove our statement, it suffices to show that a and c are adjacent vertices. Let us assume the contrary, that is, that a and c are not adjacent. Then there must be a vertex d in G so that bd is not an edge. Indeed, otherwise b would be in S.

As G is saturated non-factorizable, $G \cup ac$ has a perfect matching F_1. Since G itself does not have a perfect matching, this implies that $ac \in F_1$. Similarly, $G \cup bd$ has a perfect matching F_2. As G itself does not have a perfect matching, F_2 contains bd. Just as we did in proofs concerning matchings in bipartite graphs, let us take the symmetric difference of F_1 and F_2. This consists of alternating, (and therefore, even) cycles. Let C_1 be the cycle containing ac, and let C_2 be the cycle containing bd. We distinguish between two cases.

(a) First let us assume that $C_1 \neq C_2$. In this case, form the symmetric difference $F_3 = F_1 \bigoplus C_1$. Then we claim that F_3 is a perfect matching of G. Indeed, $ac \in (F_1 \cap C_1)$, so $ac \notin F_1 \bigoplus C_1$. On the other hand, F_3 has the same number of edges of F_1, and is a matching, so it is a perfect matching of G. This is a contradiction as G is saturated non-factorizable, and as such, has no perfect matching.

(b) Now let us assume that $C_1 = C_2$. Traverse C_1 starting at b through d, until one of a and c, say a, is reached. Let the path from b to a just traversed be P. Recall that $ab \in G$, and note that therefore, $P \cup ab$ is an alternating path (in fact, a cycle), for F_2. Form the symmetric difference $F_4 = F_2 \bigoplus (P \cup ab)$. Then we claim that F_4 is a perfect matching for G. Indeed, F_4 contains the same number of edges as F_2, but does not contain bd as $bd \in (F_2 \cap (P \cup ab))$. This shows that F_4 is a perfect matching of G, which is a contradiction.

Therefore, if ab and bc are edges in $G-S$, then so is ac, and the components of $G-S$ are complete graphs. □

The following theorem will characterize saturated non-factorizable graphs.

Theorem 11.19. *A graph G is saturated non-factorizable if and only if it has the following structure.*

(a) Either G has an odd number of vertices, and is complete, or
(b) G has an even number of vertices and consists of vertex-disjoint complete subgraphs $S_0, G_1, G_2, \cdots, G_k$ so that $k = |S_0| + 2$, each G_i has an odd number of vertices, and each vertex of each G_i is connected to each vertex of S_0.

Proof. If G has an odd number of vertices, then it does not have a perfect matching. Therefore, only the complete graph satisfies the requirement of saturated non-factorizability as that is the only graph to which no edge can be added.

If G has an even number of vertices, then let S be defined as in Lemma 11.18. Let us set $S_0 = S$. Let $G_1, G_2, \cdots, G_k$ be the connected components of $G-S$. Lemma 11.18 shows that all the G_i are complete graphs, and so is S, and by definition, each vertex of S is connected to each vertex of each G_i.

Recall that G has no perfect matching. Therefore, the number of G_i with odd components must be more than $|S|$. In fact, as G has an even number of vertices, the number of G_i with odd components must be at least $|S+2|$. On the other hand, G cannot have more than $|S+2|$ odd components, otherwise we could add a new edge connecting two of them. That would lead to a contradiction, because the resulting graph G_1 would satisfy $c_o(G_1 - S) > |S|$, and would therefore have no perfect matching. Therefore, G has exactly $|S+2|$ odd components. Finally, G has no even components, otherwise we could again add an edge connecting that component to another component without creating a perfect matching. □

Now we are in a position to prove Tutte's theorem.

Proof. (of Tutte's theorem) All we have left to do is to prove the "if" part. Assume that G satisfies the criteria but has no perfect matching. Add new edges to G until a graph with perfect matching is obtained. Let G' be the saturated non-factorizable graph that was created by this procedure.

If G has an odd number of vertices, then choosing $S = \emptyset$ we see that G does not satisfy the criteria. Thus we can assume that G has an even number of vertices. Let S' be the set of vertices of G' that are adjacent to any other vertex of G'. Theorem 11.19 then describes the structure of $G' - S'$. Let H' be the set of vertices of this graph. Then H is not empty as G was not complete (it did not have a perfect matching). Moreover,

$$H' = G_1 \cup G_2 \cup \cdots \cup G_k,$$

where the G_i are vertex-disjoint complete subgraphs, and $k = |S'| + 2$.

Our last sentence shows that $G' - S'$ has more than $|S'|$ (in fact, $|S'|+2$) components. Remove all the edges of $G' - G$ that we inserted to our original graph G. Then some of our components may split, but each of these odd components will give rise to at least one odd component of $G - S'$. This shows that $c_o(G - S') > |S'|$, so G violates the condition. This contradiction completes the proof. □

Notes

If you want to know more about matchings, you should see *Matching Theory* by László Lovász and Michael D. Plummer [28] for an extensive text.

One way to generalize our results concerning k-colorability is to ask the following question. Let G be a given graph on n vertices. At most how many edges can a graph H on n vertices have so that it does not contain a subgraph that is isomorphic to G? This leads to the area of *Extremal Graph Theory*, and you can read more about that field in the identically titled book of Béla Bollobás [6]. For an introductory treatment to Extremal Combinatorics, you may consult Chapter 6 of Introduction to Enumerative Combinatorics [7].

In Exercise 5, we define the chromatic polynomial of a graph. This polynomial tells us the number of ways the properly n-color the vertices of a given graph G. At first sight, it seems unlikely that $p(-1)$ has some direct combinatorial meaning, but amazingly, it does. In fact, $p(-1)$ is the number of *acyclic orientations* of G, that is, the number of ways to turn G into a directed graph so that no directed cycles are formed. For details, see [40] or Chapter 5 of [7].

Exercises

(1) Prove formula (11.1).

(2) Prove formula (11.2).

(3) A round robin football tournament has $2n$ participating teams. Two rounds have been played so far. Prove that we can still split the teams into two groups of n teams each so that no teams of the same group have played each other yet.

(4) Let $G = (X, Y)$ be a bipartite graph in which any vertex of X has degree at least as large as the degree of any vertex of Y. Prove that X has a perfect matching into Y.

(5) Let G be any simple graph with labeled vertices, and let $p(n)$ be the number of ways to properly n-color G. Prove that p is a polynomial function of n. What is the degree of that polynomial? We note that $p(n)$ is called the *chromatic polynomial* of G.

(6) (+) Prove that for all positive integers n, there exists a graph that does not contain any triangles and whose chromatic number is n.

(7) Prove that the number of ways to properly color an n-vertex cycle with x colors is

(a) $(x-1)[(x-1)^{n-1}+1]$ if n is even.
(b) $(x-1)[(x-1)^{n-1}-1]$ if n is odd.

(8) Let G be a bipartite graph. Prove that G has a perfect matching if and only if for all subsets X of the vertex set of G, the inequality $|X| \leq |N(X)|$ holds. Note that unlike in Philip Hall's Theorem, here we do *not* require that X be a subset of one color class.

(9) Let A be a square matrix with nonnegative integer entries in which the sum of each line, that is, each row and column, is the same positive integer r. Such a matrix is called a *doubly stochastic matrix* or *magic square*. Prove that A is a sum of permutation matrices.

(10) Let A be an $n \times n \times n$ "magic cube" with line sum 2. That is, A is a 3-dimensional matrix with nonnegative integer entries so that each line has sum 2. Is it true that $A = B + C$ where B and C are both magic cubes of line sum 1?

(11) Explain why the results of Exercise 9 and Exercise 10 are not exactly the same. Try to predict what happens in higher dimensions.

(12) Let G be a *regular* bipartite graph. Prove that G has a perfect matching.

(13) There are n children and n toys in a room. Each child wants to play

with r specific toys, and for each toy, there are r children who want to play with that toy. Prove that we can organize r playing rounds so that in each of them, each child plays with a toy he wanted to, and no child plays with the same toy twice? (Contradicting real life a little bit, but not much, we assume that only one child can play with a toy at any one time.)

(14) (-) A graph G is called *factor-critical* if $G - v$ has a perfect matching for any vertex v of G. Prove that a bipartite graph is never factor-critical.

(15) Let G_n be the bipartite graph whose color classes consist of the vertices $A_1, A_2, \cdots, A_n$ and $B_1, B_2, \cdots, B_n$, and in which A_iB_j is an edge if and only if $i + j \le n + 1$. How many matchings does G_n have? (Note that the question is not the number of perfect matchings, but the number of matchings of any size.)

(16) (Knowledge of Linear Algebra required.) Let $G(A, B)$ be a bipartite graph with $A = \{A_1, A_2, \cdots, A_n\}$ and $B = \{B_1, B_2, \cdots, B_n\}$. Let M be the $n \times n$ matrix defined by the rule $M_{i,j} = 1$ if A_iB_j is an edge of G, and $M_{i,j} = 0$ otherwise. Prove that if $\det M \neq 0$, then G has a perfect matching.

Supplementary Exercises

(17) (-) What is the chromatic number of a tree?

(18) (-) What is the chromatic polynomial of a tree?

(19) A graph is called *color-critical* if it has chromatic number k, but if we delete any vertex of the graph, we get a graph of chromatic number $k - 1$. Show an example of a color-critical graph of chromatic number three and of a color-critical graph of chromatic number four. Do not use complete graphs as examples.

(20) (-) Is there a bipartite graph with ordered degree sequence 3, 3, 3, 3, 3, 5, 6, 6, 6?

(21) (-) A school has n student clubs denoted by $c_1, c_2, \cdots, c_n$, and some students are members of more than one of them. Each club can send one representative to the general assembly, but no student can represent more than one club. Find a sufficient and necessary condition that assures that n distinct representatives $r_1, r_2, \cdots, r_n$ can be found so that r_i is a member of c_i for all $i \in [n]$.

(22) (-) Find the size of the smallest vertex cover and of a maximum matching in an odd cycle and an even cycle.

(23) Find the chromatic polynomial of $K_{3,3}$.

(24) A *wheel* is a cycle and an extra vertex that is connected to each vertex of the cycle. Find the chromatic polynomial of a wheel on $n+1$ vertices.

(25) (+) A medium-size city has three high schools, each of them attended by n students. Each student knows exactly $n+1$ who attend a high school *different from his.* Prove that we can choose three students, one from each school, so that any two of them know each other.

(26) Fix two positive integers k and n so that $k < n/2$. Let $G = (X, Y)$ be the bipartite graph in which the vertices of X are the k-element subsets of $[n]$, the vertices of Y are the $(k+1)$-element subsets of $[n]$, and there is an edge between $x \in X$ and $y \in Y$ if and only if $x \subset y$. Prove that X has a perfect matching into Y by

(a) using Philip Hall's theorem,
(b) finding a perfect matching of X to Y.

(27) Deduce Philip Hall's Theorem from Tutte's theorem.

(28) A school has various student associations. The principal wants to hold a meeting, and she wants each student association to send one representative to this meeting. No student can participate at the meeting as a representative of more than one organization. Find a sufficient and necessary condition on such a meeting being possible.

(29) Prove that G is factor-critical if and only if G has an odd number of vertices and $c_o(G - S) \leq |S|$ for all *non-empty* set S of vertices.

(30) Let G be a bipartite graph, and let uv be an edge of G. Prove that at least one of u and v have the following property.
"All maximum matchings of G contain an edge adjacent to this vertex".
Note that this is a stronger requirement than just requiring that each maximum matching contain an edge adjacent to u or v.

(31) For a graph G, let $\nu(G)$ denote the size of its maximum matching. A set of vertices S of G is called a *vertex cover*, if all edges of G have at least one of their vertices in S. Let $\tau(G)$ be the size of the smallest vertex cover of G. In other words, if you think of the edges as non-intersecting tunnels, $\tau(G)$ is the smallest number of lights we need to provide lighting for all tunnels.

(a) Prove that in any graph G, the inequality $\nu(G) \leq \tau(G)$ holds.

(b) Prove that in any *bipartite* graph G, the equality $\nu(G) = \tau(G)$ holds. (Hint: Use the result of the previous exercise, and induction on the number of vertices.)

Note that the result of part (b) is often referred to as *König's theorem*, in honor of the Hungarian mathematician Dénes König.

(32) Deduce Philip Hall's theorem from König's theorem. (The latter is stated in the previous exercise.)

(33) Deduce König's theorem from Philip Hall's theorem.

(34) For any graph G on n vertices, let $\alpha(G)$ denote the size of the largest empty subgraph of G. That is, $\alpha(G)$ is the largest number k so that G has k vertices, no two of which are adjacent. Prove that

$$\alpha(G) + \tau(G) = n,$$

where $\tau(G)$ is defined in Exercise 31.

(35) In a graph G, an *edge cover* is a set S of edges so that each vertex of G is incident to at least one edge in S. Let $\rho(G)$ be the smallest number k so that G has an edge cover consisting of k edges. Let G be a graph on n vertices so that each vertex of G has degree at least 1. Prove that then

$$\nu(G) + \rho(G) = n,$$

where $\nu(G)$ was defined in Exercise 31. What does this result imply for bipartite graphs?

Solutions to Exercises

(1) Recall that $n = tk + r$. Now we prove the statement by induction on t. For $t = 0$, the statement is true. Now let us assume that we know the statement for $t - 1$, that is, for $T(n-k,k)$, formula (1) is correct. To prove that the statement is true for t, that is, that formula (1) is correct for $T(n,k)$, it suffices to prove that the difference of the two equations given by formula (1) for $T(n,k)$ and $T(n-k,k)$ holds. That is, we have to prove that

$$\begin{aligned} T(n,k) - T(n-k,k) &= \frac{n^2 - (n-k)^2}{2k} \cdot (k-1) \\ &= \frac{(2n-k)(k-1)}{2} \\ &= \binom{k}{2} + (n-k)(k-1). \end{aligned}$$

Let us identify the edges counted by $T(n,k)$ that are not counted by $T(n-k,k)$. The graph H belonging to $T(n,k)$ has k more vertices, one in each color class, than the smaller graph H'. There are $\binom{k}{2}$ edges among these extra vertices, and each of the remaining $n-k$ vertices is connected to all of these extra vertices but one, the one in its own color class. This yields $(n-k)(k-1)$ additional edges, and the statement follows.

(2) As we have computed in the solution of the previous exercise, the definitions of $T(n,k)$ and $T(n-k,k)$ yield

$$T(n,k) - T(n-k,k) = \binom{k}{2} + (n-k)(k-1).$$

This is precisely formula (11.2).

(3) If we join teams who played with each other, we get graphs in which each vertex has degree two. In such graphs, all components must be cycles. Also, all these cycles must be of even length for no team has ever been idle. Then we can pick every other vertex of all cycles and get a set of teams with the desired property.

(4) Suppose the contrary is true. Then, by Philip Hall's theorem, there would be a set $T \subseteq X$ of vertices so that $|T| > |N(T)|$. Let $a_1, a_2, \cdots, a_t$ be the degrees of the vertices in T, and let $b_1, b_2, \cdots, b_n$ be the degrees of the vertices in $N(T)$. Our assumptions imply $t > n$, and also, $a_i \geq b_j$ for any i and j. As each edge between T and $N(T)$ has a vertex in T and one in $N(T)$, we must have

$$a_1 + a_2 + \cdots + a_t = b_1 + b_2 + \cdots + b_n.$$

However, this is impossible, as the left-hand side has more members, and they are at least as large as the members of the right-hand side.

(5) Let G have k vertices, and let $p_1, p_2, \cdots, p_k$ denote the number of ways to properly color G using *exactly* $1, 2, \cdots, k$ colors. Now let $n > k$. Then we cannot use all n colors to color G. We first have to choose the i colors ($i \in [k]$) that we will actually use, which we can do in $\binom{n}{i}$ ways. Then, we have to use the chosen i colors, which we can do in p_i ways. Therefore,

$$p(n) = \sum_{i=1}^{k} p_i \binom{n}{i}.$$

Here the p_i are constants, and the $\binom{n}{i}$ are polynomials of n, of degree i. Therefore, $p(n)$ is a polynomial of degree k.

(6) By induction on n. For $n = 2$, a single edge is such a graph. For $n = 3$, the pentagon is. Suppose we know the statement for $n - 1$, and let G be a graph with no triangles and chromatic number $n - 1$. For any vertex $x \in G$, create a new vertex x' whose neighbors are the same as those of x. Do this for all vertices of G. Then take yet another new vertex y and join it to all the vertices that we added to G. The graph obtained this way has chromatic number n and has no triangles.

(7)(a) Let n be even. The proof is by induction on n, with $n = 2$ being the initial case. If $n = 2$, then the statement is true, for an edge can be colored in $x(x-1)$ ways, and that agrees with our claim. Now suppose that the statement is true for n, and try to prove it for $n+2$. Let $A_1, A_2, \cdots, A_{n+2}$ be the vertices of our polygon. Then I have x choices for the color of A_1, $x-1$ choices for the color of A_2, $x-1$ choices for the color of A_3, and so on, $x-1$ choices for the color of A_{n+1}, and *most of the time* -we will explain this later- $x-2$ choices for the color of A_{n+2} as it cannot have the color of A_1 or A_{n+1}. This gives us $x(x-1)^n(x-2)$ colorings. The *most of the time* above refers to the possibility that A_1 and A_{n+1} can have the same color, and in this case, only that color is forbidden for A_{n+2}, so when this happens, we have $x-1$ choices for the color of A_{n+2}, not just $x-2$. So any time this happens, we have to add 1 to the number of proper colorings. To determine how many times does this happen, note that any time this happens, we can delete A_{n+2} and contract A_1 and A_{n+1} to get a properly colored n-gon. And, by the induction hypothesis, the number of such n-gons is $(x-1)[(x-1)^{n-1}+1]$.

Therefore, we get that the total number of proper colorings for our $n+2$-gon is

$$x(x-1)^n(x-2) + (x-1)[(x-1)^{n-1}+1] = (x-1)[(x-1)^{n+1}+1],$$

and the theorem is proved.

(b) If n is odd, the proof is analogous. The initial case is that of $n = 3$, and indeed, a triangle can be colored in $x(x-1)(x-2)$ ways. Then, to prove the induction step, we repeat the same argument and conclude that

$$x(x-1)^n(x-2) + (x-1)[(x-1)^{n-1}-1] = (x-1)[(x-1)^{n+1}-1],$$

and the proof follows.

(8) Let us assume that the condition does not hold, and let X be a counterexample. Let $X = A \cup B$ be the decomposition of X into two color classes. Then we have $|A| + |B| = |X| > |N(X)| = |N(A)| + |N(B)|$, and therefore, we must have either $|A| > |N(A)|$, or $|B| > |N(B)|$. Then Philip Hall's theorem shows that G does not have a perfect matching.
Now let us assume the condition holds for all X. Then in particular, it holds for all subsets that are within one color class. Philip Hall's Theorem then shows that G has a perfect matching.

(9) We prove the statement by induction on r. If $r = 1$, then our matrix is a permutation matrix, and the statement is true. Now let us assume that we know the statement for r, and prove it for $r + 1$. Let A be a magic square with line sum $r + 1$. It suffices to show that there exists an $n \times n$ permutation matrix B so that $A - B$ has nonnegative entries only.
To see this, we define a bipartite graph G in which both color classes consist of n vertices. The elements of one color class will represent the rows of A, and the elements of the other color class will represent the columns of A. Two vertices will be joined by an edge if and only if the intersection of the corresponding row and column of A is a positive entry.
Note that if we can prove that G has a perfect matching, then we are done as that perfect matching specifies n positions in A, all containing positive entries, so that no two are in the same row or column. Therefore, the permutation matrix B having its entries equal to 1 in these n positions is just the permutation matrix we were looking for. Therefore, our task is reduced to proving that G has a perfect matching. We will do this using Hall's theorem. We must show that the conditions of that theorem are satisfied, that is, any k-element subset of vertices from one color class has at least k neighbors in the other color class. Translated to the language of matrices, this means that any k rows of A must contain nonzero entries in at least k different columns. Suppose this does not hold, that is, there are k rows that contain nonzero elements only in $s < k$ columns. Then the sum of all kn elements in these k rows is kr (if you add them row by row), and at most sr (if you add them column by column). This contradicts to $s < k$, and our claim is proved.

(10) This is not true in general. A counterexample is shown below.

$$\begin{pmatrix} 0 & 1 & 1 \\ 1 & 1 & 0 \\ 1 & 0 & 1 \end{pmatrix}$$

$$\begin{pmatrix} 2 & 0 & 0 \\ 0 & 1 & 1 \\ 0 & 1 & 1 \end{pmatrix}$$

$$\begin{pmatrix} 0 & 1 & 1 \\ 1 & 0 & 1 \\ 1 & 1 & 0 \end{pmatrix}.$$

The second level can only be decomposed in one way, and its 2×2 minor in the bottom right corner makes any further decomposition impossible.

(11) The result of Exercise 9 was built on Philip Hall's Theorem for bipartite graphs. We could not use the same argument in Exercise 10 as there was no corresponding theorem for tripartite graphs. There is no corresponding theorem for general k-partite graphs either. Therefore, it is not true that a k-dimensional magic cube with line sum r is the sum of r magic cubes of dimension k having line sum 1.

(12) Let us assume that G does not have a perfect matching. By Hall's theorem, that would imply that there is a vertex set T within one color class such that $|T| > |N(T)|$. Denote by d the degree of all vertices in G. Then there are $|T|d$ edges adjacent to at least one vertex in T. The opposite endpoints of these $|T|d$ edges must be in $N(T)$. Therefore, it follows by the pigeon-hole principle that at least one vertex in $N(T)$ has degree more than d, which contradicts the assumption that G is regular.

(13) Represent children and toys with a bipartite graph G the obvious way. You get a *regular bipartite graph* with all vertices having degree r. The previous problem shows that G has a perfect matching M_1. Then M_1 defines the first playing round, and $G - M_1$ is a regular graph with all vertices having degree $r - 1$. Again, this graph has a perfect matching M_2, and so on.

This fact can also be stated as follows. It is possible to color the edges of a regular bipartite graph of degree r with r colors so that each vertex is adjacent to one edge of each color.

(14) Let G be bipartite, and let its two color classes consist of m and n vertices, with $m \leq n$. Then we cannot omit any vertex from the color class with m elements so that the resulting graph has a perfect matching. Indeed, we would get a bipartite graph with two color classes of different size.

(15) Consider a staircase Ferrers shape of row lengths $(n, n-1, \cdots, 2, 1)$, similarly to the solution of Exercise 4 of Chapter 5. Let the rows correspond to the vertices A_i, and let the columns correspond to the vertices B_j. Then each matching of G_n corresponds to a placement of non-attacking rooks on this Ferrers shape. We have seen in the solution of Exercise 4 of Chapter 5 that there are $B(n+1)$ such rook placements, where $B(n)$ is the nth Bell number.

(16) If M has a non-zero determinant, then it has a non-zero expansion term, that is, a nonzero product of the form $\prod_{i=1}^{n} M_{i,p(i)}$ for some permutation $p \in S_n$. That means that $A_i B_{p(i)}$ is an edge for each $i \in [n]$, so the set of these edges is a perfect matching in G.

Chapter 12

Do Not Cross. Planar Graphs

12.1 Euler's Theorem for Planar Graphs

Let us assume that a farming community has three houses and three wells. The families living in the three houses cannot stand each other, so they prefer not to meet when they walk to the wells. Can we build roads from each of the houses to each of the wells so that there will be no two roads among the needed nine roads that intersect?

Figure 12.1 shows a credible, but failed, attempt to build such roads.

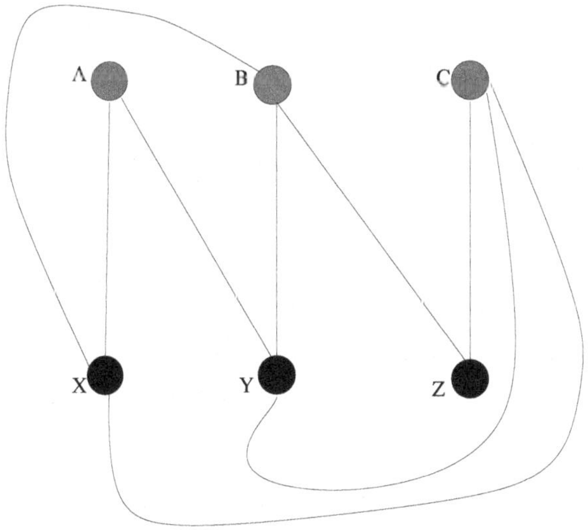

Fig. 12.1 A and Z cannot be connected.

Now you could think that maybe another attempt will succeed. Or, after many unsuccessful tries, you may think that arranging the houses and the wells differently might help. Both of these hopes are false, however. The *three houses, three wells* problem cannot be solved. In this section, we will develop a theory to prove this claim.

It is clear that we are dealing with graphs from a new aspect here. That is, we want to draw them so that their edges do not intersect. This property is central to our chapter.

Definition 12.1. Let G be a graph that can be drawn on a plane surface so that no two of its edges intersect. Then G is called a *planar* graph.

Let G be a planar graph, and draw G on a plane with no intersecting edges. Then the edges of G partition the plane into regions; we will call these regions the *faces* of G. See Figure 12.2 for an example.

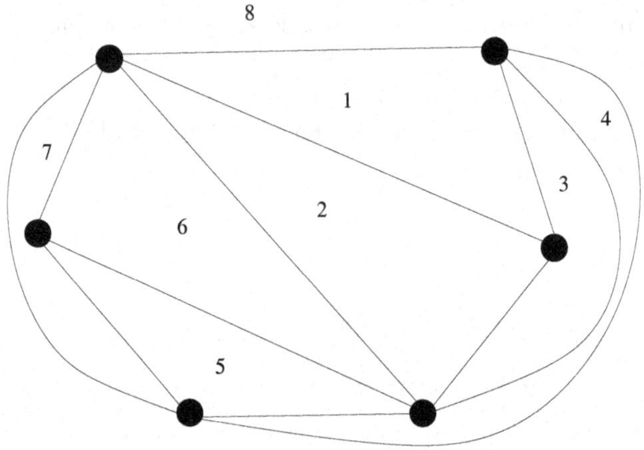

Fig. 12.2 This graph has eight faces.

The number of faces of a planar graph is just as important a parameter of that graph as the number of edges or vertices. The following theorem shows the close connection between these three parameters.

Theorem 12.2 (Euler's Theorem on Planar graphs). *Let G be a connected planar graph with V vertices, E edges, and F faces. Then $V + F = E + 2$.*

Proof. We prove the statement by induction on E, the number of edges of G. If $E = 1$, then G is either the tree of one edge, and then $V = 2$, $F = 1$, and the statement is true, or G is the one-vertex graph with a loop, and then $V = 1$, $F = 2$, and the statement is true again.

Now let us assume that we know the statement for all graphs with $E-1$ edges, and let G have E edges. We distinguish two cases.

If we can omit an edge e from G so that the new graph G' is still connected, then e is in a cycle in G, and therefore there are two different faces on the two sides of e in G. Then G' has $E-1$ edges, V vertices, and $F-1$ faces as the omission of e turned the two faces on the two sides of e into one. Therefore, $V + F - 1 = E - 1 + 2$, so $V + F = E + 2$.

If there is no e with the mentioned property, then G is a cycle-free connected graph, that is, a tree. Then we know from Theorem 10.4 that $V = E + 1$. On the other hand, $F = 1$, so the claim is again true. □

Now we are in a position to settle the problem of three houses and three wells. Indeed, that problem is equivalent to the problem of drawing $K_{3,3}$ on a plane surface without crossings.

Example 12.3. The graph $K_{3,3}$ is not planar. Therefore, there is no solution for the three houses, three wells problem.

Solution. Let us suppose that $K_{3,3}$ is planar. As it has nine edges and six vertices, it follows from Theorem 12.2 that it must have five faces. However, $K_{3,3}$ is a complete bipartite graph, so all its faces must be quadrilaterals. Five quadrilaterals need twenty edges, but in a planar graph, each edge is contained in two faces. Therefore, our graph would need ten distinct edges, but it has only nine.

Note that in particular this means that it does not matter where the houses and the wells are located with respect to each other. No arrangement will work. As $K_{3,3}$ is a subgraph of K_6, it follows from Example 12.3 that K_6 is not planar. On the other hand, K_3, the triangle is obviously planar, and so is K_4 as the reader can see by drawing a square and its two diagonals, then replacing one diagonal by an "outer" edge. It is less obvious to decide whether K_5 is planar.

Example 12.4. The graph K_5 is not planar.

Solution. Again, let us suppose that K_5 is planar. As it has five vertices and ten edges, it follows from Theorem 12.2 that it must have seven faces.

As K_5 is a complete graph, all its faces must be triangles. Seven triangles, however, would need 21 edges, which is impossible as each of the ten edges of K_5 are used in exactly two faces.

It is not by accident that we chose K_5 and $K_{3,3}$ for our examples of graphs that are not planar. Certainly, if G contains K_5 or $K_{3,3}$ as subgraph, then G cannot be planar as we cannot even draw a particular subgraph (the K_5 subgraph, or the $K_{3,3}$ subgraph) of G without crossings. The interesting fact is, however, that in some sense these two graphs are the only ones that can cause a graph to be not planar. Let us make this statement more precise. It is clear that if H is a graph that is not planar, and we remove a vertex V of degree two from H, contracting the edges AV and VB into a single edge AB, the obtained graph is still not planar. Similarly, if we split an edge CD of H into two edges by inserting a vertex F into the middle of CD, and thus replace the edge CD by the edges CF and FD, we again get a non-planar graph. If a graph T can be obtained from H by repeated applications of these two operations, then we say that H and T are *edge-equivalent.*

Then the following theorem, that we will not prove, characterizes planar graphs.

Theorem 12.5 (Kuratowski's Theorem). *A graph is not planar if and only if it contains a subgraph that is edge-equivalent to* K_5 *or* $K_{3,3}$.

12.2 Polyhedra

A *polyhedron* is a solid whose boundary is a union of polygons. We meet polyhedra every day in our lives. Common examples of polyhedra are cubes, tetrahedra, and prisms.

Polyhedra have some nice properties that are not shared by all planar graphs. Most importantly, *all their faces have at least three edges*, and *all their vertices are part of at least three edges.* It is also easy to verify that in all polyhedra, there must be at least four vertices, four faces, and six edges. We do not have to worry about loops or multiple edges in polyhedra, either.

In geometry, a polyhedron is called *regular* if it is "absolutely symmetric", that is: all its faces have the same number l of edges, all vertices are contained in the same number d of edges (d is called the *degree* of the polyhedron), all edges have the same length, all angles within the faces are equal, and all angles between the faces are equal. For example, the cube is a regular polyhedron. In combinatorics, we can disregard the conditions on the length of edges, and the size of angles, but we keep the graph-theoretical conditions that each face is a cycle with l edges, and each vertex has degree d. One could think about regular polyhedra as three-dimensional generalizations of regular polygons.

There is, however, a striking difference between regular polygons and regular polyhedra. Clearly, for all integers $n \geq 3$, there exists a regular polygon with n vertices. So the number of regular polygons that are different as graphs is infinite. In this Section we will show that this is *not* true in three dimensions. In fact, and our goal in this section will be a proof for this, there are only *five* different regular polyhedra, which is very different from the two-dimensional situation.

One of our main tools in proving this result will be Euler's theorem for planar graphs. It is not hard (see Exercise 2) to show that this theorem also holds for polyhedra by showing that polyhedra are essentially planar graphs. Nevertheless, we provide an additional proof for Euler's theorem for *polyhedra only.* The beauty of this proof lies in its simplicity as it does not use induction, or properties of trees; it only requires high school knowledge of geometry. Some of the formulae that we find on the way will be useful on their own.

Theorem 12.6. *Let P be a convex polyhedron with V vertices, F faces, and E edges. Then $V + F = E + 2$.*

Proof. Let p be a plane that is not perpendicular to any faces of P, and let us project P onto p, to get the projected image P'. As P was a convex polyhedron, the projection of a face with k edges will be a convex k-gon. Let us count the sum of angles in all the F faces of P' (the boundary B of P' is considered a face, too). There are two ways to do this, namely we can count by the vertices, or by the faces.

First we count by the vertices. Let us say that B is a convex v_1-gon, and there are v_2 vertices of P whose projected image is inside this v_1-gon. Then $v_1 + v_2 = V$.

The sum of angles around each of the v_2 interior vertices is 360 degrees, so the total sum of these angles is $360v_2$. The boundary of P' is a convex v_1-gon, so its sum of angles is $(v_1 - 2)180$. However, the sum of these angles must be counted twice as each angle is used by two different faces of P'.

Therefore, we obtain that the total sum S of angles is

$$S = (v_1 - 2)360 + 360v_2 = (V - 2)360. \tag{12.1}$$

On the other hand, we can count the angles by the faces, too. If a face of P' is a convex k-gon, then the sum of its angles is $(k-2)180$ degrees. Let $f_1, f_2, \cdots, f_F$ be the number of edges of the F faces of P. As each edge is contained in exactly two faces,

$$\sum_{i=1}^{F} f_i = 2E. \tag{12.2}$$

Therefore, the sum of the angles in all these faces is certainly

$$S = \sum_{i=1}^{F}(f_i - 2)180 = 180(\sum_{i=1}^{F} f_i) - 360F = 360(E - F). \tag{12.3}$$

Comparing (12.1) and (12.3), the proof of our theorem is immediate. □

Formula (12.2) is a useful byproduct of this proof. Note that $f_i \geq 3$ for all i as the f_i denote the number of edges of various polygons. Therefore, the left-hand side of (12.2) is at least as large as $3F$, proving the following Corollary.

Corollary 12.7. *In any convex polyhedron with F faces and E edges, $3F \leq 2E$.*

It is not too difficult to prove a similar relation between the numbers of *vertices* and edges of a convex polyhedron.

Proposition 12.8. *In any convex polyhedron with V vertices and E edges, $3V \leq 2E$.*

Proof. Let $c_1, c_2, \cdots, c_V$ denote the number of edges adjacent to each vertex. As each edge is adjacent to exactly two vertices,

$$\sum_{i=1}^{V} c_i = 2E. \tag{12.4}$$

As each vertex is contained in at least three faces, $c_i \geq 3$ for all i, so the left-hand side is at least as large as $3V$, which was to be proved. □

The reader may think that after finding relations between the number of faces and edges, as well as the number of vertices and edges, we can probably find a similarly simple relation between the number of vertices and that of faces. This is, however, not so simple. The problem is that in the two previous proofs we heavily relied on the fact that each edge is contained in *exactly two* faces, and contains *exactly two* vertices. Faces and vertices do not have such a uniform property.

We now have *lower* bounds on the number of edges in terms of the number of vertices, and also in terms of the number of faces. On the other hand, we have not proved *upper* bounds yet. It is plausible to conjecture that such an upper bound should exist in terms of the number of vertices. Indeed, if we have a simple graph on V vertices, and keep adding new edges to it, then we eventually reach K_V, which is not planar if $V > 4$. Our task is to figure out "how many edges are too many".

Lemma 12.9. *In any convex polyhedron, $E \leq 3V-6$, and also, $E \leq 3F-6$.*

Proof. We know from Corollary 12.7 that $F \leq \frac{2E}{3}$. Comparing this to Euler's theorem, we get

$$E + 2 = F + V \leq \frac{2E}{3} + V,$$

$$\frac{E}{3} \leq V - 2,$$

and the claim $E \leq 3V - 6$ follows by rearranging. Similarly, Proposition 12.8 implies $V \leq \frac{2E}{3}$, and comparing this to Euler's theorem,

$$E + 2 = F + V \leq F + \frac{2E}{3},$$

$$\frac{E}{3} \leq F - 2,$$

and again, the claim $E \leq 3F - 6$ follows by rearranging. □

The attentive reader has probably noticed the symmetric role of V and F in our results so far: these two parameters play symmetric roles in Euler's theorem, in Lemma 12.9, in Corollary 12.7 and in Proposition 12.8. Even the proofs concerning these two kinds of results were very similar. There is a deep, structural reason for this, and we will explain it shortly. First, however, we are going to use our recent results. We start with a somewhat surprising application.

Lemma 12.10. *All convex polyhedra have at least one face that has at most five edges.*

Proof. We know from Lemma 12.9 that $E \leq 3F - 6$. Comparing this to (12.2) we obtain

$$\sum_{i=1}^{F} f_i = 2E \leq 6F - 12. \tag{12.5}$$

Therefore, it cannot be that $f_i \geq 6$ for all i as that would imply $\sum_{i=1}^{F} f_i \geq 6F$. □

It should come no longer as a surprise that there is a similar result for vertices. We could tell the promised deep structural reason for this right now, but we prefer making the reader curious.

Lemma 12.11. *All convex polyhedra have at least one vertex that is contained in at most five edges.*

Proof. We know from Lemma 12.9 that $E \leq 3V - 6$. Comparing this to (12.4) we obtain

$$\sum_{i=1}^{V} c_i = 2E \leq 6V - 12. \tag{12.6}$$

Therefore, it cannot be that $c_i \geq 6$ for all i as that would imply $\sum_{i=1}^{V} c_i \geq 6F$. □

Lemmas 12.10 and 12.11 are of pivotal importance in our quest for all regular polyhedra. They show that in regular polyhedra, the degree d of each vertex can be only one of three values, namely 3, 4, or 5, and the same goes for l. That would leave us with only $3 \cdot 3 = 9$ cases to check. The following discussion will simplify that task.

Let G be any planar graph, and let us construct a new graph $G*$ as follows. The vertices of $G*$ are the centers of the faces of G. (Any interior point would do.) Two vertices A and B of $G*$ are connected by k edges if and only if the corresponding faces in G had k edges in common; in this case each common edge of those two faces will be crossed by one AB edge. This sets up a bijection between the vertices of $G*$ and the faces of G, and another bijection between the edges of $G*$ and G. Therefore, if G had E edges, V vertices and F faces, then $G*$ will also have E edges, but it will have F vertices, and V faces. The reader is invited to verify that $G*$ is also planar. See Figure 14.28 for an example.

Definition 12.12. The graph $G*$ defined in the above paragraph is called the *dual* graph of the planar graph G.

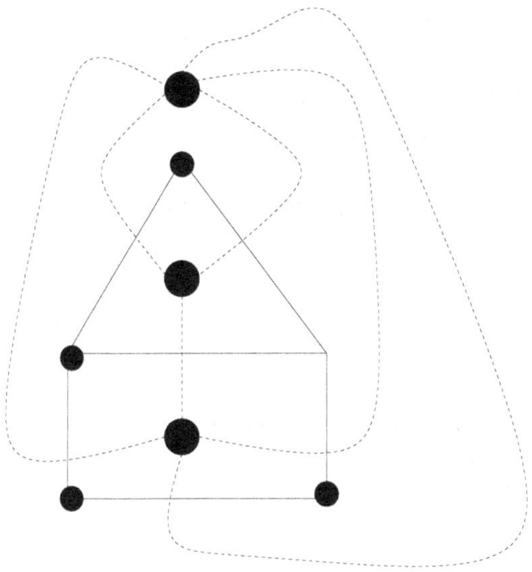

Fig. 12.3 A graph and its dual.

The reader should verify that the dual of a convex polyhedron is a convex polyhedron, and the dual of a *regular* polyhedron is a regular polyhedron.

The notion of the dual graph of a planar graph explains the similarity between results on the number of vertices and results on the number of faces. Indeed, if a theorem on parameters V and E is true for a polyhedron P, it is also true for the dual $P*$ of P, where these two parameters indicate the number of *faces* and the number of edges.

Now we are ready to find all regular polyhedra. Recall that their degrees must be 3, 4 or 5. Also remember that in a regular polyhedron, all faces have l edges, so the total number of edges is, by (12.2),

$$E = \frac{Fl}{2}. \tag{12.7}$$

(A) Let us assume first that $d = 3$. This means that $c_i = 3$ for all i, which implies by (12.4) that $3V = 2E$. Comparing this to Euler's theorem, we get $3F = E + 6$, which, together with (12.7) implies

$$3F - 6 = \frac{Fl}{2},$$

$$(6 - l)F = 12.$$

All three permitted values of l yield an integer solution to this equation.

(a) If $l = 3$, then $F = 4$, therefore $E = 3F - 6 = 6$, and $V = 4$. There indeed exists a polyhedron with these parameters, namely the tetrahedron.
(b) If $l = 4$, then $F = 6$, therefore $E = 3F - 6 = 12$, and $V = 8$. These are the parameters of the cube.
(c) If $l = 5$, then $F = 12$, therefore $E = 3F - 6 = 30$, and $V = 20$. That is, we are looking for a regular polyhedron with 12 faces, that are all pentagons. It is easy to see that such a polyhedron indeed exists: it has one pentagonal face "at the bottom", one "at the top", and to each side of these two faces we attach a new face. This polyhedron is called the *dodecahedron.*

(B) If $d = 4$, then (12.4) yields $4V = 2E$, therefore $E = 2F - 4$. Together with (12.7), this implies

$$2F - 4 = \frac{Fl}{2},$$

$$(4 - l)F = 8.$$

The only permitted value of l that leads to a positive integer solution is $l = 3$. Then we get $F = 8$, so $E = 12$, and $V = 6$. To see that such a polyhedron indeed exists, take the dual of the cube. This polyhedron is called the *octahedron.*

(C) If $d = 5$, then (12.4) yields $5V = 2E$, therefore $3E = 5F - 10$. Comparing this to (12.7) yields

$$5F - 10 = \frac{3Fl}{2},$$

$$F(10 - 3l) = 20.$$

The only permitted value of l that gives a positive integer solution to this equation is $l = 3$. Then $F = 20$, so $E = 30$, and $V = 12$. So our purported polyhedron has 20 triangular faces, 30 edges, and 12 vertices. To see that such a polyhedron indeed exists, note that we can construct one by taking the dual of the dodecahedron. This polyhedron is called an *icosahedron.* Just as the names of other discussed polyhedra referred to the number of faces, this name comes from the Greek word for twenty.

As we have examined all permitted values of d, we have proved the following theorem.

Theorem 12.13. *There are five regular polyhedra: the tetrahedron, the cube, the dodecahedron, the octahedron, and the icosahedron.*

12.3 Coloring Maps

World maps usually color the territories of neighboring countries by different colors for obvious reasons. If two countries having a common border were the same color, the viewer of the map may overlook the border between them.

This simple problem from everyday life gave rise to one of the most famous problems in Mathematics. Take any map, with the countries still uncolored, and try to color the countries so that no two neighboring countries get the same color, using as few colors as possible. What is the smallest number of colors that will suffice no matter what the map looks like?

From a graph theoretical point of view, all maps are planar graphs, so we need to find a proper coloring of the faces of a planar graph. By proper coloring, we mean the faces that have an *edge* in common must get different colors. Note that faces that only have vertices in common may get the same color. Also note that by duality, this is the same question as asking how many colors do we need to properly color the *vertices* of a planar graph. Indeed, a proper coloring of the faces of the planar graph G naturally defines a proper coloring of the vertices of $G*$, and vice versa. When coloring the vertices of $G*$, the criterion to fulfill is, of course, that adjacent vertices get different colors.

This question was probably asked first by Francis Guthrie in 1852, and got soon passed along to well-known mathematicians as A. DeMorgan, and A. Cauchy. A little bit of thinking yields that at least four colors are needed as K_4 is planar. Trying several maps, one is led to the conjecture that four colors always suffice. as long as all countries on the map are *contiguous*, that is, one can walk from any point of a country to any other point of that country without crossing into another country. For instance, the United States is *not* contiguous since one cannot walk from a point in Alaska to a point in California without crossing into Canada in between. Since four colors seem to suffice for maps with contiguous countries, this problem had been called the "Four-Color Conjecture".

For a warm-up, let us prove that six colors always suffice. We will use the dual (vertex-coloring) form of the problem as it makes induction proofs easier to describe.

Proposition 12.14. *The vertices of any planar graph can be properly colored with six colors.*

Proof. Induction on V, the number of vertices of the planar graph G. If $V = 1$, then the statement is obviously true. Let us assume that we know that the statement is true for graphs with $V - 1$ vertices. Let G have V vertices. Then we know from Lemma 12.11 that G has a vertex A of degree at most five. Remove A from G to get the graph G'. By our induction hypothesis, G' has a proper coloring with six colors. Take such a coloring of G', then color A with a color that is not the color of any of its (at most five) neighbors. $\square$

This means, by duality, that all maps can be properly colored using six colors. The situation is significantly harder if we only want to use five colors. The result, however, is the same.

Theorem 12.15. *The vertices of any planar graph can be properly colored with five colors.*

Proof. Just as in proving the previous proposition, we use induction. The only case in which the previous proof does not work is when A has five neighbors, and they are all of different colors. In this case, denote by 1, 2, 3, 4 and 5 the colors of the five neighbors y_1, y_2, y_3, y_4, y_5 of A as they follow clockwise. Let G' be the graph obtained from G by removing A and all the edges adjacent to A. If G' has a proper 5-coloring in which y_1 and y_3 are the same color, then we are done. If not, then any proper 5-coloring of G' must contain a path from y_1 to y_3 along which the vertices are alternatingly colored 1 and 3. By similar argument, if y_2 and y_4 cannot be the same color, then any proper 5-coloring of G' must contain a path from y_2 to y_4 along which the vertices are alternatingly colored 2 and 4. This, however, is a contradiction, as a path from y_1 to y_3 and a path from y_2 to y_4 must always intersect. See Figure 12.4. $\square$

Again, this means by duality that any map can be properly colored using five colors.

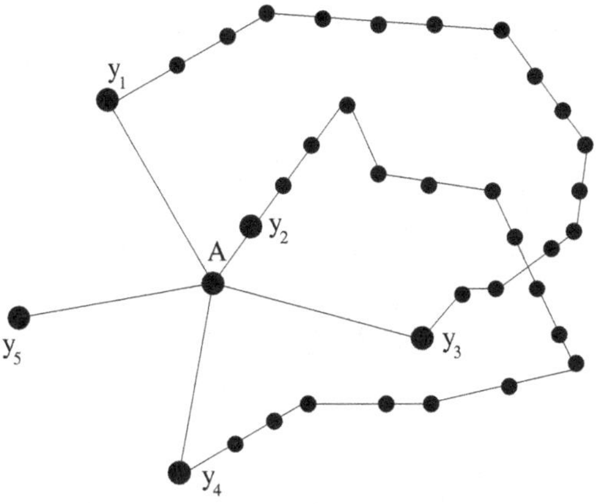

Fig. 12.4 The paths y_1y_3 and y_2y_4 intersect.

Notes

How about the big question, that of four colors? The Four-Color conjecture remained a conjecture until the 1970s. Then in 1976, Kenneth Appel and Wolfgang Haken developed a strategy to use a computer to split the problem into several cases, and check the 4-colorable property in each case. When they started running the computer program, it was not sure that the computer would ever finish. It could have happened that the cases lead to subcases, which in turn lead to subcases of subcases, and never end. This did not happen, however. After 1200 hours of running time, and the verification of 1936 cases, the computer returned the verdict *"four colors suffice"*. (Good that there were no power outages in Urbana, Illinois in those weeks!) Therefore, we can now call this statement the *Four-Color Theorem.* The proof also involved 400 pages worth of checking some other cases by humans.

A significant problem with the proof of Appel and Haken was that one did not really learn from it why the statement is true. Numerous mathematicians have kept trying to simplify the proof ever since. The 1936 cases of the Appel-Haken proof were later reduced to 1476. In 1996, a new, simpler proof was given by Neil Robertson, Daniel Sanders, Paul Seymour and Robin Thomas. It involved only 633 cases, and all of these cases could be checked by a computer. The same four researchers also found an efficient algorithm to actually give a 4-coloring of a planar graph, as opposed to simply proving that one exists. Another computer-based proof was given by Georges Gonthier and Benjamin Werner in 2005.

Exercises

(1) Generalize Theorem 12.2 for graphs that are not necessarily connected.
(2) Deduce Theorem 12.6 from Theorem 12.2.
(3) Find the only convex polyhedron for which equality holds both in Corollary 12.7 and in Proposition 12.8.
(4) Prove that in any polyhedron, there are two vertices that are adjacent to an equal number of edges.
(5) Prove that every polyhedron has two faces that have the same number of vertices.
(6) Prove the result of the previous exercise without using Euler's theorem, or its consequences.
(7) Prove that the faces of planar graph G are 2-colorable if and only if all vertices of G have even degree.
(8) Let n and k be positive integers so that the vertices of any n-vertex planar graph all of whose faces are triangles have a proper k-coloring. Prove that then the vertices of any n-vertex planar graph have a proper k-coloring.
(9) State the dual of the result of the previous exercise.
(10) Let B be a simple, bipartite, and planar graph. If each vertex of G has degree at least d, at most how large can d be?

Supplementary Exercises

(11) (-) Explain why the method we used in the text to prove Theorem 12.15 would not work if we tried to use it to prove the Four-Color Theorem.
(12) (-) The faces of a convex polyhedron are all triangles or pentagons. Prove that the number of faces is even.
(13) (-) Prove that the statement $E \leq 3V - 6$ that holds in all polyhedra does not hold in all connected simple planar graphs.
(14) How many counterexamples are there that solve the previous exercise?
(15) Let P be a polyhedron with no triangular faces. Prove that $E \leq 2V - 4$.

(16) How many connected simple planar graphs are there for which the inequality of the previous exercise does not hold?

(17) (-) Is it true that if a connected graph satisfies $E \leq 3V - 6$, then that graph is planar?

(18) (-) Take K_6, the complete graph on 6 vertices, and delete two of its edges. Prove that the obtained graph G is never planar.
What about three edges?

(19) Let P be a convex polyhedron whose faces are all either a-gons or b-gons, and whose vertices are each adjacent to three edges. Let p_a, p_b, and n respectively denote the number of a-gonal faces, b-gonal faces, and vertices of P.

(a) Express the number of edges of P in two different ways.
(b) Prove that $p_a(6-a) + p_b(6-b) = 12$.

Note that a polyhedron satisfying the conditions of this exercise is called a *trivalent, (a, b)-faced polyhedron.*

(20) Keep the notation of the previous exercise, and assume that $3 \leq a \leq b \leq 5$. Within these limits, does there exist a trivalent (a, b)-faced polyhedron for each pair (a, b)?

(21) Keeping the notation of the two previous exercises, let P be a trivalent $(5, 6)$-faced polyhedron.

(a) Prove that with these conditions, all polyhedra P will contain the same number of pentagons.
(b) Find the smallest value of n so that there exists a trivalent $(5, 6)$-faced polyhedra on n vertices in which *no two pentagonal faces share an edge.*

(22) Let G be a planar graph in which each face is either a 2-gon, or a 3-gon, or a 4-gon, and let p_2, p_3, and p_4 respectively denote the number of these faces. Let us assume furthermore that each vertex of G has degree four, and that $p_2 + p_3 = 8$, just like in an octahedron.

(a) Prove that with the given conditions, $p_2 = 0$.
(b) Prove that with the given conditions, $p_3 = 8$.

Note that a planar graph (or polyhedron, which we can now say as we know that $p_2 = 0$) satisfying the conditions of this exercise is called an *octahedrite.*

(23) Let G be a convex octogon, and let us select ten points inside G in general position (no three on the same line). Let S be the set of these ten points. Now draw some non-intersecting straight segments

so that these segments partition G into triangles, and the vertices of these triangles are the vertices of G and the elements of S.
How many triangles are formed?

(24) (+) Is it possible to partition a square into a finite number of concave quadrilaterals?

(25) (+) (Sperner's Lemma) Let T be a triangle that is partitioned into smaller triangles by line segments. Let S be the set of these triangles. Assume that none of the triangles in S that are in the interior of T contain a vertex of another triangle on the interior of their sides. Now color all the vertices of all these triangles red, blue, or green so that the three vertices of T are all different, and the vertices on the three sides of T are not colored the same as the opposite vertex of T. See Figure 12.5 for an example. Prove that there is a triangle in S whose vertices are all of different colors.

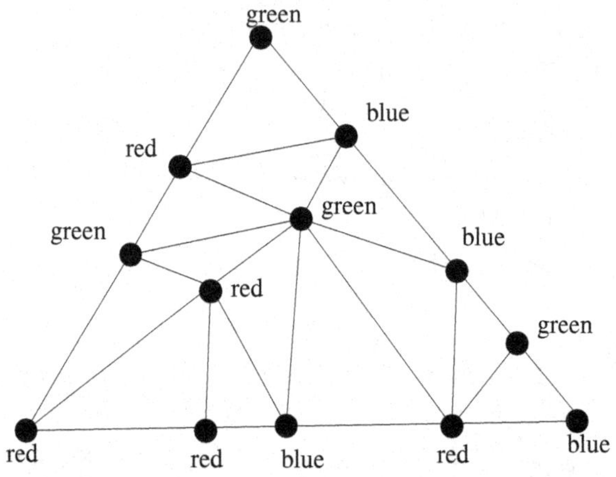

Fig. 12.5 A possible partition and coloring.

Solutions to Exercises

(1) For connected graphs, we have $F + V = E + 2$. For graphs with k connected components, we will have $F + V = E + (k + 1)$. Indeed, take the equation of the Euler theorem for each connected component, and then take the sum of these equations, each of which is of type

$V_i + F_i = E_i + 2$. As the infinite faces of the k components is common, that face is counted k times on the left-hand side. Taking that into account, our claim immediately follows.

(2) Comparing the known formulae $3F \leq 2E$ and $E = V + F - 2$, we get

$$3(E - V + 2) \leq 2E,$$

$$E \leq 3V - 6$$

as claimed.

(3) If equality holds in both formulae, then we have $V = F = 2E/3$. On the other hand, Euler's theorem forces $V + F = E + 2$. Comparing these two relations, we get $V = F = 4$, and $E = 6$. The only convex polyhedron with these parameters is the tetrahedron. (Indeed, no face can have more than three vertices.)

(4) In a polyhedron, each vertex is adjacent to at least three edges. So if our claim is not true, then there exists a polyhedron with V vertices and at least $3 + 4 + \cdots + (V + 2) = \frac{V(V+5)}{2}$ edges. On the other hand, we have seen in Lemma 12.9 that the number of edges is at most $3V - 6$. Thus we must have

$$V^2 + 5V \leq E \leq 6V - 12.$$

A routine computation shows that this is not possible as $V^2 + 5V > 6V - 12$ for all positive integers. Thus our claim is true.

(5) Our claim is equivalent to saying that every polyhedron has two faces that have the same number of edges. Assume not, and let P be a counterexample. Then the dual of P would be a counterexample for the result of the previous exercise.

(6) Let P be a polyhedron, and let L be a face of P with a maximal number n of edges. Then L shares an edge with n other faces. Each of these n faces has at least three and at most n edges. Therefore, the pigeon-hole principle implies that there must be two of them that have the same number of edges.

(7) The "only if" part is easy. If V has odd degree, then there are an odd number of faces around V, and they cannot be properly colored by two vertices.

We prove the "if" part by strong induction on F, the number of faces of G. If $F = 1$ (empty graph), or $F = 2$ (cycle), then the statement is obviously true. Now assume we know the statement for planar graphs with at most $F - 1$ faces, and let G have F faces. Take a face T of G. Omit all edges of T to get the graph G'. This decreased the number of faces of G by at least one, and decreased the degrees of vertices of T by two. Therefore, the induction hypothesis applies to G', and G' can be properly 2-colored. Let us take a proper 2-coloring of the faces of G', and assume without loss of generality that the face T' that contains the former face T is red. Let us put the edges of T back to the graph, and color T the other color, say blue. This is a proper 2-coloring of G as T shares edges with parts of T', and those are all red. See Figure 12.6 for an example.

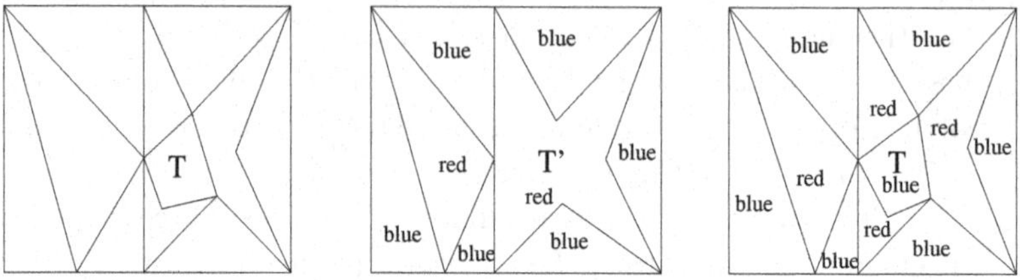

Fig. 12.6 The induction step.

(8) Induction on m, the number of non-triangular faces of our graph G. If $m = 0$, then the claim is identical to the condition, so the initial step is trivial. Now let us assume that we know the statement for $m - 1$, and prove it for m. We can assume G has no vertices of degree 1 as if it does, we can remove them without loss of generality. Therefore, G has a face that is a cycle C consisting r edges. Let $V_1, V_2, \cdots, V_r$ be the edges of this cycle. Draw (possibly curved) lines from V_1 to $V_3, V_4, \cdots, V_{r-1}$. (If all edges are straight lines, then C is a polygon, and these lines are diagonals of C cutting C into triangles.) The new graph G' we obtain has one less non-triangular faces than G, so by induction, it has a proper coloring p with k colors. Note that the set of edges of G' contains that of G, therefore p is also a proper coloring of G.

(9) For any positive integers n and k, if the faces of all n-vertex regular planar graphs with vertex degree 3 have a proper k-coloring, then the faces of all n-vertex planar graphs have a proper k-coloring.

(10) We claim that the largest possible value of d is 3. Indeed, $d = 3$ is possible, as is shown by noting that the graph give by the edges and the vertices of a cube is planar and bipartite (check this!).

On the other hand, $d = 4$ is not possible. Assume there were such a graph. Then counting the edges by their endpoints, $4V \leq 2E$, so $2V \leq E$. As our purported graph is simple and bipartite, each of its faces would have to consist of at least 4 edges, forcing $4F \leq 2E$, so $2F \leq E$. Therefore, using Euler's theorem,

$$E + 2 = V + F \leq \frac{E}{2} + \frac{E}{2} = E$$

would follow, which is a contradiction.

Chapter 13

Does It Clique? Ramsey Theory

Instead of coloring the vertices of our graphs, in this chapter we will color their edges. We will see that this leads to a completely different set of problems. Our first excursion into the land of *infinite graphs* is also part of this chapter.

13.1 Ramsey Theory for Finite Graphs

Example 13.1. Six people are waiting in the lobby of a hotel. Prove that there are either three of them who know each other, or three of them who do not know each other.

This statement is far from being obvious. We could think that maybe there is some case in which everyone knows roughly half of the other people, and in the company of any three people there will be two people who know each other, and two people who do not. We will prove, however, that this can never happen.

Solution. (of Example 13.1) Take a K_6 so that each person corresponds to a vertex. Color the *edge* joining A and B red if A and B know each other, and blue if they do not. Do this for all 15 edges of the graph. The claim of the example will be proved if we can show that there will always be a triangle with monochromatic edges in our graph.

Take any vertex V of our bicolored graph. As V is of degree five, it must have at least three edges adjacent to it that have the same color. Assume without loss of generality that this color is red. Let X, Y and Z be the endpoints of three red edges adjacent to V. (The reader can follow our argument in Figure 13.1, where we denoted red edges by solid lines.)

Now if any edge of the triangle XYZ is red, then that edge, and the two edges joining (the endpoints of) that edge to V are red, so we have a triangle with three red edges. If the triangle XYZ does not have a red edge, then it has three blue edges.

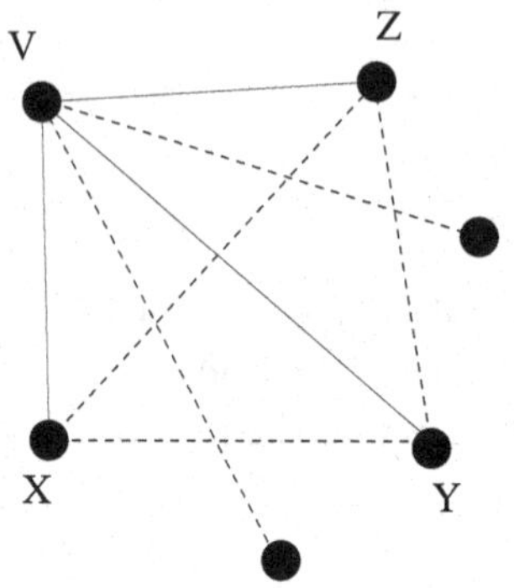

Fig. 13.1 The colors of the edges of the triangle XYZ are crucial.

This beautiful proof is our first example in *Ramsey theory.* This field is named after Frank Plumpton Ramsey, who was the first person to study this area at the beginning of the twentieth century.

We point out that the result is tight, that is, if there were only five people in the lobby of the hotel, then the same statement would be false. Indeed, take a K_5, and draw it as a regular pentagon and its diagonals. Color all five sides red, and all five diagonals blue. As any triangle in this graph contains at least one side and at least one diagonal, there can be no triangles with monochromatic edges.

Instead of taking a K_6, and coloring its edges red and blue, we could have just taken a graph H on six vertices in which the edges correspond to people who know each other. In this setup, the edges of H correspond to the former red edges, and the edges of the complement of H correspond to the former blue edges. As a complete subgraph is often called a *clique*, the statement of Example 13.1 can be reformulated as follows. If H is a simple graph on six vertices, then at least one of H and the complement of H contains a clique of size three.

The arguments used in the proof of Example 13.1 strongly depended on the parameter three, the number of people we wanted to know or not to know each other. What happens if we replace this number three by a larger number? Is it true that if there are sufficiently many people in the lobby, there will always be at least k of them who know each other, or k of them who do not know each other? The following theorem answers this question (in fact, a more general one), in the affirmative.

Theorem 13.2 (Ramsey theorem for graphs). *Let k and l be two positive integers, both of which is at least two. Then there exists a (minimal) positive integer $R(k,l)$ so that if we color the edges of a complete graph with $R(k,l)$ vertices red and blue, then this graph will either have a K_k subgraph with only red edges, or a K_l subgraph with only blue edges.*

Note that any non-empty set of positive integers has a smallest element. Therefore, if we can show that there exists at least one positive integer with the desired property, then we will have shown that a smallest such integer exists.

Example 13.3. Example 13.1, and the discussion after it shows that $R(3,3) = 6$. We also have trivial fact $R(2,2) = 2$ relating to the graph with one edge.

Proof. (Of Theorem 13.2) We prove the statement by a new version of mathematical induction on k and l. This induction will run as follows. First we prove the initial conditions that $R(k,2)$ and $R(2,l)$ exist for all k, and all l. Then we prove the induction step that if $R(k,l-1)$ exists, and also $R(k-1,l)$ exists, then $R(k,l)$ also exists.

To see that the initial conditions hold, note that $R(k,2) = k$, and similarly, $R(2,l) = l$. Indeed, either all edges of a K_k are red, and then it has a K_k subgraph with all edges red, or at least one of its edges is blue, in which case it has a K_2 subgraph with all edges blue. Analogous argument works for $R(2,l)$.

We prove the induction step by showing that

$$R(k,l) \leq R(k,l-1) + R(k-1,l). \tag{13.1}$$

Indeed, take a complete graph with $R(k,l-1)+R(k-1,l)$ vertices. Take one of its vertices, and call it V. As V has degree $R(k,l-1)+R(k-1,l)-1$, it has either at least $R(k,l-1)$ blue edges adjacent to it, or it has at least $R(k-1,l)$ red edges adjacent to it.

In the first case, let b denote the $R(k, l-1)$-element set of the other endpoints of these blue edges. Then, by the definition of $R(k, l-1)$, the set b either contains a monochromatic red K_k and we are done, or a monochromatic blue K_{l-1}, which can be completed to a monochromatic blue K_l by adding the vertex V, and we are done.

In the second case, let r denote the $R(k-1, l)$-element set of the other endpoints of these red edges. Then again, r either contains a monochromatic blue K_l and we are done, or a monochromatic red K_{k-1}, which can be completed to a monochromatic red K_k by adding the vertex V, and we are done again.

So (13.1) is proved, therefore the induction step is proved, and therefore the theorem is proved. □

Theorem 13.2 does show that the *Ramsey number* $R(k, l)$ always exists, but it does not tell us its exact value. Let us try to use this theorem to find $R(4, 3)$, the smallest Ramsey number we have not discussed yet. Formula 13.1 yields

$$R(4, 3) \leq R(4, 2) + R(3, 3) = 4 + 6 = 10.$$

The following Example shows that the upper bound obtained from Theorem 13.2 is not tight, even for such small values of k and l.

Example 13.4. The equality $R(4, 3) = 9$ holds.

Solution. As we have just seen, it follows from (13.1) that $R(4, 3) \leq 10$. To prove our claim, we have to show two things: that all 2-colorings of the edges of K_9 will result in either a red K_4 or a blue K_3, and that the same will not hold for K_8.

(1) To see the first statement, take a K_9 with two-colored edges. We claim that there has to be a vertex V so that either (i) at least six of the edges adjacent to V are red, or (ii) at least four of the edges adjacent to V are blue. If neither statements were true, then all vertices of this K_9 would have five red edges adjacent to them, which is a contradiction as the sum of the degrees in the subgraph of all red edges must be even, so it cannot be $9 \times 5 = 45$.

  (a) If there are six red edges adjacent to V, then denote by A the six-element set of their other endpoints. By Example 13.1, there is either a red triangle, or a blue triangle on A. So our K_9 either contains a blue triangle, or, together with V, a red K_4.

(b) If, on the other hand, there are four blue edges adjacent to V, then denote by B the four-element set of their other endpoints. If all edges on B are red, then there is a red K_4. If not, then there is a blue edge on B, which will form a blue triangle, together with V.

(2) In order to see that $R(4,3) > 8$, take a K_8, and label its vertices by the elements of $[n]$, in clockwise direction, say. Let the edge (i,j) (with $j > i$) be blue if $j - i$ is 1, 4, or 7, and red otherwise. This graph will not contain a blue triangle. Indeed, such a triangle would have to contain a smallest vertex i, and two of the three vertices $i+1$, $i+4$, $i+7$, but no matter which two we choose, there will be a red edge between them.
Red edges are present between vertices i and j so that $j > i$ and $j - i$ is 2, 3, 5 or 6. To get a red K_4, we would need a smallest vertex i, then three of the four vertices $i+2$, $i+3$, $i+5$ and $i+6$. This is impossible as neither $i+2$ and $i+3$, nor $i+5$ and $i+6$ can be chosen together.

This completes the proof of the equality $R(4,3) = 9$.

The following example takes the ideas seen in the preceding proof one step further.

Example 13.5. The equality $R(4,4) = 18$ holds.

Solution. Formula 13.1 shows that

$$R(4,4) \leq R(4,3) + R(3,4) = 9 + 9 = 18.$$

For an example of a 2-coloring of K_{17} without a monochromatic K_4, take the *quadratic residue graph.* That is, label the vertices from 0 to 16, and let $i - j$ be red if and only if $i - j$ is a quadratic residue modulo 17. For those not familiar with this notion, this means that if $j > i$, then the edge (i,j) is red if and only if $j - i$ is 1, 2, 4, 8, 9, 13, 15, or 16. (Since if we divide the square of an integer by 17, the remainder will always be one of these eight values.) A tedious, but conceptually not difficult, analysis of all cases shows that there will be no K_4 with monochromatic edges in this graph.

We have seen that $R(2,2) = 1$, $R(3,3) = 6$, and $R(4,4) = 18$. The exact values of $R(k,k)$ are not known if $k \geq 5$. The difficulty of this problem is illustrated by the following famous quote of Paul Erdős. "Assume an evil spirit orders us to compute $R(5, 5)$, or else he will destroy all mankind. It may then be best if all mathematicians and computers start working on the answer. If, however, he orders us to compute $R(6, 6)$, then we had better think about how to destroy him before he destroys us."

Can we at least find some bounds for the symmetric Ramsey numbers $R(k, k)$? With the methods of this section, we can mostly hope for *upper* bounds. They will be consequences of formula (13.1).

Theorem 13.6. *Let k and l be positive integers larger than one. Then*

$$R(k, l) \leq \binom{k + l - 2}{k - 1}. \tag{13.2}$$

Proof. As the reader probably guessed, we will prove this statement by the same kind of induction on k and l as we proved Theorem 13.2. If $k = 2$, our claim reduces to $R(2, l) \leq \binom{l}{1} = l$, which is trivially true. By symmetry, the statement is also true if $l = 2$.

Now let us assume that the statement is true for $R(k, l-1)$ and for $R(k-1, l)$, and prove it for $R(k, l)$. Applying formula (13.1) and the induction hypothesis, we get

$$R(k, l) \leq R(k, l-1) + R(k-1, l) \leq \binom{k + l - 3}{k - 1} + \binom{k + l - 3}{k - 2} = \binom{k + l - 2}{k - 1},$$

which is precisely what we wanted to prove. $\square$

Corollary 13.7. *For all integers $k \geq 2$, the inequality $R(k, k) \leq 4^{k-1}$ holds.*

Proof. By Theorem 13.6, we obtain

$$R(k, k) \leq \binom{2k - 2}{k - 1} \leq 4^{k-1}.$$

$\square$

A technique for proving lower bounds for Ramsey numbers will be introduced in Chapter 15.

13.2 Generalizations of the Ramsey Theorem

Example 13.8. A circle of 17 friends has the property that no matter how we choose two from these 17 friends, those two people correspond with each other on one of three given subjects. Prove that there are three friends among the circle of these 17 friends such that any two of the three of them correspond with each other on the same subject.

This example generalizes Example 13.1 in the hotel lobby in a major aspect. Now the relation between two people can be of not only two kinds (they either know each other or not), but of three kinds. So if we represent our people by a K_{17}, then we have to color the edges of this K_{17} by three colors.

Solution. (of Example 13.8) As we have just explained, we have to show that if we color each of the edges of a K_{17} either red, or blue, or green, there will always be a triangle with monochromatic edges. Choose any vertex V of our K_{17}. As V has degree 16, it follows by pigeon-hole principle that there is a color so that at least six of the edges adjacent to V have the same color, say green. Let g be the set of the other endpoints of these green edges. If there is any green edge between two vertices of g, then we are done as those two vertices of g and V span a green triangle. If not, then all the edges among the vertices of g are red or blue. However, g has at least six elements, so it follows from Example 13.1 that the vertices of g span either a red triangle, or a blue triangle.

Theorem 13.6 can be generalized to more than two colors in the following way.

Theorem 13.9. *Let $n_1, n_2, \cdots, n_k$ be positive integers, with k fixed. Then there exists a minimal positive integer $N = R(n_1, n_2, \cdots, n_k)$ so that if $n > N$, and we color all edges of $G = K_n$ with colors $1, 2, \cdots, k$, then there will always be at least one index $i \in [k]$ so that G has a K_{n_i} subgraph whose edges are all of color i.*

We only provide a sketch of a proof. After reading it, you should be able to see how to proceed in the general case. You can check your work by reading the solution of Exercise 1.

Proof. (of Theorem 13.9) We prove the statement by induction on $n_1 + n_2 + \cdots + n_k$. The initial case of $n_1 = n_2 = \cdots = n_k = 1$ is trivial. Now let us assume that we know the statement for all positive integers $n_1, n_2, \cdots, n_k$ whose sum is less than m, and prove it for the case when their sum is m.

Note that by our induction hypothesis, we know that the positive integer $R(n_1 - 1, n_2, \cdots, n_k)$ exists. Set $N = k(R(n_1 - 1, n_2, \cdots, n_k) - 1) + 2$. Let us assume that G has a vertex V so that the color that occurs most frequently among the edges adjacent to V is color 1. That means that at least $R(n_1 - 1, n_2, \cdots, n_k)$ edges adjacent to that vertex are of color 1. Let

S be the set of the endpoints of these edges (other than V), and let K_S the complete graph with vertex set S.

By the definition of $R(n_1 - 1, n_2, \cdots, n_k)$ either there exists a vertex $i \in \{2, 3, \cdots, k\}$ so that K_S has a K_{n_i} subgraph with all edges colored i and we are done, or K_S has a K_{n_1-1} subgraph with all edges colored 1, and then we are done again, adding V to this subgraph. □

Another direction in which the Ramsey theorem can be generalized is that of *hypergraphs*, or set systems. To make long story short, in that generalization, we color not the edges of K_n, but the K_r-subgraphs of K_n, for some fixed r. The special case of $r = 2$ corresponds to the traditional situation, that is, when the edges are colored. Then the following is true.

Theorem 13.10. *We color each K_r-subgraph of K_n with one of the colors $1, 2, \cdots, k$. Let $n_1, n_2, \cdots, n_k$ be positive integers. Then there exists a minimal positive integer $N = R_r(n_1, n_2, \cdots, n_k)$ so that if $n \geq N$, then there exists an index $i \in [k]$ so that K_n contains a K_{n_i} subgraph whose K_r subgraphs are all colored i.*

The proof is omitted. It is conceptually not more difficult than that of Theorem 13.9, but it involves more notations.

The following is a very surprising application of Theorem 13.10. So far our studies in Ramsey theory did not involve any geometry at all. Still, we will be able to use our last theorem to prove a result of geometric nature.

Theorem 13.11 (The Erdős-Szekeres theorem). *Let n be a positive integer. Then there exists a (minimal) positive integer $ES(n)$ so that if there are $N \geq ES(n)$ points given in the plane, no three of which are collinear, then we can choose n points from them that form a convex n-gon.*

Before reading further, you should check your understanding of the definition of $ES(n)$ by proving that $ES(4) = 5$.

Proof. We claim that $R_3(n, n)$ will always be such a positive integer (not necessarily the minimal one). Take the complete graph whose vertices are our $R_3(n, n)$ points in the plane. Color its triangles red or blue according to the following rule. Number the points from 1 to $R_3(n, n)$, and color a triangle red if the path from the smallest number via the middle one to the largest one is clockwise. Color a triangle blue if that path is counterclockwise.

As our graph has $R_3(n, n)$ vertices, there will be a K_n subgraph with monochromatic triangles. We claim that the vertices of this K_n subgraph form a convex n-gon. To see this, it suffices to show that there are no four vertices in this subgraph so that one is within the triangle spanned by the other three. In other words, we need to show that the configuration shown in Figure 13.2 does not occur.

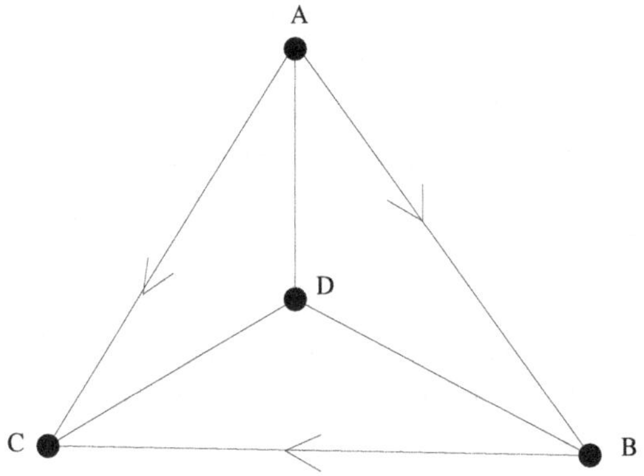

Fig. 13.2 This configuration cannot occur.

Assume without loss of generality that $A < B < C$, and that all triangles of our K_4 at hand are red. Then the fact that triangle ADB is red forces $D < A < B$. (Indeed, $A < D < B$ would mean that the triangle ADB is blue, and $A < B < D$ would mean that either the triangle BCD is blue, or $D > C$, in which case triangle ACD is blue.) Then, however, $D < A < C$, and triangle DAC is blue, which is a contradiction. This completes the proof. □

We mention that there has been a series of improvements concerning the best known upper bounds for $ES(n)$. The latest such result can be found in [43], where it is proved that $ES(n) \leq \binom{2n-5}{n-2} + 1$.

13.3 Ramsey Theory in Geometry

Example 13.12. Let us assume *all points of the plane* are colored either red, or blue. Prove that there exists a unit segment with monochromatic endpoints.

This problem is certainly different from all other problems discussed so far. The number of points in the plane is infinite, in fact, uncountably infinite. All our previously discussed problems dealt with finite graphs. Moreover, in this problem, and in what follows, we will state and prove theorems of geometric nature, making our first excursion to *combinatorial geometry*.

Solution. (of Example 13.12) Take a regular triangle T with side length one. Then by the pigeon-hole principle, T must have two vertices of the same color. Those two vertices will form a segment with the required property.

The statement of the previous example can be strengthened as follows.

Example 13.13. Let us assume that *all points of the plane* are colored either red or blue or green. Prove that there exists a unit segment with monochromatic endpoints.

Solution. Again, take any regular triangle T with side length 1, and vertices A, B, C. If A, B, and C are not all of different colors, then we are done. If they are, then append another regular triangle T' with side length 1 to one of the sides of T, say BC, as shown in Figure 13.3. Now the new vertex D of T' must be the same color as A, say red, otherwise a monochromatic unit segment is formed, either BD, or CD. Thus we have showed that the segment AD, that is of length $\sqrt{3}$, has monochromatic (red) endpoints.

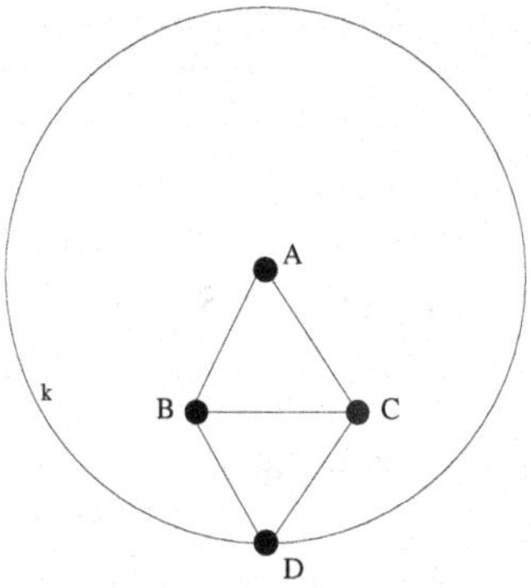

Fig. 13.3 All points of k must have the color of A.

Note that we have not used *any* special property of T other than being a regular triangle of unit side lengths. Therefore, we could repeat this argument for any regular triangle in the plane, and show this way that all segments of length $\sqrt{3}$ have monochromatic endpoints, otherwise there exists a unit segment with that property.

Finally, take any red vertex R, and take the circle k whose center is R, and whose radius is $\sqrt{3}$. Then all points of k must be red, which means that there is a unit segment with red endpoints. Indeed, k has radius $\sqrt{3}$, so k certainly has arcs of unit length.

Example 13.14. We colored all the points of the plane either red or blue. Let T be a triangle whose angles are equal to 30, 60, and 90 degrees, and whose hypotenuse is of unit length. Prove that there exists a triangle with monochromatic vertices that is congruent to T.

Solution. It follows from Example 13.12 that there exists a unit segment with monochromatic vertices. Call that segment s, and let us assume, without loss of generality, that the endpoints A and B of s are red. Now take the circle C with diameter s, and consider the four points D_1, D_2, D_3, and D_4 so that A, B and these four points divide the perimeter of C into six equal parts as shown in Figure 13.4.

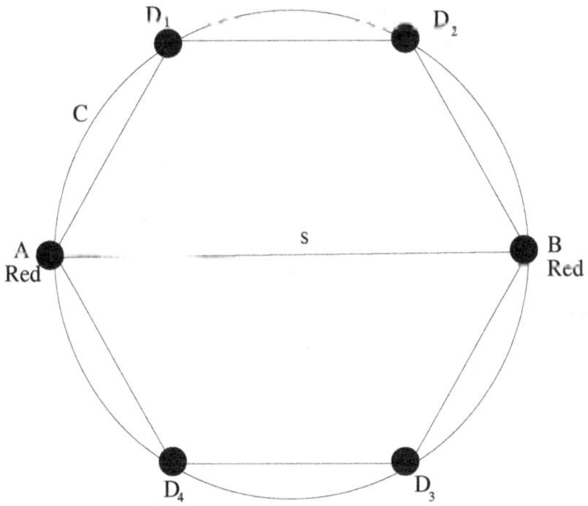

Fig. 13.4 The colors of the D_i are crucial.

If any of the D_i is red, then we are done as A, B, and this red D_i form a monochromatic (red) triangle with the required parameters. If not, then all the D_i are blue, and they form four blue triangles with the required parameters.

Notes

The first textbook on Ramsey theory was "Ramsey Theory" by Ronald Graham, Bruce Rothschild, and Joel Spencer [21]. It is an advanced book. For questions of geometric flavor, the reader is encouraged to consult "Combinatorial geometry" [31] by János Pach and Pankaj Agarwal. Finally, for questions related to coloring integers, the most comprehensive source is "Ramsey theory on the integers", by Bruce Landman and Aaron Robertson[26].

Exercises

(1) Complete the proof of Theorem 13.9.

(2) Prove that in a permutation p of length $nm+1$, there is either an increasing subsequence of length $n+1$, or a decreasing subsequence of length $m+1$. (The elements of the subsequences do not have to be in consecutive positions in p.)

(3) Each point of the space is colored either red or blue. Prove that either there is a unit square whose vertices are all blue, or there is a unit square that has at least three red vertices.

(4) Let ABC be a regular triangle, and let E be the set containing all points of the closed segments AB, AC, and BC. We color each point of E red or blue. Prove that no matter what coloring we choose, there will always be a right-angled triangle with monochromatic vertices.

(5) Eighteen teams participate at a round-robin soccer tournament. Prove that after eight rounds are played, we can still find three teams no two of which have played each other yet.

(6) Let

$$n_k = k!\left(1 + \frac{1}{1!} + \frac{1}{2!} + \frac{1}{3!} + \cdots + \frac{1}{k!}\right) + 1.$$

We color all edges of K_{n_k} with one of k colors. Prove that there will be a triangle with monochromatic edges.

(7) Let $n > 1$ be a positive integer. Prove that $R(n+2,3) > 3n$.

(8) We colored each point of the space either red, or blue, or green, or yellow. Prove that there is a segment of unit length with monochromatic vertices.

(9) Prove that it is possible to color each point of the plane either red, or blue so that there is no regular triangle with sides of unit length and monochromatic vertices.

(10) (+++) We colored each point of the plane either red, or blue. Let T be any right-angled triangle. Prove that there is a triangle that is congruent to T and has monochromatic vertices.

(11) We colored each point of the space either red or blue. Let T be a *regular* triangle. Prove that there is a triangle that is congruent to T and has monochromatic vertices.

(12) (+) We colored each point of the space either red or blue. Let T be *any* triangle. Prove that there is a triangle that is congruent to T and has monochromatic vertices.

(13) (++) We colored each point of the space either red or blue or green. Let T be as in Example 13.14. Prove that there is a triangle that is congruent to T and has monochromatic vertices.

(14) (+++) We colored each point of the space either red or blue or green. Let T be any right-angled triangle. Prove that there is a triangle that is congruent to T and has monochromatic vertices.

(15) A company has 2002 employees, from 6 different countries. Each employee has a company identification card (ID), and these cards are numbered from 1 to 2002. Prove that there is either an employee whose ID number is equal to the sum of the ID numbers of two of his compatriots, or there is an employee whose ID number is twice that of one of his compatriots.

(16) Let us color each positive integer by one of the colors $1, 2, \cdots, k$. Prove that there exists an integer $N = N(k)$ so that if $n > N$, then there are three integers a, b, c that are less than n, are of the same color, and satisfy $a + b = c$. (We allow $a = b$.)

(17) Let $N(k)$ be defined as in the previous exercise. Determine $N(2)$.

(18) Prove that $N(3) > 13$.

Supplementary Exercises

(19) The following are true for the n guests of a Christmas party.

- In any group of three guests, there are two guests who do not know each other, and
- in any groups of seven guests, there are two guests who do know each other.

At the end of the party, everyone gives a present to all the guests he or she knows. Prove that the total number of gifts given is at most $6n$.

(20) Prove that if we color the edges of K_6 red or blue, then there will be at least *two* triangles with monochromatic edges.

(21) Prove that if we color the edges of K_9 red or blue, then we will get at least *twelve* triangles with monochromatic edges.

(22) Prove that there do not exist three irrational numbers so that no matter how we choose two of them, their sum is always rational.

(23) Let k and n be positive integers satisfying $1 \leq k < n$. Prove that there do not exist n irrational numbers so that no matter how we choose k of them, their sum is always rational.

(24) There are nine passengers on a bus. Among any three of them, there are two who know each other. Prove that there are five people on the bus who know at least four of the other passengers.

(25) Continuing the previous exercise, is it true that there are five people on the bus who all know each other?

(26) Is it true that on the bus of Exercise 24 there are always six people who know at least four others?

(27) Generalize Exercise 24 for a bus with $2n+1$ passengers, keeping the condition that among any three of them, there are two who know each other.

(28) Five vertices of a regular 10-gon are colored red, and five are colored blue. Prove that there is a triangle T_1 with red vertices and a triangle T_2 with blue vertices that are congruent.

(29) Each vertex of a regular 13-gon is colored either red or blue. Prove that there exists an isosceles triangle with monochromatic vertices.

(30) We colored the edges of K_6 red or blue. Prove that there is a cycle of length four with monochromatic edges.

(31) We colored the edges of K_7 red or blue. Prove that there are at least three cycles of length four with monochromatic edges.

(32) Prove that $R(3,5) = 14$.

(33)(a) (+) Let T_m be any tree on m vertices. Let us color all vertices of $K_{(m-1)(n-1)+1}$ red or blue. Prove that there will be either a copy of T_m with all edges red, or a copy of K_n, with all edges blue.

(b) Prove that the result of part (a) is optimal.

(34) We color each vertex of the plane red or blue. Let $n \geq 3$ be an integer. Prove that there exist n points so that all these points and their centroid have the same color. Try to find a proof that only considers $2n + 1$ points. (Recall that the centroid of a set of n points in a (vector) space, viewed as the vectors $\mathbf{v_1}, \mathbf{v_2}, \cdots, \mathbf{v_n}$ is the point given by the vector $(\mathbf{v_1} + \mathbf{v_2} + \cdots + \mathbf{v_n})/n$.)

(35) We color each point of the n-dimensional plane having integer coordinates red or blue. Prove that there will be a segment with monochromatic vertices whose centroid has the same color as its two endpoints.

(36) Prove that the statement of Exercise 34 remains true even if we only color the vertices of the plane that have integer coordinates.

(37) Prove that for all integers $n \geq 1$, there exists a permutation of length n that does not contain an arithmetic progression of length 3. Note that an arithmetic progression can be increasing or decreasing.

(38) Prove that each permutation of the set of all positive integers contains an increasing arithmetic progression of length three. (A permutation of an infinite set S is an arrangement of all the elements of S in a line.)

Solutions to Exercises

(1) Proceed as in the proof provided in the text, except for the choice of N. Set $N = R(n_1 - 1, n_2, \cdots, n_k) + R(n_1, n_2 - 1, n_3, \cdots, n_k) + \cdots + R(n_1, n_2, \cdots, n_k - 1) - k + 2$. Then it follows by the Pigeon-hole Principle that there exists an $i \in [k]$ so that there are at least $R(n_1, \cdots, n_i - 1, \cdots, n_k)$ edges adjacent to V that are colored i. Then the proof is completed as in the text.

(2) Let $p = p_1 p_2 \cdots p_{nm+1}$, and let a_i denote the length of the longest increasing subsequence ending in p_i. Similarly, let b_i denote the length of the longest decreasing subsequence ending in p_i. It is then clear that if $i \neq j$, then the ordered pairs (a_i, b_i) and (a_j, b_j) are different. Indeed, either $p_i < p_j$, and then $a_i < a_j$, or $p_i > p_j$, and then $b_i < b_j$. Thus we have $nm+1$ different ordered pairs, and the statement follows by the pigeon-hole principle.

(3) First assume that there is no segment of length $b = \sqrt{2}$ whose end-

points are both red. Then take any red point, and take the sphere S of radius b that is centered at that point. Clearly, S consists of blue points only, and therefore, any unit square on S has four blue vertices. Now let us assume that there is a segment AB that is of length b and has two red endpoints. Take the circle C whose center is the midpoint of AB, whose radius is $b/2$, and that lies in the plane that is perpendicular to AB. If any point P of C is red, then the triangle ABP has three red vertices, and can be completed to a unit square. If not, then C consists of blue points only, and contains infinitely many unit squares.

(4) Denote by C_1 and C_2 the points that divide the segment AB into three equal parts. Define A_1, A_2, B_1, and B_2 analogously. There are at least two points among A_1, B_1, and C_1 that are of the same color; we can assume without loss of generality that A_1 and B_1 are both red. Now assume there is no right-angled triangle with monochromatic vertices. Then C and B_2 must both be blue. Then we cannot find a color for C_2. If C_2 is blue, then the triangle CB_2C_2 has three blue vertices, and if C_2 is red, then the triangle $A_1B_1C_2$ has three red vertices. So in any case, a triangle with monochromatic vertices is formed.

(5) Let us consider a K_{18} whose vertices correspond to the eighteen teams. After eight rounds have been played, we color the edge between two teams red if they have met, and blue if they have not. We have to show there is a blue triangle in our graph. Take any team A, and look at the nine teams A has not played yet. If there are two teams B and C among them that have not met yet, then ABC is a blue triangle, and we are done. If there were no two such teams, that would mean that any two of the nine teams that have not played A have played each other, in other words, these teams completed a round-robin tournament among themselves. However, that is impossible for nine teams in just eight rounds. Indeed, in one round, they could only play 4 games among themselves, therefore in eight rounds, they could play at most 32. That is less than the total number of $\binom{9}{2} = 36$ games needed for a round robin tournament with nine teams.

(6) We prove the statement by induction on k. If $k = 1$, then $n_k = 3$, and if we color the edges of a triangle by one color, then of course this triangle will have monochromatic edges.

Now assume that the statement is true for k, and prove it for $k+1$. Take a complete graph on n_{k+1} vertices, and select one of its vertices, say V. Then all edges adjacent to V are colored by one of $k+1$ colors.

It is easy to verify that

$$(k+1)(n_k - 1) = n_{k+1} - 2 < n_{k+1} - 1,$$

so it follows by pigeon-hole principle that at least n_k of these edges are of the same color, say black. Let B be the set of the vertices that are connected to B by a black edge. If there is a black edge joining two vertices X and Y of B, then XYV is a triangle with all edges black. If there is no such X and Y, then all edges within B are of one of the remaining k colors. As B has at least n_k edges, the statement follows by the induction hypothesis.

(7) It suffices to construct one graph G on $3n$ vertices that does not contain a K_{n+2}, but among any three of its vertices, there are two that are adjacent (so the complement of G does not contain a triangle). Such a G can be given as follows. Let the vertex set of G be $[3n]$, and draw these vertices around a cycle in increasing order. Connect i to its n left and n right "neighbors" along the cycle. This means that two vertices are joined if their difference is either at most n or at least $2n$. However, Exercise 11 of Chapter 1 shows that no matter how we choose n of these vertices, there will be two of them with difference more than n but less than $2n$. So G cannot contain a K_{n+2}.
It is easy to see that among any three vertices of G, there are two that are adjacent. Indeed, let $a < b < c$ be three vertices. If neither ab nor bc is an edge, then we must have $b - a > n$, and also $c - b > n$, which implies $c - a > 2n$, and therefore ac is an edge.

(8) This is a generalization of Example 13.13 to three dimensions. Suppose there is no such segment. Take a regular tetrahedron $ABCD$ with sides of unit length. This tetrahedron must have vertices of four different colors. Say A is red, and append another regular tetrahedron $BCDE$ to the triangle BCD. Then E must also be red, otherwise it would agree in color with one of B, C and D.
So if m is the altitude of a regular tetrahedron, then all vertices of length $2m$ have to be of the same color. In particular, the sphere whose center is A and radius is $2m$ must be red. However, there are pairs of points on that sphere whose two points are at a unit distance from each other, and therefore the claim is proved.

(9) For shortness, let $m = \sqrt{3}/2$, and note that m is the altitude of such a regular triangle. Color a point (x, y) red if $[y/m]$ is even, and blue if $[y/m]$ is odd. We get monochromatic stripes of width m so that no triangle of the required size fits within one stripe, and no triangle

of the required size is large enough to have vertices in two different stripes of the same color.

(10) This result is due to Leslie E. Shader, and can be found in the article *All right triangles are Ramsey in the plane*, Journal of Combinatorial Theory, Series A, **20** (1976), 385–390.

(11) Assume there is no such triangle. Choose the side length of T to be the unit length. Let AB be a unit segment with monochromatic vertices. We can then assume without loss of generality that A and B are both red.

Take a regular triangle ABC, and rotate it around its side AB. Then images of the vertex C form a circle c. Clearly, all points on c must be blue. The radius of c is $\sqrt{3}/2$, therefore c has pairs of points at distance 1 from each other. Let D and E be two such points. Then we can repeat the previous argument. That is, take a regular triangle DEF and rotate it around its side DE. The rotated images of the vertex F form a circle f, and they must all be red. If we do this for all possible choices of D and E, we get a torus that consists of red points only, and it is easy to see that this torus will contain a regular triangle with sides of unit length.

(12) This result can be found in P. Erdős, R. L. Graham, P. Montgomery, B. L. Rothschild, J. H. Spencer and E. G. Straus: *Euclidean Ramsey theorems I*, Journal of Combinatorial Theory, Series A, **14** (1973), 341–363.

(13) This problem can be solved with methods similar to Example 13.14. For a full solution, see M. Bóna: *A Euclidean Ramsey theorem*, Discrete Mathematics, **122** (1993), 349–352.

(14) The solution of this problem can be found in M. Bóna, G. Tóth, *A Ramsey-type problem on right-angled triangles in space.* Selected papers in honor of Paul Erdős on the occasion of his 80th birthday (Keszthely, 1993). Discrete Math. **150** (1996), no. 1-3, 61–67.

(15) Take a complete graph on 2002 vertices, and let the edge between vertices i and j (where $i < j$) be of color k if the person with ID number $j - i$ is of country k. This defines a coloring of K_{2002} by six colors. Keeping the notations of the previous exercise, $n_6 = 1958$, so there will be a triangle with monochromatic edges. Let the vertices of this triangle be $a < b < c$. Then people with ID numbers $c - a$, $c - b$, and $b - a$ are all from the same country. As $(b - a) + (c - b) = c - a$, our claim is proved. (If $b - a$ and $c - b$ are different, then the first criterion is satisfied, and if $b - a$ and $c - b$ are equal, then the second

criterion is satisfied.)

(16) Denote by C the given coloring of the positive integers. Take the complete graph K_N whose vertex set is $[N]$, and whose edges are k-colored as follows. The edge between x and y is of the color of the integer $|x - y|$ in C. It follows from Theorem 13.9 that if N is large enough, then any, and therefore, this, k-coloring of K_N contains a triangle with monochromatic edges. Let that triangle have vertices $x < y < z$. Then we know that $y - x$, $z - y$, and $z - x$ have the same color in C, so they can play the role of a, b, and c.

(17) We prove that $N(2) = 5$. Indeed, try to 2-color $[5]$ without creating a monochromatic triple so that $a + b = c$. Assume without loss of generality that 1 is red, then 2 is blue (for $1 + 1 = 2$), and 4 is red (for $2 + 2 = 4$). Then 3 must be blue (for 1+3=4), and then we cannot find a color for 5, as $5 = 1 + 4 = 2 + 3$. Therefore, $N(2) \leq 5$. On the other hand, we have just seen that R, B, B, R is a 2-coloring of $[4]$ without a monochromatic triple of the desired kind. This proves $N(2) = 5$.

(18) We show a 3-coloring of $[13]$ without a monochromatic triple of the desired kind. Denote by R, B, and G the red, blue, and green color. Then $R, B, B, R, G, G, G, G, G, G, B, B, R$ is a good coloring.

Chapter 14

So Hard To Avoid. Subsequence Conditions on Permutations

14.1 Pattern Avoidance

Let us assume that there are n children playing in our backyard, no two of whom have the same height. For the next game, they need to stand in a line so that everyone faces the back of the preceding person. Moreover, each child must be able to see all children that are shorter than him and precede him in the line. How many such lineups exist?

At this point, the reader certainly suspects that we will have to enumerate permutations of $[n]$ with some new conditions. Let $1, 2, \cdots, n$ denote the children playing in our backyard, in increasing order of height, so 1 is the shortest and n is the tallest.

Would for example 1423567 be a good lineup for $n = 7$? No, it would not, as 2 or 3 could not see 1, even if he is smaller than them and precedes them. They could not see him as their view would be blocked by 4, who is taller than them. On the other hand 6723415 would be a good lineup.

So when is a lineup good? It is good if there are no three elements a, b, c so that they are in this order (but not necessarily in consecutive positions), and $a < c < b$. Indeed, if there were three elements like that, then b could not see a.

The enumeration of permutations with subsequence conditions like this is a very active area of contemporary combinatorics. Before we continue, we make two definitions to simplify our arguments.

Definition 14.1. Let a, b, and c be three entries of a permutation that follow in this order from left to right, but are not necessarily consecutive. If $a < c < b$, then we say that the entries a, b, and c form a *132-pattern*.

Why do we call this structure a *132-pattern*? Because the entries a,

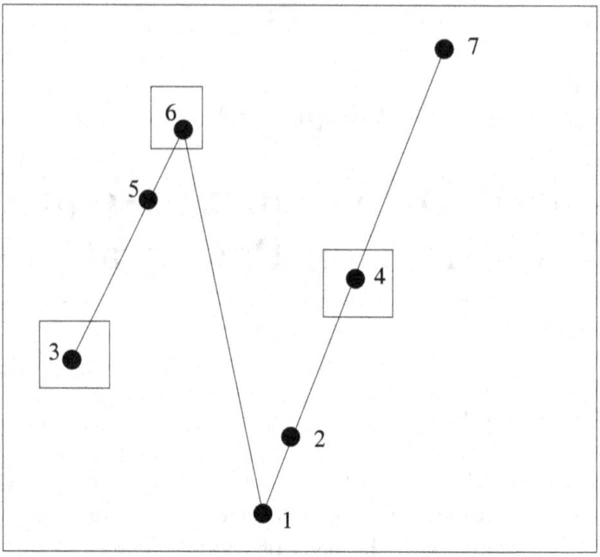

Fig. 14.1 The permutation 3561247 contains the pattern 132.

b, and c relate to each other the same way as the numbers 1, 3, and 2. That is, the leftmost one is the smallest, and the middle one is the largest. Similarly, if we had $a < b < c$, then we would say that the entries a, b, and c form a *123-pattern*, and if we had $c < a < b$, then we would say that the entries a, b, and c form a *231-pattern*.

Definition 14.2. Let p be a permutation. If there are no three entries in p that form a 132-pattern, then p is called *132-avoiding*.

Our task is therefore to find the number $f(n)$ of permutations of length n (or, in what follows, n-permutations) that are 132-avoiding.

Let us suppose that we have a 132-avoiding n-permutation in which the entry n is in the ith position. Then we claim that any entry to the left of n must be larger than any entry to the right of n. In order to see this, let us assume the contrary is true, that is, there is an entry x on the left of n and an entry y on the right of n so that $x < y$. Then the entries x, n, and y form a 132-pattern, which is a contradiction. This implies that entries $1, 2, \cdots, n-i$ are on the right of n, and entries $n-i+1, n-i+2, \cdots, n-1$ are on the left of n. Now the $i-1$ entries on the left of n must also form a 132-avoiding permutation, which they can do in $f(i-1)$ ways. Similarly, the $n-i$ entries on the right of n must form a 132-avoiding permutation, and they can do it in $f(n-i)$ ways. So there are exactly $f(i-1)f(n-i)$ 132-avoiding n-permutations in which n is in the ith position. Here we set $f(0)=1$, in order to make the recurrence work.

Summing over all $i \in [n]$ (as n can be in any position) we get the recurrence relation for

$$f(n) = \sum_{i=1}^{n} f(i-1)f(n-i), \tag{14.1}$$

if $n \geq 1$, and $f(0)=1$.

We have met this recurrence relation before, in formula (8.19), when we solved Exercise 15 of Chapter 8. We solved that recurrence relation using generating functions. The conclusion was that the solution of (8.19), and, equivalently, of (14.1) is the sequence of Catalan numbers.

Corollary 14.3. *The number of permutations of length n that avoid the pattern 132 is $c_n = \binom{2n}{n}/(n+1)$, which is the nth Catalan number.*

So we know that the number of n-permutations avoiding the pattern 132 is $c_n = \binom{2n}{n}/(n+1)$. How about the number of permutations avoiding other patterns? Before addressing that question, let us formally announce the definition of pattern avoidance for general patterns, even if we mentioned it in the text before.

Definition 14.4. Let p be an n-permutation, and let $q = q_1q_2\cdots q_k$ be a k-permutation, with $n \geq k$. Let us choose k entries of p, and denote them by $a_1, a_2, \cdots, a_k$, as they follow from left to right. If $q_i < q_j$ exactly for those indices i and j for which $a_i < a_j$, then we say that the elements $a_1, a_2, \cdots, a_k$ form a *q-pattern*.

Definition 14.5. Let p be an n-permutation, and let $q = q_1q_2\cdots q_k$ be a k-permutation, with $n \geq k$. If no k entries of p form a q-pattern, then we say that p is a *q-avoiding permutation*.

The number of q-avoiding n-permutations is denoted by $S_n(q)$.

Let us return to our main task of determining $S_n(q)$ for patterns other than 132. We start with patterns of length three as the problem is trivial for shorter patterns.

We claim that $S_n(231) = S_n(132)$. Indeed, note that 231 is precisely the *reverse* of 132. So if an n-permutation avoids 132, its reverse avoids

231, and vice versa. This sets up a natural bijection between the set of 132-avoiding n-permutations, and that of 231-avoiding n-permutations.

We also claim that $S_n(312) = S_n(132)$. To see this, define the *complement* of an n-permutation $p = p_1p_2\cdots p_n$ to be the n-permutation $\bar{p}$ whose first entry is $n+1-p_1$, whose second entry is $n+1-p_2$, and in general, whose ith entry is $n+1-p_i$. So for example, the complement of 34152 is 32514.

Now observe that 312 is the complement of 132. Moreover, note that if p avoids 312, then p_c avoids 132, and vice versa, proving $S_n(312) = S_n(132)$.

So far we have seen that $S_n(132) = S_n(231) = S_n(312)$. It is easy to extend this chain of equalities one further. Indeed, 213 is the reverse of 312, so $S_n(132) = S_n(231) = S_n(312) = S_n(213)$.

There are two more patterns of length three, namely 123 and 321. It is clear by taking reverses, or by taking complements, that $S_n(123) = S_n(321)$. This leaves us with one last question. Is it also true that $S_n(123) = S_n(132)$? If it is, that means that all permutation patterns of length three are avoided by the same number of n-permutations, and this number is c_n, the nth Catalan number. The answer to this question is in the affirmative. The proof of this is slightly harder than the previous symmetry arguments, and is the content of the following Lemma.

Lemma 14.6. *For all positive integers n, the equality*

$$S_n(123) = S_n(132)$$

holds.

We need some machinery before we start proving this Lemma. Recall that an entry of a permutation which is smaller than all the entries that precede it is called a *left-to-right minimum.* Note that the left-to-right minima of a permutation form a decreasing subsequence. For example, in the permutation 4531762, the entries 4, 3, and 1 are the left-to-right minima. Note that the leftmost entry, and the entry 1 are always left-to-right minima.

Proof. (of Lemma 14.6) We will construct a bijection f from the set of all 123-avoiding n-permutations onto the set of all 132-avoiding n-permutations which leaves all left-to-right minima fixed. (This last property is not needed for the proof of our Lemma, but it will be useful later.)

The bijection f is defined as follows. We take any 123-avoiding n-permutation p, fix all its left-to-right minima, and remove all the elements that are not left-to-right minima, leaving their places empty. Then going from the left to the right, we put the elements which are not left-to-right minima into the empty slots between the left-to-right minima so that in each step we place the smallest element we have not placed yet which is larger than the previous left-to-right minimum. In other words, in each step, we place the smallest entry that is both available (that is, it is not a left-to-right minimum) and eligible (that is, it is not smaller than the previous left-to-right minimum). The reader is invited to verify that there is always at least one such entry, so the process will never get stuck.

For example, if $p = 4\,6\,5\,1\,3\,2$, then the left-to-right minima are the entries 4 and 1, thus we leave them in the first and fourth positions. The first empty slot is the second position and we put there the smallest entry which is larger than 4, that is to say, the entry 5. Similarly, we put 6 to the third position as it is the smallest of the entries not yet used which is larger than 4 (in fact, this is the only such entry). Then by the same reasoning we put 2 into the fifth position and 3 into the sixth position. This way we get the permutation $f(p) = 4\,5\,6\,1\,2\,3$.

Note that $f(p)$ is 132-avoiding, because if there were a 132-pattern in $f(p)$, then there would be one which starts with a left-to-right minimum, but that is impossible as elements larger than any given left-to-right minimum and to the right of it are written in increasing order.

The inverse of f is even easier to describe: keep the left-to-right minima of p fixed and put all the other elements into the empty slots between them in decreasing order. Note that this procedure will not change the set of left-to-right minima of p (why?). We obtain a permutation which is the union of two decreasing subsequences and is therefore 123-avoiding. If we apply this operation to $f(p)$, then we must get p back, as the left-to-right minima have not changed, and the other elements must have been in decreasing order in p, otherwise p would not have been 123-avoiding. This completes the proof of the lemma. □

Theorem 14.7. *Let q be any permutation pattern of length three. Then for all positive integers n,*

$$S_n(q) = c_n = \frac{\binom{2n}{n}}{n+1}.$$

Proof. Lemma 14.6 and the preceding easy symmetry arguments show that $S_n(q)$ is the same for all patterns q of length three. As we know that $S_n(132) = c_n$, the statement follows. □

So we can enumerate permutations avoiding a given pattern q if the length of q is three. However, for longer patterns q, the problem becomes harder at a drastic speed. There are very few patterns q such that an exact formula is known for $S_n(q)$. To see one of the reasons for this, consider patterns of length four. There are 24 of them, but using reverses, complements, and some less obvious tricks, one can deduce that there are only three of them that are really different, namely 1234, 1342, and 1324. Computer calculations provide the following fascinating numerical evidence for these patterns (the values of $S_n(q)$, for $n \leq 8$).

- for $S_n(1342)$: 1, 2, 6, 23, 103, 512, 2740, 15485
- for $S_n(1234)$: 1, 2, 6, 23, 103, 513, 2761, 15767
- for $S_n(1324)$: 1, 2, 6, 23, 103, 513, 2762, 15793.

We see that unlike for patterns of length three, it is *no longer true* here that $S_n(q)$ does not depend on q. It also seems that for $n \geq 7$,

$$S_n(1342) < S_n(1234) < S_n(1324).$$

This is actually true. It is very surprising, and not well understood, that the monotonic pattern is in the middle of this chain. It would have been plausible to think that the monotonic pattern is the easiest, or the hardest, to avoid.

We prove the second part of this inequality. The first part follows from Exercise 5.

Theorem 14.8. *For all $n \geq 7$, the inequality $S_n(1234) < S_n(1324)$ holds.*

Proof. We are going to classify all permutations of n according to the set and position of their left-to-right minima and right-to-left maxima. This is the content of the following definition.

Definition 14.9. Two permutations x and y are said to be in the same *class* if

- the left-to-right minima of x are the same as those of y, and
- the left-to-right minima of x are in the same positions as the left-to-right minima of y, and
- the same holds for the right-to-left maxima.

For example, $x = 5\,1\,2\,3\,4$ and $y = 5\,1\,3\,2\,4$ are in the same class, but $z = 2\,4\,3\,1\,5$ and $v = 2\,4\,1\,3\,5$ are not, as the third entry of z is not a left-to-right minimum whereas that of v is.

The outline of our proof is going to be as follows: we show that each nonempty class contains *exactly* one 1234-avoiding permutation and *at least* one 1324-avoiding permutation. Then we exhibit some classes which contain more than one 1324-avoiding permutation and complete the proof.

Lemma 14.10. *Each nonempty class contains exactly one 1234-avoiding permutation.*

Proof. Suppose we have already picked a class, that is, we fixed the positions and values of all the left-to-right minima and right-to-left maxima. We claim that if we put all the remaining elements into the remaining slots in decreasing order, then we get a 1234-avoiding permutation.

Indeed, the permutation obtained this way consists of three decreasing subsequences, that is, the left-to-right minima, the right-to-left maxima, and the remaining entries. If there were a 1234-pattern in this permutation, then by the Pigeon-hole Principle two of its entries would be in the same decreasing subsequence, which would be a contradiction. On the other hand, if two of these elements, say a and b, were in increasing order, then together with the rightmost left-to-right minimum on the left of a and the leftmost right-to-left maximum on the right of b they would form a 1234-pattern. Finally, if the chosen class is nonempty, then we can indeed write the remaining numbers in decreasing order without conflicting with the existing constraints— otherwise the class would be empty. (In other words it is the decreasing order of the remaining elements that violates the least number of constraints.) □

Now comes the harder part. Recall that an *inversion* in a permutation $p = p_1p_2\cdots p_n$ is a pair (p_i, p_j) so that $i < j$ but $p_i > p_j$.

Lemma 14.11. *Each nonempty class contains at least one 1324-avoiding permutation.*

Proof. First note that if a permutation contains a 1324-pattern, then we can choose such a pattern so that its first element is a left-to-right minimum and its last element is a right-to-left maximum. Indeed, we can just take any existing pattern and replace its first (last) element by its closest left (right) neighbor which is a left-to-right minimum (right-to-left maximum). Therefore, to show that a permutation avoids 1324, it is sufficient to show that it does not contain a 1324-pattern having a left-to-right minimum for its first element and a right-to-left maximum for its last element. (Such a pattern will be called a *good pattern.*) Also note that a left-to-right minimum (right-to-left maximum) can only be the first (last) element of a 1324-pattern.

Now take any 1324-containing permutation. By the above argument, it has a good pattern. Interchange its second and third elements. Observe that we can do this without violating the existing constraints, that is, no element x goes on the left of a left-to-right minimum y such that $x < y$, and no element x goes on the right of a right-to-left maximum z such that $z < x$. The resulting permutation is in the same class as the original because the left-to-right minima and right-to-left maxima have not been changed. Let us repeat this procedure as long as we can. Note that each step of the procedure *decreases the number of inversions of our permutation* by at least 1. Therefore, we will have to stop after at most $\binom{n}{2}$ steps. Then the resulting permutation will be in the same class as the original one, but it will have no good pattern and therefore no 1324-pattern, as we claimed. □

Notation (by example): in what follows, we write $a_1 * a_2 * * b_1$ for the class of permutations of length six which have two left-to-right minima, a_1 and a_2, which are in the first and third position, and one right-to-left maximum, b_1, which is in the last position.

Finally, we must show that "at least one" in the above lemma does not always mean exactly one. If $n = 7$, then the class $3 * 1 * 7 * 5$ contains two 1324-avoiding permutations, 3 6 1 2 7 4 5 and 3 4 1 6 7 2 5. This proves $S_7(1234) < S_7(1324)$. For larger n we can extend this example in an easy way, such as taking the class $n\ (n-1) \cdots 8\ 3 * 1 * 7 * 5$. This shows that there are more 1324-avoiding permutations than 1234-avoiding ones and completes the proof of the theorem. □

As we said, there are very few patterns q that are longer than three so that an exact formula is known for $S_n(q)$. Therefore, even good approximations or upper bounds for $S_n(q)$ would be interesting. The famous Stanley–Wilf conjecture claimed that for any pattern q, there exists a constant c_q so that $S_n(q) \leq c_q^n$ for all n. This conjecture resisted numerous solution attempts in the last twenty years. Finally, the conjecture has been proved [30] using a spectacular argument, by Adam Marcus and Gábor Tardos in 2003. The best possible value of the constant c_q is still unknown. (The Marcus-Tardos proof, beautiful as it is, does not provide a constant that would seem to be close to the actually needed value of c_q.)

In some special cases, however, we can find a small constant c_q so that $S_n(q) \leq c_q^n$ for all n. The easiest case is when q is monotonic.

Theorem 14.12. *For all positive integers $k \leq n$, the inequality*

$$S_n(1234 \cdots k) \leq (k-1)^{2n}$$

holds.

Proof. Let us say that an entry x of a permutation is of rank i if x is the top of a rising subsequence of length i, but there is no rising subsequence of length $i+1$ whose top is x. Then for all i, elements of rank i must form a descending subsequence. Therefore, a q-avoiding permutation can be decomposed into the union of $k-1$ descending subsequences. There are $(k-1)^n$ ways to partition the elements into $k-1$ classes and there are less than $(k-1)^n$ ways to assign each position to one of the subsequences, completing the proof. □

Note that this result is completely in line with our earlier results, showing that $S_n(123) = c_n < 4^n$.

Additional patterns q for which an exact formula is known for $S_n(q)$ will be mentioned in the Notes. We conclude this section by presenting a recursive result. We will need the following definition.

Definition 14.13. Let $p \in S_a$, and $q \in S_b$, with $p = p_1p_2\cdots p_a$ and $q = q_1q_2\cdots q_b$. Then the *direct sum* of p and q is the pattern $p \oplus q \in S_{a+b}$ where

$$(p \oplus q)_i = \begin{cases} p_i & \text{if } i \leq a\,, \\ \\ q_{i-a} + a & \text{if } i > a. \end{cases}$$

In other words, we increase each entry of q by a before placing q after p.

Example 14.14. If $p = 132$ and $q = 2431$, then $p \oplus q = 1325764$.

Now we are in a position to announce and prove the recursive result that we promised.

Theorem 14.15. *Let q_1 and q_2 be patterns so that $S_n(q_1 \oplus 1) \leq c_1^n$ for all n, and that $S_n(1 \oplus q_2) \leq c_2^n$ for all n. Then*

$$S_n(q_1 \oplus 1 \oplus q_2) \leq (\sqrt{c_1} + \sqrt{c_2})^{2n}$$

for all n.

Example 14.16. Let $q_1 = 213$, and let $q_2 = 132$. Then Exercise 30 and Theorem 14.12 imply that $S_n(q_1 \oplus 1) = S_n(2134) < 9^n$, and also, $S_n(1 \oplus q_2) = S_n(1243) < 9^n$. Therefore,

$$S_n(q_1 \oplus 1 \oplus q_2) \leq (3+3)^{2n} = 36^n.$$

Proof. (of Theorem 14.15) Let $p \in S_n$ be a permutation that avoids $q = q_1 \oplus 1 \oplus q_2$. Color all entries of p that can play the role of the last (and largest) entry of a $q_1 \oplus 1$-pattern red, and color all other entries blue.

Then the string of all red entries must avoid $1 \oplus q_2$. Indeed, if did not, then any copy C of $1 \oplus q_2$ made up by red entries could be turned into a copy of q by using the entries on the left of C that make the leftmost entry of C red. (This is the point where we use the structure of $q = q_1 \oplus 1 \oplus q_2$, that is, the property that each entry in the first part is smaller than each entry in the second part.)

Furthermore, the string of blue entries must be $(q_1 \oplus 1)$-avoiding. Indeed, if it contained a copy D of that pattern, then the last entry of that pattern would have to be a red entry, which would be a contradiction.

Therefore, if there are k blue entries and $n-k$ red entries, then there are at most $\binom{n}{k}^2 c_1^k c_2^{n-k}$ permutations of length n that avoid q. Indeed, there are at most $\binom{n}{k}$ possibilities for the set of blue entries, and the same number of possibilities for the positions of these entries. Summing over all k, this yields

$$\begin{aligned} S_n(q) &\leq \sum_{k=1}^{n} \binom{n}{k}^2 c_1^k c_2^{n-k} \\ &\leq \left(\sum_{k=1}^{n} \binom{n}{k} \sqrt{c_1^k c_2^{n-k}} \right)^2 \\ &\leq (\sqrt{c_1} + \sqrt{c_2})^{2n}. \end{aligned}$$

We have used the fact that the sum of the squares of positive real numbers is at most as large as the square of their sum, as well as the Binomial Theorem. □

14.2 Stack Sortable Permutations

The initial setup of our topic for this section sounds similar to the well-known game of Hanoi towers. Assume we have a permutation $p = p_1 p_2 \cdots p_n$ and we want to sort its entries, to get the identity permutation $12 \cdots n$. Our only tool is a *stack*, a vertical array that can hold entries in increasing order, that is, the smallest one on top, and the largest one at the bottom.

The numbers enter the stack in the order in which they occur in the input permutation p. We take p_1, and put it in the stack. Now take p_2. If

$p_2 < p_1$, then it is allowed for p_2 to go in the stack on top of p_1, so we will put it there. If $p_2 > p_1$, however, then first we take p_1 out of the stack, and put it to the first position of the output permutation, and put p_2 into the stack. We continue this way: at step i, we compare p_i with the element $r = p_{a_{i-1}}$ currently on top of the stack. If $p_i < r$, then p_i goes on the top of the stack, if not, then r goes to the leftmost empty position of the output permutation, and p_i gets compared to the new element that is currently on the top of the stack. The algorithm ends when all n entries passed through the stack and are in the output permutation $s(p)$. See Figure 14.2 for an example of this procedure.

Example 14.17. Let $p = 2413$. Then the stages of our sorting procedure are shown in Figure 14.2.

INPUT	STACK	OUTPUT
2413		
413	2	
413		2
13	4	2
3	1 4	2
3	4	21
	3 4	21
	4	213
		2134

Fig. 14.2 Sorting 2413.

If the image $s(p)$ of p under this stack sorting operation is the identity permutation, then we say that p is *stack sortable*. So the previous example shows that 2413 is not stack sortable.

Which permutations are stack sortable? To answer this natural question, we first analyze the effect of the stack sorting operation s to pairs of entries in p.

Proposition 14.18. *Let p be a permutation, and let $a < b$ be two entries of p. Then*

(1) if a precedes b in p, then a precedes b in $s(p)$,
(2) if b precedes a in p, and there is no element c located between a and b in p so that $c > b > a$, then a precedes b in $s(p)$,
(3) if b precedes a in p, and there is an element c located between a and b in p so that $c > b > a$, then b precedes a in $s(p)$. Note that this happens when the entries a, b, and c form a 231-pattern.

Proof.

(1) As a precedes b in p, a will enter the stack before b. As $a < b$, this means that b cannot even enter (let alone, leave) the stack before a does, so a precedes b in $s(p)$.
(2) In this case the string of p between b and a is a decreasing subsequence S. The elements of S enter the stack starting with b, then they pile up on top of each other, with a entering the stack last, and getting therefore to the top of the stack. So a will be the first element of S to leave the stack. In particular, a leaves the stack before b, and thus a precedes b in $s(p)$.
(3) In this case, b has to leave the stack before c enters it. On the other hand, c has to enter the stack before a does. Therefore, b leaves the stack before a could even enter it, so b precedes a in $s(p)$. □

Theorem 14.19. *A permutation p is stack sortable if and only if it avoids the pattern 231.*

Proof. If there is a 231-pattern in p, formed by the entries $a < b < c$, then part 3 of the previous Proposition shows that b will precede a in $s(p)$, so $s(p)$ cannot be the identity permutation. If there is no 231-pattern in p, then any pair $a < b$ of entries falls either into part 1, or into part 2 of the previous proposition, and will therefore be sorted. □

So most permutations are not stack sortable. To increase the number of permutations that can be sorted using our stack, we can take $s(p)$, and pass it through the stack again, following the same rules. If the obtained permutation $s(s(p))$ is the identity permutation, then we say that p is *two-stack sortable.*

Two-stack sortable permutations are more difficult to characterize, let alone enumerate, than stack sortable permutations. One reason for this difficulty is that the two-stack sortable property is *not monotonic.* That is, there are instances when p is two-stack sortable, but a subsequence p' of p is not.

Example 14.20. Let $p = 35241$. Then $s(p) = 32145$, and $s(s(p)) = 12345$, so p is two-stack sortable. Now let $p' = 3241$. Then $s(p') = 2314$, and $s(s(p')) = 2134$, so p' is not two-stack sortable.

For this reason, we cannot hope for a characterization of two-stack sortable permutations by pattern avoidance only. However, we can still use a similar concept if we stretch the definition of pattern avoidance a little bit.

Theorem 14.21. *A permutation p is two-stack sortable if and only if it does not contain a 2341-pattern, and it does not contain a 3241-pattern, except as a part of a 35241-pattern.*

Proof. First we prove the "only if" part. Assume entries $a < b < c < d$ of p form a 2341 pattern. Then it follows from Proposition 14.18 that entries a, b, and c form a 231-pattern in $s(p)$, implying that $s(p)$ is not stack sortable.

Now assume that entries $w < x < y < z$ form a 3241-pattern in p that is not part of a 35241-pattern. Proposition 14.18 then implies that both x and y precede w in $s(p)$. If there are no entries between x and y in p that are larger than both of them, then Proposition 14.18 also implies that x precedes y in $s(p)$, and we are done as w, x, and y form a 231-pattern in $s(p)$. If there is an entry t between x and y that is larger than both of them, then, keeping in mind that the 3241-pattern $yxzu$ is not part of any 35241-pattern, the pattern $ytxzu$ must be a 34251-pattern. However, that implies that entries y, t, z, and u form a 2341-pattern, and we have seen in the previous paragraph that such a pattern prevents p from being two-stack sortable.

Now we prove the "if" part. It suffices to show that if $s(p)$ is not stack sortable, then p had to contain one of the two forbidden configurations mentioned in the theorem. If $s(p)$ is not stack sortable, then it contains a 231-pattern. Let $e < f < g$ be the entries of one such pattern. Then by Proposition 14.18, e was the rightmost of these three entries in p, and there had to be an entry h in p that separated both f and g from the entry e. If f preceded g in p, then $fghe$ was a 2341-pattern in p, and we are done.

If not, then g preceded f in p. We know that f precedes g in $s(p)$, which implies that there was no entry between g and f in p that was larger than both of them. So $gfhe$ formed a 3241-pattern in p that was not part of a 35241-pattern, completing the proof. □

The number of two-stack sortable n-permutations is known to be

$$\frac{2}{(n+1)(2n+1)} \cdot \binom{3n}{n}.$$

This formula has at least four different proofs, all of which are somewhat complicated.

We can certainly generalize our definitions. We say that a permutation p is t-stack sortable if $s^t(p)$ is the identity permutation. In other words, passing p through the stack t times, we get the identity permutation. Note that all t-stack sortable permutations will necessarily be u-stack sortable permutations, for all $u > t$.

While we are not able to enumerate t-stack sortable permutations, we will prove several interesting statements concerning them. To that end, we need to have a deeper understanding of the effects of the stack sorting operation.

Lemma 14.22. *Let $p = LnR$ be an n-permutation, where L denotes the string on the left of the entry n, and R denotes the string on the right of the entry n. Then we have*

$$s(p) = s(L)s(R)n.$$

Proof. As p passes through the stack, first the entries that belong to L enter the stack. They all leave the stack before n enters, creating $s(L)$ at the front of the output permutation. Then n enters the stack. Then all the entries belonging to R pass through the stack, creating $s(R)$ in the output permutation, while n stays in the bottom of the stack. Finally, n leaves the stack. □

We mention that the property $s(p) = s(L)s(R)n$ in fact *defines* the stack sorting operation. That is, the stack sorting operation is the only operation defined on all finite permutations that has this property.

Corollary 14.23. *All n-permutations are $(n-1)$-stack sortable.*

Proof. We prove this statement by induction on n. For $n = 1$, the statement is trivial. Now suppose the statement is true for $n-1$, and prove it for n.

Let $p = LnR$ be any n-permutation. Lemma 14.22 means in particular that $s(p)$ always ends with its largest entry, and also, if R is empty, then $s(Ln) = s(L)n$. Iterating this, $s^{n-1}(p) = s^{n-2}(s(L)s(R)n) = s^{n-2}(s(L)s(R))n$. This latter is the identity permutation as $s(L)s(R)$ is a permutation of length $n-1$, and therefore is $(n-2)$-stack sortable by the induction hypothesis. □

Corollary 14.24. *For all n-permutations p, the t-sorted image $s^t(p)$ ends in the string $(n-t+1)(n-t+2)\cdots n$.*

Proof. Immediate by induction on t. □

The property that $s(p) = s(L)s(R)n$ enables us to translate the stack sorting operation into the language of binary plane trees. If p is an n-permutation, we associate a rooted tree $T(p)$ to p as follows.

The root of $T(p)$ is a vertex labeled n, the largest entry of p. If a is the largest entry of p on the left of n, and b is the largest entry of p on the right of n, then the root will have two children, the left one will be labeled a, and the right one labeled b. If n was the first (resp. last) entry of p, then the root will have only one child, and that will be a left (resp. right) child, and it will necessarily be labeled $n-1$ as $n-1$ must be the largest of all remaining elements.

Define the rest of $T(p)$ recursively, by taking $T(p')$ and $T(p'')$, where p' and p'' are the substrings of p on the two sides of n, and affixing them to a and b.

Example 14.25. If $p = 263498175$, then $T(p)$ is the tree shown in Figure 14.3.

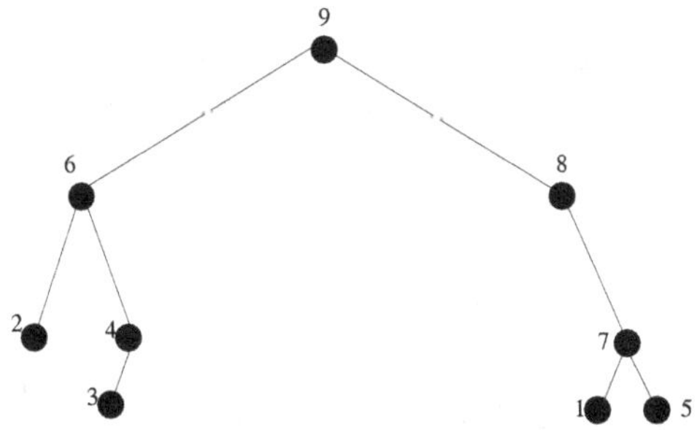

Fig. 14.3 The decreasing binary tree of $p = 263498175$.

The tree $T(p)$ is called the *decreasing binary tree* of p. It is indeed a binary tree, that is, each vertex has 0, 1, or 2 children. We repeat, for emphasis, that each child is a left child or a right child of its parent, even if that child is an only child. Given $T(p)$, we can easily recover p by reading T according to the tree traversal method called *in-order*. In other words, first we read the left subtree of $T(p)$, then the root, and then the right subtree of $T(p)$. We read the subtrees according to this very same rule.

Now let us read the tree $T(p)$ in *postorder* instead. In other words, let us first read the left subtree of $T(p)$, then the right subtree of $T(p)$, and finally the root.

Example 14.26. The tree shown in Figure 14.3 is the decreasing binary tree of $p = 263498175$. Read in postorder, it yields the permutation 234615789.

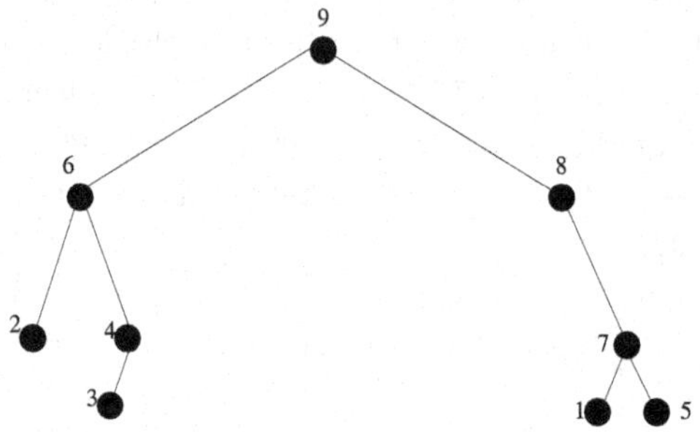

Fig. 14.4 Read in postorder, this tree yields 234615789.

The alert reader might have noted that reading $T(p)$ in postorder we precisely got the permutation $s(p)$, the image of p under the stack sorting operation. This is no accident.

Proposition 14.27. *Let p be any n-permutation. The decreasing binary tree $T(p)$ of p, read in postorder, yields the permutation $s(p)$.*

Proof. We prove the statement by induction on n, the initial case of $n = 1$ being trivial. Assume the statement is true for all positive integers less than n. Let $p = LnR$, and let us read $T(p)$ in postorder. We start with the left subtree, which is in fact $T(L)$. Reading that in postorder, we get

$s(L)$ by the induction hypothesis. Then we have to read the right subtree, which is $T(R)$. Reading that in postorder, we get $s(R)$ by the induction hypothesis. (Both L and R are shorter than p.) Finally, we read the root, which is n. So we obtain the permutation $s(L)s(R)n$, and we are done by Proposition 14.22. □

Recall that we say that i is a *descent* of the permutation $p = p_1 p_2 \cdots p_n$ if $p_i > p_{i+1}$. Similarly, we say that i is an *ascent* of the permutation p if $p_i < p_{i+1}$. Let $d(p)$ denote the number of descents of p. Note that if p is an n-permutation, then $n - 1 - d(p)$ is equal to the number of ascents of p. Moreover, if i is a descent of p, then i is an ascent of the complement of p. It follows immediately that there are as many n-permutations with k descents as there are with $n - 1 - k$ descents.

If we consider decreasing binary trees again, it is straightforward to verify that p has k descents if and only if $T(p)$ has k edges connecting a vertex to the right child of that vertex.

Let us now enumerate t-stack sortable n-permutations according to their descents. Let $W_t(n,k)$ be the number of t-stack sortable n-permutations with k descents. The following table shows the numbers $W_t(n,k)$ for small values of the parameters.

		k=0	k=1	k=2	k=3	k=4
n=4	t=1	1	6	6	1	
	t=2	1	10	10	1	
	t=3	1	11	11	1	
n=5	t=1	1	10	20	10	1
	t=2	1	20	49	20	1
	t=3	1	25	62	25	1
	t=4	1	26	66	26	1

Fig. 14.5 The numbers $W_t(n,k)$ for $n = 4$ and $n = 5$.

These data seem to suggest that $W_t(n,k) = W_t(n, n-1-k)$, for all positive integers n, k, t. If true, this would be a surprising theorem, as there seems to be nothing "symmetric" about t-stack sortable permutations, these obscure creatures. The complement, or reverse, of a t-stack sortable permutation does not need to be t-stack sortable (try 213, or 132, with $t = 1$), so these easy bijections will not work.

In the rest of this chapter, we prove this nice symmetry. We will also see the tree interpretation of the stack-sorting operation at work. The following simple map will be our main tool.

Definition 14.28. Let f be the map defined on all finite permutations as follows

- $f(1) = 1$,
- if p is an n-permutation, and $p = LnR$, and neither L nor R is empty, then $f(p) = f(L)nf(R)$,
- if p is an n-permutation and $p = Ln$, then $f(p) = nf(L)$, and
- if p is an n-permutation and if $p = nR$, then $f(p) = f(R)n$.

In words, if the maximal entry n is at neither endpoint of p, then we keep n fixed and apply f recursively on both sides of n. If n is at either endpoint, then we put n into the opposite endpoint, and apply f recursively. When we apply f recursively to L and R, then we treat L and R as permutations. This means that the maximum element of L will take over the role of n when $f(L)$ is formed, and the maximum element of R takes over the role of n when $f(R)$ is formed.

Example 14.29. If $p = 123$, then $f(p) = 321$. So if $p = 4123$, then $f(p) = 3214$.

Example 14.30. If $p = 1423$, then $f(p) = 1432$.

Example 14.31. As a consequence of the preceding examples, if $p = 412395867$, then $f(p) = 321495876$.

The following Proposition shows that the effect of f on the number of descents of a permutation is precisely what we will need.

Proposition 14.32. *For any n-permutation p, the equality $d(p)+d(f(p)) = n-1$ holds.*

Proof. We prove this claim by induction on n, the initial case being trivial. First assume that n is at neither endpoint of p, so $p = LnR$, and $f(p) = f(L)nf(R)$. Say that n is in the ith position of p. Then we have

$d(p) = d(L) + d(R) + 1$, and $d(f(p)) = d(f(L)) + d(f(R)) + 1$. So

$$d(p) + d(f(p)) = d(L) + d(R) + 1 + d(f(L)) + d(f(R)) + 1$$
$$= (i-2) + (n-i-1) + 2 = n-1,$$

which was to be proved. We used the facts that $d(L) + d(f(L)) = i - 2$ and $d(R) + d(f(R)) = n - i - 1$ by the induction hypothesis.

Now let us assume that n is in the last position, and $p = Ln$. Then clearly, $d(p) = d(L)$, while $d(f(p)) = d(nf(L)) = d(f(L)) + 1$, and the proof follows by induction. Similarly, if n is in the first position, and $p = nR$, then $d(p) = d(nR) = 1 + d(R)$, while $d(f(p)) = d(f(R)n) = d(f(R))$, and again, the proof follows by induction. □

Our f maps permutations with k descents into permutations with $n-1-k$ descents. So that we could use f to prove that the sequence $W_t(n,k)$, $0 \leq k \leq n-1$ is symmetric, we must show that f preserves t-stack sortability. The following Lemma is the key element of the proof of this.

Lemma 14.33. *For any permutation p, the equality $s(p) = s(f(p))$ holds.*

Proof. We prove the statement by induction on n, the length of p. The statement is trivially true if $n = 1$. Now let us suppose it is true for all positive integers less than n.

(1) Suppose first that the entry n is at neither end of p, and let $p = LnR$. Then

$$s(p) = s(L)s(R)n = s(f(L))s(f(R))n = s(f(L)nf(R)) = s(f(p)).$$

(2) Now suppose that the entry n is in the first position, so $p = nR$. Then

$$s(p) = s(R)n = s(f(R)n) = s(n(f(R))) = s(f(p)).$$

(3) Finally, if the entry n is in the last position, so $p = Ln$, then

$$s(p) = s(L)n = s(f(L))n = s(n(f(L))) = s(f(p)).$$

So the statement is true in all cases. Again, we used the facts that $s(L) = s(f(L))$ and $s(R) = s(f(R))$ by the induction hypothesis. □

Corollary 14.34. *The permutation p is t-stack sortable if and only if $f(p)$ is t-stack sortable.*

Proof. Both statements are true if and only if the permutation $s(p) = s(f(p))$ is $(t-1)$-stack sortable. □

Now the proof of our duality theorem is immediate.

Theorem 14.35. *For all positive integers n, k, t, the equality*

$$W_t(n,k) = W_t(n, n-1-k)$$

holds.

Proof. Corollary 14.34 and Proposition 14.32 show that f bijectively maps the set of t-stack sortable n-permutations with k descents onto that of t-stack sortable n-permutations with $n-1-k$ descents. □

In order to get a deeper understanding of this proof, let us try to go through it in terms of decreasing binary trees. A *right (left) edge* is an edge between a vertex and its right (left) child. What we want to prove is that there are as many decreasing binary trees on n vertices corresponding to t-stack sortable permutations with k right edges as there are with k left edges.

Our map f takes a tree $T(p)$, and goes through its vertices starting at the root. If the root has two children, then the two edges adjacent to the root are unchanged. However, if the root has only a left edge, then the entire left subtree of the root will be moved to the right of the root and become its right subtree. Similarly, if the root has only a right edge, then the entire right subtree of the root will be moved to the left of the root and become its left subtree. Then we proceed to the vertices immediately below the root, and apply the same rule. We continue this way until all vertices have been treated.

This procedure clearly turns vertices with only a left child into vertices with only a right child. If a vertex had two children in $T(p)$, it will have the same two children in $T(f(p))$. This proves again that $d(p)+d(f(p)) = n-1$ as the number of left edges of $T(p)$ is equal to the number of right edges of $T(f(p))$.

To see that $s(p) = s(f(p))$, we need to show that the trees $T(p)$ and $T(f(p))$ yield the same permutation when read in postorder. To see this, note that if a vertex x has only one child y, then as far as the result of the postorder reading is concerned, it does not matter whether y is a left child or a right child of x. In both cases, the postorder reading will first go through the subtree rooted at y, then go to x. On the other hand, the only effect of the map f on p is precisely this, that is, f turns each single left child into a single right child and vice versa. So f has no effect on $s(p)$, as we have claimed.

Note that we have proved a little more than we planned. We proved that *each entry* x of p has the property that the subtree of $T(p)$ rooted at x, and the subtree of $T(f(p))$ rooted at x yield the same result when read in postorder. Originally we only wanted to prove this for the full trees, that is, the subtrees of the entry n.

Example 14.36. The decreasing binary trees of $p = 356124$ and $f(p) = 536421$ yield the same permutation 351246 when read in postorder. The same is true for the subtrees of any given entry.

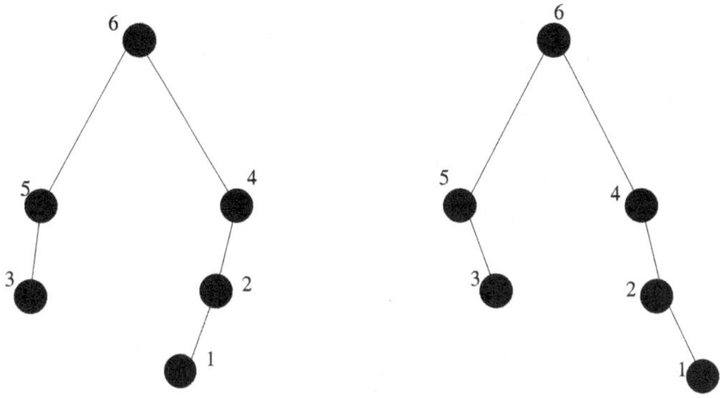

Fig. 14.6 Trees $T(p)$ and $T(f(p))$.

Notes

As pattern avoidance is the youngest of all areas covered in this book, it is also the one whose progress is the fastest. For this reason, this is the chapter that changed most since the publication of the first edition.

For a more thorough treatment of the topics discussed in this Chapter, the reader is advised to consult "Combinatorics of Permutations" by the present author [8], which devotes Chapters 4, 5, and 8 to the subject. Chapter 4 contains the proof of the Stanley-Wilf conjecture, by Adam Marcus and Gábor Tardos.

We have included several exercises that ask for the number $S_n(q_1, q_2)$ of n-permutations avoiding both patterns q_1 and q_2. Further results on this subject are available in [9] and [48].

The solution of the Stanley-Wilf conjecture implies that the limit $L(q) =$

$\lim_{n\to\infty} \sqrt[n]{S_n(q)}$ exists. This limit provides a good way of measuring the growth rate of the sequence $S_n(q)$. It was previously conjectured that if $q \in S_k$, then $L(q) \leq (k-1)^2$. However, this conjecture has recently been disproved [2]. A counterexample is given by the inequality $L(1324) > 9.35$. As far as a lower bound is concerned, it is known [24] that $L(q) \geq k^2/e^3$ for all $q \in S_k$.

Exercises

(1) Find a formula for the number of n-permutations that avoid *both* 132 and 123. We will denote this number by $S_n(132, 123)$.
(2) Find a formula for $S_n(132, 231)$.
(3) Find a formula for $S_n(132, 321)$.
(4) Find a formula for $S_n(132, 213)$.
(5) (+++) Prove that for all positive integers n,

$$S_n(1342) = \frac{(7n^2 - 3n - 2)}{2} \cdot (-1)^{n-1}$$
$$= +3\sum_{i=2}^{n} 2^{i+1} \cdot \frac{(2i-4)!}{i!(i-2)!} \cdot \binom{n-i+2}{2} \cdot (-1)^{n-i}.$$

(6) (++) Prove that for all positive integers n, we have $S_n(1423) = S_n(2413)$.
(7) Prove that the number of ways to partition a convex $n+2$-gon into triangles by non-crossing diagonals is c_n.
(8) Prove that the number of ways to partition a convex $n+1$-gon into any number of triangles and one quadrilateral by non-crossing diagonals is $\binom{2n-3}{n-3}$.
(9) (+) Let b_n be the number of n-permutations containing *exactly one copy* of the pattern 132. Find a recursive formula for b_n.
(10) Prove that $b_n = \binom{2n-3}{n-3}$, for all positive integers $n \geq 3$, where b_n is defined in the previous exercise.
(11) (+) Let d_n be the number of n-permutations containing *exactly one copy* of the pattern 123. Prove that $d_n = \frac{3}{n}\binom{2n}{n+3}$.
(12) Find a formula for $S_n(132, 123, 312)$.
(13) A partition π of $[n]$ having blocks $\beta_1, \beta_2, \cdots, \beta_k$, is called *non-crossing* if there are no four elements $1 \leq a < b < c < d \leq n$ so that $a, c \in \beta_i$ and $b, d \in \beta_j$ for some distinct blocks β_i and β_j. Prove that the number of non-crossing partitions of $[n]$ is c_n.

(14) Prove that for $k \in [n]$, the number of non-crossing partitions of $[n]$ having k blocks is equal to the number of 132-avoiding n-permutations that have $k-1$ descents.

(15) Let $N(n,k)$ be the number of 132-avoiding n-permutations with k left-to-right minima. Prove that for all $k \in [n]$, the equality

$$N(n,k) = N(n, n+1-k)$$

holds.

(16) (+) For $S \subseteq [n-1]$, let $\mathrm{Perm}_n(S)$ denote the number of 132-avoiding n-permutations with descent set S. Let $\alpha(S)$ denote its "reverse complement," that is, $i \in \alpha(S) \iff n-i \notin S$. Prove that for all $S \subseteq [n-1]$, the equality $\mathrm{Perm}_n(S) = \mathrm{Perm}_n(\alpha(S))$ holds.

(17) Let $n \geq 3$. Find all n-permutations that are not $(n-2)$-stack sortable.

(18) Find a necessary condition for a permutation to be t-stack sortable.

(19) Prove that if p does not have $t+2$ entries (not necessarily consecutive ones) so that rightmost one of them is the smallest, and the one preceding it is the largest, then p is t-stack sortable. Note that this means p avoids all $t!$ patterns of length $t+2$ that end in $(t+2)1$.

(20) Let n be an even positive integer. Find all n-permutations p for which there is no permutation $q \neq p$ so that $s(p) = s(q)$. Here s denotes the stack sorting operation.

(21) Is it true that an n-permutation is two-stack sortable if and only if there is at most one entry on the left of the entry n that is larger than the smallest entry on the right of the entry n?

(22) What is the number of *unlabeled* binary trees on n vertices? These trees are similar to decreasing binary trees in that they are rooted, each vertex has 0, 1, or 2 children, and each child is either a left child or right child of its parent, even if it is a only child. However, the vertices are *not* labeled.

Supplementary Exercises

(23) Prove that for any pattern q, and any positive integers m and n, the inequality

$$S_n(q)S_m(q) \leq S_{n+m}(q)$$

holds.

(24) (+) Let q be any pattern, and let

$$L(q) = \lim_{n\to\infty} \sqrt[n]{S_n(q)}.$$

Prove that $L(q)$ exists and is finite.

(25) Prove that $L(132) = 4$.

(26) Prove that $L(1342) = 8$.

(27) (Knowledge of basic definitions from group theory required.) Prove that if p is a q-avoiding permutation, then p^{-1} is a q^{-1}-avoiding permutation. Here t^{-1} denotes the inverse of the n-permutation t in the symmetric group S_n.

(28) Let $p = p_1p_2\cdots p_n$ be a 132-avoiding permutation. Prove that for all $i \in [2, n]$, the entry p_i is a left-to-right minimum if and only if $i - 1$ is a descent of p.

(29) Let q_1 and q_2 be two different patterns of length three. Is it true that $S_n(q_1, q_2)$ is always given by one of the formulae computed in Exercises 1 – 4?

(30) Prove that for all positive integers $k \leq n$, the equality

$$S_n(123\cdots k) = S_n(123\cdots k\,k-1)$$

holds.

(31) Find an upper bound for $S_n(3124675)$.

(32) (+) Find the ordinary generating function of the numbers $S_n(1324, 2413)$.

(33) Let q be any pattern of length k that has exactly one inversion. Prove that

$$S_n(q) \geq S_n(12\cdots k).$$

(34) A *circular translate* of the permutation $p = p_1p_2\cdots p_n$ is a permutation $p_ip_{i+1}\cdots p_np_1p_2\cdots p_{i-1}$. In other words, we get a circular translate of p by moving any initial segment of p to the end of p.
Find a formula for the number of n-permutations p so that no circular translate of p contains the pattern 132.

(35) Find a recurrence relation for the sequence $a_n = S_n(132, 4321)$. Then use that recurrence relation to prove that for all non-negative integers,

$$a_n = 2\binom{n}{4} + \binom{n+1}{3} + 1.$$

(36) (-) Show an example of a permutation of length n^2 that contains all $n!$ patterns of length n. Such a permutation is called an *n-superpattern*.

(37) (-) Show an example of an n-superpattern of length $n^2 - n + 1$.

(38) (-) Create a word of length $n^2 - n + 1$ over the alphabet $[n]$ that contains all $n!$ permutations of length n as a subword. (Letters of a subword do not have to be in consecutive positions.)

(39) Find a simple characterization for the set of permutations avoiding all of the patterns 1324, 1423, 2314, 2413, and 3412.

(40) Count the permutations of length n that avoid all five patterns listed in the previous exercise.

(41)(a) Show an example of a pair of patterns so that for all $n \geq 2$, the number $S_n(p,q)$ is even.

(b) Show that $S_n(132)$ is odd if and only if $n = 2^k - 1$.

(42) Let us assume that we have a computer program that decides whether a given m-permutation is an n-superpattern or not. We would like to use this program to find the number of m-permutations that are n-superpatterns. Let us assume for simplicity that m is odd. Prove that it suffices to test a suitably chosen set of $m!/3$ permutations with our program, and then the number of n-superpatterns of length m can be deduced.

(43) An *unlabeled plane tree* is a rooted tree that is embedded in the plane. Two unlabeled plane trees A and B are considered the same if the following hold:

(a) the roots of A and B have the same number k of children, denoted from the left to right by $A_1, A_2, \cdots, A_k$, and $B_1, B_2, \cdots, B_k$, and
(b) the subtrees rooted at A_i and B_i are isomorphic as unlabeled plane trees by this same definition.

Prove that the number of unlabeled plane trees on $n+1$ vertices is c_n.

(44) Prove that there are as many unlabeled plane trees on $n+1$ vertices with k leaves as there are with $n+1-k$ leaves.

(45) Prove that there are as many non-crossing partitions of $[n]$ with k blocks as there are with $n+1-k$ blocks.

(46) Describe the set of permutations p for which no vertex of $T(p)$ has two children.

(47) Let p and q be two n-permutations so that $T(p)$ and $T(q)$ become identical if we remove the labels of their vertices. Prove that the stack-sorting operation has the same effect on p and q. That is, prove that there is a permutation s so that $ps = qs$.

(48) A permutation p is called *sorted* if there is a permutation q so that $s(q) = p$. Is $p = 61374528$ sorted?

Solutions to Exercises

(1) We claim that $S_n(132,123) = 2^{n-1}$, and we are going to prove this by induction on n. The initial case is trivial. Assume the statement is true for $n-1$. Take any permutation of length $n-1$ that avoids both these patterns. Create two n-permutations from it by adding 1 to all its entries, then insert a new entry 1 to either its last or its next-to-last position. Clearly, these two new n-permutations avoid both 132 and 123.
We show that we obtain all n-permutations that avoid both these pattern by this procedure. We claim that such a permutation must contain the entry 1 at its last or next-to-last position. Indeed, if there are two elements on the right of 1, then they must be in either increasing or decreasing order, and must therefore form either a 123 or a 132 pattern together with the entry 1.
This proves that $S_n(132,123) = 2 \cdot S_{n-1}(132,123)$, and the proof follows by induction.

(2) Try to construct an n-permutation that avoids both 132 and 231. Then it is clear that the entry n must be either at the first or at the last position. Indeed, if there are two elements x and y bracketing n, then together with n they form either a 231-pattern, or a 132-pattern. Once n is placed, by similar argument we must place $n-1$ either in the first or the last empty position. We continue this way, having two choices at each step. Finally, we have to place 1 into the only empty spot left. So this procedure can result in 2^{n-1} different permutations. All these permutations will look like the letter V, that is, first they will decrease steadily, then they will increase steadily. Therefore, all of them will indeed avoid both 132 and 231. So we proved that $S_n(132,231) = 2^{n-1}$.

(3) Let p be an n-permutation avoiding both these patterns. In order to avoid 132, all entries on the left of n must be larger than all entries on the right of n. In order to avoid 321, all entries on the right of n must be in increasing order. Moreover, unless n is in the last position, all entries on the left of n must be in increasing order, too, otherwise two of them in decreasing order and any entry on the right of n would form a 321-pattern. So if n is in the ith position, and $i \neq n$, then there is only one such permutation, namely the permutation $(n-i+1)\ (n-i+2)\ \cdots\ n\ 1\ 2 \cdots\ n-i$. If n is in the last position, then n cannot participate in any 132- or 321-patterns, so we

can prepend it by any (132,321)-avoiding $(n-1)$-permutation. This yields the recurrence $S_n(132,321) = (n-1) + S_{n-1}(132,321)$, for $n \geq 2$, with the initial condition $S_1(132,321) = 1$. Solving this, we get $S_n(132,321) = 1 + \binom{n}{2}$.

(4) We claim that $S_n(132,213) = 2^{n-1}$. We prove this by induction on n, the initial case being trivial. Assume the statement is true for all integers smaller than n.
To avoid 132, all entries on the left of n must be larger than all entries on the right of n. To avoid 213, all entries on the *left* of n must be in increasing order. On the right of n, we must have a permutation that avoids both 132 and 213. One checks easily that these conditions together are not only necessary, but also sufficient for an n-permutation to avoid both 132 and 213.
Now assume n is in the ith position. Then the above conditions give rise to 2^{n-i-1} permutations if $i < n$, and one permutation if $i = n$. Indeed, the only freedom we have once the position of n is known is to permute the elements on the right of n, and the induction hypothesis says that we can do that in 2^{n-i-1} different ways.
So $S_n(132,213) = 1 + \sum_{i=1}^{n} 2^{n-i-1} = 2^{n-1}$.

(5) This result is due to the present author and can be found in Exact enumeration of 1342-avoiding permutations. A close link with labeled trees and planar maps, *Journal of Combinatorial Theory, Series A*, **80** (1997), 257–272.

(6) This result is due to Zvezdelina Stankova, and can be found in Forbidden Subsequences, *Discrete Mathematics*, **132** (1994), 291–316.

(7) Label the vertices of our $(n+2)$-gon by integers from 1 through $n+2$ in increasing order. Let d_n be the number of ways to partition a convex $n+2$-gon into triangles by non-crossing diagonals, and set $d_0 = 1$. We are going to find the number of partitions in which i is the smallest index in $[3, n+1]$ so that $1i$ is a diagonal in our partition π (if there is such an index).
In this scenario, $2i$ must be a diagonal π, otherwise the polygon containing 2 would have more than three sides. We have d_{i-3} possibilities for the part of π that partitions the $i-1$-gon $23\cdots i$, and we have d_{n-i+2} possibilities for the part of π that partitions the $n-i+4$-gon $1i(i+1)\cdots(n+2)$. So the number of all possibilities for such a π is $d_{i-3} \cdot d_{n-i+2}$.
Let us not forget that it can also happen that such an index i does not exist. In that case, vertex i is not part of any diagonal that is in

π, so the diagonal $2(n+2)$ must be in π. Then there are d_{n-1} ways for the part of π that partitions the $(n+1)$-gon $23\cdots(n+2)$. Summing over all cases, we get the formula

$$d_n = d_{n-1} + \sum_{i=3}^{n+1} d_{i-3} \cdot d_{n-i+2} = \sum_{j=1}^{n} d_{j-1} \cdot d_{n-j}.$$

This is identical to the recurrence relation (14.1) that we proved for 132-avoiding permutations, so the proof follows.

(8) By the previous exercise, an $(n+1)$-gon can be triangularized in any of c_{n-1} ways, using $n-2$ diagonals. The removal of any one of these $n-2$ diagonals forms a quadrilateral from two adjacent triangles. Further, there are two ways to triangularize this quadrilateral: with the diagonal we removed and the only other diagonal. Therefore, each way of partitioning the $(n+1)$-gon into one quadrilateral and $n-3$ triangles is yielded by exactly two triangularizations. Hence, the number of such ways to partition the $(n+1)$-gon is the number of triangularizations multiplied by the number of diagonals that can be chosen for removal, divided by two. This yields that the number of all such partitions is

$$\begin{aligned} c_{n-1}\frac{n-2}{2} &= \frac{1}{n}\binom{2n-2}{n-1}\frac{n-2}{2} \\ &= \frac{(2n-2)!(n-2)}{2n(n-1)!(n-1)!} \\ &= \frac{(2n-2)!}{2n(n-1)!(n-1)(n-3)!} \\ &= \frac{(2n-2)!}{2(n-1)n(n-1)!(n-3)!} \\ &= \frac{(2n-3)!}{n!(n-3)!} \\ &= \binom{2n-3}{n-3}. \end{aligned}$$

This solution is due to Christian Jones (personal communication).

(9) Clearly, $b_0 = b_1 = b_2 = 0$. Take any n-permutation p and suppose that the entry n is in the ith position in p. For shortness, call entries preceding n *front entries*, and call entries that n precedes *back entries*. Then there are three ways p can contain exactly one subsequence S of type 132.

(i) When all elements of S are front entries. Then any front entry must be larger than any back entry for any pair violating this condition would form an additional 132-subsequence with n. Therefore, the i largest entries must be front entries n (in fact, these are the entries $n-1, n-2, \cdots, n-i+1$), while the $n-i$ smallest entries must be back entries (these are the entries $1, 2, \cdots n-i$). Moreover, there can be no subsequence of type 132 formed by back entries. So all we can do is to take a 132-avoiding permutation on the $n-i$ back entries in c_{n-i} ways and take a permutation having exactly one 132-subsequence on the $i-1$ front entries. This yields $b_{i-1}c_{n-i}$ permutations of the desired property.

(ii) When all elements of S are back entries. The argument of the previous case holds here, too, we must only swap the roles of the front and back entries. In this case we have $c_{i-1}b_{n-i}$ permutations with the desired property.

(iii) Finally, it can happen that the leftmost element x of S is a front entry and rightmost element z of S is a back entry. This case is slightly more complicated. Note that here $2 \leq i \leq n-1$, otherwise either the set of front entries or that of back entries would be empty. First note that there is exactly one pair (x, z) so that x is a front entry, z is a back entry and $x < z$. (For any such pair and n form a 132-subsequence.) This implies that the front entries are $n-1, n-2, \cdots, n-i+2, n-i$ and the back entries are $1, 2, \cdots, n-i-1, n-i+1$, the only pair with the given property is $(n-i, n-i+1) = (x, z)$, and any other front entry is larger than both x and z.
Let us take these entries x and z. Clearly, all 132-subsequences of the given type must start with x and must end with z. We claim that the middle entry of S must be n. Indeed, if the middle element were some other w, then $x\ n\ z$ and $x\ w\ z$ would both be 132-subsequences. (Recall that $x < z$ and they both are smaller than any other front entry.) Moreover, we claim that x must be the rightmost front entry, in other words, it must be in the position directly on the left of n. Indeed, if there were any entry y between x and n, then $x\ y\ z$ and $x\ n\ z$ would both be 132-subsequences for y is a front entry and thus larger than x and z.
Therefore, all we can do is put the entry $n-i$ in position $(i-1)$, then take any 132-avoiding permutation on the first $i-2$ elements in c_{i-2} ways and take any 132-avoiding permutation on the $n-i$ back entries in c_{n-i} ways. This gives us $c_{i-2}c_{n-i}$ permutations of

the desired property.

Summing over all permitted i in each of these three cases we get that

$$b_n = \sum_{i=1}^{n-1} b_{i-1}c_{n-i} + \sum_{i=1}^{n-1} c_{i-1}b_{n-i} + \sum_{i=2}^{n-1} c_{i-2}c_{n-i}. \qquad (14.2)$$

Note that the first two sums are equal for they contain the same summands. Moreover, we can easily see by (8.19) or the equivalent (14.1) that the last sum equals $c_{n-1}-c_{n-2}$. Thus the above recurrence relation for b_n simplifies to

$$b_n = 2 \cdot \left(\sum_{i=1}^{n-1} b_{i-1}c_{n-i} \right) + c_{n-1} - c_{n-2}. \qquad (14.3)$$

(10) We prove that the number r_n of partitions of a convex $(n+1)$-gon P into triangles and one quadrilateral also satisfies the recurrence relation (14.3).

(I) First, we consider the case when there is no diagonal going into 1. Then it can be that $2(n+1)$ is a diagonal, and the problem is reduced to one lesser in size, with r_{n-1} partitions. Or, it can be that $2(n+1)$ is not a diagonal, and in that case, vertices $1, 2, n+1$ and a fourth vertex i form a quadrilateral. Then, to complete the partitioning, we have to triangulate the $(i-1)$-gon $2 \cdots i$ in c_{i-3} ways, and the $(n-i+2)$-gon $i(i+1) \cdots n+1$ in c_{n-i} ways. Summing this we get that in this first case there are

$$r_{n-1} + \sum_{i=3}^{n} c_{i-3}c_{n-i} = r_{n-1} + c_{n-2}$$

different partitions.

(II) Now we look at the case when there is a diagonal going into 1. Let i be *smallest* number so that $1i$ is a diagonal. Again, there are two cases: the quadrilateral is either in the part $12 \cdots i$, or in the part $i(i+1) \cdots (n+1)1$. Let us first handle the second case, as that is easier. We need to triangulate the part $12 \cdots i$, without having a diagonal touching 1 in c_{i-3} ways, (we have computed this in the solution of Exercise 7), then partition the $i(i+1) \cdots n(n+1)1$ part in r_{n-i+2} ways.

Let us return to the first case. We have to partition the first part without having a diagonal touching 1. As we have computed in case

I, there are $r_{i-2}+c_{i-3}$ possibilities, then we have to triangulate the second part in c_{n-i+1} ways. So here there are

$$\sum_{i=3}^{n-1} c_{i-3}r_{n-i+2} + \sum_{i=3}^{n}(r_{i-2}+c_{i-3})c_{n-i+1}$$

partitions.

These two cases together yield the following recurrence

$$r_n = c_{n-1} - c_{n-2} + 2\sum_{i=3}^{n-1} c_{i-3}r_{n-i+2},$$

or, writing $j = i-2$, we get

$$r_n = c_{n-1} - c_{n-2} + 2\sum_{j=1}^{n-3} c_{j-1}r_{n-i},$$

which is equivalent to (14.3) as $r_k = 0$ for $k \leq 3$.

Therefore, the sequences $\{b_n\}$ and $\{n\}$ satisfy the same recurrence relations, so they must be the same as their initial values are the same.

(11) This result is due to J. Noonan, and can be found in his article entitled "The number of permutations containing exactly one increasing subsequence of length three," *Discrete Math.* **152** (1996), no. 1-3, 307–313.

(12) Let $f(n) = S_n(123, 132, 312)$. Then in a permutation counted by $f(n)$, the entry 1 must be in one of the last two positions. If it is in the last position, then there are $f(n-1)$ possibilities for the rest of the permutation. If it is in the next-to-last position, then last position must contain the entry n, yielding $f(n-2)$ permutations. This shows that $f(n) = f(n-1) + f(n-2)$, with $f(0) = 1$ and $f(1) = 1$. This recurrence relation has been solved in Exercise 4 of Chapter 8. Recall that the numbers $f(n)$ are called *Fibonacci numbers.*

(13) The result of this exercise certainly follows from that of the next one, but we sketch a direct solution. Let π be a non-crossing partition of $[n]$, and let B be the block of π that contains the element 1. Let i be the largest element of B. Then π defines a non-crossing partition on B, and another one on $[n] - B$. It is easy to show that this decomposition leads to the same recurrence relation (14.1) that was satisfied by 132-avoiding n-permutations.

(14) This result was first published in [10]. We prove our statement by finding an appropriate bijection. Let π be a non-crossing partition of $[n]$. We construct the 132-avoiding permutation $p = f(\pi)$ corresponding to π as follows. Let k be the largest element of π which is in the same block of π as 1. Put the entry n of p in the kth position, i.e., set $p_k = n$. As p is to be 132-avoiding, this implies that the entries larger than $n-k$ are on the left of n in p, and the entries smaller than or equal to $n-k$ are on the right of n. Delete k from π and apply this procedure recursively, with obvious minor adjustments, to the restrictions of π to the sets $\{1,\ldots,k-1\}$ and $\{k+1,\ldots,n\}$, which are also non-crossing partitions. Namely, if j is the largest element in the same block as $k+1$, we set $p_j = n-k$, so that the restriction π_1 of π to $\{k+1, k+2, \ldots, n\}$ yields a 132-avoiding permutation of $\{1, 2, \ldots, n-k\}$ placed on the right of n in $p = f(\pi)$. Similarly, if in the restriction π_2 of π to the set $\{1, 2, \ldots, k-1\}$ the largest element in the same block as 1 is equal to j, we set $p_j = n-1$. Thus, recursively, π_2 yields a 132-avoiding permutation which we realize on the set $\{n-k+1, n-k+2, \ldots, n-1\}$ and we place it to the left of n in $p = f(\pi)$. In other words, with a slight abuse of notation, $f(\pi)$ is the concatenation of $f(\pi_2)$, n, and $f(\pi_1)$, where $f(\pi_2)$ permutes the set $\{n-k+1, n-k+2, \cdots, n-1\}$ and $f(\pi_1)$ permutes the set $[n-k]$. To see that this is a bijection note that we can recover the maximum of the block containing the element 1 from the position of the entry n in p, and then proceed recursively.
For example, If $\pi = (\{1,4,6\}, \{2,3\}, \{5\}, \{7,8\})$, then $f(\pi) =$ 64573812.

(15) As the reader is asked to prove in Supplementary Exercise 28, in a 132-avoiding permutation $p = p_1 p_2 \cdots p_n$ the entry p_i is a left-to-right minimum if and only if either $i = 1$ or $i-1$ is a descent. So $N(n,k)$ is also the number of 132-avoiding n-permutations with $k-1$ descents, and we need to show that this number is equal to the number $N(n, n+1-k)$ of 132-avoiding n-permutations with $n-k$ descents. For symmetry reasons, in the last sentence, the words "132-avoiding" can be replaced by "231-avoiding", and our claim then immediately follows from Theorem 14.35 by setting $t = 1$.

(16) We use induction on n. For $n = 1, 2, 3$ the statement is true. Now suppose we know it for all positive integers smaller than n. Denote by t the smallest element of S, and let p be a 132-avoiding n-permutation whose descent set is S.

(a) Suppose that $t > 1$. Then we have $p_1 < p_2 < \cdots < p_t$ and, because p avoids the pattern 132, the values of $p_1, p_2, \ldots, p_t$ are *consecutive* integers. So, for given values of p_1 and t, we have only one choice for $p_2, p_3, \ldots, p_t$. This implies

$$\mathrm{Perm}_n(S) = \mathrm{Perm}_{n-(t-1)}(S - (t-1)), \tag{14.4}$$

where $S - (t-1)$ is the set obtained from S by subtracting $t-1$ from each of its elements.

On the other hand, we have $n-t+1, n-t+2, \cdots, n-1 \in \alpha(S)$, meaning that in any permutation q counted by $\mathrm{Perm}_n(\alpha(S))$ the chain of inequalities $q_{n-t+1} > q_{n-t+2} > \cdots > q_n$ holds. In order to avoid forming a 132-pattern in q, it has to hold that $(q_{n-t+2}, \cdots, q_n) = (t-1, t-2, \cdots, 1)$. Therefore,

$$\mathrm{Perm}_n(\alpha(S)) = \mathrm{Perm}_{n-(t-1)}(\alpha(S)|n-(t-1)) \tag{14.5}$$

where $\alpha(S)|n-(t-1)$ denotes the set obtained from $\alpha(S)$ by removing its last $t-1$ elements. Then

$$\mathrm{Perm}_{n-(t-1)}(S-(t-1)) = \mathrm{Perm}_{n-(t-1)}(\alpha(S)|n-(t-1))$$

by the induction hypothesis, so equations (14.4) and (14.5) imply $\mathrm{Perm}_n(S) = \mathrm{Perm}_n(\alpha(S))$.

(b) If $t = 1$, but $S \neq [n-1]$, then let u be the smallest index which is *not* in S. Then again, to avoid forming a 132-pattern, the value of p_u must be the smallest positive integer a which is larger than p_{u-1} and is not equal to any p_i for $i \leq u-1$. So again, we have only one choice for p_u. On the other hand, the largest index in $\alpha(S)$ will be $n-u$. Therefore, in permutations q counted by $\mathrm{Perm}_n(\alpha(S))$, we must have $q_{n-u} = 1$ as q_{n-u} must be the rightmost left-to-right minimum in such permutations, and that is always the entry 1.

However, we have to be careful when we delete the entry p_u from p, and when we delete the entry 1 from q, because these deletions can have one of two different effects on the descent set of p and q. If the entries $p_{u-1}p_up_{u+1}$ form a 213-pattern, then deleting p_u will result in losing the first descent of p, while if these entries form a 312-pattern, then no descent is lost. If the entries $q_{n-u-1}q_{n-u}q_{n-u+1}$ form a 213-pattern, then deleting q_{n-u} removes the last descent of q, while if these entries formed a 312-pattern, then no descent is lost.

In order to use this information to reduce our permutations in size, we define two subsets $S', S'' \subset [n-2]$ as follows. First we define

S', the set corresponding to the case when no descents got lost. Let $i \in S'$ if and only if either $i < u$ and then, by the definition of u, $i \in S$, or $i > u$ and $i+1 \in S$. In other words, we decrease elements larger than u by 1; intuitively, we remove u from $[n-1]$, and translate the interval on its right one notch to the left. Note that $|S'| = |S|$ as we removed u, and u was not a descent anyway. If we now take $\alpha(S')$, that will consist of entries j so that $j < n-u$ and $(n-1)-(j-1) = n-j \notin S$. So in other words, we simply remove $n-u$ from $[n-1]$ (there has been nothing on the right of $n-u$ in $\alpha(S)$ to translate). Note that $|\alpha(S')| = |\alpha(S)| - 1$ as $n-u \in \alpha(S)$.

The set S'' is the set corresponding to the case when descents are lost. Therefore, we define $i \in S''$ if and only if either $i < u-1$ and then, by the definition of u, $i \in S$, or, $i > u$ and $i+1 \in S$. In other words, we decrease elements larger than $u-1$ by 1; intuitively, we remove $u-1$ from $[n-1]$, and translate the interval on its right one notch to the left. If we now take $\alpha(S'')$, that will consist of entries j so that $j < n-u+1$ and $(n-1)-(j-1) = n-j \notin S$. Note that $|S''| = |S| - 1$, and $|\alpha(S'')| = |\alpha(S)|$.

Therefore,

$$\mathrm{Perm}_n(S) = \mathrm{Perm}_{n-1}(S') + \mathrm{Perm}_{n-1}(S''),$$

and also

$$\mathrm{Perm}_n(\alpha(S)) = \mathrm{Perm}_{n-1}(\alpha(S')) + \mathrm{Perm}_{n-1}(\alpha(S'')).$$

By the induction hypothesis, the right-hand sides of these two equations agree, and therefore the left-hand sides must agree, too.

(c) Finally, if $S = [n-1]$, then the statement is trivially true as $\mathrm{Perm}_n(S) = \mathrm{Perm}_n(\alpha(S)) = 1$.

So we have shown that $\mathrm{Perm}_n(S) = \mathrm{Perm}_n(\alpha(S))$ in all cases.

(17) We prove by induction that these are precisely the permutations that end in the string $n1$. For $n = 3$, the statement is true.

Now assume the statement is true for $n-1$. Let $p = p_1 p_2 \cdots p_n$ be an n-permutation that is not $(n-2)$-stack sortable. That means that $s^{n-2}(p) = 21345 \cdots n$ as Proposition 14.24 implies that $s^{n-2}(p)$ must end in $345 \cdots n$. As each stack sorting operation moves the entry 1 up by at least one notch by Proposition 14.18, it follows that $p_n = 1$. Similarly, if $p_{n-1} \neq n$, then during the first stack sorting operation the entry 1 passes more than one entries, so in $n-2$ operations, it moves ahead more than $n-2$ notches. Therefore $p_{n-1} = n$.

To see that all such permutations are good, note that for such a p, we have $s(p) = Ln$, where L is an $(n-1)$-permutation that ends by the string $(n-1)1$. Therefore, by the induction hypothesis, $s(p)$ is not $(n-3)$-stack sortable, and the proof follows.

This result, and that of the next Exercise, is due to Julian West, who proved them in his thesis, *Permutations with forbidden subsequences; and, Stack sortable permutations,* PHD-thesis, Massachusetts Institute of Technology, 1990.

(18) Such a permutation p cannot contain the pattern $23\cdots(t+2)1$. If it did, then the entry a that plays the role of 1 in that $23\cdots(t+2)1$-pattern could move up only one notch *within the string of the entries of that pattern* during each stack-sorting operation. Therefore, after t operations, it would still be behind the first entry of that pattern.

(19) Let us denote the condition given in the exercise by C_t, and let us denote the set of the $t!$ patterns discussed in the exercise by P_t.

We are going to prove our claim by induction on t. If $t = 1$, then the condition simplifies to the 231-avoiding condition, and the statement is true. Now suppose it is true for $t-1$. Let p be as specified by the conditions of the theorem. Then $s(p)$ satisfies condition C_{t-1}. Indeed, if $s(p)$ contained a pattern q from P_{t-1}, then it follows from Proposition 14.18 that p would have to contain a pattern from P_t. (There had to be something large between the entries playing the role of $t+1$ and 1 in q.) Therefore, $s(p)$ is $(t-1)$-stack sortable by the induction hypothesis, so p is t-stack sortable.

(20) We claim that there are no such permutations. We know by Lemma 14.33 that $s(p) = s(f(p))$, where f is the map given by Definition 14.28. On the other hand, Proposition 14.32 shows that $d(p) + d(f(p)) = n-1$. Therefore, if n is even, then one of p and $f(p)$ must have an odd number of descents, and the other one must have an even number of descents. So $p \neq f(p)$, while $s(p) = s(f(p))$.

(21) No, that is not true. A counterexample is 163452. This permutation is not 2-stack sortable because of the 2341-pattern 3452.

The "only if" part is true. If there are at least two entries on the left of n that are larger than the entry c located on the right of n, then let a and b be the *leftmost* two entries with this property. If $a < b$, then $abnc$ is a 2341-pattern, and if $b < a$, then $abnc$ is a 3241-pattern that is not part of a 35241-pattern. (There is nothing between a and b that is larger than c.)

(22) We claim that the number of such trees is the Catalan number c_n

since these trees are in bijective correspondence with 231-avoiding n-permutations. Indeed, given an unlabeled binary tree B on n vertices, there is exactly one way to turn it into the decreasing binary tree $T(p)$ for a 231-avoiding permutation p. In order to see this, note that the root of B must be labeled n, and, in order to avoid 231-patterns, all vertices in the left subtree of the root must have smaller labels than all vertices in the right subtree of the root. So the sets of labels in the two subtrees are uniquely determined. This argument can be iterated for the subtrees, until we uniquely determine the label of each vertex.

Chapter 15

Who Knows What It Looks Like, But It Exists. The Probabilistic Method

We use the words "likely" or "probable" and "likelihood" or "probability" every day in informal conversations. While making these concepts absolutely rigorous can be a difficult task, we will concentrate on special cases when a mathematical definition of probability is straightforward, and conforms to common sense.

15.1 The Notion of Probability

Let us assume that we toss a coin four times, and we want to know the probability that we will get at least three heads. The number of all outcomes of the four coin tosses is $2^4 = 16$. Indeed, each coin toss can result in two possible outcomes. On the other hand, the number of *favorable* outcomes of our coin tossing sequence is five. Indeed, the five favorable outcomes, that is, those containing at least three heads, are $HHHH, HHHT, HHTH, HTHH$, and $THHH$. Our common sense now suggests that we define the probability of getting at least three heads as the ratio of the number of favorable outcomes to the number of all outcomes. Doing that, we get that the probability of getting at least three heads is $5/16$.

This common sense approach is the basis of our formal definition of probability. It goes without saying that we will have to be a little more careful. For instance, the above argument assumed, without mentioning it, that our coin is fair, that is, a coin toss is equally likely to result in a head or tail.

Definition 15.1. Let Ω be a finite set of outcomes of some sequence of trials, so that all these outcomes are equally likely. Let $A \subseteq \Omega$. Then Ω is called a *sample space*, and A is called an *event*. The ratio

$$P(A) = \frac{|A|}{|\Omega|}$$

is called the probability of A.

In particular, P is a function that is defined on the set of all subsets of Ω, and $0 \leq P(A) \leq 1$ always holds.

There are, of course, circumstances when this definition does not help, namely when Ω and A are not finite sets. An example of that situation is to compute the probability that a randomly thrown ball hits a given tree. As the ball could be thrown in infinitely many directions, and would hit the tree in an infinite number of cases, the above definition would be useless. We will not discuss that situation in this book; we will only study finite sample spaces.

Note that if A and B are disjoint subsets of Ω, then $|A \cup B| = |A| + |B|$, and therefore, $P(A \cup B) = P(A) + P(B)$. In general, we know from the Sieve formula that $|A \cup B| = |A| + |B| - |A \cap B|$, implying $P(A \cup B) = P(A) + P(B) - P(A \cap B)$. A generalization of this observation is the following simple, but extremely useful inequality.

Proposition 15.2. *Let $A_1, A_2, \cdots, A_n$ be events from the same sample space. Then*

$$P(A_1 \cup A_2 \cup \cdots \cup A_n) \leq P(A_1) + P(A_2) + \cdots + P(A_n).$$

Proof. We simply have to show that

$$|A_1 \cup \cdots \cup A_n| \leq |A_1| + \cdots + |A_n|.$$

This is true as the left-hand side counts each element of the sample space that is part of at least one of the A_i *exactly once*, while the right-hand side counts each element of the sample space that is part of at least one of the A_i *at least once*. □

The reader has already been subjected to some training in basic enumeration in Chapters 3–7. Most exercises in those chapters can be formulated in the language of probability. For example, the question "how many six-digit integers contain the digit 6" can be asked as "what is the probability that a randomly chosen six-digit integer contains the digit 6". Therefore, we do not cover these basic questions again here. Instead, we close this section by two examples that show how *counterintuitive* probabilities can be.

Example 15.3. In one of the lottery games available in Florida, six numbers are drawn from the set of numbers $1, 2, \cdots, 36$. What is the probability that a randomly selected ticket will contain at least one winning number?

Some people tend to answer $\frac{6}{36} = \frac{1}{6}$ to this question. They are wrong. That answer would be correct if *only one* number were drawn. Then the number of favorable outcomes would indeed be six, and the number of all outcomes would indeed be 36. However, when six numbers are drawn, the situation is more complicated.

Solution. (of Example 15.3) Let A be the event that a ticket contains at least one winning number, and let B be the event that a ticket does not contain any winning number. Then clearly, A and B are disjoint, and $A \cup B = \Omega$, so $P(A) + P(B) = 1$. Therefore, it suffices to compute $P(B)$. For a ticket not to contain any winning numbers, it has to contain six non-winning numbers. The number of ways that can happen is $\binom{30}{6}$. Therefore,

$$P(A) = 1 - P(B) = 1 - \frac{\binom{30}{6}}{\binom{36}{6}} = 1 - 0.3048 = 0.6952.$$

So with almost 70 percent probability, a randomly chosen ticket will contain at least one winning number! No wonder you must have more than one winning number to actually win a prize.

Note that when A and B are two disjoint events, then we say that A and B are *mutually exclusive*. In other words, it is not possible that A and B happen together. If, in addition, we also have $A \cup B = \Omega$, then we say that B is the *complement* of A. We denote this by writing $\bar{A} = B$.

Example 15.4. Forty people are present at a party, and there is nobody among them who was born on February 29. Adam proposes the following game to Bill. Each guest writes his or her birthday (just the day and month, not the year) on a piece of paper. If there are two pieces of paper with the same date on them, then Adam wins, if not, then Bill wins. When Bill heard this proposal, he looked around, and said "Fine, there are only forty people here, much less than the number of days in a year, so I am bound to win." What do we think about Bill's argument?

Solution. The problem with Bill's argument is that he fails to note the difference between one hundred percent probability and more than fifty percent probability. If we want to be one hundred percent sure that there will be two people in the room having the same birthday, then we would indeed need 366 people to be present. To have more than fifty percent chance is an entirely different issue.

In what follows, we prove that if there are at least 23 people at the party, then Adam, not Bill, has more chance of winning this game. In order to prove this, it is clearly sufficient to provide a proof for the case when there are exactly 23 people at the party as any additional person just improves Adam's chances.

Let us compute the probability that there are *no two people* at the party who have the same birthday. For that to happen, the first person's birthday can be any of the 365 possible days of the year, that of the second person could be any of 364 days, and so on. So the number of favorable outcomes is $(365)_{23}$. On the other hand, the number of all outcomes is obviously 365^{23}. Therefore, the probability that there are no two people in the room whose birthdays coincide is

$$\frac{365 \cdot 364 \cdot \dots \cdot 343}{365^{23}} = \frac{364 \cdot 363 \cdots 343}{365^{22}} < \frac{1}{2}.$$

Therefore, the probability that there are two people at the party who *do* have the same birthday is *more* than one half.

Finally, we point out that the condition that nobody was born on February 29 was only included to make the situation simpler. Indeed, February 29 exists only in leap-years, so the chance of being born on that day is 1/4 of the chance of being born on any other given day. That would make the outcomes in our sample space not equally likely, contradicting the definition of sample space. We could help this by changing our sample space from the 365-element set of dates in a year to the set of $4 \cdot 365 + 1 = 1461$ days of a 4-year cycle. That would make computations a little bit more cumbersome.

15.2 Non-constructive Proofs

If there are balls in a box, and we know that the probability that a randomly selected ball is blue is more than 0, then we can certainly conclude that there is at least one blue ball in the box. This thought seems utterly simple at first sight, but it has proved to be extremely useful in existence proofs as the following examples show.

Recall that in Chapter 13, we defined the symmetric Ramsey number $R(k, k)$. For easy reference, this was the smallest positive integer so that if we 2-color the edges of the complete graph on $R(k, k)$ vertices, we always get a K_k subgraph whose edges are all the same color.

Now we are going to find a lower bound for $R(k,k)$ by proving that $R(k,k) > 2^{k/2}$. Let us take a closer look at the statement to be proved. What it says is that if G is a complete graph on $2^{k/2}$ vertices, then *it is possible* to 2-color the edges of G so that no monochromatic copy of K_k is formed. When we proved similar statements in Chapter 13, showing that $R(3,3) > 5$, or $R(4,4) > 17$, we proved them by actually providing a coloring of K_5 or K_{17} that indeed did not contain the required monochromatic copies. However, this was more than what we strictly needed to do. To prove $R(k,k) > 2^{k/2}$, it suffices to prove that it is *possible* to 2-color the edges of G so that no monochromatic copy of K_k is formed; it is *not* necessary to *actually find* such a coloring. We will shortly see how big a difference this is.

Theorem 15.5. *For all positive integers $k \geq 3$, the inequality $R(k,k) > 2^{k/2}$ holds.*

Proof. Let $G = K_n$, and let us color each edge of G red or blue as follows. For each edge, flip a coin. If we get a head, we color that edge red, otherwise we color that edge blue. This way each edge will be red with probability one half, and blue with probability one half. We are going to show that the probability p that we get no monochromatic K_k-subgraphs in G this way is more than zero. On the other hand, $p = \frac{|F|}{|\Omega|}$, the number of favorable outcomes divided by the number of all outcomes, where Ω is the set of all possible 2-colorings of the edges of a complete graph on n vertices. So $p > 0$ implies that there is at least one favorable outcome, that is, there is at least one K_n with 2-colored edges that does not contain any monochromatic K_k-subgraphs.

Instead of proving that $p > 0$, we will prove that $1 - p < 1$, which is an equivalent statement. Note that $1-p$ is the probability that we get at least one monochromatic subgraph in our randomly colored graph $G = K_n$.

The number of ways to 2-color the edges of a given K_k-subgraph of K_n is clearly $2^{\binom{k}{2}}$ as there are two choices for the color of each edge. Out of all these colorings, only two will be monochromatic, one with all edges red, and one with all edges blue. Therefore, the probability that a randomly chosen K_k-subgraph is monochromatic is

$$\frac{2}{2^{\binom{k}{2}}} = 2^{1-\binom{k}{2}}.$$

The graph K_n has $\binom{n}{k}$ subgraphs that are isomorphic to K_k. Each of them has the same chance to be monochromatic. On the other hand, the probability that *at least one* of them is monochromatic is *at most* the sum of these $\binom{n}{k}$ individual probabilities, by Proposition 15.2. In other words, if A_S denotes the event that the K_k-subgraph S of G has monochromatic edges, then

$$P\left(\bigcup_S A_S\right) \leq \sum_S P(A_S) = \binom{n}{k} 2^{1-\binom{k}{2}}, \tag{15.1}$$

where S ranges through all K_k-subgraphs of G. Now let us assume, in accordance with our criterion, that $n \leq 2^{k/2}$. Then the last term of (15.1) can be bounded as follows.

$$\binom{n}{k} 2^{1-\binom{k}{2}} < \frac{n^k}{k!} \cdot 2^{1-\binom{k}{2}} \leq \frac{2 \cdot 2^{k^2/2}}{k! 2^{\binom{k}{2}}} = 2 \cdot \frac{2^{k/2}}{k!} < 1,$$

for all $k \geq 3$. The last inequality is very easy to prove, for example by induction. □

We have seen in Chapter 13 that $R(k,k) \leq 4^k$. Our latest result shows that $(\sqrt{2})^k < R(k,k)$. These are essentially the best known results on the size of $R(k,k)$, so there is a lot of progress to be made on Ramsey numbers.

Theorem 15.6. *Let n and m be two positive integers larger than 1, and let $m \geq 2\log_2 n$. Then it is possible to color each edge of $K_{n,n}$ red or blue so that no $K_{m,m}$-subgraph with monochromatic edges is formed.*

Proof. The number of ways to 2-color the edges of a given $K_{m,m}$ subgraph of $K_{n,n}$ is 2^{m^2}, and two of these colorings result in monochromatic subgraphs. Therefore, the probability that at least one monochromatic $K_{m,m}$ is formed is at most $\binom{n}{m}^2 2^{1-m^2}$. Therefore, all we have to prove is

$$\binom{n}{m}^2 2^{1-m^2} < 1,$$

that is,

$$2\binom{n}{m}^2 < 2^{m^2}.$$

To see this, we insert two intermediate expressions as follows.

$$2\binom{n}{m}^2 < n^{2m} \leq (2^{m/2})^{2m} = 2^{m^2},$$

where the second inequality is a simple consequence of the relation between n and m. □

Another way to formulate this same theorem is as follows. If $m \geq 2\log_2 n$, then there exists a matrix of size $n \times n$ whose entries are either 0 or 1 having no $m \times m$ minor that consists of zeros only, or of ones only.

What is amazing about this result is that *nobody knows how to construct* that matrix, or *how* to color the edges of $K_{n,n}$ so that the requirements are fulfilled. In fact, the gap between what we *can* do and what we know is *possible* is rather large. The best construction known to this day for an $n \times n$ matrix with zeros and ones, and not having $m \times m$ homogeneous minors works for $m = c\sqrt{n}$, where c is a constant. This is much more than what we know is true, that is, $m = 2\log_2 n$.

15.3 Independent Events

15.3.1 *The Notion of Independence and Bayes' Theorem*

Let us throw two dice at random. Let A be the event that the first die shows six, and let B be the event that the second die shows six. It is obvious that $P(A) = P(B) = 1/6$, and $P(A \cap B) = 1/36$. We see that $P(A) \cdot P(B) = P(A \cap B)$, and start wondering whether this is a coincidence. Now let us pick a positive integer from [12] at random. Let C be the event that this number is divisible by two, let D be the event that this number is divisible by three, and let F be the event that this number is divisible by four. Then $P(C) = 1/2$, $P(D) = 1/3$, and $P(F) = 1/4$. Furthermore, $P(C \cap D) = 1/6$, and $P(D \cap F) = 1/12$, so the "product rule" seems to hold. However, $P(C \cap F) = P(F) = 1/4 \neq P(A)P(B)$, breaking the "product rule".

Why is it that sometimes we find $P(A) \cdot P(B) = P(A \cap B)$, and sometimes we find $P(A) \cdot P(B) \neq P(A \cap B)$? As you have probably guessed, this is because sometimes the fact that A occurs makes the occurrence of B more likely, or less likely, and somctimes it does not alter the chance that B occurs at all. For example, if we choose an integer from 1 to 12, then the fact that it is divisible by two certainly makes it more likely that it is also divisible by four. Indeed, the number of all possible outcomes decreases from 12 to six, while that of favorable outcomes does not change. On the other hand, the fact that our number is divisible by two does not change its chances to be divisible by three. Indeed, the number of all outcomes decreases from 12 to six, but the number of favorable outcomes also decreases, from four to two.

This warrants the following two definitions.

Definition 15.7. If A and B are two events from the same sample space Ω, and $P(A \cap B) = P(A) \cdot P(B)$, then A and B are called *independent* events. Otherwise they are called *dependent.*

Definition 15.8. Let A and B be events from the same sample space, and assume $P(B) > 0$. Let

$$P(A|B) = \frac{P(A \cap B)}{P(B)}.$$

Then $P(A|B)$ is called a *conditional probability,* and is read *"the probability of A given B".*

That is, $P(A|B)$ is the probability of A given that B occurs. The following proposition is now immediate from the definitions.

Proposition 15.9. *The events A and B are independent if and only if $P(A|B) = P(A)$ holds.*

In other words, A and B are independent if and only if the occurrence of B does not make the occurrence of A any more likely, or any less likely.

Example 15.10. We toss a coin four times. We are not allowed to see the results, but we are told that there are at least two heads among the results. What is the probability that all four tosses resulted in heads?

Solution. Let A be the event that all four tosses are heads, and let B be the event that there are at least two heads. Then $A \cap B = A$, so $P(A|B) = P(A)/P(B)$. As the probability of getting a head at any one toss is $1/2$, we have $P(A) = \frac{1}{2^4} = \frac{1}{16}$. There is $1/16$ chance to get four heads, $4/16$ chance to get three heads and one tail, and $6/16$ chance to get two heads, two tails. Therefore, $P(B) = \frac{11}{2^4}$, and $P(A|B) = 1/11$.

Example 15.11. Let $p = p_1 p_2 \cdots p_n$ be a randomly selected n-permutation. Let A be the event that $p_1 > p_2$, and let B be the event that $p_2 > p_3$. Compute $P(A|B)$, and decide if A and B are independent.

Solution. Clearly, $P(A) = P(B) = 1/2$ as can be seen by reversing the relevant pair of entries. On the other hand, $A \cap B$ is the event that $p_1 > p_2 > p_3$, which occurs in $1/6$ of all permutations. Therefore,

$$P(A|B) = \frac{P(A \cap B)}{P(B)} = \frac{1/6}{1/2} = \frac{1}{3} \neq P(A),$$

so A and B are not independent.

Your reaction to the previous example was probably something along the lines "Of course. If $p_1 > p_2$, then p_2 is smaller than normal, so it is less likely than normal that $p_2 > p_3$." While that argument works in this case, one should be extremely careful when injecting intuition into arguments involving conditional probabilities. The following example is a striking instance of this.

Example 15.12. A University has two colleges, the College of Liberal Arts, and the College of Engineering. Each college analyzed its own admission record and each college found that last year, a domestic applicant to the college had a larger chance to be admitted than an international applicant. Can we conclude that the same is true for the entire university? (In this example, we assume that applicants can only apply to one college.)

Solution. No, we cannot. A counterexample is shown in Figure 15.1.

	Liberal Arts	Engineering	Entire University
Domestic applicants	Admitted: 10 Applied: 120 success rate:8.3%	Admitted: 10 Applied: 10 success rate: 100%	Admitted: 20 Applied: 130 success rate: 15.9%
International applicants	Admitted: 1 Applied: 15 success rate:6.7%	Admitted: 90 Applied: 100 success rate:90%	Admitted: 91 Applied: 115 success rate : 79.1%

Fig. 15.1 Not all that glitters is gold.

How is this very counterintuitive fact called *Simpson's paradox* possible? Some people do not believe it even when they see it with their own eyes. An imprecise, but conceptually correct, explanation is this. A much larger portion of the international applicants applied to Engineering, where the general rate of acceptance was higher. While it is true that the domestic students had an even higher acceptance rate in that college, it concerned only eight percent of all domestic applicants, versus more than 85 percent of international applicants. In other words, more than 85 percent of **all** international applicants got into Engineering, whereas less than 16 percent of all domestic applicants did. This is a huge difference, and the College of Liberal Arts, with relatively few applicants, cannot make up for that.

In order to find a more precise explanation, we will need Bayes' Theorem.

Theorem 15.13 (Bayes' Theorem). *Let A and B be mutually exclusive events so that $A \cup B = \Omega$. Let C be any event. Then*

$$P(C) = P(C|A) \cdot P(A) + P(C|B) \cdot P(B). \tag{15.2}$$

In other words, the probability of C is the weighted average of its conditional probabilities, where the weights are the probabilities of the conditions.

Proof. (of Theorem 15.13) As A and B are mutually exclusive, $A \cap C$ and $B \cap C$ are disjoint, and since $A \cup B = \Omega$, their union is exactly C. Therefore,

$$P(C) = P(C \cap A) + P(C \cap B),$$

and the proof follows as the first (resp. second) member of the right-hand side agrees with the first (resp. second) member of the right-hand side of 15.2. □

Now we are in a position to provide a deeper explanation for Example 15.12. Let A_1 (resp. B_1) be the event that an international (resp. domestic) applicant *applies* to the College of Liberal Arts, and define A_2 and B_2 similarly, for the College of Engineering. Let C_1 (resp. C_2) be the event that an international (resp. domestic) applicant is admitted to the *university*. Then Theorem 15.13 shows that

$$P(C_1) = P(C_1|A_1) \cdot P(A_1) + P(C_1|B_1) \cdot P(B_1),$$

and

$$P(C_2) = P(C_2|A_2) \cdot P(A_2) + P(C_2|B_2) \cdot P(B_2).$$

The criterion requiring that domestic students have larger chances to get accepted by any one college ensures that $P(C_1|A_1) < P(C_2|A_2)$, and $P(C_1|B_1) < P(C_2|B_2)$. It does not, however, say anything about $P(A_1)$ and $P(B_1)$. (We know that A_2 is the complement of A_1, and B_2 is the complement of B_1.) Therefore, we can choose A_1 and B_1 so that it is very advantageous for $P(C_1)$, and very bad for $P(C_2)$. We can do this by choosing $P(A_1)$ to be large if $P(C_1|A_1)$ is large, and by choosing $P(A_1)$ small if $P(C_1|A_1)$ is small. Similarly, we can choose $P(A_2)$ to be large if $P(C_2|A_2)$ is small, and vice versa.

In other words, weighted averages are a lot harder to control than unweighted averages. Indeed, if we impose the additional condition that $P(A_1) = P(B_1) = 1/2$, or even only the condition $P(A_1) = P(B_1)$, then the domestic students would have a greater chance to be admitted to the university.

The reader is asked to solve Exercise 33 at this point. That exercise shows a typical example when Bayes' theorem solves a problem that does not seem to be simple at the first sight, and whose results are important and counter-intuitive.

15.3.2 *More Than Two Events*

It is not obvious at first sight how the independence of three or more events should be defined. We could require the equality $P(A_1 \cap A_2 \cap \cdots \cap A_n) = P(A_1) \cdot P(A_2) \cdot \cdots \cdot P(A_n)$. This, in itself, is not a very strong requirement, however. It holds whenever $P(A_1) = 0$, no matter how strongly the other variables depend on each other. In order to have some more local conditions, we can impose the requirements that $P(A_i \cap A_j) = P(A_i)P(A_j)$ for all $i \neq j$. However, consider the following situation.

We select a positive integer from [10] at random. Let A be the event that this number is odd. Now let us select an integer from [20] at random, and let B be the event that this number is odd. Finally, let C be the event that the *difference* of the two selected integers is odd.

It is then not difficult to verify that $P(A) = P(B) = P(C) = 1/2$, and also the events A, B, and C are *pairwise* independent, that is, any two of them are independent. However, $P(A \cap B \cap C) = 0 \neq P(A)P(B)P(C) = 1/8$. Therefore, we do not want to call these events independent, either.

We resolve these possible problems by requiring a very strong property for a set of events to be independent.

Definition 15.14. We say that the events $A_1, A_2, \cdots, A_n$ are independent if, for any nonempty subset $S = \{i_1, i_2, \cdots, i_k\} \subseteq [n]$, the equality

$$P(A_{i_1} \cap A_{i_2} \cap \cdots \cap A_{i_k}) = P(A_{i_1}) \cdot P(A_{i_2}) \cdot \cdots \cdot P(A_{i_k})$$

holds.

We close this section by mentioning that Theorem 15.13 is easy to generalize to more than two conditions.

Theorem 15.15 (Bayes' Theorem, General Version). *Let $A_1, A_2, \cdots, A_n$ be events in a sample space Ω so that $A_1 \cup A_2 \cup \cdots \cup A_n = \Omega$, and $A_i \cap A_j = \emptyset$ if $i \neq j$. Let $C \subset \Omega$ be any event. Then*

$$P(C) = \sum_{i=1}^{n} P(C|A_i)P(A_i).$$

Proof. Analogous to that of Theorem 15.13. □

15.4 Expected Values

A *random variable* is a *function* that is defined on a sample space Ω, and whose range is a set of numbers. For example, if Ω is the set of all graphs on n labeled vertices, we can define the random variable X by setting $X(G)$ to be the number of edges of G, or we can define the random variable Y by setting $Y(G)$ to be the number of connected components of G.

Just as for functions, we can define the sum and product of random variables over the same sample space the usual way, that is, $(X+Y)(u) = X(u) + Y(u)$, and $(X \cdot Y)(u) = X(u) \cdot Y(u)$.

Possibly the most important and useful parameter of a random variable is its *expected value*, or, in other words, *expectation*, or *average value*, or *mean*.

Definition 15.16. Let $X : \Omega \to \mathbf{R}$ be a random variable so that the set $S = \{X(u)|u \in \Omega\}$ is finite, that is, X only takes a finite number of values. Then the number

$$E(X) = \sum_{i \in S} i \cdot P(X = i)$$

is called the *expected value*, or *expectation* of X on Ω.

Here, and throughout this chapter, $P(X = i)$ is the probability of the event that $X(u) = i$. That is,

$$P(X = i) = \frac{|\{u \in \Omega | X(u) = i\}|}{|\Omega|}.$$

In other words, $E(X)$ is the weighted average of all values X takes, with the weights being equal to the probability of X taking the corresponding value.

Remarks. Some probability variables can be defined over many different sample spaces. Our above example, the number of edges of a graph, can be defined not just over the space of all graphs on n vertices, or all *connected graphs* on n vertices, but also on all graphs on *at most* $3n$ vertices, and so on. In each case, the set $S = \{X(u)|u \in \Omega\}$ is different, therefore the expectation of X is also different. Therefore, if there is a danger of confusion, we write $E_\Omega(X)$, to denote where the expectation is taken. If there is no danger of confusion, however, we will only write $E(X)$, to alleviate notation.

Sometimes we announce both Ω and X in the same sentence as in "let $X(G)$ be the number of edges of a randomly selected connected graph G on n vertices." This means that Ω is the set of all connected graphs on n vertices, and $X(G)$ is the number of edges of the graph $G \in \Omega$.

It is possible to define the expectation of X in some cases when the set $S = \{X(u)|u \in \Omega\}$ is *not finite.* If S is a countably infinite set, we can define $E(X) = \sum_{i \in S} i \cdot P(X = i)$ as long as this infinite sum exists. See Exercise 4 for an example. If S is not countable, the summation may be replaced by integration. Details can be found in any probability textbook.

Definition 15.17. The random variables X and Y are called *independent* if for all s and t, the equality

$$P(X = s, Y = t) = P(X = s)P(Y = t)$$

holds.

15.4.1 *Linearity of Expectation*

For any real number c, we can define the random variable cX by setting $cX(u) = c(X(u))$ for all $u \in \Omega$. The following innocent-looking theorem proves to be extremely useful in enumerative combinatorics.

Theorem 15.18.

(1) Let X and Y be two random variables defined over the same space Ω. Then

$$E(X + Y) = E(X) + E(Y).$$

(2) Let X be a random variable, and let c be a real number. Then

$$E(cX) = cE(X).$$

So "taking expectations" is a linear operator. The best feature of this theorem is that it does not require that X and Y be independent! No matter how deeply X and Y are intertwined, nor how hard it is to compute, say, the probability that $X = Y$, the expected value of $X + Y$ is always given by this simple formula. This is why linearity is the most useful property of expectation, and is applied to a very wide array of problems.

Proof.   (of Theorem 15.18)

(1) Let $x_1, x_2, \cdots, x_n$ be the values that X takes with a positive probability, and let $y_1, y_2, \cdots, y_m$ be the values that Y takes with a positive probability. Then

$$E(X + Y) = \sum_{i=1}^{n}\sum_{j=1}^{m}(x_i + y_j)P(X = x_i, Y = y_j)$$

$$= \sum_{i=1}^{n}\sum_{j=1}^{m}x_i P(X = x_i, Y = y_j)$$

$$+ \sum_{i=1}^{n}\sum_{j=1}^{m}y_j P(X = x_i, Y = y_j)$$

$$= \sum_{i=1}^{n}x_i P(X = x_i) + \sum_{j=1}^{m}y_j P(Y = y_j)$$

$$= E(X) + E(Y).$$

(2) Let $r \in \Omega$, then by definition $(cX)(r) = cX(r)$. So if $x_1, x_2, \cdots, x_n$ is the range of X, then $P(cX = cx_i) = P(X = x_i)$. Therefore,

$$E(cX) = \sum_{i=1}^{n}cx_i \cdot P(cX = cx_i) = c\sum_{i=1}^{n}x_i P(X = x_i) = cE(X).$$

$\square$

In order to be able to better appreciate the surprising strength of Theorem 15.18, let $p = p_1 p_2 \cdots p_n$ be an n-permutation, and let us say that i is a *valley* if p_i is smaller than both of its neighbors, that is $p_i < p_{i-1}$, and $p_i < p_{i+1}$. We require $2 \leq i \leq n - 1$ for i to be a valley.

Theorem 15.19. *Let* $n \geq 2$ *be a positive integer. Then on average, a randomly selected permutation of length* n *has* $(n - 2)/3$ *valleys.*

Without Theorem 15.18, this would be a painful task. We would have to compute the number $v(j)$ of n-permutations with j valleys for each j, (a difficult task), then we would have to compute $\sum_j j \cdot \frac{v(j)}{n!}$. Theorem 15.18, however, turns the proof into a breeze.

Proof. (of Theorem 15.19) Take $n - 2$ different probability variables $Y_2, Y_3, \cdots, Y_{n-1}$, defined on the set of all n-permutations as follows. For an n-permutation p, let $Y_i(p) = 1$ if i is a valley, and let $Y_i(p) = 0$ otherwise. Then for $2 \leq i \leq n - 1$, every p_i has a $1/3$ chance to be the smallest of the set $\{p_{i-1}, p_i, p_{i+1}\}$. Therefore,

$$E(Y_i) = \frac{1}{3} \cdot 1 + \frac{2}{3} \cdot 0 = \frac{1}{3}.$$

Define $Y = Y_2 + Y_3 + \cdots + Y_{n-1}$. Then $Y(p)$ is the number of valleys of p. Therefore, Theorem 15.18 implies

$$E(Y) = \sum_{i=2}^{n-1} E(Y_i) = (n - 2) \cdot E(Y_1) = \frac{n - 2}{3}.$$

$\square$

Variables similar to Y_i, that is, variables that take value 1 if a certain event occurs, and value 0 otherwise, are called *indicator (random) variables*.

Theorem 15.20. *The expected value of the number of fixed points in a randomly selected n-permutation is 1.*

Proof. We define n different probability variables $X_1, X_2, \cdots, X_n$ on the set of all n-permutations as follows. For an n-permutation p, let $X_i(p) = 1$ if $p_i = i$, that is, when p has a fixed point at position i, and let $X_i(p) = 0$ otherwise.

As p_i is equally likely to take any value $j \in [n]$, it has a $1/n$ chance to be equal to i. Therefore,

$$E(X_i) = \frac{1}{n} \cdot 1 + \frac{n - 1}{n} \cdot 0 = \frac{1}{n},$$

for all $i \in [n]$. Now define $X = X_1 + X_2 + \cdots + X_n$; then $X(p)$ is precisely the number of fixed points of p. On the other hand, applying Theorem 15.18, we get

$$E(X) = \sum_{i=1}^{n} E(X_i) = n \cdot E(X_1) = n \cdot \frac{1}{n} = 1, \tag{15.3}$$

which was to be proved. $\square$

15.4.2   Existence Proofs Using Expectation

It is common sense that the average of a set of numbers is never larger than the largest of those numbers. This is true for weighted averages as well as the following theorem shows.

Theorem 15.21. *Let* $X : \Omega \to \mathbf{R}$ *be a random variable so that the set* $S = \{X(u)|u \in \Omega\}$ *is finite, and let* j *be the largest element of* S. *Then*

$$j \geq E(X).$$

Proof.   Using the definition of $E(X)$,

$$E(X) = \sum_{i \in S} i \cdot P(X = i) \leq j \sum_{i \in S} P(X = i) = j.$$

$\square$

We present two applications of this idea. The first shows that a simple graph will always contain a large bipartite subgraph.

Theorem 15.22. *Let* G *be a simple graph with vertex set* $[n]$, *and* m *edges. Then* G *contains a bipartite subgraph with more than* $m/2$ *edges.*

Proof.   Let us split the vertices of G into two disjoint nonempty subsets A and B. Then A and B span a bipartite subgraph H of G. (We remove the edges within A and within B.) Let Ω be the set of $2^{n-1} - 1$ different bipartite subgraphs we get this way. Let $X(H)$ be the number of edges in H.

On the other hand, let us number the edges of G from one through m, and let $X_i = 1$ if the edge i has one vertex in A, and one in B, and let $X_i = 0$ otherwise.

What is $P(X_i = 1)$? By our definitions, we can get a subdivision of $[n]$ leading to $X_i = 1$ by first putting the two endpoints of the edge i to different subsets, then splitting the remaining $(n-2)$-element vertex set in any of 2^{n-2} ways. Therefore, $P(X_i = 1) = \frac{2^{n-2}}{2^{n-1}-1}$, and $P(X_i = 0) = \frac{2^{n-2}-1}{2^{n-1}-1}$. This implies

$$E(X_i) = 0 \cdot P(X_i = 0) + 1 \cdot P(X_i = 1) = \frac{2^{n-2}}{2^{n-1} - 1} > \frac{1}{2}.$$

We can repeat this argument for all edges. Then we note that $X = X_1 + X_2 + \cdots + X_m$, so Theorem 15.18 implies

$$E(X) = \sum_{i=1}^{m} E(X_i) = m \cdot E(X_1) > \frac{m}{2}.$$

As the expected value of the number of edges in these bipartite subgraphs of G is more than $m/2$, it follows from Theorem 15.21 that there is at least one bipartite subgraph of G with more than $m/2$ edges. $\square$

The next example is related to a well-known problem in complexity theory, the so-called "Betweenness problem".

Example 15.23. We are given a list $L = (L_1, L_2, \cdots, L_k)$ of ordered triples $L_i = (a_i, b_i, c_i)$, so that for any i, the numbers a_i, b_i, and c_i are distinct elements of $[n]$. It is possible, however, that symbols with *different* indices i and j denote the *same* number.

Let $p = p_1 p_2 \cdots p_n$ be an n-permutation. We say that p satisfies L_i if the entry b_i is *between* a_i and c_i in p. (It does not matter whether the order of these three entries in p is $a_i b_i c_i$ or $c_i b_i a_i$.)

Prove that there exists an n-permutation p that satisfies at least one third of all L_i in any given list L.

Solution. Let Y_i be the indicator variable of the event that a randomly chosen n-permutation satisfies L_i. Then $P(Y_i = 1) = \frac{1}{3}$ as each of a_i, b_i and c_i has the same chance to be in the middle. Therefore, $E(Y_i) = \frac{1}{3}$. Now if $Y = \sum_{i=1}^{k} Y_i$, then Y is the number of L_i in L that are satisfied by p. Theorem 15.18 then implies

$$E(Y) = \sum_{i=1}^{k} E(Y_i) = \frac{k}{3},$$

and our claim follows from Theorem 15.21.

15.4.3 *Conditional Expectation*

Another way of computing the expectation of a variable is by using conditional expectations. The conditional expectation $E(X|A)$ is the expected value of X given that event A occurs. Accordingly, $E(X|A)$ is defined by replacing the absolute probabilities in the definition of $E(X)$ by probabilities conditional on the occurrence of A. In other words,

$$E(X|A) = \sum_{i} i \cdot P(X = i|A),$$

where i ranges through all values X takes with a positive probability, given that A occurs.

We can then extend Theorem 15.15 to expectations as follows.

Theorem 15.24. *Let X be a random variable, and let $A_1, A_2, \cdots, A_n$ be events in a sample space Ω so that $A_1 \cup A_2 \cup \cdots \cup A_n = \Omega$, and $A_i \cap A_j = \emptyset$ if $i \neq j$. Then*

$$E(X) = \sum_{i=1}^{n} E(X|A_i)P(A_i).$$

Proof. This follows immediately from Theorem 15.15. Just let C be the event $X = j$ in that theorem. Multiply both sides by j, and sum over all values of j taken by X with a positive probability. □

Example 15.25. We throw a die three times. Provided that the first throw was at least four, what is the expectation of the number of times a throw resulted in an even number?

Solution. If the first throw was an even number, then the expected number of times we got an even result is two as it is one over the last two throws and is one over the first throw. If the first throw was an odd number, then this expectation is 1. Therefore, Theorem 15.24 implies

$$E(X) = \sum_{i=1}^{2} E(X|A_i)P(A_i) = \frac{2}{3} \cdot 2 + \frac{1}{3} \cdot 1 = \frac{5}{3}.$$

In this problem, it was very easy to compute the probabilities $P(A_i)$. The following problem is a little bit less obvious in that aspect.

Example 15.26. Our football team wins each game with 3/4 probability. What is our expected value of wins in a 12-game season if we know that we won at least three of the first four games?

Solution. We either won three games (event A_1), or four games (event A_2) out of the first four games. If we disregard the condition that we won at least three games out of the first four (event B), then $P(A_1) = 4 \cdot \frac{1}{4}(\frac{3}{4})^3 = \frac{27}{64}$, and $P(A_2) = (\frac{3}{4})^4 = \frac{81}{256}$. That condition, however, leads us to the conditional probabilities

$$P(A_1|B) = \frac{P(A_1 \cap B)}{P(B)} = \frac{\frac{27}{64}}{\frac{27}{64} + \frac{81}{256}} = \frac{4}{7},$$

and

$$P(A_2|B) = \frac{P(A_2 \cap B)}{P(B)} = \frac{3}{7}.$$

In this problem we assume that B occurred, that is, B is our sample space. To emphasize this, we will write $P_B(A_i)$ instead of $P(A_i|B)$. We denote the expectations accordingly.

In the last eight games of the season, the expected number of our wins is certainly $8 \cdot \frac{3}{4} = 6$, by Theorem 15.18. Therefore, denoting the number of our wins by X, Theorem 15.24 shows

$$E_B(X) = E_B(X|A_1)P_B(A_1) + E_B(X|A_2)P_B(A_2) = 9 \cdot \frac{4}{7} + 10 \cdot \frac{3}{7} = 9\frac{3}{7}.$$

We see that this expectation is larger than nine, the expectation without the condition that we won at least three of the first four games. This is because that condition allowed us to win all four of those games, which is better than our general performance.

Notes

This Chapter was not as much on Probability Theory itself as on the applications of Probability in Combinatorics. While there are plenty of textbooks on Probability Theory itself, there are not as many on Discrete Probability, that is, when Ω is finite. A very accessible introductory book in that field is "Discrete Probability" by Hugh Gordon [20]. As far as the Probabilistic Method in Combinatorics goes, a classic is "The Probabilistic Method", by Alon and Spencer [3].

Exercises

(1) Let p_n be the probability that a random text of n letters has a substring of consecutive letters that reads "Probability is fun". Prove that $\lim_{n\to\infty} p_n = 1$.

(2) A big corporation has four levels of command. The CEO is at the top, (level 1) she has some direct subordinates (level 2), who in turn have their own direct subordinates (level 3), and even those people have their own direct subordinates (level 4). Nobody, however, has more direct subordinates than his immediate supervisor. Is it true that the average number of direct subordinates of an officer on level i is always higher than the average number of direct subordinates of an officer on level $i+1$?

(3) A women's health clinic has four doctors, and each patient is assigned to one of them. If a patient gives birth between 8am and 4pm, then her chance of being attended by her assigned doctor is $3/4$, otherwise it is $1/4$. What is the probability that a patient is attended by her assigned doctor when she gives birth?

(4) We toss a coin a finite number of times. Let S denote the sequence of results. Set $X(S) = i$ if a head occurs in position i first. Find $E_\Omega(X)$, where Ω is the set of all finite outcome sequences.

(5) Show that for any n, there exist n events so that any $n-1$ of them are

independent, but the n events are not.

(6) At a certain university, a randomly selected student who has just enrolled has 66 percent chance to graduate in four years, but if he successfully completes all freshmen courses in his first year, then this chance goes up to 90 percent. Among those failing to complete at least one freshmen course in their first year, the 4-year-graduation rate is 50 percent. What is the percentage of all students who cannot complete all freshmen courses in their first year?

(7) We select an element of $[100]$ at random. Let A be the event that this integer is divisible by three and let B be the event that this integer is divisible by seven. Are A and B independent?

(8) Six football teams participate in a round robin tournament. Any two teams play each other exactly once. We say that three teams *beat each other* if in their games played against each other, each team got one victory and one loss. What is the expected number of triples of teams who beat each other? Assume that each game is a toss-up, that is, each team has 50 percent chance to win any of its games.

(9) Solve the previous exercise if one of the teams is so good that it wins its games by 90 percent probability.

(10) What is the expected value of the number of digits equal to 3 in a 4-digit positive integer?

(11) Let $X(\alpha)$ be the first part of a randomly selected composition α of n. Find $E(X)$.

(12) Let $Y(\alpha)$ be the number of parts in a randomly selected composition α of n. Find $E(Y)$.

(13) Let π be a randomly selected partition of the integer n. Let $X(p)$ be the first part of π, and let $Y(p)$ be the number of parts in π. Find $E(X) - E(Y)$.

(14) Let $p = p_1 p_2 \cdots p_n$ be an n-permutation. Recall that the index i is called an *excedance* of p if $p(i) > i$. How many excedances does the average n-permutation have?

(15) Let k be any positive integer, and let $n \geq k$. Let Y be the number of k-cycles in a randomly selected n-permutation. Find $E(Y)$.

(16) Recall from Chapter 14 that $S_n(1234) < S_n(1324)$ if $n \geq 7$. Let n be a fixed integer so that $n \geq 7$. Let A be the event that an n-permutation contains a 1234-pattern, and let B be the event that an n-permutation contains a 1324-pattern. Similarly, let X, (resp. Y) be the number of 1234-patterns (resp. 1324-patterns) in a randomly selected n-permutation. What is larger, $E(X|A)$ or $E(Y|B)$?

(17) Prove that there is a tournament on n vertices that contains at least $\frac{n!}{2^{n-1}}$ Hamiltonian paths. What can we say about the number of Hamiltonian cycles?

(18) Let Y be a probability variable. Then

$$\text{Var}(Y) = E\left((Y - E(Y))^2\right)$$

is called the *variance* of Y.

(a) Prove that $\text{Var}(Y) = E(Y^2) - E(Y)^2$.

(b) Let $X(p)$ be the number of fixed points of a randomly selected n-permutation p. Prove that $\text{Var}(X) = 1$.

Note that $\sqrt{\text{Var}(X)}$ is called the *standard deviation* of X.

(19) For $i \in [n]$, define X_i as in the proof of Theorem 15.20. Are the X_i independent?

(20) Let X and Y be two independent random variables defined on the same space. Prove that $\text{Var}(X + Y) = \text{Var}(X) + \text{Var}(Y)$.

(21) We are given a list $L = (L_1, L_2, \cdots, L_k)$ of ordered 4-tuples $L_i = (a_i, b_i, c_i, d_i)$, so that for any i, the numbers a_i, b_i, c_i, and d_i are distinct elements of $[n]$. It is possible, however, that symbols with *different* indices i and j denote the same number.

Let $p = p_1p_2 \cdots p_n$ be an n-permutation. We say that p satisfies L_i if the substring of p that stretches from a_i to b_i *does not intersect* the substring of p that stretches from c_i to d_i. (It could be that a_i is on the right of b_i, or c_i is on the right of d_i.)

Prove that there exists an n-permutation p that satisfies at least one third of all L_i in any given list L.

(22) A player pays a fixed amount of n dollars to a casino for the right to participate in the following game. A fair coin is tossed several times until a tail is obtained. If the first tail is obtained as the result of the ith coin toss, then the player receives a payout of 2^i dollars, and the game ends.

(a) Assuming that the casino has unlimited resources to pay its obligations and that the player has an unlimited amount of time to pay, what is the value of n that the player should pay for the right to play this game. (For this part of the exercise, let us say that the player should pay any amount of n dollars as long as n is less than the amount of his expected winnings.)

(b) What is the probability that the player will win less than 1000 dollars? Compare your answer to the answer to part (a).

(c) What is the expected value of the player's payout if the casino has "only' 10^{14} dollars available for payouts? (This is more than the world's annual GDP.)

Supplementary Exercises

(23) What is the probability of finding two people who were born in the same month of the year in a group of six randomly selected people?
(24) Prove that it is possible to 2-color the integers from 1 to 1000 so that no monochromatic arithmetic progression of length 17 is formed.
(25) Is it true that if the occurrence of A makes B more likely to occur, then the occurrence of B also makes A more likely to occur?
(26) (-) Give an example for three events A, B, and C, so that A and B are independent, B and C are independent, but A and C are not independent.
(27) (-) Give an example for three events A, B, and C so that A and B are not independent, B and C are not independent, but A and C are independent.
(28) (-) What is the probability of the event that a randomly selected composition of n has first part 1?
(29) What is the probability of the event that a randomly selected composition of n has a second part and that second part is 1?
(30) Let $i \geq 3$. What is the probability that a randomly selected composition of n has an ith part and that part is 1?
(31) Let S be an $n \times n$ magic square (see Exercise 24 in Chapter 3) with line sum r. Let A be the event that each entry of the first row is at least $\frac{r}{2n}$, and let B be the event that each element of the second row is at least $\frac{r}{2n}$. Is the following argument correct?
"It must be that $P(B|A) < P(B)$. Indeed, if A occurs, then the entries of the first row are all larger than normal, so each entry of the second row must be smaller than normal, because the sum of each column is fixed."
(32) Can two events be at the same time mutually exclusive and independent?
(33) A medical device for testing whether a patient has a certain type of illness will accurately indicate the presence of the illness for 99 percent of patients who do have the illness, and will set off a false alarm for

five percent of patients who do not have the illness.
Let us assume that only three percent of the general population has the illness.

(a) If the test indicates that a given patient has the illness, what is the probability that the test is correct?
(b) If the test indicates that a given patient does not have the illness, what is the probability that the test is correct?
(c) What percentage of the test results provided by this device will be accurate?

(34) Adam and Brandi are playing the following game. They write each integer from 1 through 100 on a piece a paper, then they randomly select a piece of paper, and then another one. They add the two integers that are written on the two pieces of paper, and if the sum is even, then Adam wins, if not, then Brandi. Is this a fair game?
(35) Replace 100 by n in the previous exercise. For which positive integers n will the game be fair?
(36) *Note: here, and in the next several exercises, when we say that we randomly select two objects of a certain kind, we mean that we select an ordered pair (A, B) of objects of that kind. So (A, B) and (B, A) are different pairs, and $A = B$ is allowed.*

(a) Let us randomly select two subsets of $[n]$. What is the probability that they have the same number of elements?
(b) Let $f(n)$ be the probability you were asked to compute in part (a). Prove that $f(n) \sim \frac{1}{\sqrt{n\pi}}$.

(37) Let us randomly select two compositions of the integer n, and let $g(n)$ be the probability that they have the same *smallest* part. Prove that if n goes to infinity, then $g(n) \to 1$.
(38) (+) Let us randomly select two partitions of $[n]$, and let $h(n)$ be the probability that their *smallest blocks* have the same size. Prove that if n goes to infinity, then $h(n) \to 1$.
(39) Let us randomly select two permutations of length n, and let $m(n)$ be the probability that their largest cycles have the same length. Prove that

$$m(n) \geq \sum_{i=\lceil (n+1)/2 \rceil}^{n} \frac{1}{k^2}.$$

Note that the summation starts in the smallest value of i that is strictly larger than $n/2$.

(40) A dealership has n cars. An employee with a sense of humor takes all n keys, puts one of them in each car at random, then locks the doors of all cars. When the owner of the dealership discovers the problem, he calls a locksmith. He tells him to break into a car, then use the key found in that car to open another, and so on. If and when the keys already recovered by this procedure cannot open any new cars, the locksmith is to break into another car. This algorithm goes on until all cars are open.

(a) What is the probability that the locksmith will only have to break into one car?

(b) What is the probability that the locksmith will have to break into two cars only?

(c) What is the probability that the locksmith will have to break into at most k cars?

(41) (+)

(a) Let us consider the situation described in the previous exercise, but let us now assume that the manager calls two locksmiths, each of whom chooses a car and breaks into it. What is the probability that there will be no need to break into any other cars? (Make the rather conservative assumption that the two locksmiths will not break into the same car.)

(b) Same as part (a), but with k locksmiths, instead of two.

(c) Compare the result of part (a) of this exercise and part (b) of the previous exercise. Explain why the results agree with common sense.

(42) Let X be a random variable defined on the sample space of all trees on vertex set $[n]$ so that $X(t)$ equals the number of leaves of the tree t. Find $E(X)$. Explain what your result means for large values of n. (That is, explain, roughly what fraction of the vertices of a tree on $[n]$ will be leaves on average.)

(43) Find the expectation of the number of k-cycles in a randomly selected n-permutation. Then use the result to solve Exercise 7 of Chapter 6.

(44) We randomly select a cycle of an n-permutation. On average, what will be the length of this cycle?

(45) There are 16 disks in a box. Five of them are painted red, five of them are painted blue, and six are painted red on one side, and blue on the other side. We are given a disk at random, and see that one of its sides is red. Is the other side of this disk more likely to be red or

blue?

(46) There are ten disks in a basket, two of them are blue on both sides, three of them are red on both sides, and the remaining five are red on one side, and blue on the other side. One disk is drawn at random, and we have to guess the color of its back. Does it help if we know the color of its front?

(47) A pack of cards consists of 100 cards, two of them are black kings. We shuffle the cards, then we start dealing them until we draw a black king. Which is the step where this is most likely to occur?

(48) Let $p = p_1p_2 \cdots p_n$ be an n-permutation. We say that p get changes direction at position i, if either $p_{i-1} < p_i > p_{i+1}$, or $p_{i-1} > p_i < p_{i+1}$, in other words, when p_i is either a *peak* or a *valley*. We say that p has k *runs* if there are $k-1$ indices i so that p changes direction at these positions. For example, $p = 3561247$ has 3 runs as p changes direction when $i = 3$ and when $i = 4$. What is the average number of runs in a randomly selected n-permutation?

(49) What is the average number of cycles of length four in a randomly selected graph on vertex set $[n]$? (Each pair of vertices has $1/2$ chance to be connected by an edge.)

(50) Recall that a *descent* of a permutation $p = p_1p_2 \cdots p_n$ is the number of indices $i \in [n-1]$ so that $p_i > p_{i+1}$. Let X be the number of descents of a randomly selected n-permutation. Find $E(X)$ and $\mathrm{Var}(X)$.

Solutions to Exercises

(1) First, we note that the sequence $\{p_n\}$ is increasing. Indeed, $p_{n+1} = p_n + q_n$, where q_n is the probability of the event that the set of the first n letters does not contain the required sentence, but that of the first $n+1$ letters does.

It is therefore sufficient to show that the sequence $\{p_n\}$ has a subsequence that converges to 1. Such a subsequence is $r_n = p_{16n}$. (Note that the sentence "Probability is fun" contains 16 letters.)

Let a be the probability of the event that a randomly selected 16-letter string is not our required sentence. Then $a < 1$. On the other hand, $r_n \geq 1 - a^n$ as we can split a $16n$-letter string into n strings of length 16, each of which has a chance to be something else than our sentence.

So we have

$$1 - a^n \leq r_n \leq 1,$$

and our claim follows by the squeeze principle as $a^n \to 0$.

(2) That is not true. Figure 15.2 shows a counterexample. Indeed, the average number of direct subordinates of level-2 officers is $6/4 = 1.5$, while that of level-3 officers is $10/6 = 1.66$.

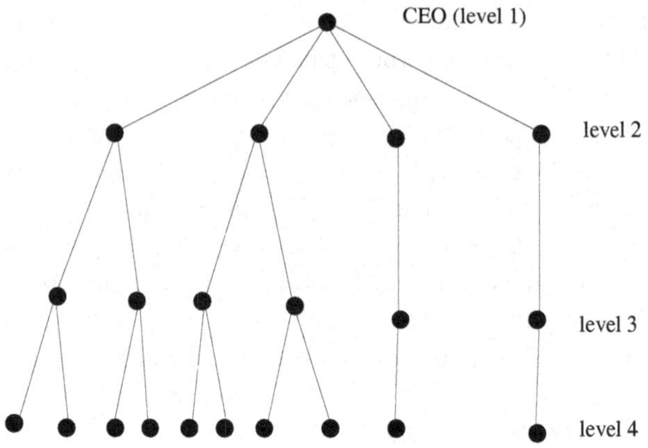

Fig. 15.2 A counterexample for Exercise 2.

(3) There is $1/3$ chance that a given patient gives birth between 8am and 4pm, and there is $2/3$ chance that she gives birth between 4pm and 8am. Therefore, Bayes' theorem shows that the answer is $\frac{1}{3} \cdot \frac{3}{4} + \frac{2}{3} \cdot \frac{1}{4} = \frac{5}{12}$.

(4) The only way for the first head to occur in position i is to have a tail in each of the first $i-1$ positions, then a head in position i. The chance of this happening is $1/2^i$. Therefore, we have

$$E(X) = \sum_{i=1}^{\infty} \frac{i}{2^i} = 2.$$

We used the fact that $\sum_{n \geq 1} nx^n = \frac{x}{(1-x)^2}$. This has been proved in two different ways in Exercise 25 of Chapter 4.

(5) Let us throw a die $n-1$ times, and for $1 \leq i \leq n-1$, denote A_i the event that throw i results in an even number. Finally, let A_n be the event that the *sum* of all the results is even. Then for any k-element subsets of these events, for $1 \leq k \leq n-1$, we have

$$P(A_{i_1}) \cdot P(A_{i_2}) \cdots P(A_{i_k}) = P(A_{i_1} \cap A_{i_2} \cap \cdots \cap A_{i_k}) = \frac{1}{2^k}.$$

However,

$$P(A_1 \cap A_2 \cap \cdots \cap A_n) = P(A_1 \cap A_2 \cap \cdots \cap A_{n-1}) = \frac{1}{2^{n-1}}$$

$$\neq \frac{1}{2^n} = P(A_1) \cdot P(A_2) \cdot \cdots \cdot P(A_n).$$

(6) Let A be the event that a randomly selected new student passes all his courses in his first year, and let C be the event that a randomly selected new student graduates in four years. Then Bayes' theorem and our data imply

$$P(C) = P(C|A)P(A) + P(C|\bar{A})P(\bar{A}),$$

$$0.66 = 0.9P(A) + 0.5(1 - P(A)),$$

yielding $P(A) = 0.4$, and $P(\bar{A}) = 0.6$. This means that sixty percent of freshmen fail to complete at least one course in their first year.

(7) No, they are not. There are 33 integers in $[100]$ that are divisible by three, and there are 14 integers in $[100]$ that are divisible by seven. Therefore $P(A) = 33/100$, and $P(B) = 14/100$. On the other hand, there are just 4 integers in $[100]$ that are divisible by 21 (so both by three and seven), which implies that $P(A \cap B) = 4/100 = 0.04$. On the other hand, $P(A)P(B) = \frac{462}{1000} = 0.0462$.

(8) Select three teams A, B, and C. Their three games within this 3-member group, that is A vs. B, B vs. C, and A vs. C can end in eight different ways. Only two of those are outcomes in which these teams beat each other. Thus the expected number of beat-each-other triples on $\{A, B, C\}$ is $1/4$. As there are $\binom{6}{3} = 20$ possible triples, it follows from Theorem 15.18 that the expected number of beat-each-other triples is 5.

(9) In this case, the ten triples not containing the strong team still have a $1/4$ chance to beat each other. In any of the other ten triples, the chances for this are $2 \cdot 0.9 \cdot 0.1 \cdot 0.5 = 0.09$. Therefore, using indicator variables and Theorem 15.18, we get that on average, there will be $10 \cdot 0.25 + 10 \cdot 0.09 = 3.4$ beat-each-other triples.

(10) Define indicator variables the usual way, that is, $X_i = 1$ if the ith digit is equal to 3, and zero otherwise. Then $E(X_1) = 1/9$, as the first digit cannot be zero, and $E(X_i) = 0.1$ if $i > 1$. Therefore, we have $E(X) = E(X_1 + X_2 + X_3 + X_4) = \frac{1}{9} + \frac{3}{10} = 0.4111$.

(11) Let A_i be the event that α has i parts. Then $P(A_i) = \frac{\binom{n-1}{i-1}}{2^{n-1}}$, and the first part of α is n/i on average. Therefore,

$$E(X) = \sum_{i=1}^{n} \frac{\binom{n-1}{i-1}}{2^{n-1}} \cdot \frac{n}{i} = \frac{1}{2^{n-1}} \cdot \sum_{i=1}^{n} \binom{n}{i} = \frac{2^n - 1}{2^{n-1}} = 2 - \frac{1}{2^{n-1}}.$$

(12) **First solution.** The number of weak compositions of n into k parts is $\binom{n-1}{k-1}$. Therefore, the probability that a randomly selected weak composition of n will have k parts is $\binom{n-1}{k-1}/2^{n-1}$. This implies

$$E(Y) = \frac{1}{2^{n-1}} \sum_{k=1}^{n} k\binom{n-1}{k-1} = \frac{1}{2^{n-1}}((n-1)2^{n-2} + 2^{n-1}) = \frac{n+1}{2}.$$

Second solution. Alternatively, we know that weak compositions of n into k parts are in bijection with $(k-1)$-element subsets of $[n-1]$. There is a natural bijection between these subsets of $[n-1]$, and $(n-k)$-element subsets of $[n-1]$, simply by taking complements. This, however, defines a bijection between weak compositions of n into k parts, and weak compositions of n into $n+1-k$ parts, and the claim follows.

(13) For all i, we have $P(X = i) = P(Y = i)$. Indeed, π has first part i if and only if its conjugate partition has i parts. Therefore, $E(X) = E(Y)$, so $E(X) - E(Y) = 0$.

(14) Let X_i the indicator variable of the event that i is an excedance of p. Then clearly, $P(X_i = 1) = \frac{n-i}{n}$, thus $E(X_i) = \frac{n-i}{n}$. Let $X(p)$ be the number of excedances of p, then $E(X) = \sum E(X_i) = \sum_{i=1}^{n} \frac{n-i}{n} = \frac{n(n-1)}{2n} = \frac{n-1}{2}$.

(15) We know from Lemma 6.19 that the probability that a given entry i of p is part of a k-cycle is $1/n$. Therefore, if Y_i is the indicator variable of i being part of a k-cycle, then $E(Y_i) = 1/n$. Now we have $Y = \frac{Y_1+Y_2+\cdots+Y_n}{k}$. Indeed, a k-cycle contains exactly k entries. Therefore, we get by Theorem 15.18 that $E(Y) = nE(Y_1)/k = 1/k$.

(16) First note that $E(X) = E(Y) = \binom{n}{4}/24$ as any four entries of p have a $1/24$ chance of forming a q-pattern, for any 4-element pattern q. Now Theorem 15.24 shows

$$E(X) = E(X|A)P(A) + E(X|\bar{A})P(\bar{A}) = E(X|A)P(A),$$

$$E(Y) = E(Y|B)P(B) + E(Y|\bar{B})P(\bar{B}) = E(Y|B)P(B).$$

Indeed, the second summands are obviously equal to zero. As $E(X) = E(Y)$, this implies $E(X|A)P(A) = E(Y|B)P(B)$, and then $P(A) > P(B)$ implies $E(Y|B) > E(X|A)$.

This makes perfect sense: a smaller number of permutations contains the same number of patterns, so on average, they must contain more patterns.

(17) Take K_n, and direct each of its edges at random, to get a tournament T. If p is an undirected Hamiltonian path in K_n, then let $X_p(T) = 1$ if p becomes a directed Hamiltonian path in T, and let $X_p(T) = 0$ otherwise. Then we have $E(X_p) = \frac{1}{2^{n-1}}$, as p has $n-1$ edges. Let $X = \sum_p X_p$, where p ranges through all $n!$ Hamiltonian paths of K_n. Then X equals the number of Hamiltonian paths of T. Theorem 15.18 then implies

$$E(X) = n!E(X_p) = \frac{n!}{2^{n-1}},$$

and our claim follows from Theorem 15.21.

For Hamiltonian cycles, the only difference is that they have one additional edge. Therefore, there exists a tournament on n vertices with at least $\frac{n!}{2^n}$ Hamiltonian cycles.

(18)(a) We get $\text{Var}(Y) = E((Y-E(Y))^2) = E(Y^2)-E(2YE(Y))+E(Y)^2$, by simply computing the square. Note that in the second term, $E(Y)$ is a number, so we get $\text{Var}(Y) = E(Y^2)-2E(Y)^2+E(Y)^2 = E(Y^2) - E(Y)^2$.

(b) Using the result computed in part (a), and the linearity of expectation, we simply have to show that $E(X^2) = 2$. For $i \in [n]$, define X_i as in the proof of Theorem 15.20. Then

$$E(X^2) = E((\sum_{i=1}^{n} X_i)^2) = \sum_{i=1}^{n} E(X_i^2) + 2\sum_{i<j} E(X_iX_j). \qquad (15.4)$$

Now note that the X_i are 0-1 variables, so $X_i = X_i^2$, and in particular, $E(X_i^2) = E(X_i) = 1/n$, by Theorem 15.20. If p is a randomly selected n-permutation, and $i < j$, then there is $1/(n-1)n$ chance that $p_i = i$, and $p_j = j$, which is the only way for X_iX_j to be nonzero (one). Therefore, $E(X_iX_j) = \frac{1}{n(n-1)}$. This, compared to (15.4), implies

$$E(X^2) = n \cdot \frac{1}{n} + 2 \cdot \binom{n}{2} \cdot \frac{1}{n(n-1)} = 1 + 1 = 2,$$

and our claim follows.

(19) No, they are not. We have computed in the proof of the previous exercise that $P(X_iX_j = 1) = \frac{1}{n(n-1)}$. On the other hand, we have computed in Theorem 15.20 that $P(X_i) = P(X_j) = \frac{1}{n}$. So $P(X_iX_j = 1) \neq P(X_i = 1)P(X_j = 1)$, and our claim is proved.

(20) It follows from part (a) of Exercise 18 that

$$\text{Var}(X+Y) = E((X+Y)^2) - E(X+Y)^2. \tag{15.5}$$

Let us express both members of the right-hand side by simpler terms as follows. On one hand,

$$E((X+Y)^2) = E(X^2) + E(Y^2) + 2E(XY) = E(X^2) + E(Y^2)$$
$$+2E(X)E(Y),$$

as X and Y are independent. On the other hand,

$$E(X+Y)^2 = (E(X)+E(Y))^2 = E(X)^2 + E(Y)^2 + 2E(X)E(Y).$$

Comparing these two equations to (15.5), we get $\text{Var}(X+Y) = E(X^2) + E(Y^2) - E(X)^2 - E(Y)^2$, and the statement is proved.

(21) Let Y_i be the indicator variable of the event that a randomly chosen n-permutation satisfies L_i. Then clearly, $P(Y_i = 1) = \frac{1}{3}$ as a_i, b_i, c_i and d_i can occur in p in 24 different ways, of which eight satisfies L_i. Indeed, we can first choose if the (a_i, b_i) pair comes first, or the (c_i, d_i) pair comes first, then we can choose the order of the elements within the pairs.
Therefore, $E(Y_i) = \frac{1}{3}$. Now let $Y = \sum_{i=1}^{k} Y_i$, then Y is the number of L_i in L that are satisfied by p. Theorem 15.18 then implies

$$E(Y) = \sum_{i=1}^{k} E(Y_i) = \frac{k}{3},$$

and our claim follows from Theorem 15.21.

(22)(a) The probability of the first tail coming at the ith toss is 2^{-i}, for all i. So the expected payout is

$$\sum_{i\geq 1} 2^{-i} \cdot 2^i = \sum_{i\geq 1} 1 = \infty.$$

So, given the (unrealistic) conditions of this part of the exercise, no finite price is too high for the player to play this game.

(b) The player will win more than 1000 dollars if and only if it takes at least ten tosses to get the first tail (since $2^{10} = 1024$ is the smallest power of 2 that is larger than 1000). That happens if and only if the first nine tosses are all heads, and the probability of that is $2^{-9} = 1/512$. So very few people would pay more than 1000 dollars to play this game, since the chances of recovering their participation fee are less than 0.002 percent.

(c) In this case, the expected value of the payout is

$$\sum_{i\geq 1} 2^{-i} \cdot \min(2^i, 10^{14}) = \sum_{i\geq 1}^{\log_2(10^{14})} 1 + 10^{14} \cdot \sum_{i>\log_2(10^{14})} 2^{-i}$$
$$= 46 + \frac{10^{14}}{2^{46}}$$
$$= 47.2.$$

So expecting a payout of more than 47.2 dollars is not realistic.

Chapter 16

At Least Some Order. Partial Orders and Lattices

16.1 The Notion of Partially Ordered Set

Let us assume that you are looking for air tickets for your upcoming trip. There are five different airlines offering service to your destination, and you know what each of them would charge for a ticket. However, price is not the only important criterion for you. The duration of the flights also matters a little bit. In other words, if airline X offers a lower price *and* a shorter flight-time than airline Y, then you say that the offer of airline X is a better offer.

Let us assume that the offers from the five airlines are as follows.

A 600 dollars, 9 hours 20 minutes,
B 650 dollars, 8 hours 40 minutes,
C 550 dollars, 9 hours 10 minutes,
D 575 dollars, 8 hours 20 minutes,
E 660 dollars, 9 hours 5 minutes.

For example, the offer of airline D is clearly better than that of airline E, but there is no such clear-cut difference between the offers of airlines C and D. You can represent the entire complex situation with the diagram shown in Figure 16.1.

In this diagram, the dots correspond to the offers. If an offer X is better than another offer Y, then X is above Y in the diagram, *and* there is a path from X to Y so that when we walk through that path, we never walk up.

This was an example of a *partially ordered set*, the main topic of this chapter. The reader probably sees the explanation for this name: some, but not necessarily all, pairs of our elements were comparable. The time has come for a formal definition.

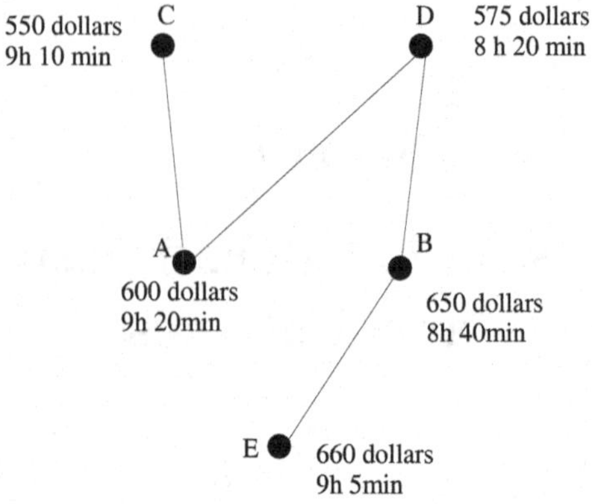

Fig. 16.1 Comparing the five offers.

Definition 16.1. Let P be a set, and let $\leq$ be a relation on P so that

(a) $\leq$ is reflexive, that is, $x \leq x$, for all $x \in P$,
(b) $\leq$ is transitive, that is, if $x \leq y$ and $y \leq z$, then $x \leq z$,
(c) $\leq$ is antisymmetric, that is, if $x \leq y$ and $y \leq x$, then $x = y$.

Then we say that $P_{\leq} = (P, \leq)$ is a *partially ordered set*, or *poset*. We also say that $\leq$ is a *partial ordering* of P.

Just as for the traditional ordering of real numbers, we write $x < y$ if $x \leq y$, but $x \neq y$. When there is no danger of confusion as to what the partial ordering $\leq$ of P is, we can write P for the poset $P_{\leq}$. If, for two elements x and y of P, neither $x \leq y$ nor $y \leq x$ holds, then we say that x and y are *incomparable*.

Example 16.2. Let P be the set of all subsets of $[n]$, and let $A \leq B$ if $A \subseteq B$. Then $P_{\leq}$ is a partially ordered set. This partially ordered set is denoted by B_n and is often called a *Boolean algebra* of degree n.

Example 16.3. The set of all subspaces of a vector space, ordered by containment, is a partially ordered set.

Example 16.4. Let P be the set of all positive integers, and let $x \leq y$ if x is a divisor of y. Then $P_{\leq}$ is a partially ordered set.

Example 16.5. Let $P = \mathbf{R}$, the set of real numbers, and let $\leq$ be the traditional ordering. Then $P_{\leq}$ is a partial order, in which there are no two incomparable elements. Therefore, we also call R a *total order*, or *chain*.

Example 16.6. Let P be the set of all partitions of $[n]$. Let α and β be two elements of P. Define $\alpha \leq \beta$ if each block of β can be obtained as the union of some blocks of α. For instance, if $n = 6$, then $\{1,4\}\{2,3\}\{5\}\{6\} \leq \{1,4,6\}\{2,3,5\}$. Then $P_{\leq}$ is a partial order, which is often called the *refinement order*, and is denoted by Π_n.

If $x \in P$ is such that there is no $y \in P$ for which $x < y$, then we say that x is a *maximal* element of P. If for all $z \in P$, $z \leq x$ holds, then we say that x is the *maximum* element of P. Minimal and minimum elements are defined accordingly. The reader should verify that all finite posets have minimal and maximal elements. Not all finite posets have minimum or maximum elements, however. The poset shown in Figure 16.1 does not have either. The minimum element of a poset, if it exists, is often denoted by $\hat{0}$, while its maximum element, if it exists, is often denoted by $\hat{1}$.

If $x < y$ in a poset P, but there is no element $z \in P$ so that $x < z < y$, then we say that y *covers* x. This notion enables us to formally define Hasse diagrams, the kind of diagrams we informally used in our introductory example.

The *Hasse diagram* of a finite poset P is a graph whose vertices represent the elements of the poset. If $x < y$ in P, then the vertex corresponding to y is above that corresponding to x. If, in addition, y covers x, then there is an edge between x and y. Alternatively, if we want to avoid the imprecise (but intuitively obvious) notion of "above", we can say that the Hasse diagram of P is the *directed* graph whose vertices are the elements of P, and in which there is an edge from x to y if x is covered by y.

Example 16.7. The Hasse diagram of B_3 is shown in Figure 16.2.

Hasse diagrams are useful to visualize various properties of posets. In particular, they can help us decide whether two small posets are isomorphic or not. We can hear the complaints of the reader that we have not even given the definition of isomorphism of posets yet. This is true, but the definition is the obvious one. That is, two posets P and Q are called isomorphic if there is a bijection $f : P \to Q$ so that for any two elements x and y of P, the relation $x \leq_P y$ holds if and only if $f(x) \leq_Q f(y)$.

It is easy to verify that up to isomorphism, there is one 1-element poset, two 2-element posets, and five 3-element posets. The Hasse diagrams of the

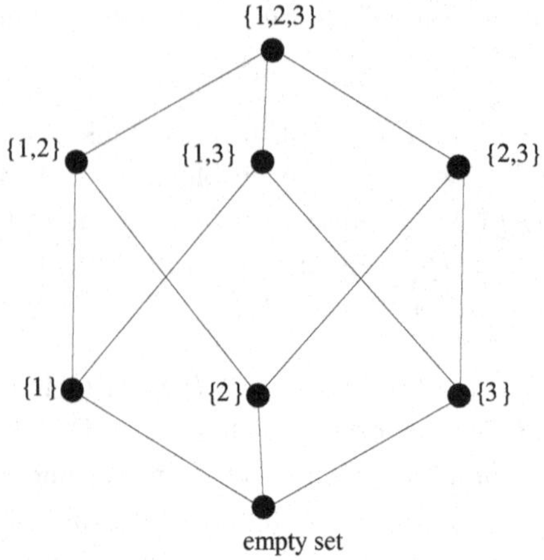

Fig. 16.2 The Hasse diagram of B_3.

latter are shown in Figure 16.3. The reader is invited to find all sixteen 4-element posets.

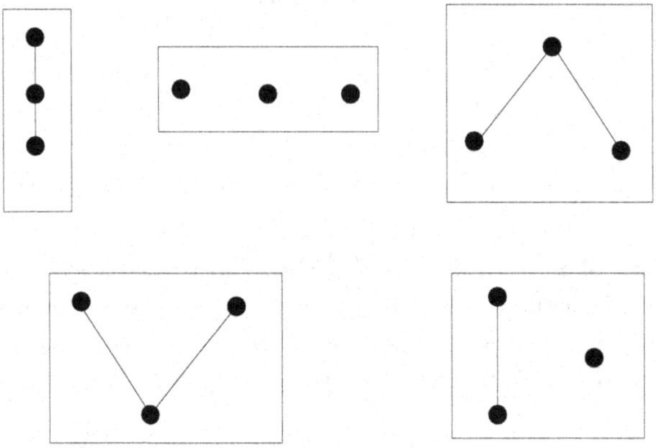

Fig. 16.3 The five three-element posets.

We have defined chains in Example 16.5. To see a finite example, in B_4, the set of subsets $\{\{2,3\},\{3\},\{1,2,3,4\}\}$ is a chain as we have $\{3\} \leq \{2,3\} \leq \{1,2,3,4\}$.

The dual notion is that of *antichains*. If the subset $S \subseteq P$ contains no two comparable elements, then we say that S is an *antichain*. For example,

$\{\{2,3\},\{1,3\},\{3,4\},\{2,4\}\}$ forms an antichain in B_4 as none of these four sets contains another one.

It is straightforward that any subset of a chain is a chain, and any subset of an antichain is an antichain. A *chain cover* of a poset is a collection of disjoint chains whose union is the poset itself. The size of a chain cover is just the number of chains in it. It seems plausible that if a poset has a large antichain, then it cannot be covered with just a few chains, and vice versa. The following classic theorem shows the precise connection between the sizes of antichains and chain covers of a poset.

Just as for matchings, a chain, (resp. antichain) X of P is called *maximum* if P has no larger chain (resp. antichain) than X, and X is called *maximal* if it cannot be extended. That is, no element can be added to X without destroying the chain (resp. antichain) property of X.

Theorem 16.8 (Dilworth's Theorem). *In a finite partially ordered set P, the size of any maximum antichain is equal to the number of chains in any smallest chain cover.*

Proof. Let a be the size of a maximum antichain A of P, and let b be the size of any smallest chain cover of P. Then it is clear that $a \leq b$. Indeed, no chain can contain more than one element of A, so at least a chains are needed in any chain cover.

We still have to prove the converse, that is, if the largest antichain of P is of size k, then P can be decomposed into the union of k chains. We prove this by induction on n, the number of elements in P. The initial case of $n = 1$ is trivial. Now let us assume that the statement is true for all positive integers less than n. We distinguish two cases.

- First let us assume that P has a k-element antichain A that contains at least one element that is not minimal, and at least one element that is not maximal. Then A "cuts P into two", that is, into the sets U and L, where U is the set of elements that are greater than or equal to at least one element in A, and L is the set of elements that are smaller than or equal to at least one element in A. Note that $U \cap L = A$. As A contains non-minimal and non-maximal elements, U and L are non-empty, and they both are partially ordered sets, with the ordering of P naturally restricted to them. Moreover, they have less than n elements, so the induction hypothesis implies that they are both unions of k chains. Each of the k chains in U has one of the

k elements of A at its bottom, and each of the k chains in L has one of the k elements of A at its top. Therefore, these $2k$ chains can be united to k chains covering P.

- Now let us assume that P does not have an antichain like the antichain A of the previous case. That is, all maximum antichains of P consist of maximal elements only, or minimal elements only. That necessarily implies that they contain all minimal elements, or all maximal elements. Let x be a minimal element of P, and let y be a maximal element of P such that $x \leq y$. (Since P is finite, such a pair of elements exist, though in the trivial special case when P itself is antichain, $x = y$ will occur.) Let P' be the poset obtained from P by removing x and y. Then the largest antichain of P' has $k - 1$ elements as it cannot contain all minimal elements or all maximal elements of P. Moreover, P' has less than n elements, so by the induction hypothesis, it can be decomposed into $k - 1$ chains. Adding the chain $x \leq y$ to this chain cover of P', we get a chain cover of P that is of size k.

$\square$

If P is an n-element poset, then a *linear extension* of P is just an *order-preserving* bijection from P onto $[n]$. That is, if $x \leq y$ in P, then $f(x) \leq f(y)$ in $[n]$.

Example 16.9. The poset shown on the left in Figure 16.4 has two linear extensions, f, and g, where $f(A) = g(A) = 4$, $f(D) = g(D) = 1$, $f(B) = g(C) = 2$, and $f(C) = g(B) = 3$. The poset shown on the right in Figure 16.4 has four linear extensions, as $\{E, F\}$ can be mapped onto $\{3, 4\}$ in two ways, and $\{G, H\}$ can be mapped onto $\{1, 2\}$ in two ways.

16.2 The Möbius Function of a Poset

In what follows we will develop some powerful computation techniques for a large class of posets. This class includes all finite posets. If the reader is only interested in finite posets, she can skip the next two paragraphs. In those paragraphs we discuss what other posets will belong to this class.

If $x \leq y$ are elements of P, then the set of all elements z satisfying $x \leq z \leq y$ is called the *closed interval* between x and y, and is denoted by $[x, y]$. If all intervals of P are finite, then P is called *locally finite*. Note that this does not necessarily mean that P itself is finite. The set of all positive integers with the usual ordering provides a counterexample.

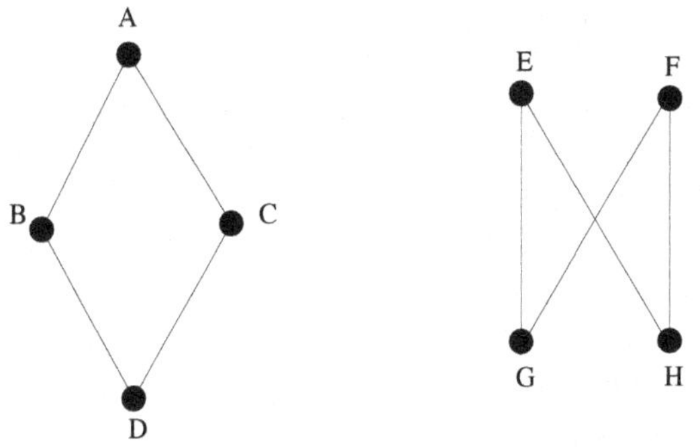

Fig. 16.4 Posets with two and four linear extensions.

A set of elements $I \subseteq P$ is called an *ideal* if $x \in I$ and $y \leq x$ imply $y \in I$. If an ideal is generated by one element, that is, $I = \{y : y \leq x\}$, then I is called a *principal ideal.* For example, if $P = B_n$, then the ideal of all subsets of $[k]$ is a principal ideal, while the ideal of all subsets that have at most four elements is not. In some of our theorems, we will have to restrict ourselves to posets in which *each principal ideal is finite.* In other words, each element is larger than a finite number of elements only. Note that this is a stronger requirement than being locally finite. The poset of all integers is locally finite, but has no finite principal ideals. Finally, we note that dual ideals, and principal dual ideals are defined accordingly.

Let $Int(P)$ be the set of all intervals of P.

Definition 16.10. Let P be a locally finite poset. Then the *incidence algebra* $I(P)$ of P is the set of all functions $f : Int(P) \to \mathbf{R}$.

Multiplication in this algebra is defined by

$$(f \cdot g)(x, y) = \sum_{x \leq z \leq y} f(x, z)g(z, y).$$

This definition of multiplication may seem odd at first sight. However, note that this is precisely the same as *matrix multiplication.* Indeed, take any linear extension $x_1 x_2 \cdots x_n$ of P, and define the $n \times n$ matrices F and G whose (i, j) entries are $f(x_i, x_j)$ and $g(x_i, x_j)$. These matrices will be upper triangular. Taking their product FG, we can see that the (i, j) entry of this product is

$$\sum_{k=1}^{n} f(x_i, x_k)g(x_k, x_j) = \sum_{k=i}^{j} f(x_i, x_k)g(x_k, x_j) = \sum_{x_i \leq x_k \leq x_j} f(x_i, x_k)g(x_k, x_j),$$

as claimed. So the incidence algebra of P is isomorphic to the algebra of $n \times n$ upper triangular matrices. We will alternatingly use the function terminology and the matrix terminology in our discussion.

Does $I(P)$ have a unit element, that is, an element u so that $uf = fu = f$ for all $f \in I(P)$? The above discussion shows that it must have as the algebra of all upper triangular matrices does have one, namely the identity matrix. The corresponding element of $I(P)$ is the function δ satisfying $\delta(x,y) = 1$, if $x = y$, and $\delta(x,y) = 0$ if $x < y$. It is straightforward to verify that indeed, this function satisfies $\delta f = f\delta = f$ for all $f \in I(P)$, so it is indeed the unit element of $I(P)$.

The following element of $I(P)$ is also a simply defined zero-one function. Nevertheless, it is surprisingly useful.

Definition 16.11. Let P be a locally finite poset. Let $\zeta \in I(P)$ be defined by $\zeta(x,y) = 1$ if $x \leq y$. Then ζ is called the *zeta function* of P.

A *multichain* in a poset is a multiset of elements $a_1, a_2, \cdots, a_m$ satisfying $a_1 \leq a_2 \leq \cdots \leq a_m$. Note that the inequalities are not strict, unlike in the definition of chains.

Proposition 16.12. *Let $x \leq y$ be elements of the locally finite poset P. Then the number of multichains $x = x_0 \leq x_1 \leq x_2 \leq \cdots \leq x_k = y$ is equal to $\zeta^k(x,y)$.*

Proof. By induction on k. If $k = 1$, then we have $\zeta^1(x,y) = 1$ if and only if $x \leq y$, and the statement is true. (In fact, the statement is even true if $k = 0$. Then $\zeta^0(x,y) = \delta(x,y) = 1$ if and only if $x = y$.)

Now let us assume that the statement is true for all positive integers less than k. Each multichain $x = x_0 \leq x_1 \leq x_2 \leq \cdots \leq x_k = y$ can uniquely be decomposed to a multichain $x = x_0 \leq x_1 \leq x_2 \leq \cdots \leq x_{k-1} = z$, and a two-element multichain $z \leq y$, where $z \in [x, y]$. Fix such a z. Then our induction hypothesis implies that the number of multichains $x = x_0 \leq x_1 \leq x_2 \leq \cdots \leq x_{k-1} = z$ is $\zeta^{k-1}(x, z)$, while the number of multichains $z \leq y$ is $\zeta(z, y)$. Summing over all z, we get that the total number of multichains $x = x_0 \leq x_1 \leq x_2 \leq \cdots \leq x_k = y$ is

$$\sum_{z \in [x,y]} \zeta^{k-1}(x, z)\zeta(z, y) = \zeta^k(x, y).$$

$\square$

The above proof shows that the number of elements of a multichain, or chain for that matter, is not always the handiest description of its size.

We will sometimes use the *length* of the chain, or multichain instead. The length of a chain (or multichain) is the number of its elements minus one. For chains, this has the following intuitive justification. If we walk up in the Hasse diagram of the poset, from the bottom of a chain of length k to its top, we will make k steps.

Lemma 16.13. *Let P be a locally finite poset. Let $[x, y] \in Int(P)$. Then the number of chains of length k that start at x and end in y is $(\zeta - \delta)^k(x, y)$.*

Proof. Analogous to that of Proposition 16.12. □

Does the zeta function of P have an inverse? That is, does there exist a function $\mu \in I(P)$ so that $\zeta\mu = \mu\zeta = \delta$? Again, resorting to our usual help, the matrix representations of the elements of $I(P)$, we see that the answer to this question should be in the affirmative. (The *zeta matrix* Z of P is just the $n \times n$ matrix whose rows and columns are labeled by the n elements of P, according to some linear extension of P, and $Z_{i,j} = \zeta_{i,j}$.) Indeed, $\zeta(x, x) = 1$ for all $x \in P$, therefore all diagonal entries of the zeta matrix Z of P are equal to 1, so $\det Z = 1$ as Z is triangular. Hence Z^{-1} exists, and those who remember the formula for the inverse of a matrix know that the matrix Z^{-1} will have integer entries only.

It turns out that the inverse of the zeta function of P is even more important than the zeta function itself. Therefore, it has its own name.

Definition 16.14. The inverse of the zeta function of P is called the *Möbius function* of P, and is denoted by $\mu = \mu_P$.

Computing the values of μ by computing the matrix Z^{-1} could be quite time-consuming. Fortunately, the triangular property of Z makes the following recursive computation possible.

Theorem 16.15. *Let P be a locally finite poset. Let $[x, y] \in Int(P)$. Then $\mu(x, x) = 1$, and*

$$\mu(x, y) = -\sum_{x \le z < y} \mu(x, z) \tag{16.1}$$

if $x < y$. In other words, μ is the only function in $I(P)$ satisfying $\mu(x, x) = 1$, and $\sum_{z \in [x,y]} \mu(x, z) = 0$ for all $x < y$.

Proof. First, we have $1 = \delta(x, x) = (\mu\zeta)(x, x) = \mu(x, x)\zeta(x, x) = \mu(x, x)$. Second, we have

$$0 = \delta(x, y) = \mu\zeta(x, y) = \sum_{z \in [x,y]} \mu(x, z)\zeta(z, y) = \sum_{z \in [x,y]} \mu(x, z)$$

if $x < y$. So the sum of $\mu(x,z)$, taken over all z in a nontrivial interval $[x,y]$ is indeed 0 as we claimed. □

Corollary 16.16. *Let P be a locally finite poset. Let $[x,y] \in Int(P)$, and let us assume that $x \neq y$. Then*

$$\mu(x,y) = -\sum_{z \in (x,y]} \mu(z,y).$$

Proof. This can be proved as Theorem 16.15, except that we have to use the equality $ZM = I$ instead of the equality $MZ = I$. □

Theorem 16.15 enables us to compute the values of $\mu(x,y)$ starting at $\mu(x,x) = 1$, and going from the bottom up.

Example 16.17. Figure 16.5 shows the computation of the values of $\mu(x,y)$. In this example, x is chosen to be the bottom element of the poset.

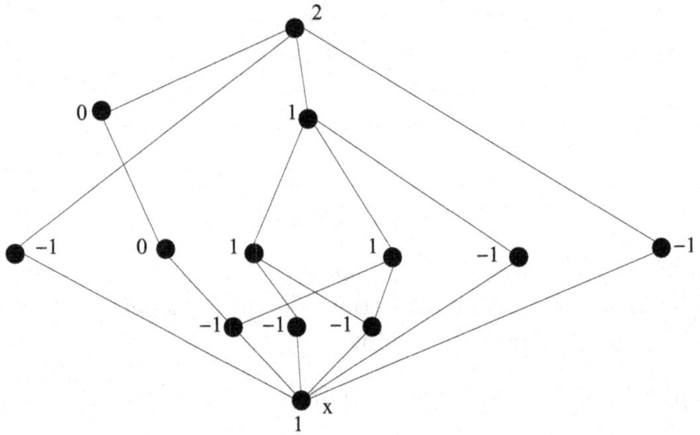

Fig. 16.5 The values of $\mu(x,y)$ when x is the bottom element.

By definition, $\mu(x,x) = 1$. Therefore, we must have $\mu(x,y) = -1$ for all y covering x. Then we compute all the other values from the bottom up, using formula (16.1).

Let us compute the value of $\mu(x,y)$ for some of the most frequently encountered posets.

Example 16.18. Let P be the poset of all nonnegative integers, and let $x < y$ be two distinct elements of P. Then $\mu(x,y) = -1$ if $x+1 = y$, and $\mu(x,y) = 0$ if $x+1 < y$.

Solution. This is straightforward by induction on $y - x$.

Example 16.19. Let $P = B_n$, and let S and T be two elements of P, that is, two subsets of $[n]$ so that $S \subseteq T$. Then

$$\mu(S,T) = (-1)^{|T-S|}.$$

Solution. Proof by induction on $k = |T - S|$. If $k = 0$, then $S = T$, so $\mu(S,T) = 1$ by definition, and the statement is true. Now let us assume that the statement is true for all nonnegative integers less than k, and let $|T - S| = k$. Then for all natural numbers i satisfying $0 \leq i \leq k-1$, the interval $[S,T]$ contains $\binom{k}{i}$ elements of P that are $|S| + i$ element subsets of $[n]$. If Z is such a subset, then it follows from the induction hypothesis that $\mu(S,Z) = (-1)^i$. Therefore, Theorem 16.15 implies

$$\mu(S,T) = -\sum_{Z \in [S,T)} \mu(S,Z) = -\sum_{i=0}^{k-1} \binom{k}{i} (-1)^i = (-1)^k.$$

The last equality is a direct consequence of Theorem 4.2. It can also be seen directly, from $(1-1)^k = 0$.

The induction step is complete, and so our statement is proved.

Example 16.20. Let P be the set of positive integers with the partial order in which $x \leq y$ if x is a divisor of y. Then

- $\mu(x,y) = (-1)^k$ if $\frac{y}{x} = p_1 p_2 \cdots p_k$, where $p_1, p_2, \cdots, p_k$ are different primes, and
- $\mu(x,y) = 0$ if $\frac{y}{x}$ is divisible by the square of a prime number.

Solution. First note that the interval $[1, \frac{y}{x}]$ and the interval $[x,y]$ are isomorphic as posets. Therefore, it suffices to prove our statements in the special case when $x = 1$. To simplify notation, we will write $\mu(y)$ instead of $\mu(1,y)$.

If $y = p_1 p_2 \cdots p_k$, where the p_i are different primes, then a little thought shows that the interval $[1,y]$ is isomorphic to the poset B_k. Indeed, a divisor of $y = p_1 p_2 \cdots p_k$ is just the product of the elements of a subset of $\{p_1, p_2, \cdots, p_k\}$. Therefore, $\mu(y) = (-1)^k$ as claimed.

We prove the second statement by strong induction on y. If $y = 4$, then the statement is true. Now assume that the statement is true for all positive integers smaller than y. Let $p_1, p_2, \cdots, p_k$ be the *distinct* prime divisors of y; it then follows that at least one of them occurs in the prime factorization of y more than once. Let us call a divisor of y *good* if it is not divisible by the square of a prime, and let us call a divisor of y *bad* if it is divisible by the square of a prime.

Then Theorem 16.15 implies

$$\mu(y) = -\sum_{z<y} \mu(z) = -\sum_{z \text{ good}} \mu(z) - \sum_{z \text{ bad}} \mu(z) = -(0+0) = 0.$$

Indeed, the set of the good elements z is precisely the interval $[1, p_1 p_2 \cdots p_k]$, and we know from Theorem 16.15 that $\sum_{z \in I} \mu(z) = 0$ for *any* interval I. On the other hand, $\mu(z) = 0$ for all bad integers z by the induction hypothesis, so it goes without saying that $\sum_{z \text{ bad}} \mu(z) = 0$ as well.

You could say "Fine, but who cares? What is the Möbius function good for?" In the following paragraphs, we will try to put our answer to this question into context.

Let $a_0, a_1, a_2, \cdots$ be a sequence of real numbers, and define the sequence $b_0, b_1, b_2, \cdots$ by

$$b_n = \sum_{i=0}^{n} a_i.$$

Then given the numbers a_i, one can certainly compute the numbers b_i. Conversely, given the numbers b_i, one can certainly compute the numbers a_i by the formula

$$a_n = b_n - b_{n-1}.$$

Now let $f : B_n \to \mathbf{R}$ be a function defined on the *subsets* of $[n]$, and let $g : B_n \to \mathbf{R}$ be another function defined on the subsets of $[n]$ by

$$g(T) = \sum_{S \subseteq T} f(S).$$

Again, given the values of f, the values of g are easy to compute. Given the values of g, however, the values of f are a little bit harder to compute. We have done this in Theorem 7.6, showing that

$$f(T) = \sum_{S \subseteq T} g(S)(-1)^{|T-S|}.$$

What was common in these two examples? In both cases, we worked in a poset. In the first case, it was the poset of all nonnegative integers (a sequence is just a function that is defined on nonnegative integers), in the second case it was B_n. We defined a function by setting its value in y to be the sum of the values of another function for *all elements of the poset that were smaller than* y. Then we showed that the values of the *original* function can be computed from the values of the new function.

The major application of the Möbius function, the *Möbius Inversion Formula,* will generalize this idea for *any* locally finite poset P.

Theorem 16.21 (Möbius Inversion Formula). *Let P be a poset in which each principal ideal is finite, and let $f : P \to \mathbf{R}$ be a function. Let the function $g : P \to \mathbf{R}$ be defined by*

$$g(y) = \sum_{x \leq y} f(x).$$

Then

$$f(y) = \sum_{x \leq y} g(x)\mu(x, y).$$

Proof. Let $x_1, x_2, \cdots$ be a linear extension of P. Let $\mathbf{f}$ be the row vector defined by $f_i = f(x_i)$, and let $\mathbf{g}$ be the row vector defined by $g_i = g(x_i)$. Let Z be the zeta matrix of P, and let M be the Möbius matrix of P. Then the equality $g(y) = \sum_{x \leq y} f(x)$ just means

$$\mathbf{g} = \mathbf{f}Z.$$

Multiplying both sides by M from the right, and using the fact that $ZM = I$, we get

$$\mathbf{g}M = \mathbf{f},$$

which is equivalent to our claim. □

Just as Theorem 16.15, this theorem also has a dual version.

Corollary 16.22. *Let P be a poset in which each principal dual ideal is finite, and let $f : P \to \mathbf{R}$ be a function. Let $g : P \to \mathbf{R}$ be defined by*

$$g(y) = \sum_{x \geq y} f(x).$$

Then

$$f(y) = \sum_{x \geq y} g(x)\mu(y, x).$$

Proof. This can be proved as Theorem 16.21, replacing the row vectors by column vectors, and right multiplication by left multiplication. □

Definition 16.23. Let P and Q be two posets. Then the *direct product* $P \times Q$ of these two posets is the poset whose elements are all the ordered pairs (p, q), where $p \in P$, and $q \in Q$, and in which $(p, q) \leq (p', q')$ if $p \leq p'$ and $q \leq q'$.

The values of the Möbius function in a direct product poset can be computed by the following Theorem.

Theorem 16.24. *Let us keep the notation of Definition 16.23. Then*

$$\mu_{P\times Q}\left((p,q),(p',q')\right) = \mu_P(p,p')\mu_Q(q,q').$$

Proof. We know that $0 = \sum_{z\in[p,p']} \mu(p, z)$, when $p \neq p'$, and also $0 = \sum_{s\in[q,q']} \mu(q, s)$, when $q \neq q'$. Multiplying these formulae together, we get

$$0 = \left(\sum_{z\in[p,p']} \mu(p,z)\right) \cdot \left(\sum_{s\in[q,q']} \mu(q,s)\right).$$

Note that we also know that $\mu_P(p,p')\mu_Q(q,q') = 1$ if and only if $p = p'$ and $q = q'$. Therefore, the function $\mu_P(p,p')\mu_Q(q,q')$ is the unique function defined on $Int(P \times Q)$ that sums to zero on all nontrivial intervals of $P \times Q$, and takes value 1 on all trivial intervals. That unique function is, by definition, the Möbius function of the poset $P \times Q$, and our statement is proved. □

Applications of this theorem will be provided in the next section, and also in the Exercises.

16.3 Lattices

There is a natural class of partial ordered sets called *lattices* for which additional techniques to compute the values of the Möbius functions are available. Let P be a poset, and let $x \in P$. If $x \leq_P a$, then we say that a is an *upper bound* for x. If $b \leq_P x$, then we say that b is a *lower bound* for x. If a is an upper bound for both x and y, then a is called a *common upper bound* for x and y. A common lower bound is defined analogously.

Now we are in a position to define lattices. Recall that the minimum (resp. maximum) element of a set, if it exists, is the element that is smaller (resp. larger) than any other element of the set.

Definition 16.25. A poset L is called a *lattice* if any two elements x and y of L have a minimum common upper bound a, and a maximum common lower bound b.

In this case, a is called the *join* of x and y, and b is called the *meet* of x and y. We denote these relations by $x \vee y = a$, and $x \wedge y = b$.

Example 16.26. The poset B_n is a lattice. Indeed, for any two subsets $S \subseteq [n]$, and $T \subseteq [n]$, the minimum subset of $[n]$ containing both S and T is $S \cup T$, and the maximum subset of $[n]$ that is contained in both S and T is $S \cap T$. Therefore, $S \cup T = S \vee T$, and $S \cap T = S \wedge T$.

Example 16.27. The poset of all subspaces of a vector space V is a lattice. If A and B are two subspaces of V, then $A \wedge B = A \cap B$, and $A \vee B$ is the subspace generated by A and B.

Example 16.28. The poset shown in Figure 16.6 is not a lattice. Indeed, elements A and B fail to have a minimum common upper bound since both C and D are minimal upper bounds for them. Similarly, C and D fail to have a maximum common lower bound since both A and B are maximal common lower bounds for them.

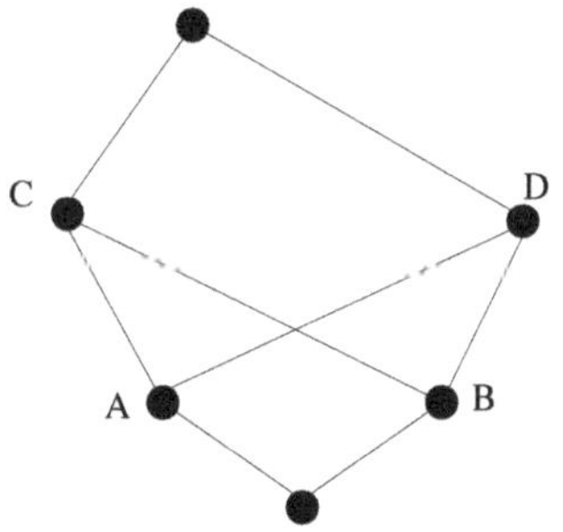

Fig. 16.6 This poset is not a lattice.

A finite lattice always has a minimum and a maximum element, as we show in Exercise 7. This is not necessarily true in infinite lattices. For example, the lattice of all finite subsets of **N** does not have a maximum element, or even a maximal element, for that matter.

The operations $\vee$ and $\wedge$ can easily be extended to more than two variables. It is straightforward to check that in a lattice, $(a \vee b) \vee c = a \vee (b \vee c)$, so we can talk about $a \vee b \vee c$, or, in more generality, $a_1 \vee a_2 \vee \cdots \vee a_n$. The same applies for the operation $\wedge$. The following simple proposition will be useful shortly. More importantly, it shows a typical lattice-theoretical argument.

Proposition 16.29. *If x, y, and t are elements of a lattice L, and $x \leq t$, and $y \leq t$, then $x \vee y \leq t$ also holds. Similarly, if $r \in L$, and $x \geq r$, and $y \geq r$, then $x \wedge y \geq r$.*

Proof. As both x and y are less than or equal to t, we know that t is a common upper bound for x and y, therefore it must be at least as large as their minimum common upper bound $x \vee y$. Similarly, r is a common lower bound for x and y, therefore it must be at most as large as their maximum common lower bound. □

We have seen that if we want to prove that a poset is a lattice, we have to show two things: the existence of the meet, and the existence of the join, for any two elements of the poset. Sometimes, one of these two claims is much easier to prove than the other. In these cases, the following lemma can help. Let us say that L is a *meet-semilattice* if, for any two elements x and y of L, the maximum common lower bound $x \wedge y$ exists.

Lemma 16.30. *Let L be a finite meet-semilattice with a maximum element. Then L is a lattice.*

In other words, if our poset is finite, and has a maximum element, then we only have to prove the existence of the meet. That of the join will automatically follow.

Proof. Let $x \in L$ and $y \in L$. Let B be the set of all common upper bounds of x and y. Then B is not empty as $\hat{1} \in B$, where $\hat{1}$ is the maximum element of L. We must show that B has a minimum element.

We know that B is a finite set as L itself is finite. Let $B = \{b_1, b_2, \cdots, b_k\}$. Then $b = b_1 \wedge b_2 \wedge \cdots \wedge b_k$ exists, and is an element of B by Proposition 16.29. Therefore, $b \leq b_i$ for all $i \in [k]$, so b is the minimum element of B. □

Example 16.31. The poset Π_n is a lattice. Indeed, we will show that it is a finite meet semilattice with a maximum element. As Π_n has $B(n)$ elements, it is finite. The maximum element of Π_n is obviously the partition consisting of one block. If α and β are two partitions of $[n]$, then $\alpha \wedge \beta$ is the partition in which the elements i and j are in the same block if and only if they are in the same block in both of α and β. Therefore, Lemma 16.30 shows that Π_n is a lattice.

Recall from Chapter 14 that a partition π of $[n]$ is called non-crossing if there are no four elements $a < b < c < d$ so that a and c belong to a block B_1 of π, and b and d belong to a block B_2 of π. As non-crossing partitions are partitions, the refinement order Π_n defines a partial order on them.

Example 16.32. The poset NC_n of non-crossing partitions of $[n]$, ordered by refinement, is a lattice.

Solution. Again, we show that NC_n is a finite meet-semilattice with a maximum element. Again, the one-block partition is the maximum element of NC_n, and our poset is finite. To see that NC_n is a meet-semilattice, note that if α and β are both non-crossing, then $\alpha \wedge_{\Pi_n} \beta$, as defined in the previous example, is also non-crossing. Therefore, $\alpha \wedge_{\Pi_n} \beta = \alpha \wedge_{NC_n} \beta$. Our claim then follows from Lemma 16.30.

We point out that it is *not* true that the *join* of two elements of NC_n is also the same in NC_n as in Π_n. Indeed, let $n = 4$, and let $\alpha = \{1\}\{2,4\}\{3\}$, and let $\beta = \{1,3\}\{2\}\{4\}$. Then $\alpha \vee_{\Pi_4} \beta = \{1,3\}\{2,4\}$, however, $\{1,3\}\{2,4\}$ is not even an element of NC_4. On the other hand, $\alpha \vee_{NC_4} \beta = \{1,2,3,4\}$.

To compute the Möbius functions of these lattices, we will need the following Theorem.

Theorem 16.33 (Weisner's theorem). *Let L be a lattice with minimum element $\hat{0}$ and with maximum element $\hat{1}$. Then for any element $a \in L - \{\hat{1}\}$,*

$$\mu(\hat{0}, \hat{1}) = - \sum_{\substack{x : x \wedge a = \hat{0} \\ x \neq \hat{0}}} \mu(x, \hat{1}).$$

In other words, for lattices, the Möbius function can be obtained by computing a significantly shorter sum than in the case of general posets. For posets, when computing $\mu(\hat{0}, \hat{1})$, in general we have to compute a sum of $n - 1$ members, where n is number of elements of the poset. However, for lattices, Theorem 16.33 shows that it is enough to sum over all elements whose meet with a is $\hat{0}$. If we choose a to be a large element, then the number of these elements will probably be small.

The remarkably simple proof we present is due to Vincent Vatter.

Proof. (of Theorem 16.33) After rearranging, our statement is equivalent to

$$0 = \sum_{x:x\wedge a=\hat{0}} \mu(x,\hat{1}) \tag{16.2}$$

as $\hat{0} \wedge a = \hat{0}$.

Now note that (16.2) is equivalent to

$$0 = \sum_{x\in L} \mu(x,\hat{1})f(x), \tag{16.3}$$

where $f(x) = 1$ if $x \wedge a = \hat{0}$, and $f(x) = 0$ otherwise.

Observe that

$$f(x) = \sum_{y\in[\hat{0},x\wedge a]} \mu(\hat{0},y). \tag{16.4}$$

Indeed, the right hand side of this equation sums the values of $\mu(\hat{0},y)$ for all y in an interval $[0, x \wedge a]$. As we know, such a sum is 0 or 1, depending on whether the interval is non-trivial or trivial, and that is exactly how f is defined as well. Comparing (16.3) and (16.4) shows that the claim to be proved is equivalent to

$$0 = \sum_{x\in L} \mu(x,\hat{1}) \sum_{y\in[\hat{0},x\wedge a]} \mu(\hat{0},y).$$

Note that $y \le x \wedge a$ if and only if $y \le x$ and $y \le a$. Therefore, reversing the order of summation in the last displayed equation, we conclude that the claim (16.2) is equivalent to

$$0 = \sum_{y\le a} \mu(\hat{0},y) \sum_{x\in[y,\hat{1}]} \mu(x,\hat{1}). \tag{16.5}$$

Finally, (16.5) clearly holds, since $y \le a < 1$, so the interval $[y,\hat{1}]$ is non-trivial, and hence the inner sum on the right-hand side is always 0. □

Now we are in a position to compute the values of $\mu_{\Pi_n}(\hat{0},\hat{1})$, and $\mu_{NC_n}(\hat{0},\hat{1})$.

Example 16.34. For all positive integers n,

$$\mu_{\Pi_n}(\hat{0},\hat{1}) = (-1)^{n-1}(n-1)!.$$

Solution. We want to use Theorem 16.33. That theorem works for any nonzero element $a \in \Pi_n$, but we want to choose an a so that the sum $\sum_{\substack{x:x\wedge a=\hat{0} \\ x\neq\hat{0}}} \mu(x,\hat{1})$ is easy to evaluate. We propose $a = \{1,2,\cdots,n-1\}\{n\}$. Then there are relatively few partitions x so that $x \wedge a = 0$. Indeed, in such partitions x, no two elements i and j of $[n-1]$ can be in the same block. Therefore, x can only be one of the $n-1$ partitions which have one doubleton block $\{i,n\}$, and $n-2$ singleton blocks. Let x be any of these partitions. We then claim that

$$[x,\hat{1}] \sim \Pi_{n-1}.$$

Indeed, if $\{i,n\}$ is the only doubleton block of x, then the elements i and n are in the same block in any partition from $[x,1]$. Therefore, Theorem 16.33 implies,

$$\mu_{\Pi_n}(\hat{0},\hat{1}) = -\sum_{\substack{x:x\wedge a=\hat{0} \\ x\neq\hat{0}}} \mu(x,\hat{1}) = -(n-1)\mu_{\Pi_{n-1}}(\hat{0},\hat{1}),$$

and our claim follows by induction on n.

The observation that $[x,\hat{1}] \sim \Pi_{n-1}$ when x is a partition of $[n]$ with one doubleton block and $n-2$ singletons can be generalized into the following statement.

Proposition 16.35. *Let y be a partition of $[n]$ that has k blocks. Then $[y,\bar{1}] \sim \Pi_k$.*

Proof. If two entries are in the same block in y, they are in the same block in all elements of $[y,\bar{1}]$. Therefore, in the poset $[y,\bar{1}]$, the blocks of y play the role of elements, and the statement follows. □

The formula obtained for the Möbius function of the partition lattice is surprisingly simple. The Möbius function of NC_n is even more surprising. Recall that the number of elements of that lattice is the Catalan number c_n.

Example 16.36. For any positive integer n,

$$\mu_{NC_n}(\hat{0},\hat{1}) = (-1)^{n-1}c_{n-1}.$$

Solution. We prove the statement by strong induction on n, the initial case being trivial. Let us assume that the statement is true for all positive integers less than n.

Let us proceed as in the previous example, with the same choice for a. What are the elements x so that $a \wedge x = \hat{0}$? As we know that $a \wedge_{\Pi_n} x = a \wedge_{NC_n} x$, it follows that these are again the partitions with $n-2$ singleton blocks, and one doubleton block, that is of the form $\{i, n\}$. What can we say about $\mu(x, \hat{1})$ if x is the mentioned partition? Since we are working in NC_n, all partitions in $[x, \hat{1}]$ are non-crossing. Since $\{i, n\}$ is a block in x, all partitions in $[x, \hat{1}]$ can naturally be decomposed as a partition of the set $[i-1]$, and a partition of the set $\{i+1, i+2, \cdots, n-1\}$.

Based on this, we claim that $[x, \hat{1}] \sim NC_i \times NC_{n-i}$. Indeed, consider first the sublattice L_1 of $[x, \hat{1}]$ in which in all partitions, each of the elements of $\{i+1, i+2, \cdots, n-1\}$ forms a singleton block (so all the "action" takes place on the partitions of $[i]$). Then L_1 is isomorphic to NC_i, because of the lattice isomorphism $f : NC_i \rightarrow L_1$ defined by $f(\{j\}) = \{j\}$ if $j \in [i-1]$, and $f(\{i, n\} = \{i\}$. Similarly, the sublattice L_2 of $[x, \hat{1}]$ in which in all partitions, each of the elements of $[i]$ forms a singleton block is isomorphic to NC_{n-i}.

As i ranges from 1 to $n-1$, using our induction hypothesis and Theorem 16.33, we get

$$\mu_{NC_n}(\hat{0}, \hat{1}) = - \sum_{\substack{x : x \wedge a = \hat{0} \\ x \neq \hat{0}}} \mu(x, \bar{1}) = - \sum_{i=1}^{n-1} (-1)^{n-2} c_{i-1} c_{n-i-1} = (-1)^{n-1} c_{n-1}.$$

How many connected simple graphs are there on the vertex set $[n]$? The difficulty here lies in enumerating *connected graphs.* It is certainly clear that there are $2^{\binom{n}{2}}$ graphs on these vertices as each of the $\binom{n}{2}$ pairs of vertices can be connected or not connected by an edge.

At any rate, the connected components of any simple graph on $[n]$ partition n in a natural way, that is, vertices that belong to the same component will belong to the same block. This partition will be called the *underlying* partition of the graph.

Now let H be any partition of $[n]$, and let us say that the blocks of H are of size $c_1, c_2, \cdots, c_h$. We cannot directly tell how many graphs on $[n]$ will have underlying partition V. However, we can easily tell how many graphs will have an underlying partition D so that $D \leq_{\Pi_n} H$. Indeed, these graphs cannot have edges between vertices that belong to different blocks of H. They can have edges within each block of H. Therefore, their number is $2^{\sum_{i=1}^{h} \binom{c_i}{2}}$.

Let $f(H)$ be the number of all graphs on $[n]$ with underlying partition H, and let $g(H)$ be the number of all graphs on $[n]$ with underlying partition

D so that $D \leq_{\Pi_n} H$.

Then $g(H) = 2^{\sum_{i=1}^{h} \binom{c_i}{2}}$, and

$$g(H) = \sum_{D \leq_{\Pi_n} H} f(D),$$

so the Möbius Inversion Formula implies

$$f(H) = \sum_{D \leq_{\Pi_n} H} g(D)\mu_{\Pi_n}(D, H).$$

We wanted to compute the number of *connected* graphs on $[n]$, that is, graphs whose underlying partition is the one-block partition N. Substituting N in the last equation, and using Proposition 16.35, we get

$$f(N) = \sum_{D \in \Pi_n} 2^{\sum_{i=1}^{d} \binom{d_i}{2}}(-1)^{d-1}(d-1)!,$$

where d is the number of blocks of D, and $d_1, d_2, \cdots, d_d$, are the sizes of the blocks in D.

Notes

We recommend [41] for further information on Möbius functions. For a different approach (dimension theory) on posets, see "Combinatorics and Partially Ordered Sets. Dimension Theory" by William T. Trotter [44]. Finally, we mention that Dilworth's theorem has a far-reaching generalization, the Greene–Fomin–Kleitman theorem. See [22] for details.

Exercises

(1) Let p be a permutation, and let d be the smallest integer so that p can be written as the union of d increasing subsequences. Prove that the longest decreasing subsequence of p consists of d elements.

(2) The *dimension* of a partial ordered set P is the minimum number of linear orders of the vertex set of P so that the intersection of these linear orders is precisely the partial order of P. Find a natural way to associate a poset of dimension two to each permutation. Will this mapping be injective?

(3) Let P be the set of all finite permutations, and let $p \leq_P p'$ if p is contained in p' as a pattern. Does this poset contain an infinite antichain?

(4) Let P be any locally finite poset and let $x_1, x_2, \cdots, x_n$ be a linear extension of P. Find a formula for the number of all chains from x_i to x_j, using the zeta function, or zeta matrix of P.

(5) We define the *covering matrix* C of a poset P as follows. The rows and columns are indexed by the vertices, listed according to some linear extension. $C_{i,j} = 1$ if x_j covers x_i, and $C_{i,j} = 0$ otherwise. Prove that the (i, j)-th entry of the matrix $(I - C)^{-1}$ is equal to the number of maximal chains of the interval $[x, y]$.

(6) Let P be any locally finite poset, let $x_i, x_j \in P$, and assume $x_i < x_j$. Prove that $\mu(x_i, x_j) = c_0 - c_1 + c_2 - c_3 + \cdots$, where c_i is the number of chains of length i from x_i to x_j. (So $c_0 = 0$ and $c_1 = 1$.)

(7) Prove that a finite lattice always has a minimum element and a maximum element.

(8) Find an example for a lattice that does not have a minimum element.

(9) Find a proof for the formula of the Möbius function of B_n using Theorem 16.24.

(10) Prove that in any lattice, we have $(x \wedge y) \vee y = y$.

(11) Prove that it is *not true* in every lattice that if $x \leq z$, then

$$x \vee (y \wedge z) = (x \vee y) \wedge z,$$

for all $y \in L$. A lattice in which this is true is called *modular.*

(12) Prove that it is *not true* in every lattice, not even in every modular lattice, that

$$x \wedge (y \vee z) = (x \wedge y) \vee (x \wedge z)$$

for all $x, y, z \in L$. A lattice in which this is true is called *distributive.*

(13) Prove that the condition of the previous exercise, that is, $x \wedge (y \vee z) = (x \wedge y) \vee (x \wedge z)$, for all $x, y, z \in L$, is equivalent to the condition

$$x \vee (y \wedge z) = (x \vee y) \wedge (x \vee z),$$

for all $x, y, z \in L$ so the latter can also be used to define distributive lattices.

(14) Prove that all distributive lattices are modular.

(15) In a lattice, we say that a is a *complement* of b if and only if $a \wedge b = \hat{0}$, and $a \vee b = \hat{1}$. Prove that if L is a distributive lattice, and $b \in L$ has a complement, then b has a unique complement $a \in L$.

(16) Show an example for a distributive lattice L in which each element has a complement, but $L \neq B_n$ for any n.

(17) Decide whether B_n, Π_n, and NC_n are distributive lattices.

(18) Let x and y be two given elements of Π_n so that $x \leq y$. Compute $\mu_{\Pi_n}(x, y)$.

(19) Is NC_n a modular lattice?

(20) A poset P is called *self-dual* if there exists a bijection $f : P \to P$ so that $f(x) \geq f(y)$ if and only if $x \leq y$. In other words, the Hasse diagram of P is invariant to the "turn upside down" operation. The bijection f is called an *anti-automorphism* of P.
Decide if the posets B_n, D_n, and Π_n are self-dual.

(21) Prove that NC_n is self-dual.

(22) Let Q_n be the poset of non-crossing partitions of $[n]$ in which $\pi \leq \pi'$ if the set of minimal elements of π is a proper subset of the set of minimal elements of π'. (By minimal element, we mean an element that is minimal in its block.) Prove that Q_n is self-dual.

(23) Keep the notation of the previous two exercises. Prove that if $x \leq_{NC_n} y$, then $y \leq_{Q_n} x$.

Supplementary Exercises

(24) (-) Find all sixteen 4-element posets.

(25) (-) Prove that B_n is a distributive lattice.

(26) (-) Is it true that every finite poset has as many antichains as ideals?

(27) (-) Let P be a finite poset, and let $J(P)$ be the poset whose elements are the *ideals* of P, ordered by inclusion. Prove that P is a lattice.

(28) Let P be a finite poset, and let $J(P)$ be defined as in the previous exercise. Prove that $J(P)$ is a *distributive* lattice.

(29) (-) What is P if $J(P) = B_n$?

(30) (-) Let $p = p_1 p_2 \cdots p_n$ be a permutation. Let P_p be the poset whose elements are the elements of $[n]$, and in which $i < j$ if $i < j$ in the usual ordering of natural numbers *and* i is on the left of j in P.

 (a) What is p if P_p is the n-element chain?
 (b) What is p if P_p is the n-element antichain?
 (c) What are the minimal and maximal elements of P_p?

(31) Find the number of all 2-element antichains in B_n.

(32) Find the number of all 2-element chains in B_n.

(33) Let P be the product of a k-element chain and an n-element chain. What is the size of the largest chain and the largest antichain of P?

(34) Let m and n be two distinct positive integers, and let D_m and D_n be the (respective) lattices of their divisors. Under what conditions is it true that the product of D_m and D_n equals D_{mn}?

(35) Let $M(n,k)$ be the multiset consisting of k copies of each element of $[n]$. Define a partial ordering $P(n,k)$ on the set of all sub-multisets of $M(n,k)$ as follows. Let $x \leq y$ if for all $i \in [n]$, the multiset x contains at most as many copies of i as y.
Find a general formula for $\mu_{P(n,k)}(x,y)$. Explain the connection between this exercise and Example 16.20.

(36) Prove that the poset $\mathbf{N^k}$ does not have infinite antichains for any k. (Recall that this is the poset of vectors with nonnegative integer coordinates, $x \leq y$ if and only if $x_i \leq y_i$ for all i.)

(37) Let P be a poset that has a minimum element $\hat{0}$, and let x be an element of P that covers one single element y. Let us assume that $y \neq \hat{0}$. Prove that $\mu(0,x) = 0$.

(38) Let m be any positive integer, and let P be a fixed poset. Let $\Omega_P(m)$ be the number of order-preserving maps f from P to the set $\{1, 2, \cdots, m\}$. In other words, if $x \leq_P y$, then $f(x) \leq f(y)$. Prove that $\Omega_P(m)$ is always a polynomial in m. This polynomial is called the *order polynomial* of P.

(39) What is $\Omega_P(m)$ if P is a k-element chain?

(40) What is $\Omega_P(m)$ if P is a k-element antichain?

(41) What is $\Omega_P(m)$ if P is the three-element poset consisting of one maximum element and two minimal elements?

(42) A chain in a poset is called maximal (or saturated) if it cannot be extended. Let B_n be the poset of all subsets of $\{1, 2, \cdots, n\}$. How many maximal chains does B_n have?

(43) How many linear extensions does the following poset have?

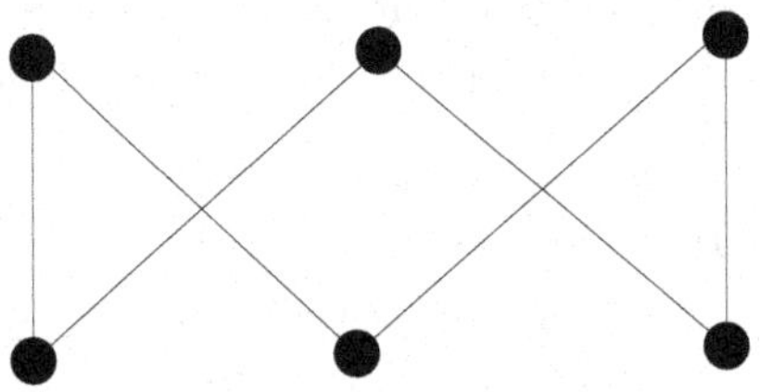

Fig. 16.7 How many linear extensions does this poset have?

(44) Find the number of linear extensions of the direct product of a 2-element chain and an n-element chain.

(45) Prove that for any finite poset P, the number of elements in any maximum chain equals the number of antichains in the smallest antichain cover.

(46) Let P be a poset having n elements. Prove that P contains either a chain of at least $\sqrt{n}$ elements, or an antichain of at least $\sqrt{n}$ elements.

(47) Let us define a partial order on the set of all partitions of the integer n as follows. If $a = (a_1, a_2, \cdots, a_k)$ and $b = (b_1, b_2, \cdots, b_t)$, then we say that $a \geq b$ if for all $i \in [k]$ the inequality

$$\sum_{j=1}^{i} a_i \geq \sum_{j=1}^{i} b_i$$

holds. Note that if $k \leq t$, then we set $a_{k+1} = a_{k+2} = \cdots = a_t = 0$. This order D_n is called the *dominance* order.

(a) Is D_n a lattice?
(b) Is D_n self-dual?

(48) Let Y be the poset of all partitions of all nonnegative integers ordered lexicographically. That is, $a \leq_Y b$ if $a_i \leq b_i$ for all i. Note that this automatically implies that if a has more (positive) parts than b, then a cannot be smaller than b in Y.

(a) Explain what this ordering means in terms of Ferrers shapes.
(b) Is Y a lattice?
(c) Prove that if an element $x \in Y$ covers k elements, then x is covered by $k + 1$ elements.

(49) Is it true that every interval of NC_n is self-dual?

(50) An *interval order* is a poset P that is isomorphic to a poset Q whose elements are closed intervals of real numbers, with the *precedence* ordering. That is, $[a, b] < [c, d]$ if $b < c$.
Prove that an interval order cannot contain two chains $c_1 < c_2$ and $d_1 < d_2$ so that for any $i, j \in [2]$, the elements c_i and d_j are incomparable. (This condition is often expressed by saying that P is **2+2**-avoiding.)
We point out that the converse is also known: if P does not contain four elements like that, then P is an interval order.

(51) A *unit interval order* is a poset P that is isomorphic to an interval order Q whose elements are closed intervals of *unit length*.

Prove that a unit interval order cannot contain a chain $c_1 < c_2 < c_3$ and an element d so that d is incomparable with c_i, for all $i \in [3]$. (These conditions are often expressed by saying that P is both **2+2**-avoiding and **3+1**-avoiding.)
We point out that the converse is also known: if P does not contain four elements like that, then P is a unit interval order.

Solutions to Exercises

(1) We show that this is a special case of Dilworth's Theorem. Indeed, let us introduce a partial ordering P on the elements of our permutation $p = p_1p_2\cdots p_n$ as follows. Let $p_i <_P p_j$ if $p_i < p_j$ as integers *and* $i < j$. Then chains in P are the increasing subsequences of p, and antichains of P are the decreasing subsequences of p.

(2) See the partial ordering defined in the previous exercise. This map is not one-to-one. For instance, p and p^{-1} are mapped into the same poset.

(3) Yes, it does. There are several ways to find an infinite antichain in this poset, and one of them is this.
Let $a_1 = 13, 12, 10, 14, 8, 11, 6, 9, 4, 7, 3, 2, 1, 5$. We view a_1 as having three parts: a decreasing sequence of length three at its beginning, a long alternating permutation starting with the maximal element of the permutation and ending with the entry 7 at the fifth position from the right (in this alternating part odd entries only have even neighbors and vice versa. Moreover, the odd entries and the even entries form two decreasing subsequences so that $2i$ is between $2i+5$ and $2i+3$), and a terminating subsequence 3 2 1 5.
To get a_{i+1} from a_i, simply insert two consecutive elements right after the maximum element m of a_i, and give them the values $(m-4)$ and $(m-1)$. Then make the necessary corrections to the rest of the elements, that is, increment all old entries on the left of m (m included) by two and leave the rest unchanged (see Figure 16.8). Thus the structure of any a_{i+1} is very similar to that of a_i—only the middle part becomes two entries longer.
We claim that the a_i form an infinite antichain. Assume by way of contradiction that there are indices i, j so that $a_i < a_j$. How could that possibly happen? First, note that the rightmost element

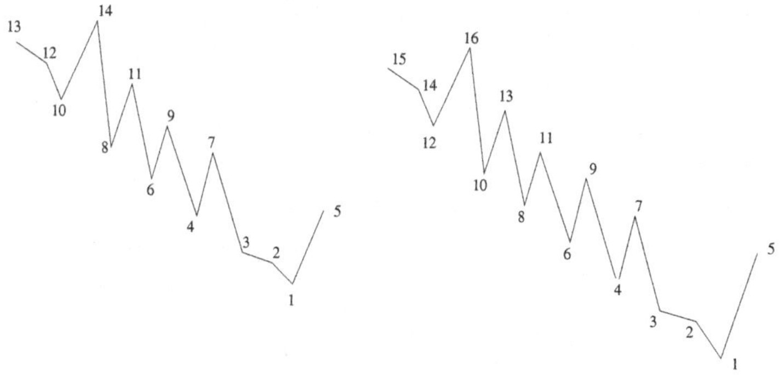

Fig. 16.8 Elements of our antichain.

of a_j must map to the rightmost element of a_i, since this is the only element in a_j preceded by four elements less than itself. Similarly, the maximal element of a_j must map to the maximal element of a_i, since, excluding the rightmost element, this is the only element preceded by three smaller elements. This implies that the first four and the last six elements of a_j must be mapped to the first four and last six elements of a_i, thus none of them can be deleted.

Therefore, when deleting elements of a_j in order to get a_i, we can only delete elements from the middle part, M_j. We have already seen that the maximum element cannot be deleted. Suppose we can delete a set D of entries from M_j so that the remaining pattern is a_i. First note that D cannot contain three consecutive elements, otherwise every element before those three elements would be larger than every element after them, and a_i cannot be divided into two parts with this property. Similarly, D cannot contain two consecutive elements in which the first is even. Thus D can only consist of separate single elements (elements whose neighbors are not in D) and consecutive pairs in which the first element is odd. Clearly, D cannot contain a separate single element as in that case the middle part of resulting permutation would contain a decreasing 3-subsequence, but the middle part, M_i, of a_i does not. On the other hand, if D contained two consecutive elements x and y so that x is odd, then let z be the element that immediately precedes x. Then all elements preceding z in the remaining permutation are larger than all elements on the right of z, including z. This is again a contradiction, as our permutations

a_i cannot be divided into two parts with this property.

This shows that D is necessarily empty, thus we cannot delete any elements from a_j to obtain some a_i where $i < j$. We have shown that no two elements in $\{a_i\}$ are comparable, so $\{a_i\}$ is an infinite antichain.

Note that all elements of our antichain avoid the pattern 123.

(4) Let Z be the zeta matrix of P. We claim that the number of all chains from x_i to x_j is equal to the (i,j)-th entry of the matrix $(2I - Z)^{-1}$. Note that $2I - Z = I - (Z - I)$. Therefore, $(2I - Z)^{-1} = (I - (Z - I))^{-1} = I + (Z - I) + (Z - I)^2 + (Z - I)^3 + \cdots$. As discussed in Lemma 16.13, the element of $(Z - I)^k$ in position (i,j) is the number of all k-element chains from x_i to x_j, and the proof follows.

(5) Note that $(I - C)^-1 = I + C + C^2 + \cdots$. It follows from the definition of C that the (i,j)-th element of C^k is the number of all k-element maximal chains from x_i to x_j, and the proof follows.

(6) We know from Lemma 16.13 that $c_k = (\zeta - \delta)^k(x_i, x_j)$. In other words, c_k is the (i,j)-entry of the matrix $(Z - I)^k$. Therefore $c_0 - c_1 + c_2 - c_3 + \cdots$ is the (i,j)-entry of the matrix $\sum_k (-1)^k (Z - I)^k = (I + Z - I)^{-1} = Z^{-1} = M$. Therefore,

$$c_0 - c_1 + c_2 - c_3 + \cdots = \mu(x_i, x_j)$$

as claimed.

(7) Let L be a finite lattice, and assume it does not have a minimum element. Then it has at least two different minimal elements x and y. Take $x \wedge y$; it has to be smaller than or equal to both x and y. As both x and y are minimal, this forces $x = x \wedge y$ and $y = x \wedge y$. This contradicts to $x \neq y$. The existence of a maximum element can be proved analogously.

(8) Take all subsets of **N** that have a *finite complement.* These subsets are partially ordered by containment, and form a lattice where the meet is the intersection, and the join is the union. There is no minimum element, however. Indeed, if there were such an element K, with complement size k, then we could take any subset of **N** whose complement is of size $k + 1$ to reach a contradiction. This lattice does not even have a minimal element.

(9) Note that $B_n = I_2^n$, where I_2 is the chain of two elements.

(10) The left-hand side is an upper bound of y, so it is at least y. On the other hand, y is a common upper bound for y and $x \wedge y$, so it is indeed their lowest common lower bound.

(11) The lattice shown in Figure 16.9 is a counterexample. Indeed, in that lattice, $x \vee (y \wedge z) = x \vee \hat{0} = x$, and $(x \vee y) \wedge z = \hat{1} \wedge z = z$.

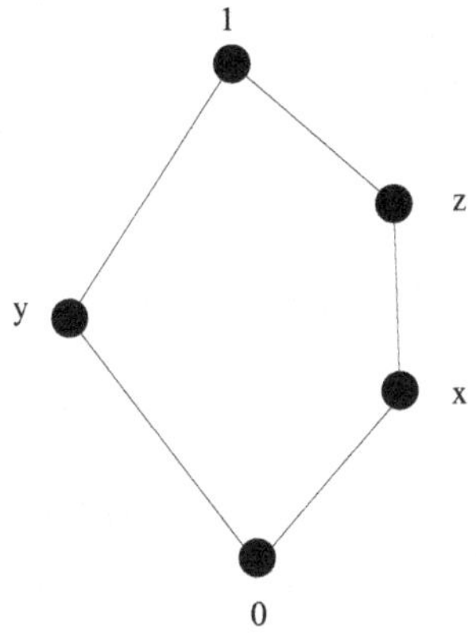

Fig. 16.9 A lattice that is not modular.

(12) The lattice shown in Figure 16.10 is a counterexample. Indeed, in that lattice, $x \wedge (y \vee z) = x \wedge \hat{1} = x$, and $(x \wedge y) \vee (x \wedge z) = \hat{0} \wedge \hat{0} = \hat{0}$.

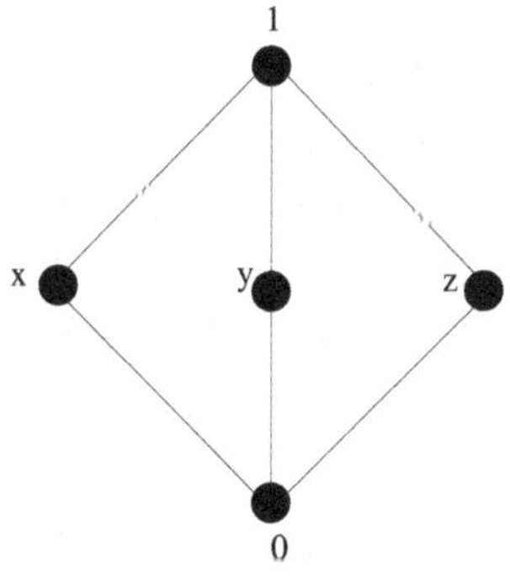

Fig. 16.10 A lattice that is not distributive.

(13) Let us assume that the condition of Exercise 12 holds. Apply this to the right-hand side of our new condition, considering $x \vee y$, the expression in the first set of parentheses, as one element. We get

$$(x \vee y) \wedge (x \vee z) = [(x \vee y) \wedge x] \vee [(x \vee y) \wedge z] = x \vee [(x \vee y) \wedge z]$$
$$= x \vee [(x \wedge z) \vee (y \wedge z)] = [x \vee (x \wedge z)] \vee (y \wedge z) = x \vee (y \wedge z),$$

as claimed. The other implication can be proved in an analogue way.

(14) Let L be distributive, and let $x \leq z$. Then

$$(x \vee y) \wedge z = (x \wedge z) \vee (y \wedge z) = x \vee (y \wedge z),$$

which was to be proved.

(15) Let us assume that the opposite is true, that is, that there is another element $c \in L$ so that c is the complement of b. In that case, we have $(a \vee b) \wedge (c \vee b) = \hat{1} \wedge \hat{1} = \hat{1}$, and also, $(a \vee c) \wedge b \leq b$. So this lattice could only be distributive if $b = \hat{1}$ held, but then b would only have $\hat{0}$ for its complement.

(16) Let L be the set of all subset of $\mathbf{N}$ that are either finite, or co-finite (have a finite complement). Then the complement of x is its set-theoretical complement. As our lattice is infinite, it is not isomorphic to B_n for any n. (It can be shown that there is no finite example for L.)

(17) If $n \geq 3$, then Π_n and NC_n are not distributive. Indeed, let a, (resp. b, and c) be three partitions with $n-2$ singleton blocks, and the only doubleton block $\{1,2\}$ (resp. $\{1,3\}, \{2,3\}$). Then the distributivity axioms do not hold for these three elements.
On the other hand, B_n is always distributive. Indeed, for all three subsets $A, B, C \subseteq [n]$, we have

$$A \cap (B \cup C) = (A \cap B) \cup (A \cap C).$$

This is because both sides consist of the elements of $[n]$ that are elements of A, and at least one of B and C.

(18) As $x \leq y$, the blocks of y are unions of the blocks of x. Say that y has k blocks, and they are unions of $u_1, u_2, \cdots, u_k$ blocks of x. Then it is straightforward to see that

$$[x,y] \sim \Pi_{u_1} \times \Pi_{u_2} \times \cdots \times \Pi_{u_k}.$$

Therefore, Theorem 16.24 and Example 16.34 imply

$$\mu(x,y) = \Pi_{i=1}^{k}(-1)^{u_k - 1}(u_k - 1)!.$$

(19) No, NC_n is not modular if $n \geq 3$. Let $n = 4$, and let $x = \{1,3\}\{2\}\{4\}$, $y = \{1\}\{3\}\{2,4\}$, and $z = \{1,2,3\}\{4\}$. Then $x \leq z$, but $x \vee (y \wedge z) = x \vee \hat{0} = x$, and $(x \vee y) \wedge z = \hat{1} \wedge z = z$. If $n \geq 4$, then the same example will work, by adding all the other elements as singleton blocks.

(20) The poset B_n is self-dual as the map defined by $f(S) = S^c$ is an anti-automorphism. The poset D_n is self-dual as the map $g(k) = n/k$ is an anti-automorphism. However, Π_n is not self-dual if $n \geq 3$. If it was, it

would have as many elements covering $\hat{0}$ (atoms) as elements covered by $\hat{1}$ (coatoms). That is not the case, as Π_n has $\binom{n}{2}$ atoms, namely the partitions that have one doubleton block, and $n-2$ singletons, and $2^{n-1}-1$ coatoms, namely the 2-block partitions.

(21) This result was first proved in [37]. Write the elements $1, 2, \cdots, n$ clockwise around a circle, and write elements $1', 2', \cdots, n'$ interlaced in counterclockwise order, so that $1'$ is between 1 and n, $2'$ is between n and $n-1$, and so on, i' is between $n+2-i$ and $n+1-i$. For $\pi \in NC_n$, join by chords cyclically successive unprimed elements belonging to the same block of π. Then define $g(\pi)$ to be the coarsest non-crossing partition on the elements $1', 2', \cdots, n'$ so that the chords joining primed elements of the same block do not intersect the chords of π. See Figure 16.11 for an example.

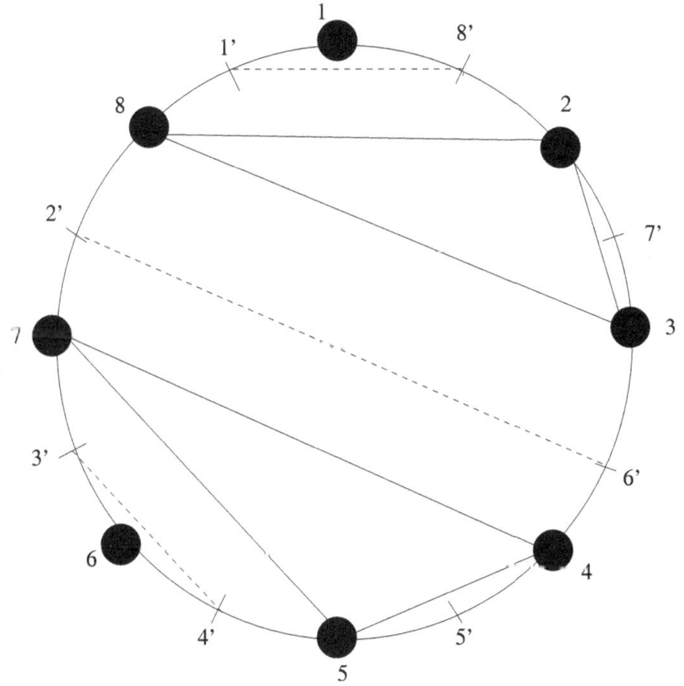

Fig. 16.11 The partition $\pi = (\{1\}, \{2, 3, 8\}, \{4, 5, 7\}, \{6\})$ and its image $g(\pi)$.

The map g is certainly a bijection, and it is order-reversing in NC_n

since merging two blocks of π subdivides a block of $g(\pi)$.

(22) In Exercise 14 of Chapter 14, we have seen that there is a bijection between non-crossing partitions of $[n]$ with a given set of minimal elements, and 132-avoiding n-permutations with a given descent set. Then Exercise 16 of Chapter 14 shows that the latter form a self-dual poset when ordered by the strict containment of their descent set.

(23) If $x \leq_{NC_n} y$, then each block of y is the union of some blocks of x. This means however, that the minimal element of each block of y is also a minimal element of x. So the set of minimal elements of y is strictly contained in that of x, and the statement follows.

Chapter 17

As Evenly As Possible. Block Designs and Error Correcting Codes

17.1 Introduction

17.1.1 *Moto-cross Races*

The following is a real-life example from moto-cross competitions. There are sixteen drivers competing for a main prize and various other prizes. The racetrack can safely accommodate four drivers in any single course, called a heat. So there will be numerous heats, and at the end of each heat, participants get points based on their rank in that heat. At the end of the day, the driver with the highest number of total points is declared the winner, the driver with the second highest number of points is declared the runner-up, and so on.

The question is how to schedule the various heats. If there were only a first prize to be awarded, then we could simply split the set of 16 contestants into four heats of four contestants each, and then have the four heat-winners compete in a final. That would be a short and fair race in that the best driver would win. It would not be fair if a second prize, a third prize, or additional prizes were awarded. Indeed, if the second-best driver is unlucky, she could be put in the same heat as the best driver, and so would not qualify for the final.

The other extreme solution is a race that is very fair, but unreasonably long. Schedule each of the $\binom{16}{4}$ possible four-driver heats to run once, and rank the drivers at the end based on their total number of points earned. The problem is that such a race would consist of $\binom{16}{4} = 1820$ heats. Even if one heat took only ten minutes, and there were no down-time between consecutive heats, this would take more than 12 days. The attention span of most fans is shorter than that.

Therefore, the challenge is to design the heats so that the race is both fair and reasonably short (it can be completed in an afternoon). The length of the race is very easy to measure. The fairness of the race is a little bit more complex, but participants can agree on a few simple principles. As drivers earn points in each heat they participate, each driver should compete in the same number of heats. Each heat should use the available space, that is, it should contain four drivers. Finally, to avoid situations in which some drivers had stronger opponents than others, we can require that for any two drivers, there is at least one heat in which they both competed. If this "at least one" can be replaced by "exactly one", that is even better. This has the additional benefit of being a way to break two-way ties: if, at the end of the race, two drivers have the same aggregate score, let the one who beat the other in their only shared heat be ranked ahead of the other.

If such a race can be designed, then each driver needs to participate in at least five heats. Indeed, each driver needs to compete against 15 others in at least one heat, but can compete only against three in any one heat. At least five heats for each of sixteen drivers means at least 80 ordered pairs (d, H), where d is a driver competing in heat H. Since each heat consists of four drivers, this means that there would have to be at least 20 heats.

So the question now is whether these minimal values can actually be attained. That is, can we plan 20 heats of four drivers each so that each driver participates in exactly five heats, and for any two drivers d_1 and d_2, there is a heat in which they both competed? (It follows from the preceding paragraph that if such a set of 20 heats exists, then for any two drivers d_1 and d_2, there is a *unique* heat H in which they both competed.)

It is very far from obvious, but it is possible to design such a race. An example is shown below. If each heat takes ten minutes to complete, and there is a two-minute down-time in the 19 breaks between heats, then it takes only 238 minutes, or slightly less than four hours, for the entire contest to take place.

Example 17.1. *The following collection of heats satisfies all the requirements. Every driver is in five heats, every heat consists of four drivers, and for any two drivers, there is exactly one heat containing both of them.*

(1) 1, 2, 3, 4
(2) 5, 6, 7, 8
(3) 9, 10, 11, 12
(4) 13, 14, 15, 16

(5) 1, 5, 9, 13
(6) 2, 6, 10, 14
(7) 3, 7, 11, 15
(8) 4, 8, 12, 16

(9) 1, 6, 11, 16
(10) 2, 5, 12, 15
(11) 3, 8, 9, 14
(12) 4, 7, 10, 13

(13) 1, 7, 12, 14
(14) 2, 8, 11, 13
(15) 3, 5, 10, 16
(16) 4, 6, 9, 15

(17) 1, 8, 10, 15
(18) 2, 7, 9, 16
(19) 3, 6, 12, 13
(20) 4, 5, 11, 14

At this point, the reader hopefully wants to know how this collection of heats was built, and for what set of requirements will such a collection exist. By the end of this chapter, we will find some interesting answers to these questions.

17.1.2 *Incompatible Computer Programs*

Let us assume that we have downloaded seven new programs to our computer. We have enough memory to run any three of them at the same time, but not more. We want to test that the seven programs are pairwise compatible with each other. That is, we want to make sure that there are no programs A and B among our seven programs so that running A and B at the same time always results in a system error (for reasons unrelated to space or memory availability). Let us assume that all incompatibilities are caused by a *pair* of programs, that is, if a subset S of programs cannot run together, then there is a 2-element subset $T \subset S$ so that even the elements of T cannot run together. What is the most efficient way to do this if testing the simultaneous operation of any k programs ($k \leq 3$) takes one minute?

Before we start looking for the most efficient way to test all pairs of programs, let us discuss some obvious upper bounds for the needed time. First, it is clear that $\binom{7}{3} = 35$ minutes will suffice, since we can simply test all three-element subsets of the seven-element set $\{A, B, C, D, E, F, G\}$ of programs. There is a tremendous waste in this method, since every pair of programs will run together five times (since there are five choices for the third element of the triple that is being tested). A somewhat better upper bound is 21 minutes. Indeed, we can just test each of the $\binom{7}{2} = 21$ pairs of programs. However, it is easy to see the waste in this method as well. In each test run, we are only testing a pair, and not a triple, of programs, and hence we gain information about the compatibility of just one pair, and not three pairs.

What is the best testing time that we can achieve? First, let us note that we may assume that in an optimal testing scheme, each run consists of testing a *triple* of programs, and not a pair. Indeed, we were told that testing a triple takes one minute, the same amount of time that it takes to test a pair, so we do not lose any time by completing all pairs of our testing scheme to triples.

Let b be the number of all triples used in our testing scheme. In each of these triples, three pairs are tested, so our scheme tests at most $3b$ pairs, exactly that many if all these $3b$ pairs are different. On the other hand, we need to test $\binom{7}{2} = 21$ pairs. Therefore, for any testing scheme, $3b \geq 21$, and so $b \geq 7$. So we need at least seven minutes to test all 21 pairs.

There remains the question whether we can actually *construct* a testing system that consists of only seven triples and still tests for every pair. The following example shows that the answer is in the affirmative.

Example 17.2. The following family F of subsets of the set $S = \{A, B, C, D, E, F, G\}$ has the property that for any two elements x and y of S there is a (necessarily unique) element of F that contains both x and y.

- $\{A, B, D\}$,
- $\{B, C, E\}$,
- $\{A, F, C\}$,
- $\{A, E, G\}$,
- $\{B, F, G\}$,
- $\{C, D, G\}$,
- $\{D, E, F\}$.

One way to verify that the preceding example is correct is by checking that if $x \in S$, then x occurs in exactly three subsets belonging to F, and that no element other than x occurs in more than one of those three subsets.

So we can test all pairs using the above scheme in just seven minutes. Note that the above scheme is free of any waste; each pair is tested once, but no pair is tested more than once. Usually, it is not possible to completely eliminate waste like that. For instance, if we had to test all pairs of an eight-element set of programs using a scheme of triples, then we would have had to test $\binom{8}{2} = 28$ pairs. As 28 is not divisible by three, we would have needed 10 triples, meaning that there would have been pairs tested more than once. If we had to test all pairs of a six-element set of programs using a scheme of triples, that would have meant testing $\binom{6}{2} = 15$ pairs. While 15 is divisible by three, complete elimination of waste is still not possible for the following reason. Each program has to be part of at least three triples, since each triple contains two other programs. So placing program A in just two triples will test it in only four pairs. However, placing A in three pairs tests A in six pairs. There are only five programs other than A, so at least one of the pairs containing A gets tested twice.

As we see, there are several reasons for which a completely waste-free testing scheme may not exist. From a more high-brow point of view, just because in some cases we may not be able to prove that such a scheme does not exist, it does *not* follow that it exists; it could be that it does not exist for some reason that is unknown to us. This raises a whole family of questions, such as what the sufficient and necessary conditions of the existence of a waste-free testing scheme are, and what the most interesting generalizations of this problem are.

17.2 Balanced Incomplete Block Designs

Let S be a finite set of v elements called *vertices*. Let $\mathcal{B}$ be a collection of b non-empty subsets of S called *blocks*. Then the pair $(S, \mathcal{B})$ is called a *block design* or just a *design*. Note that this definition does not exclude the possibility of some blocks appearing in $\mathcal{B}$ more than once, so the repetition of blocks is allowed. That said, most of our examples will not contain repeated blocks.

If a design $(S, \mathcal{B})$ contains at least one block that does not contain all vertices of S, then the design is called *incomplete*, otherwise it is called *complete*. It goes without saying that complete designs are not very interesting, unless some additional structure is placed on them. Our examples will all be incomplete designs.

If in a design, each block consists of k vertices, then the design is called *uniform* or k-uniform. Note that simple graphs are 2-uniform designs, with the edges of the graph being the blocks. If each vertex of a design occurs in exactly r blocks, then the design is called *regular* or r-*regular*.

Finally, if in a k-uniform, r-regular incomplete design $(S, \mathcal{B})$, each pair of vertices occurs together in exactly λ blocks, then we say that $(S, \mathcal{B})$ is a *balanced incomplete block design* or BIBD, of parameters (b, v, r, k, λ). Alternatively, we may simply call such a design a (b, v, r, k, λ)-design, since the mere existence of the parameters r, k, and λ shows that $(S, \mathcal{B})$ is regular, uniform, and balanced, and if $k < v$, then $(S, \mathcal{B})$ is incomplete.

Example 17.3. The design of Example 17.1 is a balanced incomplete block design of parameters $(20, 16, 5, 4, 1)$. That is, it has 20 blocks, 16 vertices, each vertex occurs in five blocks, each block consists of four vertices and each pair of vertices occurs together in exactly one block.

Example 17.4. The design of Example 17.2 is a balanced incomplete block design of parameters $(7, 7, 3, 3, 1)$. That is, it has seven blocks, seven vertices, each vertex occurs in three blocks, each block consists of three vertices, and each pair of vertices occurs together in exactly one block.

The following is an even smaller example of a BIBD.

Example 17.5. Let $S = \{a, b, c\}$, and let $\mathcal{B}= \{\{a, b\}, \{b, c\}, \{a, c\}\}$. Then $(S, \mathcal{B})$ is a balanced incomplete block design with parameters $(3, 3, 2, 2, 1)$.

The preceding example can be easily generalized as follows.

Example 17.6. Let $1 < k < n$. Then the family of all k-element subsets of $[n]$ is a BIBD. The reader is invited to verify that the parameters of this block design are $\left(\binom{n}{k}, n, \binom{n-1}{k-1}, k, \binom{n-2}{k-2}\right)$.

We point out that while all BIBDs have a lot of symmetries built in them, some have more than others. The BIBDs of Examples 17.2 and 17.5 both satisfy the equalities $b = v$, and, following from the latter, $r = k$. If a BIBD satisfies either (or as we will see very soon, equivalently, both) of these equalities, then it is called *symmetric*. So the BIBDs of Examples 17.2 and 17.5 are symmetric, those of the other examples are not.

At this point, you may say, "OK, so there exist a lot of BIBDs. However, can we have idea about how difficult it is to create one? What are the necessary conditions on the parameters for the existence of a BIBD?" The following two propositions give a partial answer to those questions.

Proposition 17.7. *If a (b, v, r, k, λ)-design exists, then $bk = vr$.*

Proof. Let $(S, \mathcal{B})$ be such a design. Then both sides count all ordered pairs (w, B), where $w \in S$, $B \in \mathcal{B}$, and $w \in B$. In other words, these are ordered pairs consisting of a vertex and a block containing that vertex. The left-hand side counts these pairs according to the blocks (there are k such pairs for each of b blocks), and the right-hand side counts these pairs according to the vertices (there are r such pairs for each of v vertices)¿ □

Note that in the proof of Proposition 17.7, the parameter λ did not play any role, and even the fact that λ existed, that is, that the design was balanced, was irrelevant. So the equality $bk = vr$ holds for all regular, uniform designs.

Proposition 17.8. *If a (b, v, r, k, λ)-design exists, then $r(k-1) = \lambda(v-1)$.*

Proof. Let $(S, \mathcal{B})$ be such a design. Let $x \in V$ be a fixed vertex. Then both sides count all ordered pairs (w, B), so that x and w are both vertices in the block B, with $x \neq w$. The left-hand side counts these pairs by first choosing one of the r blocks that contain x, then choosing any of the remaining $k - 1$ vertices of B for the role of w. The right-hand side counts these pairs by first choosing w in one of $v - 1$ ways, then choosing one of the λ blocks that contain both x and w for the role of B. □

Note that Propositions 17.7 and 17.8 show that the three parameters v, k, and λ determine the other two in any BIBD. Therefore, it is correct to refer to BIBDs as (v, k, λ)-designs. In this terminology, the BIBD of Example 17.1 is a $(16, 4, 1)$-design, and the BIBD of Example 17.2 is a $(7, 3, 1)$-design.

17.3 New Designs From Old

There are several ways to create new designs from ones that we already have at hand. Perhaps the simplest of these is the *complementary design.* For two sets X and Y, let $X \setminus Y$ denote the set of elements that are in X but not in Y.

Definition 17.9. Let $\mathcal{D} = (S, \mathcal{B})$ be a design. The *complementary design* of $(S, \mathcal{B})$ is the design $\mathcal{D}^c$ whose set of vertices is S, and whose blocks are the complements of the blocks in $\mathcal{B}$ in S. That is, B is a block of $\mathcal{D}^c$ if and only if $S \setminus B$ is a block of $\mathcal{D}$.

Example 17.10. If $\mathcal{D}$ is the design of Example 17.2, then $\mathcal{D}^c$ is the design on vertex set $\{A, B, C, D, E, F, G\}$ that has blocks

- $\{C, E, F, G\}$,
- $\{A, D, F, G\}$
- $\{B, D, E, G\}$,
- $\{B, C, D, F\}$,
- $\{A, C, D, E\}$,
- $\{A, B, E, F\}$,
- $\{A, B, C, G\}$.

The reader should spend a moment on verifying that the complementary design of a regular, uniform design is also regular and uniform. It is a little bit less obvious, but still straightforward to see that the complementary design of a BIBD is also balanced, and so it is also a BIBD. You are asked to prove this in Exercise 5.

We present the next way to define a new design from an old one because it allows us to talk about the *incidence matrix* of a design, a concept which will be useful in the near future.

Definition 17.11. Let $\mathcal{D}$ be a design with blocks $B_1, B_2, \cdots, B_b$ and vertices $v_1, v_2, \cdots, v_v$. The *incidence matrix* of $\mathcal{D}$ is the $v \times b$ matrix A defined by

$$A_{i,j} = \begin{cases} 1 \text{ if } v_i \in B_j, \\ \\ 0 \text{ if } v_i \notin B_j. \end{cases}$$

Example 17.12. Let $\mathcal{D}$ be the design whose blocks are the two-element subsets of $[4]$. Then the incidence matrix of $\mathcal{D}$ is

$$A = \begin{array}{c} \\ 1 \\ 2 \\ 3 \\ 4 \end{array} \begin{array}{c} \begin{array}{cccccc} \{1,2\} & \{1,3\} & \{1,4\} & \{2,3\} & \{2,4\} & \{3,4\} \end{array} \\ \begin{pmatrix} 1 & 1 & 1 & 0 & 0 & 0 \\ 1 & 0 & 0 & 1 & 1 & 0 \\ 0 & 1 & 0 & 1 & 0 & 1 \\ 0 & 0 & 1 & 0 & 1 & 1 \end{pmatrix} \end{array}.$$

Now we are in a position to present the next example of constructing new designs from old.

Definition 17.13. Let $\mathcal{D}$ be a design with incidence matrix A. Then the *dual* of $\mathcal{D}$ is the design $\mathcal{D}^d$ whose incidence matrix is A^T, that is, the transpose of A.

Example 17.14. The dual $\mathcal{D}^d$ of the design of Example 17.12 is the design whose incidence matrix is

$$A = \begin{array}{c} \\ a \\ b \\ c \\ d \\ e \\ f \end{array} \begin{array}{c} \begin{array}{cccc} \{a,b,c\} & \{a,d,e\} & \{b,d,f\} & \{c,e,f\} \end{array} \\ \left(\begin{array}{cccc} 1 & 1 & 0 & 0 \\ 1 & 0 & 1 & 0 \\ 1 & 0 & 0 & 1 \\ 0 & 1 & 1 & 0 \\ 0 & 1 & 0 & 1 \\ 0 & 0 & 1 & 1 \end{array} \right) \end{array}.$$

That is, if we relabel the rows of A^T by the letters a, b, c, d, e, f, then the blocks of $\mathcal{D}^d$ are $\{a,b,c\}$, $\{a,d,e\}$, $\{b,d,f\}$, and $\{c,e,f\}$.

Note that $\mathcal{D}^d$ of Example 17.14 is *not* a BIBD since it is not balanced. This is because some pairs of blocks of $\mathcal{D}$ intersect, while some others do not. Hence some pairs of vertices of $\mathcal{D}^d$ do appear together in a block, while some others do not.

The incidence matrix is a seminal tool in using techniques of linear algebra to prove facts about designs. Let us start with a simple observation.

Proposition 17.15. *Let A be the incidence matrix of a design $\mathcal{D}$ with parameters (b, v, r, k, λ). Then*

$$AA^T = (r - \lambda)I_v + \lambda J_v, \tag{17.1}$$

where I_v is the identity matrix of size $v \times v$ and J_v is the matrix of size $v \times v$ whose entries are all equal to 1.

Proof. If $i \neq j$, then the (i, j)-entry of AA^T is the dot product of the ith and jth rows of A. This dot (scalar) product is the sum of 0s and 1s, with a 1 for every block that contains both the ith and the jth vertex. There are λ such blocks. If $i = j$, then the (i, i)th entry of AA^T is the dot product of the ith row of A by itself. This dot product is a sum of 0s and 1s, a 1 for each time the ith vertex appears in a block. That happens r times. □

Recall from linear algebra that the number m is called an *eigenvalue* of the square matrix M if there exists a nonzero vector $\mathbf{x}$ so that $M\mathbf{x} = m\mathbf{x}$. In that case, $\mathbf{x}$ is called an *eigenvector*. A subspace spanned by eigenvectors of M is called an *eigenspace* of M.

Corollary 17.16. *If A is the incidence matrix of a BIBD with parameters (b, v, r, k, λ), then the eigenvalues of AA^T are $r - \lambda$ with multiplicity $v - 1$, and $r + \lambda(v - 1) = rk$ with multiplicity one. In particular, since none of these eigenvalues is 0, we have* $\det AA^T \neq 0$.

Proof. We have seen in Proposition 17.15 that $AA^T = (r - \lambda)I + \lambda J$. On the one hand, all nonzero vectors are eigenvectors of $(r - \lambda)I$, with eigenvalue $r - \lambda$. On the other hand, it is easy to verify that any vector with coordinate sum 0 is an eigenvector of λJ with eigenvalue 0. Furthermore, vectors of the form $(x, x, \cdots, x)$ also form an eigenspace (a one-dimensional eigenspace) of J, with eigenvalue λv.

Therefore, the eigenvectors of $AA^T = (r - \lambda)I + \lambda J$ are the eigenvectors of λJ, and their associated eigenvalues are the sums of their associated eigenvalues for the matrix $(r - \lambda)I$ and the matrix λJ. This proves our claim. □

We have pointed out immediately after Example 17.14 that the dual of a BIBD is *not* necessarily a BIBD because it is not necessarily balanced. Indeed, it follows from the definition of dual designs that for $\mathcal{D}^d$ to be balanced, the following would have to hold. For any two distinct blocks B_i and B_j of $\mathcal{D}$, the size of the intersection $B_i \cap B_j$ should be a *fixed* integer l, that is, it should not depend on the choice of B_i and B_j. In that case, we say that $\mathcal{D}$ is a *linked* BIBD. If $\mathcal{D}$ is linked, and only then, $\mathcal{D}^d$ will be balanced, since any two of its vertices will occur together in l blocks.

Proposition 17.15 makes it easy to prove that all symmetric designs are linked.

Proposition 17.17. *All symmetric BIBDs are linked.*

Proof. Let $\mathcal{D}$ be symmetric with parameters (v, k, λ). Then the adjacency matrix A of $\mathcal{D}$ is a *square* matrix, so the product $A^T A$ exists. Note that A^T is the adjacency matrix of $\mathcal{D}^d$. So if we can prove that $A^T A = AA^T$, then we will be done, since all non-diagonal entries of AA^T are equal to λ, which then implies that any two blocks of A intersect in exactly $l = \lambda$ vertices.

Given that $AA^T = (r-\lambda)I+\lambda J$, the proof of the equation $AA^T = A^T A$ is purely algebraic. Note that $JA = AJ$, where J is the $v \times v$ matrix whose entries are all equal to 1 . We know from Corollary 17.16 that A^{-1} exists, otherwise $\det AA^T$ would be 0. Multiply both sides of (17.1) by A^{-1} from the left, then by A from the right, to get

$$\begin{aligned} A^T A &= A^{-1}((r-\lambda)I)A + \lambda A^{-1}JA \\ &= (r-\lambda)I + \lambda J. \end{aligned}$$

So $AA^T = A^T A$ as claimed, implying that $\mathcal{D}$ is linked. □

In particular, since a symmetric BIBD is linked, its dual is also a BIBD. The following theorem shows that the opposite is true as well. That is, if the dual of a BIBD is a BIBD, then that BIBD must be symmetric. This is because, as we will see, all BIBDs have at least as many blocks as vertices.

Theorem 17.18 (Fisher's inequality). *If $\mathcal{D}$ is a BIBD on v vertices and b blocks, then $v \leq b$.*

Proof. Let us consider the incidence matrix A of $\mathcal{D}$. We know from Proposition 17.15 that $\det AA^T \neq 0$, so in particular, this $v \times v$ matrix has rank v. On the other hand, the rank of the product of two matrices is never more than the rank of either matrix, so

$$v = \text{rank}(AA^T) \leq \text{rank}(A).$$

Finally, the rank of any matrix is the number of its linearly independent columns, so at most the number of its columns. Hence

$$\text{rank}(A) \leq b.$$

The last two displayed inequalities imply our claim by transitivity. □

See Exercise 10 for a very interesting variation of this Theorem.

The following result provides another necessary condition for the existence of symmetric BIBDs.

Theorem 17.19. *If $\mathcal{D}$ is a symmetric BIBD with parameters (v, r, λ), and v is even, then $(r-\lambda)$ is a perfect square.*

Proof. In the proof of Corollary 17.16, we have determined the eigenvalues of the matrix AA^T. It is therefore easy to compute $\det AA^T$ as the product of these eigenvalues, namely

$$\det AA^T = (r+(v-1)\lambda) \cdot (r-\lambda)^{v-1}.$$

Clearly, the left-hand side is a perfect square since it is equal to $(\det A)^2$. On the right-hand side, using Proposition 17.8 and the fact that our design is symmetric, we have $(r + (v-1)\lambda) = r + r(k-1) = rk = r^2$. Therefore, $(r-\lambda)^{v-1}$ has to be a perfect square as well. For that to happen, $r - \lambda$ has to be a perfect square since $v - 1$ is odd. □

Note that Theorem 17.19 has a counterpart for the case of odd v which is much more difficult to prove. We present that theorem without a proof. In the Notes section, we give some pointers to more specialized books that contain the proof of this and other theorems that go beyond the scope of this book.

Theorem 17.20. *If there exists a symmetric BIBD with parameters (v, r, λ), and v is odd, then there exist integers x, y, and z that are not all equal to zero so that*

$$x^2 = (k-\lambda)^2 + (-1)^{(v-1)/2} \cdot \lambda z^2.$$

Theorems 17.19 and 17.20 are called the *Bruck-Ryser-Chowla* theorem.

We mention two more ways of creating new BIBDs from existing ones. For these methods to work, the existing BIBD has to be *symmetric.*

Definition 17.21. Let $\mathcal{D}$ be a symmetric BIBD, and let B be a block of $\mathcal{D}$. The *residual design* of $\mathcal{D}$ with respect to D is obtained from $\mathcal{D}$ by removing B from $\mathcal{D}$, and removing all vertices of B from all remaining blocks of $\mathcal{D}$.

In Exercise 8, you are asked to verify that this is a correct definition, that is, that the block design defined by Definition 17.21 is indeed a BIBD.

Definition 17.22. Let $\mathcal{D}$ be a symmetric BIBD, and let B be a block of $\mathcal{D}$. The *derived design* of $\mathcal{D}$ with respect to B is obtained from $\mathcal{D}$ by removing B from $\mathcal{D}$, and replacing every other block B_i by $B_i \cap B$.

You will be asked to justify this definition in Exercise 9.

17.4 Existence of Certain BIBDs

In Section 17.2, we have seen examples of *necessary* conditions for the existence of balanced incomplete block designs. Propositions 17.7 and 17.8 provided such conditions in terms of the parameters b, v, r, k, and λ. Later, these were supplemented by Theorems 17.18, 17.19, and 17.20. In Section

17.3, we saw examples showing how to obtain new BIBDs from old ones. However, we have not seen many examples of constructing BIBDs, especially infinite families of them, from "scratch", as opposed to constructing them from other designs. An exception was the rather easy Example 17.6.

The reason for this is that sufficient conditions on the existence of certain designs are typically more difficult to prove than necessary conditions. We will mention some of them here without complete proofs.

Theorem 17.23. *Let $n = p^k$ be a power of a prime number p, with $k \geq 1$. Then a symmetric BIBD with parameters $(n^2 + n + 1, n + 1, 1)$ exists.*

Reading this, the reader should wonder why the fact that n is a power of a prime matters. The answer is that if n is a power of a prime, then a finite field F_n with n elements exists. We can then take the $n^3 - 1$ ordered triples of the form (x, y, z) in which at least one of the three coordinates is non-zero, and identify two such triples if they are constant multiples of each other. As there are $(n - 1)$ non-zero constants in F_n, this creates $(n^3 - 1)/(n - 1) = n^2 + n + 1$ non-identical triples. These will be the vertices of the BIBD that we are constructing.

Interestingly, the blocks will also be defined as ordered triples (a, b, c) of the same kind. So $(1, 2, 3)$ labels both a block and a vertex. (One should think of the block $(1, 2, 3)$ and the vertex $(1, 2, 3)$ as *different* objects with the same label– say one is red and the other is blue.) Then we say that block (a, b, c) contains vertex (x, y, z) if

$$xa + yb + zc = 0.$$

It can then be shown that these rules define a BIBD with the aforementioned parameters.

A BIBD described in Theorem 17.23 is called a *projective plane* of order n. So Theorem 17.23 shows that if n is a prime power, then a projective plane of order n exists. This, of course, does not imply that no projective plane can exist if n is not a prime power. However, no such plane is known, and it is conjectured that no such projective planes exist.

Note that if we apply Theorem 17.23 with $n = 2$, then we conclude that a symmetric BIBD with parameters $(7, 3, 1)$ exists. We have seen one such design, that of Example 17.2. We say that designs $\mathcal{D}$ and $\mathcal{H}$ are isomorphic if there is a bijection f from the vertex set of $\mathcal{D}$ to the vertex set of $\mathcal{H}$ so that $\{v_1, v_2, \cdots, v_k\}$ is a block that appears in $\mathcal{D}$ exactly m times if and only if $\{f(v_1), f(v_2), \cdots, f(v_k)\}$ is a block in $\mathcal{H}$ exactly m times. In Exercise 16, you are asked to prove that *all* BIBDs with parameters $(7, 3, 1)$ are isomorphic. So the design of Example 17.2 is *the* projective plane of order two. It is called the the *Fano plane*, in honor of the Italian mathematician Gino Fano. The Fano plane is often represented by the diagram shown in Figure 17.1.

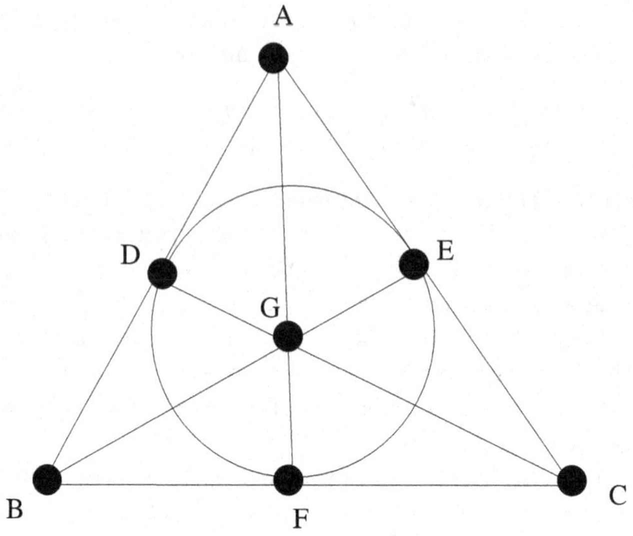

Fig. 17.1 The Fano plane.

In this representation, the blocks are the six straight lines, *and* the circle. The fact that a circle is used is not by accident. See Exercise 30 for an explanation.

17.4.1 *A Residual Design of a Projective Plane*

As an application, consider a derived design of a projective plane with parameters $(n^2 + n + 1, n + 1, 1)$. In Exercise 9, you are asked to compute the parameters of such a design. For now, we just state that if $n = 4$, then the residual design of a $(21, 5, 1)$-projective plane is a BIBD with parameters $(20, 16, 5, 4, 1)$. If that 5-tuple seems familiar, then that is because we have seen such a design, namely in the very first example of this chapter, Example 17.1, that discussed moto-cross races. The design presented there was a residual design of a projective plane of order four. It can be shown that such a residual design can be constructed *directly*, without constructing the projective plane, as follows. Consider the 4-element finite field F_4, with elements $\{0, 1, a, a^2\}$. Let the vertices of our design be the 16 ordered pairs (x, y), where x and y are elements of F_4. Let the blocks be the solutions of the 16 equations of the form $y = mx + c$, where m and c are also elements of F_4, and the solutions of the four equations of the form $x = c$. It is straightforward to show that each of these equations has four solutions of the form (x, y), so we get 20 blocks of size four. It can be shown that the other conditions of being a BIBD are also satisfied by the design we have just defined. You will be asked to fill in the gaps (in a more general form) in Exercises 31 and 32. A residual design of a finite projective plane is called a *finite affine plane.*

17.5 Codes and Designs

17.5.1 *Coding Theory*

The quest for ways to encode a message reliably, securely, and economically goes back to at least the Roman empire. Here security means that intercepted messages are only readable by the intended recipients. This excludes commonly used ways of communication, such as natural languages. Economy means that a message should not take unreasonably long to encode or decode, that is, both the number of symbols used to encode a message and the number of different kinds of symbols used to encode the message should be kept as low as possible without hurting the other goals of the encoding process. Third, the coding process should be reliable. In other words, the recipient should not misinterpret our message. This means that our coding system should be *injective*, that is, different messages should be encoded differently. We may impose the stronger requirement that the recipient be able to decode our message *even if there are a few mistakes* in the coded message that he receives, as long as the number of those mistakes is not too large.

Coding theory is the huge discipline that studies the above problems from various aspects. It is the subject of independent graduate and undergraduate courses. We will not even attempt to give an overview of the topic here; we will just show a few connections between coding theory and the theory of designs.

17.5.2 *Error Correcting Codes*

Let us assume that in a crucial moment of a football game, the coach wants to send a message to one of his players, who is standing far away from him. Before the game, the coach and a player agreed on signals as follows. If the coach lifts his left hand, that means that the player should try to run a (previously discussed) very risky play next time he gets the ball; if the coach lifts his right hand, then the player should not try to run that play.

Given the importance of this decision, but also the short time that is left to relay the message, the coach will send the same signal three times, with short pauses in among those times. In the heat of the moment, the player may misinterpret one of the signals, or may mistake someone else's hand for the coach's hand. Both of these are rather unlikely, but not impossible, sources of error. It is extremely unlikely that any of these errors would occur more than once out of the three signals. Assuming that the coach will not change his mind between his first and last signal, there are only two possible series of signals he can send, namely LLL or RRR (L for raising his left hand, and, yes, R for raising his right hand). If the player sees one of these sequences, he will immediately know what to do. However, if he mistakes a signal for its opposite, and he gets a sequence like LRL or RRL, then he is most likely to take a *majority vote.* For instance, if the player reads the sequence LLR, he can argue as follows. It is extremely unlikely that I misread the signal *more than once.* Therefore, the two times I saw the left hand of the coach raised, could not both be in error. So the correct signal is L, and I will run the risky play.

The previous example was simple since there were only two possible correct outcomes. Now let us assume that we want to send a message in the form of a 0-1 string that will encode which of the four friends Anna, Benjamin, Catherine and David plans to attend a party. This message can be described by a 0-1 string of length four, in which the first, second, third, and fourth digit is 1 if, respectively, Anna, Benjamin, Catherine and David plans to attend the party. So the number of possible correct messages is $2^4 = 16$. If the receiver gets the message 1001 and knows that there is at most one erroneous digit, then she does not know which digit is erroneous and needs to be corrected. However, if we send the intended message three times, and the receiver still knows that at most one digit is erroneous, then she can easily figure out which digit that is. Indeed, for each of $k = 1, 2, 3, 4$, consider the 3-tuple of the kth, $k+4$th and $k+8$th digits. If they are all identical, then they are correct, if they are not all identical, then the digit that occurs only once has to be changed to the opposite digit.

The preceding examples show the important expectations that we have towards *error correcting codes*. They should be *compact*, that is, the words encoding the messages should be as short as possible, errors should be recognizable if we know that there are not too many of them, and errors should be correctable in an unambiguous way. This last expectation suggests that in some sense, our legal codewords (the possible correct messages) should be quite dissimilar, that is, changing a digit in a legal codeword should not lead to another legal codeword, since then we would be unable to tell when that digit is correct. These concepts will be formalized in the next subsection.

17.5.3 *Formal Definitions About Codes*

Let S and T be two finite alphabets. Let T^* be the set of all finite sequences whose elements are in T. A *code* c is an *injective* function $c : S \to T^*$. If $t \in T^*$ is in the range of $c : S \to T^*$, then t is called a *codeword* for the code C. The set C of all codewords for c is also called the code c. If $T = \{0,1\}$, then the code is called *binary*. The injection $c : S \to T^*$ can easily be extended to the set S^* of all finite sequences (or words) over S by setting

$$c(s_1 s_2 \cdots s_n) = c(s_1)c(s_2)\cdots c(s_n).$$

The fact that $c : S \to T^*$ is an injective function does not in itself guarantee that the extended function $c : S^* \to T^*$ is injective.

Example 17.24. Let $S = \{x, y, z\}$, let $T = \{0,1\}$, and let $c(x) = 0$, $c(y) = 1$, and $c(z) = 01$. Then $c : S^* \to T^*$ is not injective since $c(xy) = c(z) = 01$.

That is, a recipient receiving the string 01 could not know whether the original message was xy or z. If this problem does not occur, that is, if the extended function $c : S^* \to T^*$ is injective, then we say that c is *uniquely decodable.*

A good way to make sure that c is uniquely decodable is by making sure our code $c : S \to T^*$ is *prefix-free.*

Definition 17.25. We say that a code $c : S \to T$ is *prefix-free* if there are no two codewords $c(x)$ and $c(y)$ so that $c(x) = c(y)q$ for some string $q \in T^*$.

Example 17.26. Let $S = \{x, y, z\}$, let $T = \{0,1\}$, and let $c(x) = 0$, $c(y) = 11$, and $c(z) = 10$. Then c is prefix-free.

The following theorem shows the main advantage of prefix-free codes.

Theorem 17.27. *If c is prefix-free, then it is uniquely decodable.*

Proof. Let $c : S \to T^*$ be a prefix-free code, and let us say that

$$c(x_1 x_2 \cdots x_k) = c(y_1 y_2 \cdots y_m) = t_1 t_2 \cdots t_n,$$

where the x_i and the y_j are elements of S, and $t_r \in T$.

We prove that then $k = m$, and $x_1 x_2 \cdots x_k = y_1 y_2 \cdots y_k$, so $x_i = y_i$. We prove this by strong induction on n. For $n = 1$, the statement is obvious since $c : S \to T$ is an injective function by definition. Now let us assume that the statement is true for all positive integers less than n. Observe that $c(x_1) = c(y_1)$ must hold, otherwise one of $c(x_1)$ and $c(y_1)$ would be a prefix of the other. As c is injective on S, this implies that $x_1 = y_1$. Hence

$$c(x_2 \cdots x_k) = c(y_2 \cdots y_m) = t_h \cdots t_n$$

for some $h > 1$. In particular, $t_h \cdots t_n$ has less than n letters, and as such, is uniquely decodable by the induction hypothesis. So $x_i = y_i$ for all i as claimed. □

The converse of Theorem is not true as you are asked to show in Exercise 33.

One way to make absolutely sure that a code is prefix-free is to choose all codewords to be of the same length. These codes have many additional advantages in addition to being uniquely decodable. Therefore, *for the rest of this chapter, all codes will consist of codewords of the same length.*

Definition 17.28. Let v and w be two n-letter words over the same finite alphabet. The *Hamming distance* of v and w, denoted by $d(v, w)$, is the number of positions in which v and w differ.

Example 17.29. Let $v = (0, 1, 1, 0, 1, 0)$, and let $w = (1, 0, 1, 0, 1, 1)$. Then $d(v, w) = 3$, since v and w differ in their first, second, and sixth entries.

The Hamming distance satisfies the *triangle inequality.* This is the content of the next theorem.

Theorem 17.30. *Let u, v, and w be three n-letter words over the same alphabet. Then*

$$d(u, w) \leq d(u, v) + d(v, w).$$

Proof. We can turn u to v by changing at most $d(u, v)$ letters, then we can turn v to w by changing at most $d(v, w)$ letters. So we can turn u to w by changing no more than $d(u, v) + d(v, w)$ letters. □

If we are to create a code that can correct e errors, or, in what follows, e-error correcting codes, then it suffices to choose the codewords in such a way that the distance between any two of them is at least $2e + 1$. That makes sure that if a codeword is transmitted by no more than e errors, then it will be closer to one codeword w than all others. The receiver can then conclude that w was the intended message.

A geometric way to describe the above idea is as follows. Let us say that the *ball* $B(w, e)$ whose center is the codeword w and whose radius is e is the set of all words v so that $d(v, w) \leq e$. A code c is e-error correcting if the balls of radius e around any two of its codewords w and w' are disjoint. In other words, a code is e-error correcting if and only if the distance of any two codewords is at least $2e + 1$.

Definition 17.31. An (n, m, d)-code is a code that consists of m codewords of length n so that the Hamming distance of any two codewords is at least d.

Intuitively, the key of creating efficient e-error correcting codes is to "pack" the codewords as closely as possible while making sure that the balls of radius e centered at each codeword remain disjoint. It is plausible to think that highly symmetric structures will be useful in the creation of such codes. For instance, one can try to build an error correcting code using the incidence matrix A of the BIBD of Example 17.2, that is, the matrix

$$\begin{array}{c} \\ A \\ B \\ C \\ D \\ E \\ F \\ G \end{array} \begin{array}{c} \begin{array}{ccccccc} \{ABD\} & \{BCE\} & \{ACF\} & \{AEG\} & \{BFG\} & \{CDG\} & \{DEF\} \end{array} \\ \left(\begin{array}{ccccccc} 1 & 0 & 1 & 1 & 0 & 0 & 0 \\ 1 & 1 & 0 & 0 & 1 & 0 & 0 \\ 0 & 1 & 1 & 0 & 0 & 1 & 0 \\ 1 & 0 & 0 & 0 & 0 & 1 & 1 \\ 0 & 1 & 0 & 1 & 0 & 0 & 1 \\ 0 & 0 & 1 & 0 & 1 & 0 & 1 \\ 0 & 0 & 0 & 1 & 1 & 1 & 0 \end{array} \right) \end{array}.$$

First, we note that the Hamming distance of any two distinct rows is four, since any two vertices of the $(7, 3, 1)$-design at hand occur together in exactly one block. So for any two rows, there are exactly one position in which both rows contain a 1, and two positions in which both rows contain a 0. Therefore, the two rows differ in four positions. Hence, the seven rows of the above matrix form a $(7,7,4)$-code. In particular, this code is 1-error correcting.

At this point, the reader should feel some discomfort since the code we have just created is wasteful. Indeed, having the codewords four apart is not any better than having them three apart for error correction purposes. Both constructions lead to 1-error correcting codes. So it is natural to ask whether the (7,7,4) code above can be extended in some natural way.

The answer is yes. This is the content of the next theorem. If $\mathbf{v}$ is a vector of length n whose entries are all 0 or 1, then the *complement* of $\mathbf{v}$ is the 0-1 vector of the same length whose coordinates disagree with the corresponding coordinates of $\mathbf{v}$ in every position. In other words, the complement of $\mathbf{v}$ is equal to $\mathbf{1}-\mathbf{v}$ in the binary arithmetic, where $\mathbf{1}$ is the vector of length n whose entries are all equal to 1.

Theorem 17.32. *Let T be the incidence matrix of a $(7,3,1)$-design $\mathcal{D}$. The following 16 codewords form a $(7,16,3)$-code over the binary alphabet.*

(1) The seven rows of the matrix T,
(2) the complements of the seven rows of T,
(3) the word 0000000, *and*
(4) the word 1111111.

Proof. All we need to show is that the distance between any two codewords is at least three. Let v and w be two codewords. If they both correspond to the same row of T, then they both correspond to a vertex in $\mathcal{D}$. As $\mathcal{D}$ is linked with link number 1, this implies that there is exactly one position in which v and w both have a 1. Hence there are two positions in which they both have a 0, so $d(u,v)=4$. If u and v are both complements of rows of T, then the same argument, applied to the complement of u and v, proves that again, $d(u,v)=4$.

If u is a row of T, and v is the complement of another row of T, then u and v together contain seven 1s. Note that $d(u,v)$ is the number of columns with exactly one 1, therefore $d(u,v)$ is an odd number. So it suffices to show that $d(u,v)\neq 1$. As u contains three 1s and v contains four 1s, $d(u,v)=1$ is only possible if v contains a 1 in each position where u does. That implies that the vertices corresponding to u and the complement of v do not appear together in any block of $\mathcal{D}$, which is a contradiction.

In all other cases, the statement $d(u,v)$ is obviously true. □

The reader is probably wondering why we stopped at 16 codewords instead of trying to extend our code even further. The answer is that the addition of further codewords is not possible without losing the 1-error correction property.

Indeed, if c is a binary code whose codewords are of length seven, then the ball $B(w,1)$ centered at the codeword w and of radius 1 consists of eight words, namely w itself, and the seven words obtained from w by changing exactly one letter of w. If c is 1-error correcting, then the m balls $B(w,1)$ are all disjoint, showing that their total volume is at most the total number of 7-letter binary words, that is, $2^7 = 128$. As the volume of each such ball is eight, there can be no more than $128/8 = 16$ balls, and so, there can be at most 16 words in such a code.

The following example was a special case of some general phenomena. The first is a simple upper bound for the number of codewords in an e-error correcting code.

Proposition 17.33. *Let c be an r-error correcting code over the binary alphabet in which all codewords are of length n. Then the number $|C|$ of codewords in c is at most*

$$\frac{2^n}{\sum_{i=0}^{r}\binom{n}{i}}.$$

You are asked to provide the proof in Exercise 11.

Second, the (7,16,3)-code we created from the (7,3,1)-BIBD was *as good as it could be* in that the balls of radius 1 centered at the codewords filled the entire vector space B^7 (the seven-dimensional vector space over the binary field) with no gaps. This is an important concept that has its own name.

In what follows, when we mention a code *over* B^n, we mean a code over B whose words are of length n.

Definition 17.34. Let c be an r-error correcting code over B^n. We say that c is *perfect* if each word $v \in B^n$ belongs to a ball $B(w,r)$ for some codeword w.

So the code of Theorem 17.32 is perfect. For perfect codes, the statement of Proposition 17.33 can be strengthened as follows.

Lemma 17.35. *Let c be a perfect code over B^n that is r-error correcting. Then*

$$|c| = \frac{2^n}{\sum_{i=0}^{r}\binom{n}{i}}.$$

You are asked to provide the proof in Exercise 12.

In particular, a perfect r-error correcting code over B^n does not exist unless 2^n is divisible by $\sum_{i=0}^{r}\binom{n}{i}$. In other words, for such a code to exist, it is necessary for $\sum_{i=0}^{r}\binom{n}{i}$ to be a power of 2.

So a perfect 1-error correcting code over B^n can exist only if $\binom{n}{0}+\binom{n}{1} = n+1$ is a power of 2, that is, when $n = 2^k - 1$ for some positive integer k. The code we created in Theorem 17.32 was the special case $k = 3$. The special case $k = 2$ also yields a perfect 1-error correcting code, namely the somewhat unexciting code consisting of the two words 000 and 111.

The last paragraph raises two interesting questions. First, we have seen that a perfect 1-error correcting code over B^n can only exist in very limited circumstances, namely when $n = 2^k - 1$. Is this necessary condition sufficient?

Question 17.36. *Is it true that for all integers $k \geq 2$, there exists a 1-error correcting code over $B_n = B_{2^k-1}$ that consists of*

$$\frac{2^n}{n+1} = \frac{2^{2^k-1}}{2^k} = 2^{2^k-k-1}$$

codewords?

Second, are there other r-error correcting binary codes that are perfect? Note that this question can be broken down into the following two parts.

Question 17.37. *Are there any pairs of positive integers (n, r) so that $n > r > 1$ and $\sum_{i=0}^{r}\binom{n}{r}$ is a power of 2?*

If yes, are there any r-error correcting codes over B^n that are perfect?

The answer to Question 17.36 is in the affirmative. In order to be able to prove that, we need a few simple definitions. A code c over B^n is called *linear* if the set of codewords forms a subspace of the vector space B^n. In other words, c is a linear binary code if the binary sum of any two codewords is also a codeword. Note that the sum of two codewords is computed letter by letter, in the binary arithmetic, without carries. So the sum of 101 and 011 is 110. The *weight* of a codeword is simply the sum of its letters. If the code is binary, then this means the number of letters equal to 1 in a given codeword. We are now ready to state the next theorem.

Theorem 17.38. *For all positive integers $k \geq 2$, there exists a perfect binary $(2^k - 1, 2^{2^k-k-1}, 3)$ code.*

The codes described in Theorem 17.38 are called *Hamming codes*, to honor their inventor, Richard Hamming. It follows from the parameters of these codes that they are 1-error correcting codes.

Proof. (of Theorem 17.38) Let A be any $(2^k - k - 1) \times k$ matrix whose rows are the $2^k - k - 1$ binary strings of length k that contain at least two digits equal to 1. Note that in particular, the rows of A are all different. Now consider the matrix $H = I|A$, where I is the identity matrix of size $(2^k - k - 1) \times (2^k - k - 1)$, and $I|A$ simply means that A is placed on the right of I. So H is a $(2^k - k - 1) \times (2^k - 1)$ matrix over the binary alphabet, and each row of H contains at least three digits equal to 1.

Now let c be the binary code whose codewords are the $2^{2^k - k - 1}$ possible linear combinations of the rows of H with coefficients 0 or 1. Note that all these linear combinations are different, since even their $(2^k - k - 1)$-prefixes are different. Indeed, for $i \leq 2^k - k - 1$, a linear combination will have a 1 in the ith position if and only if the coefficient of the ith row in that linear combination is 1.

The rows of H are obviously of length $2^k - 1$ as claimed, hence so are all their linear combinations, so all the codewords of c. There remains to show that the distance of any two codewords is at least three.

We claim that every nonzero codeword of c has weight at least three. Indeed, every codeword obtained by adding at least three rows of H contains at least three 1s among its first $2^k - k - 1$ entries, every codeword obtained by adding two rows of H contains two 1s among its first $2^k - k - 1$ entries and at least one 1 after that, while the rows of H contain, by definition, one 1 among their first $2^k - k - 1$ entries and at least two 1s after that.

Now let us assume that v and w are two distinct codewords so that $d(v, w) < 3$. Then $w - v = w + v$ is also a codeword, and the weight of $w - v$ is less than 3, since $w - v$ can have a 1 only in the positions where v and w differ. That contradicts the fact proved in the previous paragraph, namely that all codewords in c have weight at least three.

Therefore, c is indeed a $(2^k - 1, 2^{2^k - k - 1}, 3)$ binary code as claimed. □

The answer to Question 17.37 is not as positive as that to the previous question. It is easy to notice that if n is odd, and $r = (n - 1)/2$, then

$$\sum_{i=0}^{r} \binom{n}{i} = 2^{n-1}.$$

However, this is not too exciting, since this equality simply means that there is a perfect binary $(n, 2, n)$ code, in other words, a perfect binary code that is $(n-1)/2$-error correcting, but which consists of only *two* codewords of length n. This is obvious, since any word and its opposite will do, such as $11\cdots1$, and $00\cdots0$.

Interestingly, there are not many other pairs (n,r) so that $1 < r < n$ and $\sum_{i=0}^{r}\binom{n}{r}$ is a power of two. It turns out that the only such pairs are $(23,3)$ and $(90,2)$. As far as these pairs of parameters are concerned, there does exist a perfect binary code, called the *Golay code* with $n = 23$ and $r = 3$, but there does not exist a perfect binary code with parameters $n = 90$ and $r = 2$.

Notes

The topic of design theory is huge, and is the subject of many books at different levels. At the introductory level, a recent and readable volume is *Introduction to Combinatorial Designs* [46] by W. D. Wallis. For a comprehensive reference, we recommend the *Handbook of Combinatorial Designs* [14].

The theory of designs is closely connected to other areas in combinatorics. The reader got a glimpse of one of those areas, coding theory, in this chapter. For further reading in that subject, a very reader-friendly possibility is *Codes* by Norman Biggs [5].

Another related topic is that of *Latin squares.* These interesting structures have not been discussed in our text, but they are the subject of Exercises 18-22, and of Exercise 43 of this chapter.

Exercises

(1) (-) Does there exist a BIBD with parameters $(120, 10, 36, 3, 9)$?
(2) (-) Does there exist a BIBD with parameters $(120, 10, 36, 3, 8)$?
(3) (-) Does there exist a BIBD with $b = 35$, $k = 3$, and $r = 4$?
(4) (-) Recall from Chapter 9 that a regular graph is a simple graph in which each vertex has the same degree. So, regular graphs are regular, 2-uniform designs, with the edges of the graphs playing the role of blocks. Which regular graphs are BIBDs?
(5) (-) Prove that the complementary design of a BIBD is also a BIBD. What are the parameters of this BIBD?
(6) What is the largest possible number of blocks in a uniform design on v

vertices if no block is repeated?

(7) Describe all symmetric BIBDs with $3 \leq v \leq 12$ and $\lambda = 1$.

(8) Prove that the residual design of a symmetric BIBD is a BIBD. What are the parameters of a residual design of a symmetric design with parameters (v, k, λ)? (See Definition 17.21 for the definition of the residual design of a symmetric BIBD.)

(9) Prove that the derived design of a symmetric design is a BIBD. What are the parameters of a derived design of a symmetric BIBD of parameters (v, k, λ)? (See Definition 17.22 for the definition of derived designs.)

(10) Let $\mathcal{D}$ be a block design without repeated blocks with b blocks and v vertices. Let us assume that each vertex is in more than one block, and that every pair of distinct vertices appears together in exactly one block. Prove that $v \leq b$.
Note that unlike in Theorem 17.18, $\mathcal{D}$ is not assumed to be regular or uniform. On the other hand, the number of times each pair of vertices appears together in a block is not just equal to some fixed λ, but to 1.

(11) Prove Proposition 17.33.

(12) Prove Lemma 17.35.

(13) Let $c : S \rightarrow \{0, 1\}^*$ be a prefix-free code in which b_i codewords have length i. Prove that $\sum_i \frac{b_i}{2^i} \leq 1$.

(14) Let c be a Hamming code with parameters $(2^k - 1, 2^{2^k - k - 1, 3}, 3)$ as constructed in the proof of Theorem 17.38. Let us assume that $k \geq 3$. Let A be the matrix whose columns are the weight-3 codewords of c. Consider the design $\mathcal{D}$ whose incidence matrix is A.

  (a) Prove that $\mathcal{D}$ is uniform and incomplete.
  (b) Prove that $\mathcal{D}$ is balanced.
  (c) Conclude that $\mathcal{D}$ is a BIBD, and determine the parameters of $\mathcal{D}$.

(15) A t-design is a design in which every t-element set of vertices appears together in exactly λ blocks. So BIBDs are 2-designs. Prove that if $\mathcal{D}$ is a t-design with parameters (b, v, r, k, λ), then

$$b\binom{k}{t} = \lambda\binom{v}{t}.$$

(16) Prove that all $(7, 7, 3, 3, 1)$-designs are isomorphic.

(17) If $\mathcal{D}$ is a design, then an *automorphism* of $\mathcal{D}$ is a bijection f from the vertex set of $\mathcal{D}$ into the vertex set of $\mathcal{D}$ so that if $\{v_1, v_2, \cdots, v_k\}$ is a block with multiplicity m, then $\{f(v_1), f(v_2), \cdots, f(v_k)\}$ is also a block with multiplicity m, and vice versa.

How many automorphisms does the Fano plane have?

(18) A *Latin square* is an $n \times n$ matrix in which each row and column contain exactly one copy of each element of $[n]$. Let $L(n)$ be the number of Latin squares of side length n (or *order* n). Find $L(n)$ for all positive integers $n \leq 4$.

(19) Magic cubes were defined in Exercise 10 of Chapter 11. Prove that the number of magic cubes of side length n having line sum 1 is equal to $L(n)$.

(20) Two Latin squares A and B are called *orthogonal* if for each ordered pair $(x, y) \in [n] \times [n]$, there is exactly one position (i, j) so that $A_{i,j} = x$ and $B_{i,j} = y$. A set S of several $n \times n$ Latin Squares is called *mutually orthogonal* if every pair (A, B) of Latin squares with $A \in S$, and $B \in S$, is orthogonal. Let $N(n)$ be the highest number so that there exist $N(n)$ mutually orthogonal Latin squares of side length n. Find $N(n)$ for all positive integers $n \leq 4$.

(21) (+) Prove that if a k-element set of mutually orthogonal Latin squares of order n exists, then $k < n$.

(22) The affine plane of order four was defined in Subsection 17.4.1. The general definition is analogous. If q is a power of a prime, then the affine plane A_q of order q is the BIBD whose vertices are the ordered pairs (x, y), where x and y are elements of the q-element field F_q, and whose blocks are the solutions (x, y) of the q^2 equations of the form $y = mx + c$, where m and c are elements of F_q, and the solutions of the q equations of the form $x = c$.

  (a) Prove that the $q^2 + q$ blocks A_q can be classified into $q + 1$ classes so that each class has q blocks in it, blocks in the same class are pairwise disjoint, and if two blocks are in two distinct classes, then they have exactly one vertex in common.

  (b) (+) Prove that if $q > 1$ is a power of a prime, then $q - 1$ mutually orthogonal Latin squares of side length q do exist.

Supplementary Exercises

(23) (-) Prove that every 3-element subset S of the 5-element set $\{b, v, r, k, \lambda\}$ of parameters of a BIBD determines the value of the two parameters that are not in S.

(24) (-) Let $\mathcal{D}$ be a BIBD with parameters $(b, v, r, 3, 1)$. Prove that either $v = 6i + 1$ or $v = 6i + 3$ for some non-negative integer i.

(25) (-) Prove that a balanced, uniform design is regular. (A balanced design is a design in which every pair of vertices belongs to the same number of blocks.)

(26) (-) Construct a BIBD with parameters $(12, 9, 4, 3, 1)$.

(27) (-) Give an example of a BIBD with no repeated blocks in which $\lambda > k$.

(28) How many different (7,7,3,3,1)-BIBDs are there on vertex set [7]? Note that two such designs are different if their sets of blocks are different. That is, we do *not* require that the BIBDs be non-isomorphic; we simply require that they be non-identical.

(29) We are given five distinct positive integers, and we are told that they are the five parameters of a BIBD. We are not told which number corresponds to which parameter. Can we find it out?

(30) (+) Let S be a finite set of points in the plane so that not all points of S are on the same straight line. Prove that there exists a straight line that contains exactly two points of S.

(31) In Exercise 22, we define a design on q^2 vertices and $q^2 + q$ blocks, where q is a power of prime. Why is that design $\mathcal{D}$ uniform? Why is it regular?

(32) Why is the design $\mathcal{D}$ discussed in the previous exercise balanced?

(33) Show an example of a code that is not prefix-free but still uniquely decodable.

(34) State and prove the version of Proposition 17.33 for an alphabet of size k instead of the binary alphabet.

(35) Prove that in a linear binary code, either all codewords have even weight, or at most half of the words have even weight.

(36) Let c be a linear code. Prove that the minimum distance between any two codewords of c is equal to the minimum weight of nonzero codewords in c.

(37) Let c be a code over the binary alphabet in which every codeword has the same length. Let $\bar{c}$ be a code obtained by the rule

$$\bar{c}(s) = \begin{cases} c(s)0 \text{ if } c(s) \text{ has even weight,} \\ \\ c(s)1 \text{ if } c(s) \text{ has odd weight.} \end{cases}$$

Let us assume that the minimum distance between any two codewords in c is d. What is the minimum distance between any two codewords of $\bar{c}$?

(38) Recall that t-designs are defined in Exercise 15. Prove that if $\mathcal{D}$ is a t-design with parameters (b, v, r, k, λ), then

$$r\binom{k-1}{t-1} = \lambda\binom{v-1}{t-1}.$$

(39) (+) Prove that if $\mathcal{D}$ is a t-design, then it is also a u-design for all integers $u \in [2, t]$.

(40) (+) Let $\mathcal{D}$ be a design with no repeated blocks. A *subdesign* of $\mathcal{D}$ is a design $\mathcal{F}$ so that the set of vertices of $\mathcal{F}$ is a subset of the set of vertices of $\mathcal{D}$, and the set of blocks of $\mathcal{F}$ is a subset of the set of blocks of $\mathcal{D}$.
Let $\mathcal{D}$ be a (v, k, λ)-design, and let $\mathcal{F}$ be a (w, k, λ)-subdesign of $\mathcal{D}$, with $w < v$. Prove that then $w \leq \frac{v-1}{k-1}$.

(41) Show two non-isomorphic BIBDs with parameters $(14, 7, 6, 3, 2)$.

(42) (+) Show four non-isomorphic BIBDs with parameters $(14, 7, 6, 3, 2)$.

(43) (+) Prove that if $n-1$ mutually orthogonal Latin squares exist, then there exists an $(n^2+n, n^2, n+1, n, 1)$-design.

Solutions to Exercises

(1) No. The first four parameters force the equality $\lambda = 8$ by Proposition 17.8.

(2) Yes. The design of all 3-element subsets of $[10]$ is a BIBD with these parameters.

(3) No. By Proposition 17.7, if bk is odd, then so is vr.

(4) If a simple graph is a BIBD, then any pair of its vertices must occur together in the same number of edges (0 or 1). Only the complete graph and the empty graph have this property.

(5) Let $\mathcal{D}$ be a BIBD with parameters (b, v, r, k, λ). Then the blocks of $\mathcal{D}^c$ are all of size $v-k$ since they are complements of k-element blocks. Each vertex x of $\mathcal{D}^c$ is part of $b-r$ blocks (the complements of those that did not contain x in $\mathcal{D}$). Finally, if x and y are two vertices, then in $\mathcal{D}$, they each belonged to r blocks, and there were λ blocks that contained both. So, by the Priciple of Inclusion-Exclusion, there were $2r-\lambda$ blocks that contained at least one of them, hence there were $b-2r+\lambda$ blocks that contained neither. The complements of these

blocks will contain both of them, so any two vertices appear together in exactly $b-2r+\lambda$ blocks in $\mathcal{D}^c$. So $\mathcal{D}^c$ is a BIBD with parameters $(b, v, b-r, v-k, b-2r+\lambda)$.

(6) If the design is k-uniform, and has no repeated blocks, then it cannot have more than $\binom{v}{k}$ blocks, and we have seen in Chapter 4 that for a given v, the binomial coefficient $\binom{v}{k}$ is maximal when $k = \lfloor v/2 \rfloor$. So we have $b \leq \binom{v}{\lfloor v/2 \rfloor}$.

(7) For symmetric BIBDs, the claim of Proposition 17.8 reduces to the equality $\lambda(v-1) = k(k-1)$. We know that $v \leq 12$, so $v - 1 \leq 11$. That means that if $\lambda = 1$, then $k(k-1) = v-1$ has to be 2 or 6, since these are the only allowed numbers that are of the form $k(k-1)$. In the first case, $v = 3$, and $k = 2$, and we recognize the design of all 2-element subsets of $[3]$. In the second case, $k = 3$ and $v = 7$, and we recognize the design of Example 17.2.

(8) Let $\mathcal{D}$ be a symmetric BIBD with parameters (v, k, λ). The residual design $\mathcal{R}$ of $\mathcal{D}$ will have one less block than $\mathcal{D}$. The size of these blocks will be $k - \lambda$, since any two blocks of $\mathcal{D}$ intersect in λ vertices, so in particular, B intersects any other block in λ vertices. (See the proof of Proposition 17.17 for this fact.) The number of vertices decreases by k, to $v-k$, since the vertices of B are removed. The remaining vertices still occur in all r blocks they occurred before, and any pair of them still occurs together λ times. So we conclude that $\mathcal{R}$ has parameters $(b-1, v-k, r, k-\lambda, \lambda)$.

(9) As $\mathcal{D}$ is linked with link number λ, any given block intersects B in $\mathcal{D}$ in λ vertices. So the blocks of the derived design $\mathcal{D}'$ will be of size λ. The number of vertices will be k since vertices outside B will disappear. The number of blocks will be $b-1 = v-1$ since B disappears, all other blocks intersected B, so their intersection with B will be in $\mathcal{D}'$. Any two vertices will appear together in $\lambda - 1$ blocks, (all those blocks that contain both them in $\mathcal{D}$, except for the block B). Finally, all vertices will appear in $r - 1 = k - 1$ blocks (all blocks in which they appeared before, except B). So $\mathcal{D}'$ has parameters $(v-1, k, k-1, \lambda, \lambda-1)$.

(10) Proceed as in the proof of Theorem 17.18. We get that $AA^T = D + J$, where D is a $v \times v$ matrix whose non-diagonal entries are all 0, (since any pair (v_1, v_2) of vertices appears together in exactly one block), and whose ith diagonal entry is $r_{v_i} - 1$, where r_{v_i} is the number of blocks in which vertex v_i appears.
Now we show that $\det(AA^T) \neq 0$ by showing that it has linearly independent rows. Indeed, let us assume that the rows of AA^T have

a non-trivial linear combination with coefficients $c_1, c_2, \cdots, c_v$ that is equal to 0. It then follows from routine algebra that

$$c_1(v_1 - 1) = c_2(v_2 - 1) = \cdots = c_v(v_v - 1).$$

As the numbers v_i are larger than 1, the numbers $v_i - 1$ are larger than 0. No c_i can be 0 since that would imply that all c_i are equal to 0, and the linear combination at hand is non-trivial. Therefore, either all c_i are positive, or all c_i are negative, both of which contradicts the assumption that the c_i are coefficients of a linear combination of the all-positive rows that equals 0. So $\det(AA^T) \neq 0$, and therefore, AA^T has rank v. As $\text{rank}(AA^t) \leq \text{rank}(A)$, this implies that $\text{rank}(A) = v$. The rank of a matrix cannot be more than the number of its columns. Applying this fact for the matrix A, we get $v \leq b$.

(11) As c is r-error correcting, the balls centered at the codewords and having radius r must all be disjoint. Therefore, their total volume is the sum of their individual volumes, which is $|C| \cdot \sum_{i=0}^{r} \binom{n}{i}$. This cannot be more than the total number of strings of length n over the binary alphabet.

(12) Continuing the line of thought of the previous exercise, if c is perfect, then the union of the balls centered at the codewords and having radius r must be the B_n.

(13) Let t be a randomly selected binary string of length n, where n is at least as long as all the codewords of c. Let w be a given codeword of c of length i. Then the probability of the event A_w that w is a prefix of t is 2^{-i}. The probability of the event that *any* codeword of c is a prefix of t is

$$P\left(\bigcup_w A_w\right) = \sum_w P(A_w) = \sum_i \frac{b_i}{2^i}.$$

This number cannot be more than 1 since it is a probability. The fact that the code is prefix-free is crucial, since that fact assures that $\bigcup_w A_w$ is a *disjoint* union of events, hence its probability is equal to the sum of the probabilities of the A_w.

(14)(a) As the columns of A are the weight-3 codewords, each column contains three 1s, and so each block contains three vertices, hence $\mathcal{D}$ is uniform. As $k \geq 3$, the number of v of vertices of $\mathcal{D}$ is at least $2^k - 1 \geq 7$, so $\mathcal{D}$ is incomplete.

(b) We show that any two vertices of $\mathcal{D}$ appear together in exactly one block. In other words, we show that for any $i \neq j$, there is exactly one position in which both the ith and jth row of A contain a 1. There cannot be more than one such position, since if the ath and bth position were both like that, then the ath and bth columns would differ in at most two positions, contradicting the fact that any two columns (codewords) differ in at least three positions. Now we show that there is indeed column that contains a 1 in positions i and j. Let w_{ij} be the word of weight two in B^{2^k-1} that has a 1 in positions i and j and 0 everywhere else. As c is a perfect code, w_{ij} belongs to the ball of radius 1 of exactly one codeword x of c. By the definition of Hamming codes, x has weight at least 2. However, as x is at distance 1 from $w_{i,j}$, the weight of x cannot be exactly two, so it is at least three. As x has weight three, it must contain a 1 in positions i and j, otherwise it would be at distance at least two from $w_{i,j}$. This proves that $\mathcal{D}$ is balanced, and $\lambda = 1$.

(c) We know that $\mathcal{D}$ is a balanced, uniform design, hence by Supplementary Exercise 25 it is regular. As the codewords are of length $2^k - 1$, we have $v = 2^{k-1}$. Then Proposition 17.8 implies

$$r = \frac{\lambda(v-1)}{k-1} = \frac{2^k - 2}{2} = 2^{k-1} - 1,$$

and then 17.7 implies

$$b = \frac{vr}{k} = \frac{(2^k - 1)(2^{k-1} - 1)}{3}.$$

(15) Both sides count the same objects, namely pairs (S, B), where S is a t-element subset of vertices that are all part of the block B. The left-hand side first chooses B in b ways, then $S \subset B$. The right-hand side first chooses S in $\binom{v}{t}$ ways.

(16) We show that every design with those parameters has to be built up the same way. This will imply a stronger statement. Let $\mathcal{D}$ be a design with those parameters, and let A, B, and C be three vertices in $\mathcal{D}$ that do *not* form a block. Then any pair of these three vertices appears together in a block, and each of those three blocks has a separate third vertex. Let us say that $\{A, B, D\}$ is a block, $\{B, C, E\}$ is a block, and $\{A, C, F\}$ is a block. Finally, $\mathcal{D}$ has a seventh vertex, which we will call G. As $\lambda = 1$, G has to appear together with each of the other six vertices in exactly one block. What can be the third vertex of the block that contains A and G? It cannot be anything that already shares a block with A, so it can only be E. So $\{A, E, G\}$ is a block. Similarly, the third vertex of the block containing B and G must be

F, so $\{B, G, F\}$ is a block. Finally, the third vertex containing C and G must be D, so $\{C, D, G\}$ is a block.

We have found six of the seven blocks of $\mathcal{D}$. The seventh block must be formed by the vertices that have not been used in three blocks yet, so $\{D, E, F\}$ is a block.

Note that we have not used anything other than the parameters of $\mathcal{D}$ in this argument. So any $(7, 7, 3, 3, 1)$-design can be built up this way, from starting by any three vertices A, B, and C that do not form a block. So we have in fact proved that if $\mathcal{D}$ and $\mathcal{H}$ are two $(7, 7, 3, 3, 1)$-designs, then they are isomorphic, and an isomorphism can be found by choosing three vertices A, B, and C that do not form a block, and mapping them to any three vertices A', B', and C' of $\mathcal{H}$ that are not in the same block. These three images will determine the isomorphism.

(17) By the previous argument, f is determined by the images of any three vertices A, B, and C that do not form a block. There are seven possible choices for $f(A)$, six possible choices for $f(B)$, then four possible choices for $f(C)$ since we cannot choose the vertex that forms a block with $f(A)$ and $f(B)$. Hence the Fano-plane has $7 \cdot 6 \cdot 4 = 168$ automorphisms.

(18) It is obvious that $L(1) = 1$ and $L(2) = 2$. In order to construct a Latin square of side length 3, first we need to place the three 1s on the board in one of $3! = 6$ ways, then the three 2s in one of two ways, and fill the remaining boxes by 3s. So $L(3) = 12$. In order to determine $L(4)$, note that in any Latin square of side length 4, we can permute the columns so that the top row becomes 1234, then we can permute the top three rows so that the leftmost column becomes 1234. There remains to fill out the remaining 3×3 grid, and it is easy to verify that there are four ways to do that. Recalling the operations that made sure that the first row and column were in increasing order, this implies that $L(4) = 4! \cdot 3! \cdot 4 = 576$.

Note that a Latic square in which both the first row and first column is increasing is called a *reduced* Latin square.

(19) If L is a Latin square of side length n, then let its entries 1 correspond to the 1s at the bottom floor of the magic cube M, let the entries 2 of L correspond to the 1s on the second floor of M, let the entries 3 of L correspond to the 1s on the third floor of M, and so on.

(20) Clearly, $N(1) = 1$ and $N(2) = 1$. We claim that $N(3) = 2$. On the one hand, there exist two orthogonal Latin squares of side length three,

as can be seen from the examples of

$$A = \begin{pmatrix} 1\,2\,3 \\ 2\,3\,1 \\ 3\,1\,2 \end{pmatrix} \quad \text{and} \quad B = \begin{pmatrix} 1\,2\,3 \\ 3\,1\,2 \\ 2\,3\,1 \end{pmatrix}.$$

Now we prove that $N(3) < 3$. Let us assume that P, Q, and R are mutually orthogonal Latin squares of side length three. As P and Q are orthogonal, there is a position in which they both contain a 1. Let us assume without loss of generality that this is the position $(1, 1)$. Then, again without loss of generality, P has a 1 in $(2, 2)$ and $(3, 3)$, while Q has a 1 in $(2, 3)$ and $(3, 2)$. The set of positions in which R has a 1 must intersect the set of positions in which P has a 1 in exactly one position, and the set of positions in which Q has a 1 in exactly one position. This is impossible.

(21) Permute the *symbols* of each of the mutually orthogonal Latin squares $S_1, S_2, \cdots, S_k$ so that each of these squares has an increasing top row. This does not change orthogonality. Now consider the first entry of the second row of each of these squares. This is a collection T of k entries, none of which is equal to 1, since the entry just above them is 1. No two entries of T can be equal, say b, since if S_i and S_j both contained b in the $(2, 1)$-position, then S_i and S_j would not be orthogonal. Indeed, the pair (b, b) would occur in both the $(2, 1)$-position of both of these squares, and in their $(1, b)$-positions.

(22) The reader may want to use Example 17.1 to illustrate the concepts explained here. Note that in that example, $q = 4$.

(a) Let all blocks that were defined by keeping m fixed and by letting c vary over all elements of F_q belong to the class C_m. For instance, C_0 is the class of all blocks defined by the equations $c = y$. This defines q classes indexed by the q elements of F_q; finally let C_∞ be the class of blocks defined by the equations $x = c$, where c varies over all elements of F_q. First, it is straightforward to see that two blocks of the same class are always disjoint. Indeed, if two blocks of C_m (for $m \neq \infty$) had a non-empty intersection containing the vertex (x, y), that would mean that $c_1 = mx + y = c_2$ for two different values of c_1 and c_2, which is absurd. Similarly, if two blocks of C_∞ had a non-empty intersection containing the vertex (x, y), then $c_1 = x = c_2$ would have to hold for two distinct values c_1 and c_2.

Now let B and B' be blocks from two different classes. First, consider the case when they are respectively defined by the equations $m_1x + c_1 = y$ and $m_2x + c_2 = y$, with $m_1 \neq m_2$. Then it is straightforward to check that the system of those two equations has a unique solution (x, y), so $|B \cap B'| = 1$. If B' is from the class F_{q+1}, then the second equation has to be replaced by $x = c_2$, and then $y = m_1c_2 + c_1$ from the first equation. So again there is one (x, y) that is in $B \cap B'$.

(b) If q is a power of a prime, then the finite field F_q does exist, and the affine plane over F_q can be created as explained in part (a) of this exercise. This affine plane is a $(q^2 + q, q^2, q + 1, q, 1)$-design, and its set of blocks can be partitioned into classes as explained in part (a).

Now we create $q - 1$ mutually orthogonal Latin squares from $q - 1$ of these classes. The classes that will not correspond to a Latin square will be C_0 and C_∞. In each class C_a, number the blocks in some arbitrary way $C_{a1}, C_{a2}, \cdots, C_{aq}$. Let $(i, j) \in [q] \times [q]$. Find the unique vertex v in the intersection of blocks C_{0i} and $C_{\infty j}$. This v lies in exactly one block of each of the remaining $q - 1$ classes. If, in the class C_a, the vertex v lies in the block C_{ak}, then we put the number k in the (i, j)-position of the Latin square S_a that we are creating.

This procedure defines $q - 1$ square matrices of side length q. Each of these squares $S_1, S_2, \cdots, S_{q-1}$ is a Latin square. Indeed, the jth column of S_a will contain all the numbers k of blocks C_{ak} in the class C_k that contain a vertex of $C_{\infty j}$. Given that each vertex of $C_{\infty j}$ is part of one block of C_a, each element of $[q]$ will occur once in this column. A dual argument shows that the ith row of S_a also contains each element of $[q]$ once.

Finally, we must show that S_a and S_b are orthogonal if $a \neq b$. Consider the q squares where S_a has an entry x. These squares were filled in instances when the vertex v of the above algorithm was part of the block C_{ax}. The block C_{ax} has a one-element intersection with the block C_{b1}, and one-element intersection with the block C_{b2}, and so on, a one-element intersection with each block C_{bk}, for $k \in [q]$. Therefore, in the q positions where S_a has an entry x, the Latin square S_b contains the q different elements of $[q]$. So S_a and S_b are indeed orthogonal.

Chapter 18

Are They Really Different? Counting Unlabeled Structures

18.1 Enumeration Under Group Action

18.1.1 *Introduction*

Let us assume that we have a garbage can whose base has the shape of a regular hexagon. We see this garbage can every day, and one day, we start thinking about the following problem. How many different ways are there to paint the six sides of the can using only the colors red, blue, white, and green?

This question is easy if the faces are distinguishable. Indeed, in that case, we have four choices for each of the six sides, resulting in a total of $4^6 = 4096$ paint jobs. However, what can we say if the sides are indistinguishable as is often the case in practice? That is, what if two paint jobs are considered identical if one can be obtained from another by simply rotating the can around a vertical axis going through its center?

This is a much more difficult problem than the problem of counting the ways of coloring six *distinguishable* objects using a finite set of colors. Clearly, some general theory would be helpful, since a naive way of counting would probably consist of considering too many cases. In this section, we will learn such a technique. First, we need to learn about a structure from Abstract Algebra that will enable us to precisely describe how one paint job can be turned into another.

18.1.2 *Groups*

18.1.2.1 *Groups in General*

Definition 18.1. A *group* G is a set endowed with an operation called *multiplication* defined on ordered pairs of elements of G so that all of the following conditions hold.

(1) The set G is closed under multiplication. That is, for all elements a and b of G, the product $a \cdot b$ and the product $b \cdot a$ are elements of G.
(2) There exists an identity element $id \in G$ so that for all $g \in G$, the equalities
$$id \cdot g = g \cdot id = g$$
hold.
(3) Multiplication is associative, that is, $(a \cdot b) \cdot c = a \cdot (b \cdot c)$, for all elements a, b, and c of G.
(4) Each element $g \in G$ has an *inverse*, that is, there exists a unique element g^{-1} so that $g \cdot g^{-1} = g^{-1} \cdot g = id$.

Example 18.2. The set $\mathbf{Z}$ of all integers is a group if the usual *addition* of integers is defined as the operation on $\mathbf{Z}$.

It is easy to see why $\mathbf{Z}$ is *not* a group with the usual multiplication of integers as the operation. No elements other than 1 and -1 would have an inverse if we used that definition, so the last criterion in the definition of groups would not be satisfied.

Example 18.3. The set $\mathbf{R} - \{0\}$ of *non-zero* real numbers forms a group with the usual multiplication of the real numbers as the operation.

Note that it is important that we exclude 0 from our set. Indeed, 0 does not have an inverse under usual multiplication (there is no real number x so that $x \cdot 0 = 1$), so the last criterion would not be fulfilled.

We point out that the definition of groups does *not* require that in a group, multiplication be commutative. That is, it is possible in a group that $ab \neq ba$. In fact, it turns out that *most* groups are non-commutative.

Example 18.4. Let $SL_2(\mathbf{R})$ be the set of 2×2 matrices with real entries that have determinant 1. Then $SL_2(\mathbf{R})$ is a non-commutative group.

Solution. It is easy to verify that all the group axioms are satisfied by $SL_2(\mathbf{R})$, so $SL_2(\mathbf{R})$ is indeed a group. Indeed, the fact that the elements of $SL_2(\mathbf{R})$ have determinant 1 assures that their set is closed under multiplication. The fact that their determinant is nonzero assures that all elements have an inverse. Matrix multiplication is known to be associative and the identity matrix serves as the identity element of the group.

On the other hand, let $A = \begin{pmatrix} 1 & 0 \\ a & 1 \end{pmatrix}$, and let $B = \begin{pmatrix} 1 & b \\ 0 & 1 \end{pmatrix}$. Then we have

$$AB = \begin{pmatrix} 1 & b \\ a & 1+ab \end{pmatrix}, \qquad \text{and} \qquad BA = \begin{pmatrix} 1+ab & b \\ a & 1 \end{pmatrix},$$

and so $AB \neq BA$ if $ab \neq 0$.

Two groups G and M are called *isomorphic* if there exists a bijection $f : G \to M$ that preserves the group operation. That is, if g and g' are elements of G, then $f(g \cdot g') = f(g) \cdot f(g')$.

18.1.2.2 *Subgroups and Cosets*

If a subset H of the elements of a group G forms a group with the same operation of G restricted to H as the operation, then H is called a *subgroup* of G. This is denoted by the symbols $H \leq G$.

Equivalently, H is a subgroup of G if it contains the identity element of G, and is closed under the operation of G and under taking the inverse of each element of H in G.

Example 18.5. Let $GL_2(\mathbf{R})$ be the set of all 2×2 matrices with real entries and *non-zero* determinants. Then $SL_2(\mathbf{R}) \leq GL_2(\mathbf{R})$.

If H is a subgroup of G, and $a \in G$, then we define

$$aH = \{ah | h \in H\}.$$

In other words, aH is the set of elements of G that can be obtained by left-multiplying the elements of H by a. The set aH is then called a *coset* of H in G. We point out that HH denotes the set of all two-term products formed by elements of H, and that clearly, $HH = H$. The following property of cosets is what makes them interesting.

Proposition 18.6. *Let $H \leq G$. Let $a \in G$ and $b \in G$. Then either $aH = bH$, or $aH \cap bH = \emptyset$.*

Proof.   Let us assume that $c \in aH \cap bH$. Then there exist elements $h_1 \in H$ and $h_2 \in H$ so that $ah_1 = c = bh_2$. In other words, $ah_1h_2^{-1} = b$, so, since $h_1h_2^{-1}$ in H, we have that $b \in aH$. However, that implies that $bH \subseteq aHH = aH$. In an analogous way, $aH \subseteq bH$, proving our claim. $\square$

Corollary 18.7. *If G is a finite group and H is a subgroup of G, then $|G|$ is divisible by $|H|$.*

Proof.   We claim that the *different* cosets of H partition the set of elements of G. Indeed, each element g of G belongs to a coset of H, namely to the coset gH. It follows from Proposition 18.6 that the different cosets of H do not overlap. As each of these cosets has size $|H|$, the Corollary is proved. $\square$

Note that the ratio $|G|/|H|$ is called the *index* of H in G, and is denoted by $G : H$.

18.1.3   Permutation Groups

In Chapter 6, when we first defined permutations as bijections on a finite set, we mentioned how to multiply two permutations together. For easy reference, if $f : [n] \to [n]$ and $g : [n] \to [n]$ are permutations, then the permutation fg is simply the permutation that is obtained by first applying the bijection f to $[n]$, then the bijection g to $[n]$. In other words, fg is the *composition* of f and g, that is, $(fg)(x) = g(f(x))$.

The following proposition shows that permutations of length n form a group. This will be the most important group for us in this chapter.

Proposition 18.8. *The set of $n!$ permutations of $[n]$ forms a group with the multiplication of permutations as its operation.*

Proof.   The identity map id of $[n]$ (the permutation $12 \cdots n$) has the property that $\mathrm{id} \cdot p = p \cdot \mathrm{id} = p$ for all permutations p of length n. Each permutation is a bijection from $[n]$ to $[n]$, and so has an inverse permutation p^{-1} that satisfies $p \cdot p^{-1} = p^{-1} \cdot p = \mathrm{id}$. Multiplication of permutations is associative, since if p, q, and r are permutations, $p(i) = j$, $q(j) = k$, and $r(k) = m$, then

$$((p \cdot q) \cdot r)(i) = r((p \cdot q)(i)) = r(q(p(i))) = r(q(j)) = r(k) = m,$$

while

$$(p \cdot (q \cdot r))(i) = (q \cdot r)(p(i)) = r(q(p(i))) = r(q(j)) = r(k) = m. \quad \square$$

The group of $n!$ permutations of length n is called the *symmetric group* of degree n and is denoted by S_n.

Definition 18.9. Let H be a subgroup of S_n for some n. Then H is called a *permutation group.*

Recall that we defined automorphisms of graphs in Chapter 9. For easy reference, an automorphism of a simple graph G is a bijection f from the set of vertices of G onto that same set so that uv is an edge in G if and only if $f(u)f(v)$ is an edge in G.

Example 18.10. The set of all automorphisms of a graph G, with composition of functions as the operation, forms a permutation group, denoted by $Aut(G)$, and called the *automorphism group* of G.

Solution. By definition, the elements of $Aut(G)$ are permutations of the set of vertices of G. As the elements of $Aut(G)$ are multiplied as permutations, multiplication in $Aut(G)$ is associative. The identity map of the vertex set of G is always an automorphism. Let f^{-1} be the inverse of $f \in Aut(G)$ as a permutation. If uv is an edge of G, then $f^{-1}(u)f^{-1}(v)$ must be an edge as well, otherwise f would map the non-edge $f^{-1}(u)f^{-1}(v)$ into the edge uv. So $f^{-1} \in Aut(G)$, proving our claim.

Permutation groups are of seminal importance for theory we present in this chapter. Therefore, we spend a paragraph to put them in context. In some sense, permutation groups are not different objects from groups in general (which are sometimes called *abstract groups*). Indeed, each permutation group is a group, and it can be shown that each group of n-elements is isomorphic to a subgroup of S_n, that is, each finite group is isomorphic to a permutation group. However, what is different between abstract groups and permutation groups is the way we view them. An abstract group, such as the set of all real numbers with addition as the operation, stands on its own. There is no need to consider any other objects to understand and describe this group. On the other hand, a permutation group, such as S_n or one of its subgroups, consists of bijections (permutations) on the set $[n]$. So the elements of the set $[n]$, that is, the numbers $1, 2, \cdots, n$ are present. The elements of S_n *act* on $[n]$; they *permute* these elements among each other. They move them around. So a permutation group always *acts* on some set of objects. For instance, the automorphism group of a graph acts on the set of vertices of that graph (and, for that matter, also on the set of edges of that graph).

At this point, the reader may well start seeing where we are going with this. Recalling the introductory example of this chapter, we point out that the rotations of the hexagon-based garbage can that turn a paint job into another one will be part of a permutation group. We do need some additional machinery before we can make this observation useful in counting.

Definition 18.11. Let G be a permutation group acting on a set S, and let $i \in S$. Then the set

$$G_i = \{g \in G | g(i) = i\}$$

is called the *stabilizer* of i.

Example 18.12. Let $G = S_6$, acting on the set $S = [6]$. Let $i = 1$. Then G_i is the set of all permutations in S_6 that keep 1 fixed. In other words, this is the group of all permutations on the set $\{2, 3, 4, 5, 6\}$, which is isomorphic to S_5.

Example 18.13. Let $G = Aut(C_n)$ be the automorphism group of a cycle on $n \geq 3$ vertices. Let i be any vertex of C_n. Then $G_i = S_2$ is the two-element group.

Solution. Indeed, let j and k be the neighbors of i. Then any automorphism f of C_n that fixes i must either fix both of j and k, or interchange them. In either case, the rest of f is determined by $f(j)$ and $f(k)$. So G_i is isomorphic to the 2-element group S_2 acting on j and k.

Proposition 18.14. *For any group G acting on the set S, and any $i \in S$, we have $G_i \leq G$. That is, the stabilizer of i in G is a subgroup of G.*

Proof. We verify that all conditions listed in the definition of groups hold for G_i. As $G_i \subseteq G$, the associtiavity of the operation automatically holds. Furthermore, G_i has an identity element, namely the identity element of G, which fixes all elements of S, including i. If both f and g fix i, then $g(f(i)) = g(i) = i$, so G_i is closed under multiplication. Finally, if $f(i) = i$, then applying $f^{-1} \in G$ to both sides of the preceding equation, we get $i = f^{-1}(i)$, so $f^{-1} \in G$ as well. □

A notion that is somewhat complementary to that of the notion of stabilizers is that of *orbits*.

Definition 18.15. Let G be acting on a set S and let $i \in S$. Then the *orbit* of i by G, denoted by i^G, is the set of all elements of S to which i can be moved by an element of G. That is

$$i^G = \{g(i) | g \in G\}.$$

The following lemma connects the notions of the orbit and the stabilizer of G. Roughly speaking, if the orbit of i is big, then the stabilizer of i is small, and vice versa.

Lemma 18.16. *Let G be a finite permutation group acting on a set S, and let $i \in S$. Then*

$$\frac{|G|}{|G_i|} = \left|i^G\right|.$$

Example 18.17. If G and i are as in Example 18.12, then $|G| = 6! = 720$, $G_1 = 5! = 120$, and $i^G = [6]$ and so $\left|i^G\right| = 6$.

Example 18.18. If G and i are as in Example 18.13, then $|G| = 2n$, $|G_i| = 2$, and $\left|i^G\right| = n$ since i^G is the entire vertex set of the cycle C.

Proof. We have seen in the proof of Corollary 18.7 that the left-hand side is the number of different cosets of G_i in G. We show that these cosets are in bijection with the elements of i^G.

Indeed, let C be a coset of G_i in G. Then $C = gG_i$ for some $g \in G$. We then set $f(C) = g(i) \in i^G$.

Before we prove that f is a bijection, we need to prove that f is well-defined. That is, we need to prove that for each C, it defines one and only one object $f(C)$. This is not obvious since there could be several candidates for g that satisfy $C = gG_i$. However, Exercise 3 shows that any two such candidates will only differ by an element of G_i, so they will have the same effect on i. That is, if $gG_i = g'G_i = C$, then $g'^{-1}g \in G_i$, so $g'^{-1}g(i) = i$, which implies that $g(i) = g'(i)$. So indeed, f is well-defined, that is, it is a function on the *cosets* of G_i in G, and it does not depend on the choice of the element g of a given coset.

It is straightforward to prove that f is a bijection. Indeed, if $j \in i^G$, then there exists an element $g \in G$ so that $g(i) = j$. Then it is a direct consequence of the definition of f that $j = f(gC)$. Again, there are several candidates for g, but if $g(i) = g'(i)$, then $g'^{-1}g(i) = i$, so $g'^{-1}g \in G_i$, and by Exercise 3 the cosets gG_i and $g'G_i$ are identical. □

The following notion is complementary to the notion of stabilizers in a different sense. It collects the objects that are fixed by a *given element of the permutation group* as opposed to the stabilizer, which collects the elements of the permutation group that fix a *given object.*

Definition 18.19. Let G be a permutation group acting on a set S, and let $g \in G$. Let

$$F_g = \{i \in S | g(i) = i\}.$$

The following theorem will be our fundamental tool in counting objects that cannot be moved into one another by a given permutation group. It is sometimes called *Burnside's lemma*, or *Burnside's theorem.*

Theorem 18.20. *Let G be a permutation group acting on a set S. Then the number of orbits of S under the action of G is equal to*

$$\frac{1}{|G|}\sum_{g\in G}|F_g|.$$

Proof. As we have defined two distinct notions, G_i and F_g, that related to pairs (g, i) so that $g \in G$, $i \in S$, and $g(i) = i$, it is perhaps not surprising that the proof of this important result consists of counting these pairs in two different ways, and equating the resulting formulae.

Let n be the number of orbits of S under the action of G. It then suffices to show that

$$|G| \cdot |n| = \sum_{g\in G}|F_g|. \tag{18.1}$$

Here the right-hand side is clearly the number of ordered pairs (g, i) so that $g \in G$, $i \in S$, and $g(i) = i$, summed first for each fixed g, then summed as g ranges over G.

Summing the number of these same pairs first for each fixed i, then summing as i ranges over S, we get

$$\begin{aligned}\sum_{g\in G}|F_g| &= \sum_{i\in S}|G_i|\\ &= \sum_{i\in S}\frac{|G|}{|i^G|}\\ &= |G|\sum_{i\in S}\frac{1}{|i^G|}\\ &= |G|\cdot n.\end{aligned}$$

Here we applied Lemma 18.14 in the second step. In the last step, we noted that each orbit of size k contributes k fractions equal to $1/k$ each to the sum, for a total contribution of 1 from that orbit. So the sum evaluates to the number of orbits. □

Corollary 18.21. *Let the finite permutation group G act on the finite set S. Then the average number of fixed points of this action is equal to the number of orbits of this action.*

Proof. The displayed chain of equations in the proof of Theorem 18.20 shows that $|G_i| = n|G|$, where n is the number of orbits of the action of G on S. Dividing both sides by $|G|$, we get our claim. □

The following example illustrates Theorem 18.20 in a very simple case.

Example 18.22. Let us color the edges of K_3 red or blue. Let us consider two colorings equivalent if they can be moved into each other by an automorphism. How many inequivalent colorings are there?

Solution. The key is to put the problem into a context where Burnside's lemma can be applied, that is, where the number we are looking for is precisely the number of orbits of a set under an appropriate group action.

In this case, this is easy. Let S be the set of all colorings of the edges of K_3 with colors red or blue, and let G be the automorphism group of K_3. Then G acts as a permutation group on the set S (it moves colorings into other colorings). Under this group action, two colorings belong to the same orbit precisely when they are equivalent, so the number of inequivalent colorings is precisely the number of orbits.

Note that $G = Aut(K_3)$ consists of the identity permutation, two 3-cycles, and three permutations that have a fixed point and a 2-cycle.

All we need to do now is to find the numbers $|F_g|$ for all six elements of G. Clearly, the identity permutation fixes all eight colorings, the two 3-cycles fix only the two colorings in which all edges have the same color, and the remaining three permutations fix four colorings each. Indeed, the permutation $(AB)(C)$ fixes all four colorings in which the color of AB and the color of BC agrees.

Therefore, the number of orbits, and so the number of inequivalent colorings is

$$\frac{1}{|G|}\sum_{g\in G}|F_g| = \frac{8+2+2+4+4+4}{6} = 4.$$

This result is very easy to verify since the four inequivalent colorings are precisely those with 0, 1, 2, or 3 blue edges. Once the number of blue sides is determined, the coloring is determined. This, of course, would not be true for regular polygons with more than three sides.

We can now see that in order to answer the question asked in the introduction of this chapter, we have to compute F_g for each of the 6 elements of the group of rotations of the regular hexagon, viewed as a permutation group acting on the set of the 4^6 colorings of the six sides of the hexagon using the four available colors.

The identity permutation (rotation by zero degrees) will keep all 4^6 colorings fixed. Rotations by 60 or 300 degrees will only fix the four colorings that use one color each. Indeed, such a rotation moves a side of the hexagonal can into a neighboring side, so if a coloring is fixed by such a rotation, then in that coloring, each side must have the same color as its neighbors.

By a similar argument, if a coloring is fixed by a rotation by 120 or 240 degrees, then in that coloring *second* neighbors must have the same color. There are $4^2 = 16$ colorings like that, since there are four possible choices for the color of the first, third, and fifth sides, and four possible choices for the color of the second, fourth, and sixth sides.

An analogous argument shows that rotation by 180 degrees fixes $4^3 = 64$ colorings. Therefore, Theorem 18.20 shows that the number of inequivalent colorings is

$$\frac{1}{|G|}\sum_{g\in G}|F_g| = \frac{4^6+2\cdot 4+2\cdot 4^2+4^3}{6} = 700.$$

Example 18.23. Let H be the graph shown in Figure 18.1. Find the number of ways to color the edges of H using only the color red, blue, and green if two colorings are considered identical if H has an automorphism that takes the first coloring into the second one.

Solution. It is easy to see that H has six automorphisms, and that all these automorphisms keep all vertices that have degree more than one fixed. Therefore, the automorphisms keep a coloring fixed if and only if it keeps the coloring of the five edges leading to the degree-one vertices fixed. The other three edges will never be moved by an automorphism, so all their 27 colorings will always be fixed.

As far as the five edges leading to the degree-one vertices go, the identity will keep all their $3^5 = 243$ colorings fixed, and the two automorphisms of degree six keep nine of their colorings fixed (one color for the three edges on

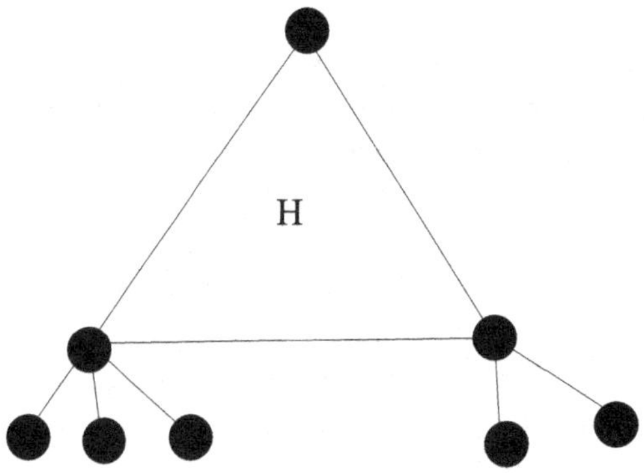

Fig. 18.1 Count the ways to color the edges of this graph with three colors.

the left, one color for the three edges on the right). The two automorphisms of degree three keep 27 colorings of these five edges fixed (one color for the three edges on the left, one color for each of the two edges on the right). Finally, the only automorphism of degree two keeps 81 coloring of these five edges fixed, since the two edges on the right must have the same color.

Therefore, by Theorem 18.20, the edges of the graph H have

$$\frac{27(243 + 2 \cdot 9 + 2 \cdot 27 + 81)}{6} = \frac{27 \cdot 396}{6} = 1782$$

non-identical colorings.

Classic applications of Theorem 18.20 include the enumeration of colorings of the five regular polyhedra that were discussed in Chapter 12. We will encounter some of those problems in the Exercises section.

18.2 Counting Unlabeled Trees

18.2.1 *Counting Rooted Non-plane 1-2 trees*

In Chapter 8, we have learned various techniques to use generating functions in order to solve enumeration problems. These techniques were applicable in many situations. However, there was something common in all the circumstances in which our methods were useful. The structures to be enumerated were always *labeled* structures. That is, there was an n-element *set* of *distinct* objects, and the task was to compute the number of ways to carry out a task on that set. Using different objects for a given role resulted in new structures. Examples of these structures were permutations, set partitions, and graphs with labeled vertices.

In this section, we will learn a technique to enumerate certain *unlabeled* structures. The tools we present will not always lead to exact enumeration formulae, but they will provide a way to obtain the number of our structures recursively. Techniques of Analytic Combinatorics, which are beyond the scope of this book, can be used to obtain the *approximate* number of the studied structured from the formulae that we deduce. Unless otherwise stated, in the rest of this section all graphs are unlabeled graphs.

As our first example, we consider the problem of enumerating rooted trees on n vertices in which each non-leaf vertex has at most two children. Let us call these trees *non-plane 1-2 trees*. The word "non-plane" emphasizes the fact that our trees are *not* embedded in the plane, that is, there is no notion of the "left child" and the "right child" of a vertex.

Let b_n be the number of non-plane 1-2 trees on n vertices. Set $b_0 = 1$. The reader is invited to verify that $b_1 = b_2 = 1$, $b_3 = 2$, and $b_4 = 3$. The six non-plane 1-2 trees on five vertices are shown in Figure 18.2, and we conclude that $b_5 = 6$.

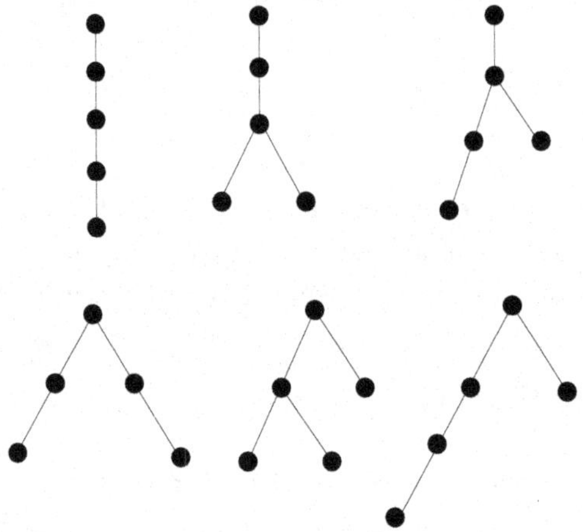

Fig. 18.2 The six non-plane 1-2 trees on five vertices.

Cutting off the root of an n-vertex non-plane 1-2 tree, we get a *multiset* of two non-plane 1-2 trees (one of which may be a 0-vertex tree) that together have $n-1$ vertices. The word *multiset* is of double importance in the preceding sentence. First, as our trees are non-plane, the order of the obtained two trees is insignificant. This is why we do not get an *ordered* pair of two trees. Second, we cannot even say that we get a *set* of two trees, since it is possible that the two obtained trees are identical. This is why we get a *multiset* of two trees.

Note that the advantage of setting $b_0 = 1$ was that now we do not have to treat the case when the root has only one child separately. When the root has one child, we can still say that the root has two subtrees, one of which is empty.

Let $B(x) = \sum_{n \geq 0} b_n x^n$ be the ordinary generating function of the sequence b_n.

If the order of the two subtrees of the root mattered, then we could use the Product Formula (Theorem 8.5) to get the generating function $xB^2(x)$. However, in our setup the two subtrees of the root are unordered. It would be tempting to say that since the two subtrees of the root form an *unordered pair*, the number of possible such pairs is half of what it would be if they formed an ordered pair, and so the generating function for non-plane 1-2 trees is $\frac{1}{2}xB^2(x)$. The problem with this argument is that sometimes the two subtrees of the root are *identical*. See Figure 18.3 for an illustration. Trees with this property are counted by $xB^2(x)$ only once and not twice, hence $\frac{1}{2}xB^2(x)$ slightly undercounts our trees.

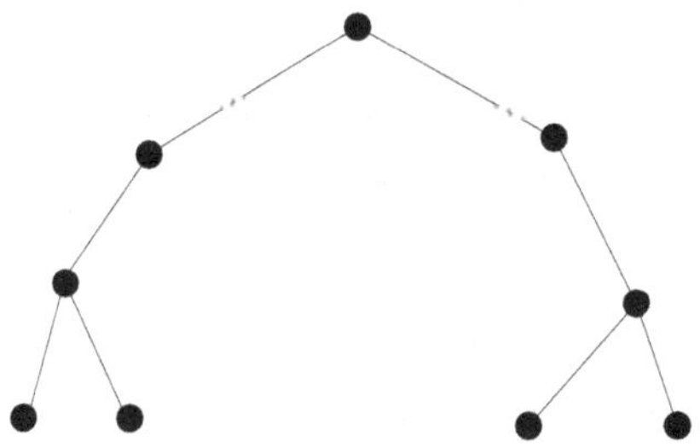

Fig. 18.3 A non-plane 1-2 tree in which the root has two identical subtrees.

How many non-plane 1-2 trees are there on n vertices that fall in this exceptional class, that is, how many non-plane 1-2 trees are there on n vertices in which the two subtrees of the root are identical? If n is even, then there are no such trees (since there are $n-1$ non-root vertices), and if $n = 2k+1$, then there are b_k such trees, since there are b_k choices for one subtree of the root, and the other subtree must be identical to the first. Hence the generating function of trees in this exceptional class is $\sum_{k\geq 0} x \cdot b_k x^{2k} = xB(x^2)$.

So we have proved the following theorem.

Theorem 18.24. *Let $B(x)$ be defined as above. Then we have*

$$B(x) = 1 + \frac{x}{2}\left(B^2(x) + B(x^2)\right). \tag{18.2}$$

You could ask what the use of this theorem is, given that we cannot obtain an explicit formula for the numbers b_n from (18.2). A recurrence relation can certainly be deduced from (18.2), but that could be done without introducing generating functions. You are asked to prove such a relation in Exercise 28.

The real advantage of Theorem 18.24 is that it makes it possible to evaluate, in the asymptotic sense, how large the numbers b_n are. The methods needed to do this are beyond the scope of this book. The result is that there exists a constant c so that

$$b_n \sim c \cdot \frac{a^n}{n^{3/2}},$$

where $a = 2.4832535....$ The constant c is close to 0.791.

Compare this to the case of unlabeled *binary plane* trees on n vertices. Their number is the Catalan number

$$c_n = \frac{\binom{2n}{n}}{n+1} \sim \pi^{-1/2} \cdot \frac{4^n}{n^{3/2}}.$$

So the binary plane trees are about 1.6^n times as numerous as the non-plane 1-2 trees. This agrees with our intuition. Indeed, in a binary plane tree, if a vertex is the only child of its parent, it can be its left or right child; in a non-plane tree, there is no such issue. For vertices with a sibling, this issue comes up only if the two siblings have non-identical subtrees, which happens often if the subtrees are large, but not so often if the subtrees are small, and never if they are empty. So we expect the set of binary plane trees be more numerous than the set of non-plane 1-2 trees, by a number between 1 and 2^n, and that is indeed the case.

18.2.2 *Counting Rooted Non-plane Trees*

What if we drop the condition that each vertex has at most two children? In that case, our method needs some significant modification. Indeed, as long as a vertex has two children, there are only two cases, in that the subtrees rooted at those children are either identical or not. However, if a vertex can have 17 children, then it could happen that out of the 17 subtrees rooted at those children, there is a set of five subtrees that are identical to each other, there are another two subsets of three subtrees each that are identical to each other, and the rest of the subtrees are all different. We need a method that can keep track of all these cases.

The method we present will be somewhat similar to the method we used in Chapter 8 when we proved that the generating function of the sequence $p(n)$ counting partitions of the integer n is

$$\sum_{n\geq 0} p(n)x^n = \prod_{i\geq 1} \frac{1}{1-x^i} = (1+x+x^2+\cdots)\cdot(1+x^2+x^4+\cdots)\cdots. \quad (18.3)$$

Indeed, if a partition of n has m_i parts equal to i, then it is accounted for on the right-hand side by the summand $x^{m_i i}$ of the ith infinite sum.

The task of finding the ordinary generating function of the sequence T_n, where T_n is the number of rooted non-plane trees is somewhat similar to finding the above generating function for the sequence $p(n)$. Indeed, cutting of the root of such a tree that has n vertices, we get a *multiset* of trees whose numbers of vertices add up to $n-1$. However, the difference is that among these smaller trees, there could be some that have the same size but are different. This did not happen for integer partitions; indeed all parts of size i were identical.

For instance, there are two rooted non-plane trees on three vertices. (In one, the root has one child, in the other one, the root has two children.) Therefore, a formula for $T(x) = \sum_{n\geq 1} T_n x^n$ similar to (18.3) will contain not one but *two* factors of the form $(1+x^3+x^6+\cdots) = \frac{1}{1-x^3}$. One of these factors will account for the number of 3-vertex subtrees of the first kind, and the other one will account for the number 3-vertex subtrees of the second kind.

Extending this argument, we see that

$$T(x) = x\prod_{i\geq 1}\left(\frac{1}{1-x^i}\right)^{T_i}. \quad (18.4)$$

Again, the factor x on the right-hand side accounts for the root. Equation (18.4) seems to be interesting, but somehow needing more work. In particular, the right-hand side contains all T_i, but in a way which does not make it clear how $T(x)$ can appear on the right-hand side. Fortunately, some purely algebraic manipulations will resolve this issue. First, using the fact that $y(x) = \exp(\ln y(x))$, and the rule stating that the logarithm of a product equals the sum of the logarithms of the factors, (18.4) can be written in the form

$$T(x) = x \exp\left(\sum_{n\geq 1} T_n \ln\left(\frac{1}{1-x^n}\right)\right). \tag{18.5}$$

The reader is invited to verify that the first few numbers T_n, starting with $n = 1$, are 1, 1, 2, 4, 9, and 20.

At this point, the reader might point out that (18.5) is even worse than (18.4), especially if we note that $\ln\left(\frac{1}{1-x^n}\right)$ is an infinite sum, so the right-hand side involves an infinite sum of infinite sums. This is true, but further simplification is possible. Indeed, we know that $\ln(1/(1-x)) = \sum_{k\geq 1} \frac{x^k}{k}$, so

$$\log\left(\frac{1}{1-x^n}\right) = \sum_{k\geq 1} \frac{x^{nk}}{k},$$

and therefore,

$$\begin{aligned}\sum_{n\geq 1} T_n \ln\left(\frac{1}{1-x^n}\right) &= \sum_{n\geq 1} T_n \sum_{k\geq 1} \frac{x^{nk}}{k} \\ &= \sum_{k\geq 1}\sum_{n\geq 1} T_n \frac{x^{nk}}{k} \\ &= \sum_{k\geq 1} \frac{T(x^k)}{k}.\end{aligned}$$

Comparing the last displayed chain of equations to (18.5), we see that we have proved the following theorem.

Theorem 18.25. *Let $T_0 = 0$, and let T_n be the number of all rooted non-plane trees on n vertices. Let $T(x) = \sum_{n\geq 1} T_n x^n$ be the ordinary generating function of the sequence $\{T_n\}_{n\geq 1}$. Then*

$$T(x) = x \exp\left(\sum_{k\geq 1} \frac{T(x^k)}{k}\right). \tag{18.6}$$

Again, using methods of analytic combinatorics, it is possible to show from (18.6) that

$$T_n \sim c \cdot \frac{a^n}{n^{3/2}}, \tag{18.7}$$

where $c = 0.439924\cdots$ and $a = 2.95576\cdots$.

Compare this with Cayley's formula, proved in Chapter 10. That formula implies that the number of rooted *labeled* trees on n vertices is n^{n-1}. If every rooted unlabeled non-plane tree on n vertices could be labeled in $n!$ different ways, then the number of labeled trees would be $n!$ times larger than T_n. However, there are a few trees that can be labeled in less than $n!$ ways (think of a star rooted at its center). So we expect this ratio to be a little bit less than $n!$. This intuition proves to be correct. Indeed, by Stirling's formula (see formula (3.1)),

$$\frac{n^{n-1}}{n!} \sim \frac{e^n}{\sqrt{2\pi}n^{3/2}},$$

which is a little bit less than T_n since $e = 2.718 < a = 2.95576$. We also point out that the number of rooted unlabeled *plane* trees on n vertices is the Catalan number c_{n-1}, as you were asked to prove in Exercise 43 of Chapter 14. Catalan numbers grow roughly as fast as powers of 4, so somewhat faster than the numbers T_n that grow roughly as fast as powers of 2.95. The difference in growth rates is the effect of the difference between plane trees and non-plane trees.

18.2.3 *Counting Unrooted Trees*

What can we say about the number of all *unlabeled trees*, that is, unlabeled trees that are not rooted? This seems to be a difficult question since some trees can be rooted in many non-equivalent ways (the path is an example for this), while some others can only be rooted in a few ways (the star can only be rooted in two ways).

The following famous result of Richard Otter establishes a surprisingly close connection between the number of rooted and unrooted trees on n unlabeled vertices.

Theorem 18.26. *Let $t_0 = 0$, let t_n be the number of unrooted trees on n unlabeled vertices for $n \geq 1$, and let $t(x) = \sum_{n\geq 1} t_n x^n$. Then we have*

$$t(x) = T(x) - \frac{1}{2}\left(T(x)^2 - T(x^2)\right). \tag{18.8}$$

Theorem 18.26 is surprising for more than just its simple form. The right-hand side of (18.8) contains the number 2 three times, in ways that are similar to what we have seen in formula (18.2) for rooted non-plane 1-2 trees. This is unexpected, since the trees counted by Theorem 18.26 do not have limits on their vertex degrees.

The reader can get familiar with the problem at hand by verifying that the first few values of the sequence of the numbers t_n, starting with $n = 1$, are 1, 1, 1, 2, 3, and 6.

Proof. (of Theorem 18.26) In order to prove Theorem 18.26, we need a few definitions. If v and w are two vertices of the tree T, and T has an automorphism f so that $f(v) = w$, then we call v and w *similar* vertices. In other words, v and w are similar if they are in the same orbit of the action of $Aut(T)$ on T. The number of orbits of this action will be called the *number of non-similar vertices* of T. In an analogous way, $Aut(T)$ acts on the set of all edges of T. Edges of T that are in the same orbit of this action are called *similar edges*, and the number of *non-similar edges* is the number of orbits of the action of $Aut(T)$ on the set of edges of T.

A *symmetry edge* of T is an edge that connects two similar vertices of T. Let v and w be the two endpoints of a symmetry edge L of the tree T. In Exercise 30, the reader is asked to prove that each $f \in Aut(T)$ that satisfies $f(v) = w$ must also satisfy $f(w) = v$. It then follows easily (see Exercise 31) that no tree can have more than one symmetry edge.

The following, somewhat technical lemma is (almost) the only graph-theoretical part of the proof of Theorem 18.26.

Lemma 18.27. *Let T be a tree, and let T_0 be a subtree of T so that no two vertices of T_0 are similar in T. Let P be a vertex of T that is not in T_0, and let P be adjacent to a vertex Q in T_0.*

Let $f \in Aut(T)$ so that $P' = f(P) \in T_0$. Set $Q' = f(Q)$. Denote by e the edge between P and Q, and denote by e' the edge between P' and Q'. Then either $Q = Q'$ or $e = e'$.

See Figure 18.4 for an illustration. It is easy to see that in that particular example, the function f interchanging P and P' and keeping all other vertices fixed is an automorphism of T.

Proof. (of Lemma 18.27) We show that if $e \neq e'$, then $Q = Q'$. If we remove the edge $e = PQ$ from T, we get the trees T_P and T_Q, where T_P is the tree containing P. In a similar manner, if we remove e' from T, we get the trees $T_{P'}$ and $T_{Q'}$.

We will now prove that the edge e' is in T_0. That, in turn, implies that $Q' \in T_0$ since Q' is an endpoint of e'. As $f(Q) = Q'$, this forces $Q = Q'$, since T_0 does not contain two distinct vertices that are similar in T.

Let us assume by way of contradiction that $e' \notin T_0$. That means that

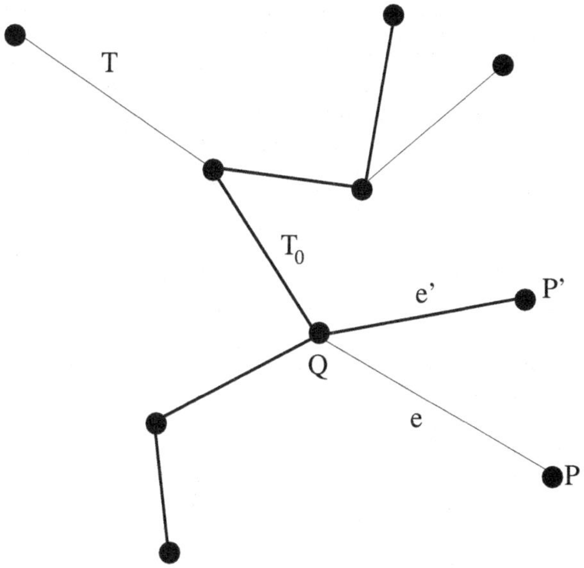

Fig. 18.4 An instance of the situation in Lemma 18.27.

the removal of e' leaves T_0 intact, so T_0 is part of $T_{P'}$ or $T_{Q'}$. However, by definition, $P' \in T_0$, so $T_0 \subseteq T_{P'}$ since P' is an endpoint of e'. Since by definition, $Q \in T_0$, this implies that $Q \in T_{P'}$. As Q is adjacent to P (by way of e), this implies that $e \in T_{P'}$, since we know that e is *not* identical to the removed edge e'. The fact that $e \in T_{P'}$ implies that $e \notin T_{Q'}$ since $T_{P'}$ and $T_{Q'}$ are disjoint.

In a similar manner, $e = PQ$ is by definition not part of T_0 (since P is not part of T_0), so its removal leaves T_0 intact. This means T_0 is a subtree of T_P or T_Q, but since $Q \in T_0$, therefore $T_0 \in T_Q$. In particular, $P' \in T_Q$ since $P' \in T_0$. As the removed edge e is not identical to the edge $e' = P'Q'$, this implies that $Q' \in T_Q$, and also $e' \in T_Q$.

Finally, because $Q' \in T_Q$, and the removed edge $e = PQ$ is not in $T_{Q'}$, it follows that $T_{Q'} \subseteq T_Q$. This containment is strict, since T_Q contains the edge e', while $T_{Q'}$ by definition does not. This is a contradiction, since $f(T_Q) = f(T_{Q'})$, so T_Q and $T_{Q'}$ have the same number of edges. So $e' \in T_0$, and, as we explained in the second paragraph of this proof, that forces $Q = Q'$. □

Note that if $e = e'$, that means that the edge PQ and the edge $P'Q'$ are identical. As $P \neq P'$, this means that $P = Q'$ and $P' = Q$. In particular, P and Q are similar, and the edge PQ is a symmetry edge. Also note that $f(P) = P' = Q$, and $f(Q) = Q' = P$, so f reverses the edge e.

The following, perhaps surprising, lemma shows that the number of non-similar vertices and the number of non-similar edges of a tree are very closely connected, almost in the same way as the number of all vertices and the number of all edges.

Lemma 18.28. *In any tree T, the number of non-similar vertices is one larger than the number of non-similar edges (symmetry edge excluded).*

The reader should verify this claim for the tree T shown in Figure 18.4 before reading further.

Proof. (of Lemma 18.28) Let T_0 be a maximal (that is, non-extendible) subtree of T that contains no two vertices that are similar in T. Let T_0 consist of m vertices. See Figure 18.5 for an illustration.

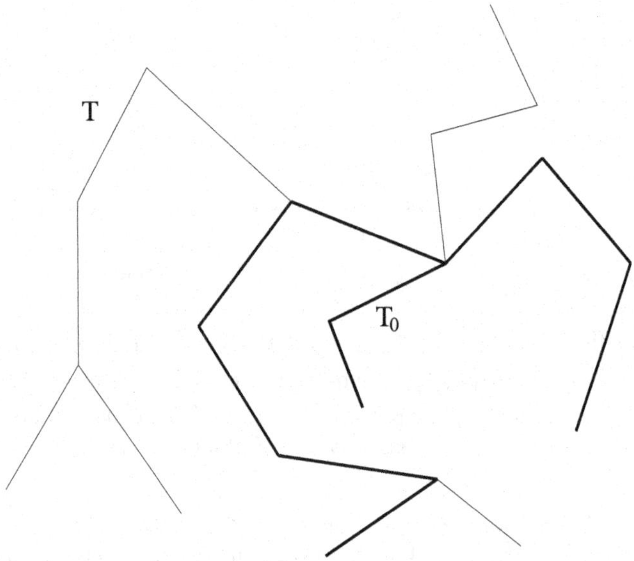

Fig. 18.5 The tree T_0 has no two vertices that are similar in T, but no other vertices of T can be added to T_0 without destroying that property of T_0.

We will prove two claims, which will together imply the statement of Lemma 18.28.

(1) We claim that T has exactly m non-similar points. We claim that this is true since each vertex v of T is similar to a vertex of T_0.
We prove this claim by induction on k, the distance of a vertex $v \in T$ from T_0. For $k = 1$, the claim is obvious. Indeed, in that case, v is a neighbor of a vertex of T_0, so if v were not similar to a vertex of T_0, then v could be added to T_0, contradicting the maximality of T_0.
Now let us assume that we know that our claim holds for k and prove it for $k+1$. Let v be at distance $k+1$ from T_0, and let u be the neighbor of v on the unique path from v to T_0. Then u is at distance k from T_0, so by the induction hypothesis, u is similar to a vertex $u_0 \in T_0$. So the neighbors of u (including v) are similar to the neighbors of u_0, which are vertices at distance at most one from T_0. By transitivity, (and because the claim holds for $k = 1$), this implies that all neighbors of u, including v, are similar to a vertex in T_0. This completes the induction step and the proof of our claim.

(2) We claim that T has $m - 1$ non-similar edges that are not symmetry edges. We claim that this is true since each non-symmetry edge of T is similar to an edge of T_0.
Let pq be an edge of T. By the previous claim, q is similar to a vertex in T_0, so we may assume that $q \in T_0$ (since there is an automorphism of T that maps q into a vertex of T_0). If $p \in T_0$ as well, then our claim is obviously true. If $p \notin T_0$, then p is similar to a vertex $p' \in T_0$, otherwise p could be added to T_0, contradicting the maximality of T_0. Now apply Lemma 18.27, with the vertices p, p', and q playing the roles of the vertices P, P' and Q, and $f \in Aut(T)$ being an automorphism such that $f(p) = p'$. Then Lemma 18.27, and the remark after it imply that either $f(q) = q$, or pq is a symmetry edge. So if pq is not a symmetry edge, then $f(pq) = p'q$, which is an edge of T_0 since both p' and q are in T_0, and the automorphism f of T preserves edges. So pq is similar to the edge $p'q$ of T_0.

The statement of the Lemma is now obvious since $m \quad (m-1) - 1$. □

Now let v_n be the total number of non-similar vertices of all n-vertex unlabeled trees, and let e_n be the total number of all non-similar edges (that are not symmetry edges) of all n-vertex unlabeled trees. In other words, v_n is the sum of the number of non-similar vertices taken over all n-vertex unlabeled trees, and e_n is the sum of the number of all non-similar edges (excluding symmetry edges) of all n-vertex unlabeled trees.

The following proposition expresses the number of our unrooted trees in a way that will be useful in our enumeration efforts.

Proposition 18.29. *For all positive integers $n \geq 2$, we have*

$$t_n = v_n - e_n.$$

Proof. (of Proposition 18.29) Let $v(t)$ (resp. $e(t)$) denote the number of non-similar vertices (resp. edges) of the unrooted tree t. Then by Lemma 18.28, we have

$$\begin{aligned} v_n - e_n &= \sum_t (v(t) - e(t)) \\ &= \sum_t 1 \\ &= t_n. \end{aligned}$$

Here t ranges all unrooted trees on n unlabeled vertices. □

In order to use Proposition 18.29 to compute t_n, we need to determine both v_n and e_n. First, we have

$$v_n = T_n, \tag{18.9}$$

since rooting the unlabeled tree t at its vertex x will result in a rooted tree different from the rooted tree obtained from rooting t at its vertex y if and only if x and y are non-similar vertices.

Second, e_n is equal to the number of rooted trees t on n unlabeled vertices in which one edge that is not a symmetry edge is colored blue. We claim that the number of the latter is

$$e_n = \frac{1}{2} \sum_{\substack{i \neq j \\ i+j=n}} T_i T_j + \binom{T_{n/2}}{2},$$

where $T_{n/2} = 0$ if n is odd.

Indeed, removing the colored edge (but not its vertices) from t will result in two smaller *rooted* trees, the endpoints of the removed edges being the roots. If these trees have a different number of vertices, then they are counted by the first summand of the right-hand side, and if they have the same number of vertices, then they are counted by the second term of the right hand side. (Since the removed edge was not a symmetry edge, the two obtained smaller trees are always different.)

A routine rearrangement of the last displayed equation yields

$$e_n = \frac{\sum_{i+j=n} T_i T_j}{2} - \frac{T_{n/2}}{2}. \tag{18.10}$$

Substituting the expressions that we obtained for v_n and e_n in formulae (18.9) and (18.10) into the formula of Proposition 18.29, we get

$$t_n = T_n - \frac{1}{2}\left(\left(\sum_{i+j=n} T_i T_j\right) - T_{n/2}\right),$$

which is equivalent to the claim of Theorem 18.26. □

The asymptotics of t_n are not surprising, given that each unrooted tree on n vertices can be rooted in at least one and at most n different ways. So $t_n \leq T_n \leq nt_n$ clearly holds. This implies that t_n must be of the same exponential order as T_n. A more precise analysis, whose scope is beyond this book, leads to the result

$$t_n \sim \frac{a^n}{n^{5/2}} b, \tag{18.11}$$

where $a = 2.95576$ just as in formula (18.7), and $b = 0.534949$.

Notes

For an introducton to Analytic Methods in Combinatorics, we refer the reader to Chapter 5 of *Generatingfunctionology* by Herb Wilf [49]. For a more detailed treatment, we recommend *Analytic Combinatorics* [18] by Philippe Flajolet and Robert Sedgewick.

A central part of these methods is a theorem that claims that (with suitable conditions), the exponential order of the coefficients of the power series $A(x)$ equals $1/a$, where a is the *singularity* of $A(x)$ that is closest to 0. *Singularity* is an important concept in complex analysis, and the reader is encouraged to read about it in the aforementioned books. For instance, if $a = 1/4$, then the coefficients of $A(x)$ grow roughly as fast as 4^n.

Exercises

(1) (-) Prove that the group of all real numbers with addition as the operation is isomorphic to the group of all positive real numbers as multiplication as the operation.

(2) (-) Prove Proposition 18.14.

(3) (-) Let H be a subgroup of G, and let a and b be two elements of G. Prove that $aH = bH$ if and only if $b^{-1}a \in H$.

(4) We say that the action of the group G on the set S is *transitive* if for any two elements $s, t \in S$, there is a $g \in G$ so that $g(s) = t$. Give an example of a regular graph R such that the action of $Aut(R)$ on the vertex set of R is *not* transitive.

(5) Show an example of a graph G such that the action $Aut(G)$ on the set of vertices of G is transitive, but the action of $Aut(G)$ on the set of edges of G is not transitive.

(6) Let H be a regular hexagon. Let us color the six sides of H using only the colors red, white, blue, and green, and consider two colorings equivalent if one can be moved into the other by rotating H around its center, or reflecting H through one of its six symmetry lines (three of which are diagonals, and the other three of which connect midpoints of opposing sides). How many non-equivalent colorings are there?
Note that in the next three exercises, we consider two colorings equivalent if there is an automorphism of the graph at hand that moves one coloring into the other.

(7) Find the number of non-equivalent ways to color the vertices of a tetrahedron using only red, blue or green.

(8) Find the number of non-equivalent ways to color the vertices of an octahedron using only red or blue.

(9) Find the number of non-equivalent ways to color the vertices of a cube using only red or blue.

(10) Let us say that two n-permutations q and r are *c-equivalent* if there exists a natural number $j \leq n-1$ and a bijection between the cycles of q and the cycles of r so that each cycle of r can be obtained from the corresponding cycle of q by adding j (modulo n) to each entry of the corresponding cycle of q. For instance, $q = (413)(52)$ and $r = (241)(35)$ are c-equivalent, as can be seen by selecting $j = 3$.

   (a) Describe, in the language of group theory, what it means for q and r to be c-equivalent. Your description should be in terms of multiplying permutations within the symmetric group.
   (b) (+) Analyze the sizes of the equivalence classes created by the c-equivalence relation to prove that if n is a prime, then $(n-1)! + 1$ is divisible by n. (This statement is called *Wilson's theorem.*)

(11) Let P be any of the five regular polyhedra. For $f \in Aut(P)$, let $X(f)$ be the number of vertices fixed by f. Find $E(X)$.

(12) A non-plane 2-tree is a rooted tree in which each non-leaf vertex has exactly two children.

(a) (-) Prove that all non-plane 2-trees have an odd number of vertices.
(b) Let d_n be the number of non-plane 2-trees on $2n+1$ vertices. Prove that $d_n = b_{2n+1}$, where b_m is the number of non-plane 1-2 trees on m vertices.

(13) Recall from Chapter 6 that a permutation p is called an *even* permutation if it has an even number of inversions, or, equivalently, if $\det A_p = 1$. See Exercise 11 of Chapter 6 for a definition of A_p.
What is the average number of fixed points in all *even* permutations of length n?

(14) We have seen in Chapter 14 that the number of *decreasing binary trees* on vertex set $[n]$ is $n!$. Recall that those trees were plane trees. Now let N_n be the number of *decreasing binary non-plane* trees on vertex set $[n]$. In such a tree, the root has label n, and each non-leaf vertex v has at most two children, and the labels of these children are less than the label of v. The order of the children of v is insignificant, so there is no "left child" and "right child".
Let $N_0 = 1$, and let $N(x) = \sum_{n\geq 0} N_n \frac{x^n}{n!}$. Prove that $N^2(x) + 1 = 2N'(x)$.

(15) Deduce from the result of the previous exercise that $N(x) = \tan x + \sec x$.

(16) Let E_n be the number of permutations of length n whose descent set is $\{2, 4, \cdots\}$.

(a) Characterize the decreasing binary trees (the plane version) of such permutations.
(b) Prove that $E_n = N_n$, where N_n is defined in the previous exercise.

(17) (+) Find a bijective proof of the result of part (b) of the previous exercise.

(18) Let $d_0 = 0$, let d_n be the number of all *decreasing non-plane trees* on vertex set $[n]$ if $n \geq 1$. In these trees, each non-leaf vertex v can have any number of children, as long as the labels of these children are smaller than the label of v. The children of v are unordered.
Let $D(x) = \sum_{n\geq 1} d_n \frac{x^n}{n!}$.

(a) Let $D_k(x)$ be the exponential generating function for the sequence counting decreasing non-plane trees in which the root has exactly k children. Prove that

$$D_k'(x) = \frac{D^k(x)}{k!}.$$

(b) Use the result of part (a) to find a closed form for $D(x)$, and then for d_n.

(19) Find a combinatorial proof of the result of part (b) of the previous exercise.

(20) Let g_n be the number of *plane* 1-2 trees on n unlabeled vertices. In such trees, each non-leaf vertex v has either one or two children. If v has two children, then the order of those children matters, that is, v has a left child and a right child. However, if v has only one child, that child is not a left child or right child (unlike in decreasing binary trees).

Set $g_0 = 0$, and verify that $g_1 = 1$, $g_2 = 1$, $g_3 = 2$, $g_4 = 4$, and $g_5 = 9$. Find a closed form for the generating function $G(x) = \sum_{n\geq 0} g_n x^n$.

Supplementary Exercises

(21) A permutation group G acting on the set S is called *transitive* if for any two elements $s, t \in S$, there exists $g \in G$ so that $g(s) = t$.
Let R be a *regular* simple graph (meaning that all vertices of R have the same degree. Is $Aut(R)$ a transitive permutation group as it acts on the vertex set of R?

(22) A permutation group G acting on the set S is called *primitive* if there is no non-trivial partition of S that is preserved under the action of G. That is, there is no partition π of S other than the 1-block partition and the $|S|$-block partition with the property that for all $g \in G$ and all blocks B of π, the set $g(B)$ is a block of π.
Now let S be the set of vertices of a cube, and let G be the automorphism group of the cube. Is G primitive over S?

(23) Let H be a simple graph on n vertices. Let $l(H)$ be the number of different ways to label the vertices of l using each element of $[n]$ exactly once. Two labelings are considered identical if the labeled graphs defined by them have identical sets of edges. That is, for all $i, j \in [n]$, the edge ij is either part of both labeled graphs, or of neither labeled graphs.
Find a formula for $l(H)$ in terms of $Aut(H)$.

(24) Euler's theorem states that if p is a prime and x is a positive integer, then $x^p - x$ is divisible by p. Use Theorem 18.20 to prove Euler's theorem.

Note that in the next three exercises, we consider two colorings equivalent if there is an orientation-preserving automorphism *of the graph at hand that moves one coloring into the other.*

An automorphism f of a graph G is called orientation preserving if the orientation (positive or negative, clock-wise or counter-clockwise) of each cycle $A_1A_2\cdots A_k$ of G agrees with the orientation of the cycle $f(A_1)f(A_2)\cdots f(A_k)$. Note that the orientation-preserving automorphisms of regular polyhedra are precisely those automorphisms that correspond to three-dimensional movements. This group is sometimes called the *rotation group* of the polyhedron.

(25) Find the number of non-equivalent ways to color the vertices of a cube using only red or blue.

(26) Find the number of non-equivalent ways to color the vertices of an octahedron using only red or blue.

(27) Find the number of non-equivalent ways to color the vertices of a tetrahedron using only red, blue or green.

(28) Let b_n be the number of unlabeled non-plane 1-2 trees on n vertices. Prove that

(a) $b_n = \sum_{k=0}^{i-1} b_k b_{n-1-k}$ if $n = 2i$, and
(b) $b_n = \frac{1}{2}b_i(b_i+1) + \sum_{k=0}^{i} b_k b_{n-1-k}$ if $n = 2i+1$.

(29) (-) Show an example of a tree on six vertices with a symmetry edge, and of a tree on six vertices without a symmetry edge.

(30) Let v and w be the endpoints of a symmetry edge L of the tree T. Let $f \in Aut(T)$. Prove that if $f(v) = w$, then $f(w) = v$.

(31) Prove that a tree can have at most one symmetry edge.

(32) Let $p = p_1p_2\cdots p_n$ be a permutation. Let us define two families of operations on p.

(a) For $1 \leq i \leq n$, let $H_i(p) = p_ip_{i+1}\cdots p_np_1\cdots p_{i-1}$, and,
(b) for $0 \leq i \leq n-1$, let $V_i(p) = (p_1+i)(p_2+i)\cdots(p_n+i)$, where integers larger than n are taken modulo n.

In other words, $H_i(p)$ shifts p "horizontally" and $V_i(p)$ shifts p "vertically", by i units.

Let us call two permutations *equivalent* if one can be obtained from the other by a series of operations H_i and V_j, where i and j are allowed to change. Find the number of equivalence classes of n-permutations under this equivalence relation.

(33) (-) Let P be any of the five regular polyhedra. Let $f \in Aut(P)$. Let $Y(f)$ be the number of *edges* of P fixed by Y. Find $E(Y)$. (We say

that $f \in Aut(P)$ fixes the edge e if f either leaves both endpoints of e fixed, or interchanges them.)

(34) (+) Recall that in Chapter 16, we proved that if Π_n is the lattice of partitions of $[n]$ ordered by refinement, then

$$\mu_{\Pi_n}(\hat{0}, \hat{1}) = (-1)^{n-1}(n-1)!.$$

Use this fact to give an alternate proof of Wilson's theorem, which is stated in Exercise 10, part (c).

(35) (-) Recall that t_n is the number of unlabeled trees on n vertices. An *unlabeled forest* is a forest whose components are unlabeled trees. Let f_n be the number of such forests on n vertices, and let $f(x) = \sum_{n \geq 0} f_n x^n$, with $f_0 = 1$. Express $f(x)$ in terms of the numbers t_n.

(36) (-) A *rooted unlabeled non-plane forest* is a forest on unlabeled vertices whose components are rooted non-plane trees. Let F_n be the number of such forests on n vertices, and set $F_0 = 0$. Find a simple expression for $F(x) = \sum_{n \geq 0} F_n x^n$ in terms of one of the generating functions discussed in this chapter.

(37) Let $h_0 = 0$, and let h_n be the number of rooted non-plane trees on n unlabeled vertices in which each vertex has at most *three* children if $n \geq 0$. Find a functional equation satisfied by the generating function $H(x) = \sum_{n \geq 0} h_n x^n$.

(38) Let $n \geq 3$, and let u_n be the number of graphs on n unlabeled vertices that contain exactly one cycle and that cycle is a triangle. Express the generating function $U(x) = \sum_{n \geq 3} u_n x^n$ in terms of some generating functions introduced in this chapter.

(39) Let a_n be the number of rooted plane 1-2 trees on vertex set $[n]$ in which every non-leaf vertex v has a label that is larger than the label of its children. If v has two children, then one of them is the left child of v and the other one is the right child of v, but if v has only one child, that child is not a left child or right child (unlike in decreasing binary trees). Let $a_0 = 1$, and let $A(x) = \sum_{n \geq 0} a_n \frac{x^n}{n!}$. Prove that $A'(x) = A^2(x) - A(x) + 1$.

(40) Let a_n be the number defined in the previous exercise. Find a (simply defined) class of permutations of length n that has a_n elements.

Solutions to Exercises

(1) The map $f(x) = e^x$ is an isomorphism between these two groups.

(2) If g and g' both fix i, then so does gg' since $(gg')(i) = g'(g(i)) = g'(i) = i$. The inverse of g reverses the action of g, so if g takes i to i, then g^{-1} takes i to i as well. The identity permutation fixes every element of S, including i. Finally, multiplication is automatically associative, since it is associative in the group G.

(3) If $aH = bH$, then multiplying both sides by b^{-1} from the left, we get $b^{-1}aH = H$. If $b^{-1}a \notin H$, then this cannot hold, since H contains the identity element of G. So $aH = bH$ implies that $b^{-1}a \in H$.
Now let us assume that $b^{-1}a \in H$. Let $x \in bH$. Then $x = bh$ for some $h \in H$. Note that
$$x = b(b^{-1}a)(a^{-1}b)h = a(a^{-1}b)h \in aH.$$
Indeed, if $b^{-1}a \in H$, so is its inverse, $a^{-1}b$. So $bH \subseteq aH$. Applying the same argument for the $a^{-1}b$, which is in H if and only if its inverse, $b^{-1}a$ is, we get the inclusion $aH \subseteq bH$.

(4) Let R be an octogon, with its vertices labeled 1 through 8 in circular succession, and with the diagonals 15, 28, 37, and 46. Then R is a regular graph since each of its vertices has degree three. However, the action of $Aut(R)$ on the vertex set of R is not transitive, since some vertices of R are *not* part of any triangles (3 and 7), while other vertices are.

(5) Let G have vertices $ABCA'B'C'$, where ABC and $A'B'C'$ are triangles, and AA', BB' and CC' are edges. One can think of G as a triangular prism. It is clear that $Aut(G)$ acts transitively on the set of vertices. However, $Aut(G)$ does not act transitively on the edges, since edges AA', BB' and CC' are not part of any triangles, while the other edges are.

(6) Let G be the automorphism group of a regular hexagon. It follows from Exercise 8 (c) of Chapter 9 that G has 12 elements, which can be identified as six rotations around the center of the hexagon (by angles $i\pi/3$, for $i = 0, 1, \cdots, 5$) and six reflections (through three maximal diagonals and three lines connecting the midpoints of opposite faces).
Let us start with the easy cases. If g is the identity element of G, then it of course fixes all colorings, so $F_g = 4^6 = 4096$. If g is a rotation by $\pi/3$ or $5\pi/3$, then it only fixes colorings in which all sides have the same color (four colorings each). If g is a rotation by π, then g fixes the colorings in which opposite faces have the same color (there are $4^3 = 64$ such colorings). If g is a rotation by $2\pi/3$ or $4\pi/3$, then g fixes the colorings in which second neighbors have the same color. There are $4^2 = 16$ such colorings.

If g is a reflection, then we distinguish two cases. If g is a reflection through a maximal diagonal, then it fixes $4^3 = 64$ colorings, and if g is a reflection through a line connecting the midpoints of two opposite sides, then g fixes $4^4 = 256$ colorings.
So

$$\frac{1}{|G|}\sum_{g\in G}|F_g| = \frac{4^6 + 2\cdot 4 + 4^3 + 2\cdot 4^2 + 3\cdot 4^3 + 3\cdot 4^4}{12} = 430.$$

(7) The automorphism group of the tetrahedron is S_4, the group of all permutations of length 4. The elements of this group can be of five different cycle types.

  (a) Four 1-cycles. This permutation fixes all 81 colorings.
  (b) Two 1-cycles and one 2-cycle. Such permutations fix 27 colorings, since the two colors in the 2-cycle have to be identical.
  (c) Two 2-cycles. By the same argument as in the preceding case, such permutations fix nine colorings.
  (d) One 1-cycle and one 3-cycle. Such permutations fix nine colorings.
  (e) One 4-cycle. Such permutations fix three colorings.

The number of permutations of each type is, 1, 6, 3, 8, and 6 as we learned in Chapter 6. Therefore, Theorem 18.20 yields that the number of inequivalent colorings is

$$\frac{81 + 6\cdot 27 + 3\cdot 9 + 8\cdot 9 + 6\cdot 3}{24} = 15.$$

(8) Burnside's theorem is not the simplest way to solve this problem. Let us consider the possible distributions of colors. There is one coloring in which all vertices are red, one coloring in which all vertices are blue, one coloring in which five vertex is red and one is blue, and one coloring in which five vertices are blue and one is red.
If four vertices are red and two are blue, then those two blue vertices may be at distance one or two from each other, leading to two inequivalent colorings. The same holds for four blue and two red vertices. Finally, if there are three red and three blue vertices, then the three red vertices may or may not contain an opposite pair, leading to two more colorings. So there are altogether ten inequivalent colorings.

(9) There is one such coloring with zero, one, seven, or eight red vertices. There are three inequivalent colorings with two red vertices, since the two red vertices may be at distance 1, $\sqrt{2}$, or $\sqrt{3}$. The same goes for colorings with two blue vertices.

Now consider colorings with three red vertices. One possibility is that all three red vertices are on the same face. Another one is when two red vertices are endpoints of the same edge, and the third is the unique vertex at distance $\sqrt{3}$ from one of them. The only remaining case is the one in which no edge has two red endpoints, that is, the red vertices are the three neighbors of a given vertex. This means there are three inequivalent colorings with three red vertices, and three inequivalent colorings with three blue vertices.

Finally, there is the case of four red vertices and four blue vertices. There is one coloring in which all red vertices are on one face. There are three colorings in which three red vertices and a blue vertex are on one face. The only remaining case is when each face has two red and two blue vertices. That can happen in two ways, namely either there will be an edge with monochromatic vertices, or not. So there are a total of six colorings with four vertices of each color.

So altogether, there are $1+1+1+1+3+3+3+3+6 = 22$ inequivalent colorings.

(10) (a) Let q and r be c-equivalent by way of the integer j, and let $q(a) = b$. Then $r(a+j) = b+j$. Now let $g_j(i) = i+j$ modulo n (that is, $n+t$ is identified with t for positive t). Then g_j is a permutation (a "cyclic translation") on $[n]$. Furthermore, $r(a+j) = b+j$ just means that $r(g_j(a)) = g_j(q(a))$. So we have that $g_j \cdot r = q \cdot g_j$ as permutations, or $r = g_j^{-1} \cdot q \cdot g_j$. This fact is referred to by saying that r and q are *conjugates* by way of a cyclic translate g_j.

(b) The n cyclic translates g_j form a group G that acts on the set of all n-permutations by the conjugate relation. That is, the action of g_j on q sends q to $r = g_j^{-1} \cdot q \cdot g_j$. The size of this group is n, and so, by Lemma 18.16, the sizes of the orbits of its action, that is, the sizes of c-equivalence classes, are divisors of n. If n is a prime, that means that the orbit sizes are 1 or n.

It is easy to characterize the c-equivalence classes of size 1. The reader is invited to verify that these correspond to powers of the n-cycle $(12\cdots n)$. In particular, if a permutation p is not the identity, and not an n-cycle, then the size of its equivalence class is n. The number of such permutations is $n! - [(n-1)!+1]$, so this number is divisible by n, but then so is $(n-1)!+1$.

(11) By Corollary 18.21, the average number of fixed points of the action of a finite group G on the finite set S is the number of orbits of that action. The automorphism group of each regular polyhedron acts

transitively on the vertex set of that polyhedron, which means that it has only one orbit. So $E(X) = 1$.

(12)(a) This is true for the one-vertex tree, then it follows by induction, since each non-plane 2-tree consists of a left subtree, a right subtree, and a root.

(b) Let T be a tree counted by d_n. Remove all $n+1$ leaves of T. This leaves us with a tree $f(T)$ counted by b_n. We are going to show that f is a bijection by showing that it has an inverse. Let U be a tree counted by b_n. Let us add two leaves to be the children of each leaf of U, and one extra child to each vertex of U that has one child. Let $g(U)$ be the obtained non-plane 2-tree with n non-leaf vertices. Then g and f are inverses of each other.

(13) Even permutations of length n form a group (called the alternating group and denoted by A_n). It is straightforward to see that A_n is transitive over the set $[n]$ if $n \geq 3$. So by Corollary 18.21, the average number of fixed points is 1 if $n \geq 3$. By trivial considerations, it is 1 if $n = 1$, and 2 if $n = 2$.

(14) By Theorem 8.21, $N^2(x)$ is the generating function counting two-element sequences binary non-plane trees (on a combined vertex set $[n-1]$) obtained by cutting of the root of such a tree on vertex set $[n]$. If, instead of sequences, we count 2-element *sets* of such trees, this generating function turns into $(N^2(x)+1)/2$. (The reader should explain why the addition of 1 is needed.) On the other hand, the number of such 2-element sets of trees on $[n-1]$ is equal to the number of such trees on $[n]$, that is, N_n, so their generating function is also equal to $\sum_{n\geq 1} N_n \frac{x^{n-1}}{(n-1)!} = N'(x)$.

(15) Note that $N(x) = \tan x + \sec x$ does solve the initial value problem $2N'(x) = N^2(x) + 1$, with $N(0) = 1$. It is well-known in the theory of differential equations that the solution of a (correctly stated) initial value problem is unique, so $N(x)$ must equal $\tan x + \sec x$.

(16)(a) If n is odd, then these are precisely the permutations whose decreasing binary tree has the property that every non-leaf vertex has exactly two children. If n is even, then the right-most vertex has one (left) child, other non-leaf vertices have two children. These claims are easy to prove by induction on n.

(b) We show that the sequences E_n and N_n satisfy the same recurrence relation, namely $2N_{n+1} = \sum_{i=0}^{n} \binom{n}{i} E_i E_{n-i}$ for $n \geq 0$, with $N(0) = 1$. For the sequence N_n, this is equivalent to the result of Exercise 14. For the sequence E_n, we proceed as follows. It is clear that

E_n is also the number of permutations of length n with descent set $\{1, 3, \cdots\}$. So the number of permutations of length $n+1$ with descent set $\{1, 3, \cdots\}$ plus the number of permutations of length $n+1$ with descent set $\{2, 4, \cdots\}$ is $2E_{n+1}$. On the other hand, we get such a permutation by placing the entry $n+1$ into the $(i+1)$st position, selecting the i entries that will be on the left of that position, then placing one of the E_{n-i} permutations of length $n-i$ with descent set $\{2, 4, \cdots\}$ on the right of $n+1$, and placing the *reverse* of one of the E_i permutations of length i with descent set $\{2, 4, \cdots\}$ on the left of $n+1$.

(17) The first such proof was found by R. Donaghey [17].

(18)(a) By the Product Formula of Exponential Generating Functions (Theorem 8.21), we know that $D^k(x)$ is the generating function of the sequence counting ordered lists of k such trees so that the vertex set of their union is $[n]$. Therefore, $D^k(x)/k!$ is the generating function for the sequence counting k-element *sets* of such trees.

On the other hand, by cutting off the root n of a decreasing non-plane tree in which the root has k children, we get a set of k decreasing binary trees so that the union of their vertex sets is $[n-1]$. This shows that the coefficient of $x^n/n!$ in $D^k(x)/k!$ agrees with the coefficient of $x^{n+1}/(n+1)!$ in $D_k(x)$, proving our claim.

(b) Summing the result of part (a) over all k, we get $D'(x) = e^{D(x)}$, or, equivalently, $D'(x)e^{-D(x)} = 1$. Integrating both sides, we obtain $-e^{-D(x)} = x - 1$, since the constant of integration must be chosen to -1 to conform with the fact $D(x) = 0$. So $e^{-D(x)} = 1 - x$, hence

$$D(x) = \ln\left(\frac{1}{1-x}\right) = \sum_{n\geq 1} \frac{x^n}{n!}.$$

So $d_n = (n-1)!$.

(19) Let $p = p_1 p_2 \cdots p_{n-1}$ be a permutation. Define the decreasing non-plane tree D_p by making p_i the parent of p_j if, when walking from p_j to the right, p_i is the first entry we meet that satisfies $p_i > p_j$. If there is no such p_i, then let n be the root of p_j. It is straightforward to verify that the map $p \to D_p$ is a bijection.

(20) As the root has either one or two children, the Product formula (Theorem 8.5) implies that $G(x) = 1 + xG(x) + xG^2(x)$. This leads to

$$G(x) = \frac{1 - x - \sqrt{1 - 2x - 3x^2}}{2x}.$$

Note that these numbers are the *Motzkin numbers* that we have seen in Exercise 16 of Chapter 8. The indexing is shifted by 1, so $g_n = M_{n-1}$.

Chapter 19

The Sooner The Better. Combinatorial Algorithms

19.1 In Lieu of Definitions

In all preceding parts of this book, when we considered a problem, we were interested in enumerating certain structures, finding the number of ways in which a certain task could be carried out, or deciding whether a structure with a certain set of properties can exist.

In this chapter, we will consider combinatorial problems from a new aspect. Instead of finding the number of ways in which we can carry out a task, we will be asking *how fast* we can carry out that task.

For our purposes, an *algorithm* is a finite sequence of unambiguously defined steps that carries out a task. We will not attempt to define an algorithm better than that sentence as that would be a topic for a logic course. Let us nevertheless point out that one could question what "unambiguously defined" means. Consider for instance the following definition.

"Let N be the smallest positive integer that cannot be defined using the English language and writing less than one thousand letters."

Now is N defined or not? The above sentence took less than one thousand letters to write, so it would seem that after all, it does define N within the allowed limits. However, N, by definition, cannot be defined with those limits.

The above paradox, which is sometimes called the *typewriter paradox*, is caused by the fact that the meaning of the word "defined" is not precise. As we said, we will not attempt to resolve that problem in this class, we will simply work with algorithms in which each step will be defined with no room left for ambiguity.

The above "pseudo-definition" of an algorithm nevertheless made it clear that an algorithm must consist of a *finite* number of steps. So if a procedure ever gets into an infinite loop, then that procedure is *not* an algorithm.

Example 19.1. The following procedure is not an algorithm as it contains an infinite loop.

```
# start with a_1=1
# for i larger than 0 do
# a_{n+1} = -a_n
```

The data that the algorithm is given at the beginning is called the *input* of the algorithm, and the data that the algorithm returns is called the *output*.

19.1.1 *The Halting Problem*

In order to further illustrate the difficulties of properly defining an algorithm, consider the following. Let us assume that we formally defined what an algorithm is. Then if somebody gives us a text T, we can surely decide whether T is an algorithm or not, can we not? Even more strongly, we can surely find a generic way, that is, an *algorithm* that decides whether T is an algorithm or not, can we not? In particular, we can surely decide that if we give a specific input t to T, then T will eventually halt or go into an infinite loop, can we not?

It turns out that no, we cannot. Let us assume that we can, that is, that there exists an algorithm $Halt(T,t)$ so that

$$Halt(T,t) = \begin{cases} \text{"Yes" if } T \text{ halts when given } t \text{ as input,} \\ \\ \text{"No" if } T \text{ goes into an infinite loop when given } t \text{ as input.} \end{cases}$$

What we do next will sound familiar to readers who took a course on Set Theory. We will present a proof by the so-called *diagonalization* method.

Write a program *Diagonal* so that

$$Diagonal(s) = \begin{cases} \text{returns "Yes" and halts if } Halt(s,s) \text{ is "No",} \\ \\ \text{goes into an infinite loop if } Halt(s,s) \text{ is "Yes".} \end{cases}$$

Now we are making one more step of this strange, self-referring kind. We feed *Diagonal* to itself as input. Will *Diagonal*(*Diagonal*) stop or not? Let us consider both possible answers.

(1) Let us assume first that *Diagonal*(*Diagonal*) halts. By the definition of *Diagonal*, that means that *Halt*(*Diagonal*, *Diagonal*) is "No". However, by the definition of *Halt*, that means that *Diagonal* does not halt on itself. This is a contradiction.
(2) Let us now assume that *Diagonal*(*Diagonal*) goes into an infinite loop. By the definition of *Diagonal*, that means that *Halt*(*Diagonal*, *Diagonal*) is "Yes". However, by the definition of *Halt*, that means that *Diagonal* does halt on itself. This, again, is a contradiction.

So our original assumption that *Halt* exists led to a contradiction, therefore *Halt* does not exist.

It is important to point out that all we proved is that there is no algorithm that would decide whether *any given* text T is an algorithm, that is, whether T will halt on an arbitrary input t. For a *specific* text T, we can very often decide whether T halts on t or not.

19.2 Sorting Algorithms

One of the classes of algorithms used most often in real life are *sorting* algorithms. These arrange certain objects in a line according to a specified property of the objects. In our examples, we will most often sort sets of real numbers. In order to simplify the discussion, we will assume that all the real numbers to be sorted are all *distinct*, but if we allowed multisets of real numbers, the algorithms would still work, with minor modifications.

19.2.1 *BubbleSort*

There are n children standing in a line, and they are of all different heights. We would like to rearrange the line so that the children are in increasing order of their height. What is the best way to achieve that goal?

The question at the end of the last paragraph is very imprecise. What makes it imprecise is the word *best*, that is, we have not said what we mean by the best way. We will revisit this problem in the next chapter, when we will formalize our ways of describing the efficiency of various algorithms. However, for now, let us say that we measure efficiency by the number of pairwise comparisons an algorithm makes. That is, the less pairwise comparisons an algorithm makes, the better it is. So the best algorithm is the one that makes the smallest number of comparisons.

One idea that naturally comes to mind is the following. Let $a_1, a_2, \cdots, a_n$ denote the heights of the n children, with a_i being the height of the child who is currently in the ith place of the line. Let us compare a_1 and a_2. If $a_1 < a_2$, then the relative order of the first two children is the desired one, and we do nothing. If $a_1 > a_2$, then the relative order of the first two children is not the desired one, and we ask them to change places. Note that in either case, we have made one comparison so far.

After making sure that the relative order of the first two children in the original line was the desired one, we compare the heights of the two children who are currently in the second and the third position of the line. If it is the desired one, that is, the second child is shorter than the third, then we do nothing, otherwise we ask them to change places.

We then continue the same way, that is, we compare the third and the fourth children of the current line, and if they are in the wrong order, we ask them to change places, then we compare the fourth and the fifth children, and if they are in the wrong order, we ask them to change places, and so on. The first part of the algorithm will end after we compared the two last children of the then-current line, and made sure they were in the right order. After that is done, we can be sure that the tallest child is indeed in the last place of the line. Indeed, no matter where he was in the line, once our swapping procedure reached him, he moved back on place in each step, until he reached the end of the line.

Example 19.2. If $n = 5$, and originally, the children's line corresponded to 4, 1, 5, 2, 3, then this first round of comparisons will take place as follows.

(1) Start with 4, 1, 5, 2, 3.
(2) As $4 > 1$, interchange these two children, to get 1, 4, 5, 2, 3.
(3) As $4 < 5$, do nothing.
(4) As $5 > 2$, interchange these two children, to get 1, 4, 2, 5, 3.
(5) As $5 > 3$, interchange these two children, to get 1, 4, 2, 3, 5.
(6) End of first round.

Unfortunately, the tallest child is the only one who is surely in his right place after this part of our algorithm. Indeed, it could even happen that the second-tallest child is in the first place! The reader is invited to check that this could happen if a_1 is the largest of all a_i, and a_2 is the second largest, or if a_2 is the largest and a_1 is the second largest. The reader is also invited to check that for any positions i and $i+1$, with $i \leq n-2$, we cannot know for sure that the child in position i is shorter than the child in position $i+1$.

Therefore, we will now repeat almost all the steps of the first part of the algorithm. That is, we compare the first two children of the current line, and if they are in the wrong order, we ask them to change places, then we compare the second and the third children of the then-current line, and proceed in an analogous way, and so on. The last pairwise comparison we will make in this round is comparing the $(n-2)$nd and $(n-1)$st children of the then-current line. Indeed, there is no need to compare the $n-1$st and nth children of the line, since we already know that the latter is the tallest of all n children.

After this second round of comparisons, we can be sure that the second-tallest child is at her right place (since she was taller than anybody who got compared to anyone in this round). Again, nothing more is assured. Therefore, we need to run another round of comparisons on the first $n-2$ children of the current line again. That will make sure that the third-tallest child gets in his right place, and so on. When we run the ith round of comparisons, the ith-tallest child will get in his right place. Therefore, when we run the $n-1$st run, of comparisons, the $(n-1)$st-tallest (or second-shortest) child gets in his right place. At that point, our task is done since the remaining child is automatically in his right place, namely the first place.

Example 19.3. Continuing the process started in Example 19.2, we would proceed as follows.

(1) Our starting point for the second round would be the line 1, 4, 2, 3, 5.
(2) As $1 < 4$, we would do nothing.
(3) As $4 > 2$, we would interchange these two children, to get 1, 2, 4, 3, 5.
(4) As $4 > 3$, we would interchange these two children, to get 1, 2, 3, 4, 5.
(5) This would end the second round of comparisons. In this particular case, no further comparisons would result in any changes, since we have reached the increasing order.

How many comparisons will we have to make while rearranging the line of n children? The first round will take $n-1$ comparisons, the second round will take $n-2$, comparisons, and so on, with the ith round taking $n-i$ comparisons. Therefore, the total number of comparisons that we will have to make is $\sum_{i=1}^{n-1}(n-i) = \sum_{i=1}^{n-1} i = \binom{n}{2}$. Another way of seeing this is by considering that every pair of elements gets compared exactly once.

As we mentioned before, the procedure of arranging n objects in a line in a previously specified order is called *sorting.* The sorting algorithm we presented above that used subsequent comparisons of adjacent elements is called *Bubble Sort.* (If we imagine that the objects are arranged vertically, then the largest, second-largest, third-largest elements, will rise to their correct positions as bubbles in water.)

In a generic programming language often called *pseudo-code*, the Bubble Sort algorithm can be described as follows.

```
#  for i := 1 to n - 1
#      do for j := 1 to n - i
#             do if a_j > a_{j+1}
#                    then t :=  a_j
#                         a_j := a_{j+1}
#                         a_{j + 1} := t
```

Here the variable i tells in which round of comparisons we are, and the variable j tells which comparison of that round we are currently carrying out. The temporary variable t is needed so that while we declare the new a_j to be equal to the old a_{j+1}, we do not lose the value of the old a_j before we assign it as the new value of a_{j+1}.

A "generic programming language", or *pseudo-code* is a loosely defined concept. It describes algorithms in a way programming languages do, but without the formal constraints. It helps to get a quick overview of what the algorithm does.

19.2.2 *MergeSort*

We have seen in the previous section that BubbleSort can sort an array of n real numbers in $\binom{n}{2}$ steps, even in the worst case. It is natural to ask whether we can find a better algorithm, that is, an algorithm that uses less pairwise comparisons, even in the worst case. Every element needs to get compared to another element at least once throughout any sorting procedure, otherwise we will have no information about the size of that element. So we cannot expect to find an algorithm that uses less than $n/2$ comparisons. This leaves a rather big gap between the two bounds we currently have, that is, the (trivial) lower bound $n/2$ and the proved upper bound $\binom{n}{2}$.

It turns out that the truth is much closer to the lower end. There exist several sorting algorithms that can sort n elements in no more than $cn \log_2 n$ steps, for some positive constant c.

One of these algorithms is called *MergeSort.* This is because this algorithm will first *split* the list of n objects in two parts which are as equal in size as possible, then *sort* both parts, and then **merge** the two sorted lists together. OK, you could say, but *how* will this algorithm sort those two partial lists? The answer to this is the most self-contained answer possible. MergeSort will sort those two lists by Mergesort again, that is, by splitting each of them into two sublists, sorting each of them by MergeSort, and then merging each pair of ordered lists into an ordered list. In other words, MergeSort is a recursive algorithm that calls unto itself in each step.

There is one more detail that we must discuss. How do we merge two sorted lists, say $a_1 < a_2 < \cdots < a_m$ and $b_1 < b_2 < \cdots < b_k$? We can do this efficiently as follows. Compare a_1 to b_1. If $a_1 < b_1$, then a_1 is the smallest of all $m + k$ elements at hand, and we can put it to the front of the merged list. Then we can continue with the lists $a_2 < a_3 < \cdots < a_m$ and $b_1 < b_2 < \cdots < b_k$ and repeatedly use the merging procedure we are describing. If $a_1 > b_1$, then b_1 is is the smallest of all $m + k$ elements at hand, and we can put it to the front of the merged list. Then we can continue with the lists $a_1 < a_2 < \cdots < a_m$ and $< b_2 < \cdots < b_k$, and use the same procedure again.

Example 19.4. The following shows MergeSort at work on the list 3, 1, 4, 2.

(1) Start with the list 3, 1, 4, 2.
(2) *Split* the list into the partial lists 3, 1 and 4, 2.
(3) *Sort* the partial lists, to get the sorted partial lists 1, 3 and 2, 4.
(4) *Merge* the partial lists 1,3 and 2, 4 to get 1, 2, 3, 4.
(5) End.

In the above example, the procedure worked in a very "symmetric" way since the number of elements to sort, four, was a power of two. This does not have to be the case for MergeSort to work.

Example 19.5. The following shows MergeSort at work on the list 4, 2, 1, 5, 6, 3.

(1) Start with the list 4, 2, 1, 5, 6, 3.
(2) *Split* the list into the partial lists 4, 2, 1 and 5, 6, 3.
(3) *Sort* the partial lists as follows.
    (a) *Split* them into the partial lists 4, 2, and 1; and 5, 6, and 3.
    (b) *Sort* the partial lists, to get the sorted lists, 2, 4, and 1; and 5, 6, and 3.
    (c) *Merge* the partial lists 2, 4, and 1; and 5, 6, and 3, to get the sorted lists 1, 2, 4 and 3, 6, 5.
(4) *Merge* the sorted partial lists 1, 2, 4 and 3, 6, 5 to get the sorted list 1, 2, 3, 4, 5, 6.
(5) End.

In pseudo-code, MergeSort can be implemented as follows.

```
# MergeSort(i=1..n)
# if 1<n do
# m=[(1+n)/2];
                 # Mergesort(1,m);
                 # Mergesort(m+1,n);
# merge(1,m,n)
# end
```

Here $merge(1, m, n)$ is the subalgorithm that merges two sorted partial lists. It can be implemented for instance by copying the two ordered partial lists into a temporary list (so that the original lists are not overwritten), and then by moving the smallest elements still in the two lists into the new, sorted list. In pseudo-code, this can be done as follows.

```
# merge(1,  m,  n)
# for i=1..n do
#        b_i=a_i;
#    i=1; j=m+1; k=lo;
#    while (i<=m && j<=n)
#        if (b_i<b_j)
#            a_{k+1}=b_{i+1};
#        else
#            a_{k+1}=b_{j+1};
# end
```

How many comparisons will MergeSort make when it sorts an n-element list? Let $M(n)$ denote this number. Then $M(1) = 0$ and $M(2) = 1$. For the general case, we claim that if $n = 2k$, then

$$M(2k) = 2M(k) + 2k - 1, \tag{19.1}$$

and when $n = 2k + 1$, then

$$M(2k+1) = M(k) + M(k+1) + 2k. \tag{19.2}$$

Both of these formulae are easy to explain. Indeed, the first two terms of the right-hand side are the number of comparisons needed to sort the two partial lists, and the last term is the number of comparisons needed to merge the two sorted partial lists. Indeed, after each comparison, we are able to place one element to its right place in the merged list, so $n - 1$ comparisons will place the first $n - 1$ elements in their right place, which then will force the last element into its right place.

Note that (19.1) and (19.2) can be comprised in the formula

$$M(n) = M(\lfloor n/2 \rfloor) + M(\lceil n/2 \rceil) + n - 1. \tag{19.3}$$

These cumbersome divisibility issues suggest that we first try to find an exact formula for $M(n)$ in the special case when $n = 2^t$. In that case, set $m_t = M(2^t)$. Then (19.3) translates to

$$m_t = 2m_{t-1} + 2^t - 1 \tag{19.4}$$

for $t \geq 1$, and $m_0 = 0$. Let $m(x) = \sum_{t \geq 0} m_t x^t$ be the ordinary generating function of the sequence m_t. Multiplying both sides of (19.4) by x^t and summing over $t \geq 1$, we get

$$m(x) = 2xm(x) + \frac{2x}{1-2x} - \frac{x}{1-x}.$$

This implies

$$\begin{aligned} m(x) &= \frac{2x}{(1-2x)^2} - \frac{x}{(1-x)(1-2x)} = \frac{x}{(1-x)(1-2x)^2} \\ &= \frac{1}{(1-2x)^2} + \frac{1}{1-x} - \frac{2}{1-2x}. \end{aligned}$$

Therefore, $m_t = (t-1)2^t + 1$. This means that if $n = 2^t$, then

$$M(n) = M(2^t) = n(\log_2 n - 1) + 1.$$

If n is not a power of 2, then we can add new objects to the list which are larger than all existing objects until n does become the closest power of 2 that is not smaller than n, that is, $n_1 = 2^{\lceil \log_2 \rceil}$. We can then sort the new list in $n_1(\log_2 n_1 - 1) + 1$ steps as above. The obtained sorted list will contain the original n elements in the right order, at the beginning of the list. Finally, note that $n_1 < 2n$, and $\log_2 n_1 < \log_2 n + 1$, therefore, in terms of n, the sorting algorithm will never take more than $1 + 2n \log_2 n$ comparisons.

At this point, we would like to stress that MergeSort is a very significant improvement compared to BubbleSort. Indeed, the ratio of the numbers of steps in the two algorithms is not more than

$$g_n = \frac{1 + 2n \log_2 n}{\binom{n}{2}} = \frac{1}{\binom{n}{2}} + 4\frac{\log_2 n}{n-1}.$$

Therefore $\lim_{n\to\infty} g_n = 0$, so for large n, the number of comparisons MergeSort needs is negligible when compared to the number of comparisons that BubbleSort needs.

19.2.3 *Comparing the Growth of Functions*

In the rest of this chapter, we will define ways to describe good estimates of the number of steps an algorithm makes. As the example of MergeSort shows, these estimates can often be obtained much faster than a precise formula, and will still provide a good measurement of the efficiency of the algorithm. In order to facilitate discussion of these estimates, we make the following three definitions, which are all very widely used in approximation theory.

Definition 19.6. Let $f : \mathbf{Z}^+ \to \mathbf{R}$ be a function, and let $g : \mathbf{Z}^+ \to \mathbf{R}$ be another function. We say that $f(n) = O(g(n))$ (read "f is big O of g") if there exists a positive constant c so that

$$f(n) \leq cg(n)$$

for all $n \in \mathbf{Z}^+$.

In other words, $f(n) = O(g(n))$ means that $f(n)$ is at most a constant factor larger than $g(n)$, for all n. In other words, $g(n)$ approximates $f(n)$ up to a constant factor.

Example 19.7. Let $M(n)$ be defined as above. Then $M(n) = O(n \log_2 n)$.

Solution. We have seen that $M(n) \leq 1 + 2n \log_2 n$ for all n. Furthermore, $M(1) = 0$, and $1 \leq n \log_2 n$ if $n \geq 2$. Therefore, $M(n) \leq 3n \log_2 n$.

Example 19.8. Let $f(n) = 100\binom{n}{2} + \binom{n}{3}$. Then $f(n) = O(n^3)$.

Solution. Use $c = 51$.

Example 19.9. Let $f(n) = \binom{n}{2}$ and let $g(n) = 1000n$. Then $f(n) \neq O(g(n))$.

Solution. No matter what c we choose, $f(n) > cg(n)$ will hold when $n > 2000c + 1$.

Let us return for a minute to the function $M(n)$ that counted the number of comparisons MergeSort needed to make in order to sort an n-element list. Then the statement that $M(n) = O(n^2)$ is certainly correct since n^2 grows much faster than $2n \log_2 n + 1$. However, this statement is not very informative since it is not very sharp. It simply says that $M(n)$ is smaller than another function, but it does not say how much smaller. There are other notions that can make this statement more precise.

Definition 19.10. Let $f : \mathbf{Z}^+ \to \mathbf{R}$ and $g : \mathbf{Z}^+ \to \mathbf{R}$ be two functions. We say that $f(n) = \Omega(g(n))$ (read "f is omega of g") if there exists a positive constant c so that

$$f(n) \geq cg(n)$$

for all $n \in \mathbf{Z}^+$.

Example 19.11. Let $f(n) = 0.001n$ and let $g(n) = 100 \log_2 n$. Then $f(n) = \Omega(g(n))$.

Solution. Choose $c = 10^{-5}$.

Finally, our last notation brings the previous two together.

Definition 19.12. Let $f : \mathbf{Z}^+ \to \mathbf{R}$ and $g : \mathbf{Z}^+ \to \mathbf{R}$ be two functions. We say that $f(n) = \Theta(g(n))$ (read "f is theta of g") if $f(n) = O(g(n))$ and $f(n) = \Omega(g(n))$.

Example 19.13. Let $f(n) = n^2 + n\log^3(n)$. Let $g(n) = n^2$. Then $f(n) = \Theta(g(n))$.

Solution. On the one hand, $f(n) = O(g(n))$ as can be seen by choosing $c = 2$. On the other hand, $f(n) = \Omega(g(n))$ as can be seen by choosing $c = 1$.

19.3 Algorithms on Graphs

19.3.1 *Minimum-cost Spanning Trees, Revisited*

We saw an algorithm on graphs in Chapter 10. That algorithm, called Kruskal's algorithm, took a connected simple graph whose edges were assigned a cost (or weight) as the input and returned a minimum-cost spanning tree of G as an output. In Chapter 10, we were concerned about the *greedy* property of the algorithm, that is, the fact that in each step, the algorithm took the edge that increased the short-term costs the least. In each step, the algorithm chose an edge that did not create a cycle and had the minimum cost of all edges with that property.

Now we will consider that algorithm from a different aspect. Our goal is to decide *how many steps* the algorithm takes. Foretelling the need for a unified approach that we will introduce in the next section, we point out that *what a step is* needs a little bit of explanation. In the sorting algorithms of the previous section, we simply counted comparisons. However, in Kruskal's algorithm, it is not so clear what we should count. Indeed, choosing an edge from a graph is easy, but choosing an edge that does not create a cycle is more difficult (in a very large graph), because we need to make sure that indeed, no cycle is created, and that in itself can take a long time if we do not have an efficient method to do it.

Let us discuss an efficient way of running the Kruskal algorithm. As we said, each round of that algorithm will look for the lowest-cost edge that can be added to the set S of edges already selected without creating a cycle. This means that if there are several edges that can be added without creating a cycle, then we have to look for the one with minimum cost. Finding a minimum-cost edge in each round, and then forgetting the results of all comparisons made in the process seems wasteful. It is therefore sensible to sort *all edges* of G at the beginning of the algorithm. As we have seen in the previous section, this can be done in $O(E \log_2 E)$ steps, where E is the number of edges of G. Let $edges = \{e_1, e_2, \cdots, e_E\}$ be the obtained list of all edges of G in non-decreasing order of their costs.

Now in the first round of Kruskal's algorithm, we choose the edge e_1, and in the second round, we choose e_2. As G is simple, e_1 and e_2 never form a cycle. The third round is more complicated as e_1, e_2, and e_3 could form a cycle. If that happens, e_3 is rejected, and e_4 is selected. However, as we proceed further, we need an efficient approach to decide whether the next edge of *edges* is eligible to be chosen or not, that is, whether its addition would create a cycle or not. It would take very long to consider every possible subset of edges containing the newly chosen edge and verify that they do not contain a cycle. Instead, we propose the following. From the beginning of the algorithm, let us keep track of the *connected components* of the graph T of selected edges.

That is, when we choose e_1, let us put the two endpoints of e_1 into a new set C_1, indicating that they are in the same component of T. After selecting e_2, put its endpoints in C_1 if the two-edge graph with edges e_1 and e_2 is connected, and put them into a new set C_2 if that graph is not connected.

Continue this way. That is, in round i, scan the still unused edges of *edges* until you find the first edge whose endpoints are not in the same component C_i. (It could be that they are in different components, or it could be that one or both of them are not in any components yet.) Add that edge e_h to T. Discard the edges preceding e_h from *edges*. If they could not be added to T before without forming a cycle, they cannot be added to T now without forming a cycle.

Then update the list of components. That is, if neither endpoint of e_h was included in any C_i before, create a new component with the endpoints of e_i. If one of them was in C_j, and the other one was in no component, add that other one to C_j. Finally, if one endpoint of e_h was in C_i and the other one in C_j, then unite C_i and C_j, and add both e_i and e_j to the obtained component. Rename that component so that it inherits the label of the *larger* of its predecessors, that is, the component which had more vertices in it.

This assures that the graph T remains cycle-free since we never join two of its vertices in the same component by an edge.

Before counting the steps in this second part of the procedure, let us consider an example.

Example 19.14. The above implementation of Kruskal's algorithm applied to the graph of Figure 19.1 runs as follows.

(1) Start with the graph shown in Figure 19.1 (with the costs assigned to the edges, but not yet the labels e_i).
(2) Sort the edges according to their cost. Obtain the list *edges* $= \{e_1, e_2, \cdots, e_{11}\}$.
(3) Select e_1. Create the component $C_1 = \{A, B\}$.
(4) Select e_2. Create the component $C_2 = \{G, H\}$.

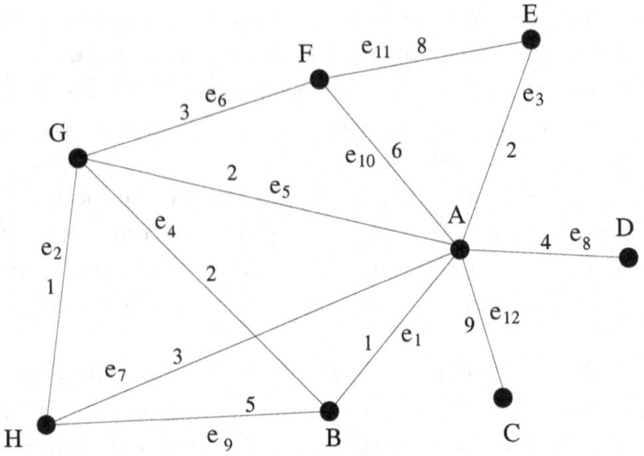

Fig. 19.1 The graph G with its cost function and sorted edges.

(5) Select e_3. Add E to C_1, to get $C_1 = \{A, B, E\}$.
(6) Select e_4. Unite C_1 and C_2 to get $C_1 = \{A, B, E, G, H\}$.
(7) Select e_6. (Note that e_5 is ineligible since its endpoints both belong to C_1.) Add F to C_1 to get $C_1 = \{A, B, E, F, G, H\}$.
(8) Select e_8. Note that e_7 is ineligible since its endpoints both belong to C_1. Add D to C_1 to get $C_1 = \{A, B, D, E, F, G, H\}$. Select e_{12}, since e_9, e_{10}, and e_{11} are ineligible as their endpoints both belong to C_1.
(9) End with the tree whose edges are e_1, e_2, e_3, e_4, e_6, e_8 and e_{12}.

Returning to the question of how many steps the Kruskal algorithm takes, let us say that for this algorithm, a step is whenever we do something, that is, put a vertex in a component, unite two components, or check whether two vertices are in the same component. Note that in each round, we have to scan at most E edges before finding the minimal-cost edge that is eligible. After finding this edge e, there are two possibilities. If e will not unite two existing components, but create a new component of two vertices, or add one vertex to an existing component, then we can record that in a constant number of steps. Indeed, we spend at most two steps adding a vertex to one or two components. If e unites two components, say C_i and C_j, then we change the label of the vertices in the *smaller* component to the label of the other component. This may take as many as $n/2$ steps. However, if x is a vertex whose label changed this way, then the component containing x at least doubled in size. This cannot happen more than $\log_2 n$

times for any x. So each x will change labels no more than $\log_2 n$ times, therefore the number of all changes of labels is not more than $n \log_2 n$.

To summarize, it takes $O(E \log_2 E)$ steps to sort the edges according to their costs, then it takes $O(E + n \log_2 n)$ steps to find the minimum-cost tree. Therefore, the total number of steps needed is $O(E \log_2 E)$ since our graph is connected, so $E \geq n - 1$.

19.3.2 *Finding the Shortest Path*

The next problem we consider is one that everyone with a driver's license has faced before. Given a starting point s, an endpoint t, and a network of one-way streets, find the *shortest* path from s to t.

We will present an algorithm that will find not simply the shortest path to t, but the shortest path to *any* point on the map. The algorithm is called the *Dijkstra algorithm*, after its inventor, the Dutch mathematician Edsger W. Dijkstra.

Our input is a *directed* simple graph G. The edges of G all have a positive cost; the cost of e_i will be denoted by $d(e_i)$. One can think of $d(e_i)$ as the "length" of e_i.

While looking for the shortest path from s to any given vertex t, we will associate a number $\delta(v_i)$ to each vertex v_i. This number can be thought of as the "length of the shortest discovered path" from s to v_i. Originally, we set $\delta(s) = 0$ and $\delta(t) = \infty$ for all $t \neq s$, since we have not yet discovered any paths from s to t.

In what follows, we split the vertex set $V(G)$ of G into two parts, the set S vertices to which we already have a path from s, and the set T of vertices to which we do not yet have a path. So at the beginning, $S = \{s\}$ and $T = V(G) - s$.

For all edges sv, set $\delta(v) = d(s, v)$ replacing the original $\delta(t) = \infty$. This makes perfect sense, since it expresses the fact that if there is an edge from s to v, then the minimum distance from s to v is the length of that edge.

Now we describe a generic step of the algorithm. This step will be applied several times, following the initial step described in the previous paragraph.

Find a vertex $v \in T$ for which $\delta(v)$ is minimal. Put v into S, and proceed with all the edges leaving v and going to a vertex outside S as you proceeded with the edges leaving s. More precisely, if vr is an edge with $r \in T$, and $\delta(v) + d(v, r) \geq \delta(r)$, then do nothing. Otherwise replace $\delta(r)$ by $\delta(v) + d(v, r)$, corresponding to the fact that we have just found a shorter path to r, namely the path that consists of a shortest path to v, and the edge vr. This step is often described by saying that we *relax* the edge vr.

When this is done, iterate this procedure. That is, find the vertex $v' \in T$ for which $\delta(v')$ is minimal, and start over. Stop when all vertices are in S, and therefore, all edges are relaxed, or when there are no edges going from S to T (the latter can happen when G is not strongly connected).

Throughout the algorithm, ties can be broken in any way. An example is shown below.

Example 19.15. For the graph shown in Figure 19.2, Dijkstra's algorithm works as follows.

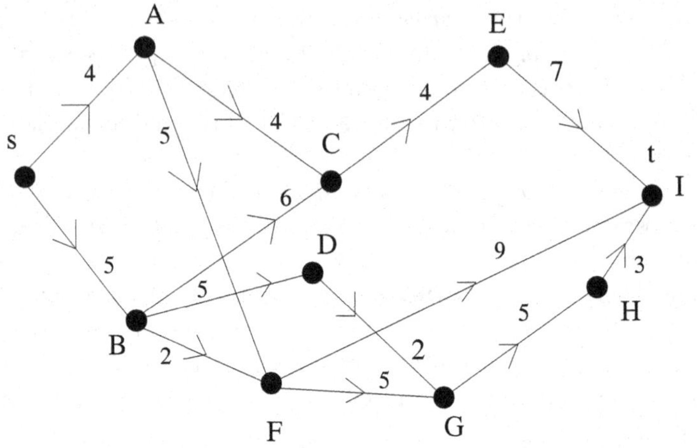

Fig. 19.2 We will use the Dijkstra algorithm on this graph.

(1) Start with the graph shown, and set $S = \{s\}$.
(2) Relax the edges leaving s. Set $\delta(A) = 4$ and $\delta(B) = 5$. Put A in S.
(3) Relax the edges leaving A. Set $\delta(C) = 8$ and $\delta(F) = 9$. Put B in S.
(4) Relax the edges leaving B. Set $\delta(D) = 10$, and $\delta(F) = 7$ (so $\delta(F)$ is being reset). Put F in S.
(5) Relax the edges leaving F. Set $\delta(G) = 12$ and $\delta(I) = 16$. Put C in S.
(6) Relax the only edge leaving C. Set $\delta(E) = 12$. Put D in S.
(7) Relax the only edge leaving D. This happens to have no effect, since $\delta(G) = 12$. Put G in S.
(8) Relax the only edge leaving G. Set $\delta(H) = 17$. Put H in S.
(9) Relax the only edge leaving H. This has no effect. Put I in S.

(10) End.

Figure 19.3 shows the graph of Figure 19.2 with the values of δ written next to the vertices in italics, and the weights of the edges written on the edges in Roman font.

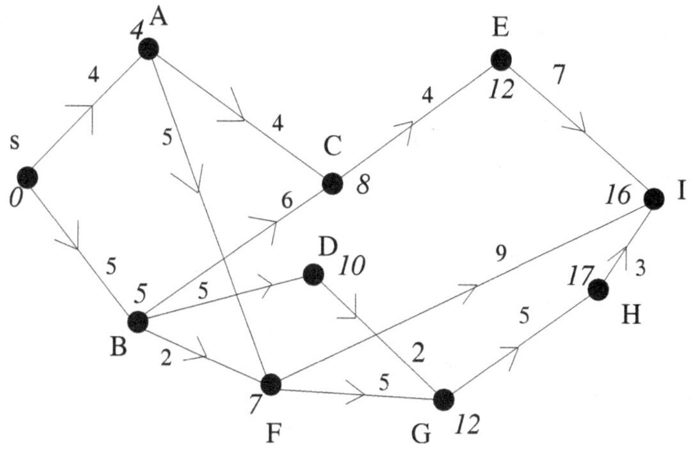

Fig. 19.3 The Output of the Dijkstra algorithm.

Several questions are in order. First, how do we read off a shortest path from s to some t from the output of this algorithm? (We say *a* shortest path, not *the* shortest path, since there could be times when there are several paths of minimum length.) Second, why is it that the final value of $\delta(t)$ is indeed the length of a shortest path from s to t. Third, how many steps does it take to run this algorithm?

We are going to answer the first two questions at once, by one theorem. Before we can announce the theorem, we need one more notion. Let us say that for a given vertex t, every time an edge vt is relaxed and the value of $\delta(t)$ decreases, we set $prev(t) = v$. This expresses the fact that at that point, there is a shortest path from s to t that ends in the edge vt. Note that $prev(t)$ is always a vertex that got placed into S before t, and that when the Dijkstra algorithm is finished, $prev(t)$ is defined for all $t \neq s$. Therefore, as G is finite, for all vertices $t \neq s$, there exists a positive integer k so that $prev^k(t) = s$. Here $prev^k$ simply means k successive applications of $prev$.

Theorem 19.16. *For any simple graph G, and for any pair of distinct* *vertices s and t, the Dijkstra algorithm will either produce a shortest path from s to t and compute its length, or show that there is no path from s to t, as follows.*

(a) Once the algorithm is finished, the path whose edges listed from the end are $(prev(t), t)$, $(prev(prev(t)), prev(t))$, and so on, $(s, prev^k(t))$ is a shortest path from s to t. If $prev(t)$ is not defined, then there is no path from s to t.

(b) Furthermore, the length of any shortest path from s to t is equal to the value of $\delta(t)$ when the algorithm is finished. If there is no path from s to t, then this value will be ∞.

Proof. We first prove part (b). In fact, we prove the following, even stronger statement. We claim that at each stage of the Dijkstra algorithm,

(i) if $t \in S$, then $\delta(t)$ is the length of a shortest path from s to t, and
(ii) if $t \notin S$, then $\delta(t)$ is the length of a shortest path from s to t whose last edge is an edge from S to t.

We prove these claims by induction on the size of S. If $|S| = 1$, that is $S = \{s\}$, then (i) holds since $\delta(s) = 0$. Claim (ii) holds since if (s, t) is an edge, then $\delta(t) = d(s, t)$ is the length of a shortest path from (s, t) with the desired property, and if (s, t) is not an edge, then $\delta(t) = \infty$.

Now let us assume that the claims hold for the case of $|S| = k$, and prove them for the case of $|S'| = k + 1$. Let $S' = S \cup x$, that is, x is the new vertex added to S in this step of the Dijkstra algorithm. We first show that (i) holds for x. Before this step, x was outside S, so by the induction hypothesis, $\delta(x)$ was the length of a shortest (s, x) path q whose last edge was from S to x. Note that every path p from s to x has to first leave S and then arrive at x. If the first vertex of p outside S is some $y \neq x$, then p is not a shortest (s, x) path. See Figure 19.4 for an illustration.

Indeed, the part of p that is between s and y cannot be shorter than q since if it were, then y, not x, would be selected to be added to S in this step. (Recall that by the induction hypothesis, we know that $\delta(x)$ is minimal for $x \notin S$, and that $\delta(x)$ is the length of a shortest (s, x) path ending in an edge from S to x.) We would then have to get from y to x, which would make p longer than q.

Also note that adding x to S will not change the label of the vertices that are already in S since, by the definition of the Dijkstra algorithm, edges within S are not being relaxed. Therefore, (i) is proved.

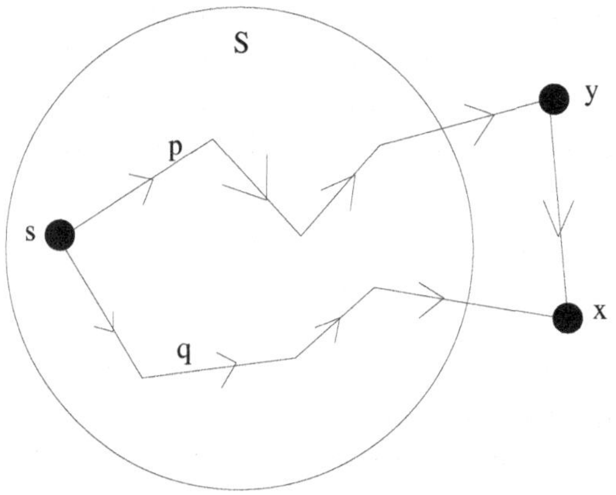

Fig. 19.4 The path p is longer than the path q.

In order to prove (ii), let $h \in V(G) - S'$. Note that if there is a shortest (s,h) path ending in an (S',h) edge that does not end in an (x,h) edge, then the placement of x into S' did not change anything, so our claim holds by the induction hypothesis. If all the shortest (s,h) paths ending in an (S',h) edge do end in an (x,h) edge, then the Dijkstra algorithm sets $\delta(h) = \delta(x) + d(x,h)$, and our claim is proved since in all such paths, the path from s to x must be a shortest path from s to x, and so have length $\delta(x)$.

This proves part (b) of the theorem. Part (a) is now not difficult to see. Indeed, now that we know that the Dijkstra algorithm correctly computes the minimum distances to all vertices, we see that part (a) simply describes how we can keep track of the way those minimum distances are actually achieved. A minimum distance is achieved by a shortest path, and the path described in part (a) is a path achieving the minimum distance from s to t, therefore it is a shortest path. □

Note that the Dijkstra algorithm takes $O(n^2)$ steps. Indeed, in each stage, we add one vertex to S, so there are at most n stages, and in each of those stages, we must find the vertex $v \notin S$ for which $\delta(v)$ is minimal. This can be done in $O(n)$ steps, (as you are asked to prove in Supplementary Exercise 14) proving our claim.

The Dijkstra algorithm has several refinements and enhanced versions. Perhaps the most widely used special case is *breadth first search.* This is the special case when all edges have weight one, and the task is reduced to finding the path from s to t that contains the minimum number of edges. The name "breadth first search" refers to the fact that the algorithm will first reach all the neighbors of s, before going deeper into the graph. This is in contrast to another approach, *depth first search*, which we define in Exercise 8.

Notes

A very readable introduction to the topics of this chapter and the next one is Herb Wilf's book *Algorithms and Complexity* [50]. A classic comprehensive textbook on algorithms is *Introduction to Algorithms*, by Cormen et al. [16]. In this book, the reader will find several sorting algorithms which are roughly as effective as MergeSort, as well as a detailed analysis of the two graph traversal algorithms that we mentioned here, breadth first search, and depth first search.

Exercises

(1) Consider the following sorting algorithm. First, sort $n-1$ objects recursively with the algorithm we are defining. Then insert the nth object a_n to its correct place as follows. First, compare a_n to the middle element of the sorted list L of $n-1$ elements. Depending on the result of that comparison, a_n needs to be inserted into the first or second half of L. Whichever half it is, insert a_n into it by the same procedure. That is, compare a_n to the element in the middle of that half of L, and conclude in which quarter of L the correct place of a_n is. Let $b(n)$ be the number of steps this sorting algorithm will take in the worst case. Prove that $b(n) = O(n \log_2 n)$.

(2) Prove that if A is any sorting algorithm that uses only pairwise comparisons, and $f(n)$ is the number of comparisons that A makes in the worst case when sorting n elements, then $f(n) = \Omega(n \log_2 n)$. Conclude that the best sorting algorithms based on comparison make $\Theta(n \log_2 n)$ comparisons.

(3) Let us assume that we have a machine that can do k-wise comparisons in one step, for a fixed positive integer k. That is, if we give k distinct real numbers to the machine as input, it will output the sorted list of those numbers in one step.

Let $g(n)$ be the number of times we have to run this machine in order to sort n distinct real numbers. Is it true that $g(n) = \Omega(n \log_2 n)$?

(4) Construct an algorithm that finds the largest and the second largest elements of an n-element set of real numbers using at most $\frac{3}{2}n + 2$ pairwise comparisons.

(5) Let k be a positive integer. Construct an algorithm that finds the kth largest of an n-element set of real numbers using $O(n)$ pairwise comparisons.

(6) Construct an algorithm that will list all $n!$ permutations of length n. Each element of the list should be obtained from the previous one by at most $n - 1$ steps.

(7) Let G be a directed graph which contains no directed cycles. Prove that then the vertices of G can be listed in some order $v_1, v_2, \cdots, v_n$ so that if $i > j$, then there is no directed path from v_i to v_j.

(8) Let G be a directed graph. The following algorithm, called *depth first search* obtains all vertices t that are reachable from vertex s of G by a directed path. First, go from s to a vertex s_1 using an edge (s, s_1), then go from s_1 to vertex s_2 different from s and s_1 using an edge (s_1, s_2), and so on, as long as this is possible. Let us assume that we are forced to stop after k vertices $s, s_1, \cdots, s_{k-1}$, that is, there is no edge leaving s_{k-1} that ends in a vertex that we have not reached before. Then we go back to the predecessor of s_{k-1}, the vertex s_{k-2}, and continue the algorithm from there the same way. (This is called *backtracking.*) Each time we get stuck at some vertex, we backtrack to the predecessor of that vertex.

Now let G be a directed graph with no loops so that each vertex of G is reachable from vertex s by a directed path. Find a sufficient and necessary condition for G not containing any directed cycles, in terms of the depth first search algorithm, starting at s.

(9) Decide if the following statements are true or false.

(a) If $a > 0$, then $n \log n = O(n^{1+a})$.
(b) $2^{n^2} = O(n!)$.
(c) $n! = \Theta\left(\frac{n^n}{e^n}\right)$.

(10) Consider the following algorithm. Let G be a connected simple graph whose edges have non-negative costs assigned to them. Start with the one-vertex subgraph v, for any $v \in G$. Build a graph from v as follows. In each step, if T is the vertex set of the graph that has already been built, find the lowest-cost edge between T and $V(G) - T$ whose addition

will not form a cycle in the graph that is being built. Add that edge to the graph being built. Stop when $T = V(G)$.
Prove that this algorithm constructs a minimum-cost spanning tree for G.

(11) Let S be a finite set of n elements. Consider the following sorting algorithm. Pick an element $s \in S$ at random. Compare the remaining $n-1$ elements of S to s, and based on the results, break up $S - s$ into the sets A and B, where $A = \{i \in S | i < s\}$, and $B = \{i \in S | i > s\}$. Now S is "partially sorted", as AsB. Next, use the same procedure recursively, first on A and B, and then on the obtained smaller blocks, until a completely ordered list is obtained.

   (a) Prove that sorting by this algorithm will take no more than $\binom{n}{2}$ comparisons.
   (b) When will this sorting procedure take *exactly* $\binom{n}{2}$ comparisons?

(12) (+) Keep the definitions of the preceding exercise. Let $X(p)$ denote the number of comparisons it takes to sort S using a given set p of picks. Prove that $E(X) = O(n \log n)$.

Supplementary Exercises

(13) Give a simple proof using graph theory for the fact that there is no algorithm that sorts n objects with less than $n-1$ steps.

(14) Find two different algorithms to find the largest element of an n-element list of real numbers. Both algorithms should use $n-1$ comparisons.

(15) A group of 64 ping-pong players want to find the best and second best players among themselves. Show that they can achieve this by playing a total of 68 games, in the following sense. After 68 games, there will be two players A and B so that all other 62 players lost a game to A or B, and neither A nor B lost a game to any of the other 62 players.

(16) Let X be a random variable defined on the set of all n-permutations by setting $X(p)$ to be the number of times that BubbleSort interchanges two elements while sorting the entries of p. Compute $E(X)$.

(17) The depth first search algorithm, defined in Exercise 8 can be applied to undirected graphs as well. Note that if G is a connected undirected

graph, then the algorithm will in fact find a spanning tree T of G that will be rooted at the vertex s in which the algorithm started.
Let us say that vertex a is a descendant of vertex b in T if the only path in T connecting b to the root s goes through a.
Prove that if e is an edge of G, then one endpoint of e is a descendant of the other.

(18) Prove that if a connected graph on n vertices does not contain a path of length k, then it has at most $(k-1)n$ edges.

(19) Explain how the Dijkstra algorithm can be used to decide whether an undirected graph G is connected.

(20) Decide whether the following statements are true or false.

(a) $\sin n = O(1)$,
(b) $\sin n = \Theta(1)$,
(c) $\left(\frac{n}{\log_2 n}\right)^n = O(n!)$.

(21) The *diameter* of a graph was defined in Exercise 29 of Chapter 10. Find an algorithm that computes the diameter of a graph G on n vertices in $O(n^3)$ steps.

(22) Write a pseudo-code for Kruskal's algorithm.

(23) Write a pseudo-code for the Dijkstra algorithm.

(24) Let M be a list of n positive integers which are not necessarily distinct, such that no element of M is larger than k. Give an algorithm that lists the n elements of M in non-decreasing order that uses $O(n+k)$ steps.

(25) Set $n = k$ in the preceding exercise. Then the result of that exercise is an algorithm that sorts an input of size n in $O(n)$ steps.

Solutions to Exercises

(1) Let us first assume that $n = 2^t$. Let us compute how many comparisons it takes to find the correct place of the nth element a of the list once the $(n-1)$ other elements are sorted. The reader is invited to verify that in the first step, we compare a to the middle element of a list of length $2^t - 1$, in the second step, we compare a to the middle element of a list of length $2^{t-1} - 1$, and so on, and in step i, we compare a to the middle element of a list of length 2^{t+1-i}. Therefore,

in the tth step, we compare a to the "middle" element of a "list" of length one, after which we know the correct place of a. So the correct place of a could be found in $t = \log_2 n$ steps.
If $n \neq 2^t$, then there exists a positive integer u so that $2^u + 1 \leq n \leq 2^{u+1}$. In this case, we complete our list by adding extra elements to its end so that it has 2^{u+1} elements. We can then find the correct place of a in the new list in $u + 1 \leq 2\log_2 n$ steps. So it never takes more than $2\log_2 n$ comparisons to find the correct place of the nth element a. Then, by the same argument, it takes at most $2\log_2(n-1)$ comparisons to find the correct place of the $(n-1)$st element, at most $2\log_2(n-2)$ comparisons to find the correct place of the $(n-2)$nd element, and so on. Therefore, the total number of comparisons is at most

$$\sum_{i\leq n} 2\log_2 i \leq 2n\log_2 n.$$

(2) There are $n!$ possible orders of n distinct elements, and in the worst case, each pairwise comparison will eliminate at most half of the orders that were possible before that comparison. So in the worst case, after one comparison, there will be $n!/2$ possible orders, after two comparisons, at least $n!/4$ possible orders, after 3 comparisons, at least $n!/8$ possible orders, and so on. Therefore, if after m comparisons, there is only one possible order left, then $n! \leq 2^m$, or $\log_2 n! \leq m$. From Stirling's formula,

$$m \geq \log_2 n! = n\log_2(n/e) + \log_2(\sqrt{2\pi \cdot n}) = \Omega(n\log_2 n)$$

proving our claim.
On the other hand, we have seen that it is possible to sort n elements by only $O(n\log_2 n)$ pairwise comparisons, so indeed, the best sorting algorithms will take $\Theta(n\log_2 n)$ steps.

(3) Analogous to the solution of the previous exercise. The only difference is that now each step has $k!$ possible outcomes, so if there are a possible orders before a step, then if we are unlucky, then there could be at least $a!/k!$ possible orders after that step. As $k!$ is just a constant, like 2 in the previous exercise, the rest of the solution unchanged, except that $k!$ plays the role of 2.

(4) Let us split our set of elements into two blocks of equal size, or as equal as possible size. In each set, find the maximal element, then compare the two maximal elements. This takes $n-1$ comparisons. Say that we find that the maximal element a of block A is larger than

the maximal element b of block B. Then a is the maximal element of our set, and the second maximal element is either B, or the maximal element of $A - a$. Find the maximal element of $A - a$ in at most $(n+1)/2$ steps, then compare it to b in one steps. This will provide the desired output with at most $n + \frac{n+3}{2}$ comparisons.

(5) Let $a_1, a_2, \cdots, a_n$ be our elements. First, order the first k elements of the list using MergeSort in $O(k \log_2 k) = O(1)$ steps. Then find the place of a_{k+1} in this list in at most k steps, and discard the last element. Continue this way. In each stage, find the place of the new element in the k-element list that we keep, and discard the last $(k+1)$st element of that list. Each stage takes at most k steps, so the whole procedure takes no more than $k \log_2 k + nk = O(n)$ steps. (Note that we could find the place of the new element in $O(\log k)$ steps as opposed to k steps, but that would not be a significant improvement, since k is a constant.)

(6) We will list the permutations of length n in lexicographic order. That is, $p = p_1p_2 \cdots p_n$ will precede $q = q_1q_2 \cdots q_n$ in the order if, for the smallest index i for which $p_i \neq q_i$, the inequality $p_i < q_i$ holds.
In order to get the permutation immediately following $p = p_1p_2 \cdots p_n$ in this order, first find the largest ascent of p, that is, the largest i so that $p_i < p_{i+1}$. If there is no such i, then p is the decreasing permutation, which is the last permutation of our the list, and we stop. Otherwise, let p be followed by q, defined by $q = p_1p_2 \cdots p_{i-1}q_iq_{i+1} \cdots q_n$, where q_i is the smallest element in $\{p_{i+1}, p_{i+2}, \cdots p_n\}$ that is larger than p_i, and the string $q_{i+1} \cdots q_n$ contains the remaining entries of $p_ip_{i+1} \cdots p_n$ in *increasing* order. The reader is invited to prove that each permutation occurs exactly once in this list since each permutation (other than the increasing one) has a unique predecessor.

(7) We use induction on n. For $n = 2$, the statement is true. Now let us assume that the statement is true for n, and prove it for $n+1$. Let G have $n+1$ vertices, and let $G' = G - v_{n+1}$. As G' contains no cycles, its vertices can be listed the right way by the induction hypothesis. Let L be this list. Let A be the set of vertices $a \in G'$ so that there is a directed path from a to v_{n+1}. Let B be the set of vertices $b \in G'$ so that there is a directed path from v_{n+1} to b. As G has no directed cycles, this implies that there can be no directed path from B to A. So all vertices of A precede all vertices of B in the list. Then v_{n+1} can be inserted anywhere between the end of A and the start of B in L.

(8) The depth first search algorithm creates a directed spanning tree of

G. In this tree, the parent of each vertex v is its unique predecessor, that is, the vertex from which v was first reached. We claim that G is acyclic if and only if there is no edge from G that goes from a vertex v to one of the ancestors of v. If there is such an edge (v, u), then G contains a cycle since u is an ancestor of v, so there is a path from u to v.

If there is a cycle C with vertices $c_1, c_2, \cdots, c_k$ in G, then let c_i be the first vertex of C reached by depth first search. Then all the other c_j, including c_{i-1}, are descendents of c_i in the depth first search tree (since the algorithm will not backtrack from c_i before reaching all vertices reachable from c_i, which includes all of C). Therefore, (c_{i-1}, c_i) is an edge of the desired kind.

(9) Let us assume that our algorithm (called *Prim's algorithm*) creates the tree T with edges $e_1 \leq e_2 \leq \cdots \leq e_{n-1}$, while there is another, cheaper tree F. If there are several candidates for F, choose one so that the number of edges that are part of both T and F is *maximal*. Then there is a smallest index i so that $e_i \notin F$. Let A be the vertex set of edges $e_1, e_2, \cdots, e_{i-1}$. Then e is an edge between A and $V(G) - A$. Let x and y be the endpoints of e_i. Then there is a unique path from x to y in F. Let f be the edge of F along that path that connects a vertex in A to a vertex in $V(G) - A$. As in step i, we added e and not f to our tree T, the inequality $w(f) \geq w(e_i)$ must hold.

Now remove f from F and add e_i to F instead. This creates another spanning tree of G with at most as large a cost as F. Indeed, the new graph F' has $n-1$ edges and is connected (why?). As F had minimal cost, it follows that $w(F) = w(F')$, but F' and E have one more edge in common than F and E, which is a contradiction.

(10)(a) True. After simplifying by n, the statement is reduced to $\log n = O(n^a)$, and that is true since $\lim_{n\to\infty} \frac{\log n}{n^a} = 0$ by the l'Hospital rule.

(b) False. In fact $\lim_\infty n!/2^{n^2} = 0$ as can be seen by taking logarithms. Using Stirling's formula, $\log n! = n(\log n - 1) + (\log n + \log \pi)/2$, while $\log 2^{n^2} = n^2 \log_2$. Now use part (a).

(c) False. By Stirling's formula, $n! \sim \left(\frac{n^n}{e^n}\right)\sqrt{2\pi n}$, and that extra $\sqrt{2\pi n}$ factor will outgrow any constant.

(11)(a) No pair will get compared more than once.

(b) Each pair will get compared once if we never pick an element that properly partitions the set of remaining elements, that is, if each time we make a pick, we choose the largest or the smallest available element.

This sorting procedure is called *QuickSort*.

(12) Let $X_i(p)$ be the number of times the element a_i gets compared during the sorting sequence defined by p. It suffices to show that $E(X_i) = O(\log n)$, and the claim will follow by linearity of expectation.
In order to see that $E(X_i) = O(\log n)$, note that the *size* of the block containing a_i decreases, on average by at least half by each pick involving that block (the reader should prove this fact), so on average, this block will shrink into a singleton after $O(\log n)$ splits.

Chapter 20

Does Many Mean More Than One? Computational Complexity

The wide variety of problems in which algorithms are used suggests that we look for a unified approach that measures how efficient various algorithms are. In the previous chapter, we did that by counting the number of steps the algorithms used, but that meant that for each problem, we had to specify what counted as a step. Our goal now is to have standards that can be applied to every algorithm.

20.1 Turing Machines

A *Turing machine* is an idealized computer named after the English mathematician Alan Turing. It is meant to simulate how a human being would carry out an algorithm step by step, moving from one stage to the next, according to some well-defined rules. Formally, a Turing machine T consists of the following four parts.

(1) A *tape*. This is a one-dimensional array of cells, which is infinite at both ends, so that we never run out of tape. Each cell contains a symbol from a finite alphabet A. Two of these symbols have to be *blank* and *start*. If we have not written anything to a cell yet, then we assume it contains the *blank* symbol. The *start* symbol is the one that the machine will read first. It will tell the machine to start. The tape is often called the *input* of the machine.

(2) A *head*. Fair enough, if the tape contains a lot of information, the machine should be able to read it. The head can move both ways along the tape, and can read the symbol in the cells, and can replace a symbol in the cells. This is often expressed by saying that the head is *read-write*.

(3) A set S of *states*, containing the *start state* s. As the head moves along the tape, the machine changes from one state to another. How T reacts to a certain symbol it is reading depends on the state it is in. That is, it can happen that when the machine reads symbol a and is in state t, it will react differently from when it reads symbol a is in state u.

(4) A *transition function* (or *program*)

$$f : S \times A \to (S \cup \text{“}Yes\text{”} \cup \text{“}No\text{”}) \times A \times \{\leftarrow, \rightarrow, stay\}.$$

This function describes how T works. The definition is not nearly as difficult as it may look. The domain of f is $S \times A$, which makes perfect sense since the action of the machine must depend on the state in which it is, and the symbol it is currently reading. The range of f is a direct product of three factors, and we will survey them separately. The first factor, $S \cup \text{“}Yes\text{”} \cup \text{“}No\text{”}$ means that when T reads the input of the given cell in its current state, it may go to a “Yes” state (often called the accepting state), or to a “No” state, (the rejecting state). Note that it follows from the above definition that the machine will always halt immediately after reaching the “Yes” state or the “No” state. Also note that the “Yes” state and the “No” state are so special that they are not part of S.

Some enhanced versions of Turing machines can simply halt without saying “Yes” or “No”, and these machines have a “Halt” state for stopping like that, but we will not use that model. We will concentrate on Turing machines that are used to test “yes or no” questions, hence the accepting and rejecting states.

The second factor A of the right-hand side is needed since T can write another symbol into the cell it is reading. Finally, the third factor $\{\leftarrow, \rightarrow, stay\}$ is needed since after writing into the current cell, the head may move one notch to the left, one notch to the right, or it may stay where it was.

While this definition may seem too cumbersome, or too broad, it comprises almost everything an algorithm can possibly do. Therefore, most algorithms we encounter can be executed by Turing machines.

There are several versions of enhanced Turing machines, and a few simplified versions as well. The machines described above are often called *deterministic* Turing machines since knowing the state in which the machine is, the position of the head, and the content of the cell the machine is currently reading means knowing what the machine will do next. The adjective *deterministic* will be further explained when we put these machines into contrast with different machines.

Example 20.1. We can use a Turing machine to decide whether a certain positive integer a is divisible by 3 or not. This Turing machine will have the following parameters.

(1) The set of states

$$S = \{start, 0, 1, 2, Yes, No\}.$$

(2) The set of symbols of the alphabet

$$A = \{start, blank, 0, 1, 2, 3, 4, 5, 6, 7, 8, 9, end\}.$$

(3) The tape containing, from left to right, *start*, the digits of a in order, and *end*.

(4) The program f defined as follows.

(5)(a) When the head reads *start*, it moves to the first digit of a in the next cell on the right. It stays in state *start*.

(b) When the machine reads the ith digit of a, it reacts as follows. If that digit is 0, 3, 6, or 9, it stays in its current state. If that digit is 1, 4, or 7, it moves one state up, (that is, if it was in state 0, it goes into state 1, if it was in state 1, it goes into state 2, and if it was in state 2, it goes in state 0). Finally, if that digit is 2, 5, or 8, the machine moves two states up.
The head then moves to the next cell on the right of the current cell.

(c) If the machine is in state 0 when the head reaches the cell containing the symbol *end*, the machine goes to "Yes" state and halts. If the machine is in state 1 or 2 when the head reaches the cell containing the symbol *end*, the machine goes to "No" state and halts.

The above program used the fact that a is divisible by three if and only if the sum of its digits is divisible by three.

The reader should not be horrified. In the rest of the chapter, we will not analyze every single algorithm so painfully. The goal of the above example was to show how to translate an algorithm into the terminology of Turing machines. The main advantage of this model is that now it is absolutely clear what a *step* is (a step of the head, either $\rightarrow$, or $\leftarrow$, or *stay*), and it is also clear what the running time of an algorithm is (the number of steps of the head). This is why Turing machines are so appropriate for analyzing the efficiency of a very wide array of algorithms.

20.2 Complexity Classes

In this section, we will encounter some of the most intriguing problems of modern mathematics. They are related to attempts of describing which questions can be decided by Turing machines in an efficient way.

20.2.1 *The Class* P

A *decision problem* is a "yes-or-no" question asked about a combinatorial object, such as "Is this graph bipartite?" or "Is this graph connected?" or "Is this integer prime?" or "Does this permutation contain an even number of cycles of length seven?". A language L is the set of all objects for which the answer of a given decision problem is "Yes". So, following up on the above examples, the class of all bipartite graphs, the class of all connected graphs, the set of all prime numbers, and the set of all permutations with an even number cycles of length seven each form a language.

We will say that a Turing machine T *accepts* the language L if given input x, T stops in the accepting state if $x \in L$, and T stops in the rejecting state if $x \notin L$.

We are now ready for the first major definition of this section.

Definition 20.2. We say that a language L is in **P** if there exists a Turing machine T and a positive integer k so that T accepts L in $O(n^k)$ time, where n is the size of the input.

That is, if an input x of length n is given to T, then $O(n^k)$ moves of the head are enough for T to decide whether $x \in L$ or $x \notin L$.

If a language L is in **P**, we often say that membership in L can be *tested in polynomial time.*

The reader might think that we are too imprecise here since **P** does not discriminate between languages that can be accepted in $O(n)$ time or in $O(n^{20})$ time. We have two answers to that, the first one of which will be clearer after the next example.

Example 20.3. Let L be the language consisting of all simple graphs that contain a triangle. Then $L \in \mathbf{P}$.

Solution. A Turing machine can simply go through all $\binom{n}{3}$ triples of the n-element vertex set of the input graph and check whether all three pairs of vertices in any given pair are adjacent. There are only $\binom{n}{3} = O(n^3)$ triples to check, and in each of them, there are only three edges to check. Finally, the head of the Turing machine never needs to travel more than n^2 cells between checking two edges, so the statement follows.

Note that it paid off that in the definition of **P**, we did not specify what k had to be. This, for instance, obviates the question of what the size of the input should really be, the number of vertices of the graph, n, or number of entries of the adjacency matrix of the graph, n^2. (The adjacency matrix is needed to describe which vertices are connected by an edge.)

There is also no need to figure out clever ways to send the head from one tape to another, since even sending it from one end to the other will not hurt.

Example 20.4. Let L be the language consisting of permutations p (given in the one-line notation) for which p^2 is the identity permutation. Then $L \in \mathbf{P}$.

In this example, the size of the input is clearly the length n of p.

Solution. Let $p = p_1 p_2 \cdots p_n$. For each $i \in [n]$, if $p_i = j$, check whether $p_j = i$. If this always holds, accept, otherwise reject. There are n equalities to check, and between checking two entries, the head never needs to travel more than n cells, so there will be no more than $O(n^2)$ steps.

The class **P** of problems is an example of a *complexity class*, that is, a class of problems that are roughly equally difficult to solve. While the reader might object by saying that there is quite some difference between a problem that takes n steps to solve and a problem that takes n^{100} steps to solve, this difference is still much smaller than the difference between the latter and a problem that takes 2^n steps to compute. Indeed, if we have a computer that can solve a problem with input size m in $\log m$ time, then the first two problems will take $\log n$ and $100 \log n$ time for this computer to solve, respectively. These times will only differ by a constant factor. The last problem will take $n \log 2$ time, which is an order of magnitude higher. More precisely, as n goes to infinity, the first two problems will take a negligible amount of time to compute when compared to the last problem. This is our second answer to the question as to why it makes sense to put problems solvable in $O(n)$ time and in $O(n^{20})$ time into the same class.

Loosely speaking, **P** is the set of languages that can be decided by an effective algorithm. Indeed, polynomial time is in some sense the best that we can expect, since it takes n steps just to read the input.

20.2.2 *The Class* NP

There is a wide array of decision problems for which no polynomial-time algorithm is known. To be more precise, there are languages L for which there is no polynomial-time algorithm known to *test* whether $x \in L$, for an arbitrary input x. Quite often, there is a weaker algorithm which will not decide whether x is in L or not, but if someone claims that $x \in L$ for a specific reason, the algorithm will *verify* that reasoning in polynomial time, and decide whether that reasoning is correct. If it is, then $x \in L$. If not, then it does *not* follow that $x \notin L$ since it could still be that $x \in L$ for some other reason.

For instance, let L be the set of all pairs (S, m) so that S is a set of positive integers that have a subset T so that the sum of the elements of T is m. Now let $x = (A, t)$ be a pair so that A is a set of positive integers, t is a positive integer, and let us see what we can say about the membership of x in L. We could certainly take all $2^{|A|}$ subsets of A and check if any of those have sum t, but that would take more than a polynomial amount of time. Indeed, $2^{|A|}$ is an exponential function of the size of the input. On the other hand, if someone claims that a certain subset $B \subseteq A$ has sum t, then we can *verify* that claim in $O(n)$ steps, by simply taking the sum of all elements of B. Of course, if the claim turns out to be false, we are out of luck, since it could well be that $x \in L$ thanks to some other subset $B' \subseteq A$.

This set of decision problems, that is, the problems for which we can *verify* (but not necessarily *test*) membership in polynomial time, turns out to be extremely important. This warrants the following formal definition.

Definition 20.5. We say that a language L is in **NP** if there exists a positive integer k and a Turing machine T so that the following hold.

- For each $x \in L$, there exists a *witness* $W(x)$ so that when T is given input $(x, W(x))$, it will recognize that $x \in L$ in $O(n^k)$ time.
- For each $x \notin L$, no such witness exists. That is, no matter what input $(x, W'(x))$ we give to T, we cannot "trick" T into falsely saying that $x \in L$.

In other words, $L \in$ **NP** if witnesses for the claim that $x \in L$ can be verified in polynomial time (but not necessarily *found* in polynomial time). We point out that the witness is often called a *certificate*.

So the introductory example of this subsection says that the language of pairs (S, m), where S is a set of positive integers that has a subset summing to m is in **NP**. This is actually a version of a well-known decision problem, called the *subset sum problem.* We will take a second look at other versions of this problem shortly.

Let us consider a few other classic examples.

Example 20.6. Let L be the language of all undirected graphs that have a Hamiltonian cycle. Then L is in **NP**.

Solution. An ordered list $v_1, v_2, \cdots, v_n$ of vertices of G can play the role of the witness $W(G)$. Then all we need is to check whether v_1v_2 is an edge, v_2v_3 is an edge, and so on, up to $v_{n-1}v_n$, and, at the end, v_nv_1. This means that a Turing machine T only needs to check the existence of n edges. As the head of T never needs to move more than n^2 cells between two checks, T can verify in $O(n^3)$ time whether $v_1, v_2, \cdots, v_n, v_1$ is a Hamiltonian cycle.

Example 20.7. Let L be the language of all pairs of simple graphs (G, H) so that G is isomorphic to H. Then L is in **NP**.

Solution. A bijection $f : V(G) \rightarrow V(H)$ can play the role of the witness $W(G, H)$. Then all that a Turing machine T needs to do is to check whether it holds for all $u, v \in V(G)$ that if uv is an edge, then $f(u)f(v)$ is an edge. As this means checking at $\binom{n}{2} = O(n^2)$ edges, and the head of T never travels more than $O(n^2)$ cells between two checks, our statement is proved.

The following proposition compares the two complexity classes we defined so far.

Proposition 20.8. *We have* $\mathbf{P} \subseteq \mathbf{NP}$.

Proof. If $L \in \mathbf{P}$, then there exists a Turing machine T that can test membership in L in polynomial time. So if we give input $(x, W(x))$ to T, then T can simply ignore $W(x)$ and can still verify $x \in L$ in polynomial time. □

At this point, it seems very natural to ask whether the containment in Proposition 20.8 is strict.

Question 20.9. *Does the equality* $\mathbf{P} = \mathbf{NP}$ *hold?*

This turns out to be one of the most intriguing open problems in mathematics today, and probably the single most intriguing open problem of theoretical computer science. It is one of the seven *Millennium Prize Problems.* These are seven particularly difficult open problems designated by the Clay Mathematics Institute in Cambridge MA in 2000. There is a one million dollar prize offered for the solution of each of them. The interested reader can learn more about these problems in the Notes section of this chapter.

It may sound very surprising that this Question 20.9 is still open. After all, *verifying* a witness seems to be a much simpler task than finding one. However, there are several other points to consider. First, for **P** to be equal to be **NP**, we would not need a Turing machine T that can test membership as fast as another machine T' can verify membership. It would be enough to have T test membership in $O(n^{10000})$ time while T' could verify membership in $O(n)$ time. Second, in order to prove that $\mathbf{P} \neq \mathbf{NP}$, one would need to find a language $L \in \mathbf{NP}$ so that $L \notin \mathbf{P}$. And how do you prove that a certain language is *not* in **P**?

There are other methods that could possibly be used to find the answer to Question 20.9. We will mention a few of them in the rest of this section.

20.2.2.1 *The Class* coNP

There is a subtle way of taking the complement of a complexity class. It is given by the following definition.

Definition 20.10. We say that the language L is in **coNP** if there exists a Turing machine T and a positive integer k so that the following hold.

- For each $x \notin L$, there exists a *witness* $W(x)$ so that when T is given input $(x, W(x))$, it will recognize that $x \notin L$ in $O(n^k)$ time.
- For each $x \in L$, no such witness exists. That is, no matter what input $(x, W'(x))$ we give to T, we cannot "trick" T into falsely saying that $x \notin L$.

In other words, **coNP** is the class of languages for which we can *verify non-membership* in polynomial time.

The following is a classic example of a naturally defined problem which is easily seen to be in **coNP**, but requires more work to be seen in **NP**.

Example 20.11. Let *PRIMES* be the set of all prime numbers. Then $L \in \mathbf{coNP}$.

Solution. Let x be the positive integer for which we want to show that $x \notin PRIMES$. A proper divisor $d = W(x)$ of x can play the role of witness for $x \notin L$. Indeed, then T can simply divide x by d and verify that there is no remainder. If x has n digits, then this can be done in $O(n^2)$ time.

Note that as we said before, n must be the size of the input, that is, the number of digits of x. Therefore, the following argument would be wrong. "The language *PRIMES* is in **P** since we could simply check for each integer i satisfying $2 \leq i \leq \sqrt{x}$ whether i divides x. This takes $\sqrt{x}$ steps, which is less than a polynomial function of x." The problem with this argument is that we need an algorithm that is polynomial *in terms of* n, not in terms of x.

The reader may ask why we defined **coNP** before defining **coP** as the set of languages for which we can test non-membership in polynomial time by a Turing machine.

We encourage the reader to spend a moment trying to figure that out before reading further. The answer is that **coP** = **P** since if T can test for non-membership in L in polynomial time, then the same T can test for membership in L in polynomial time, by simply interchanging the accepting and rejecting states at the end. This line of thinking leads to the following proposition.

Proposition 20.12. *We have* $\mathbf{P} \subseteq \mathbf{NP} \cap \mathbf{coNP}$.

Proof. On the one hand, Proposition 20.8 shows that $\mathbf{P} \subseteq \mathbf{NP}$. On the other hand, by the same argument, $\mathbf{coP} \subseteq \mathbf{coNP}$. As $\mathbf{coP} = \mathbf{P}$, our claim is proved. □

We would like to point out that it is somewhat more difficult to prove that the language L consisting of all prime numbers is in **NP**. That result is called *Pratt's theorem*, and can be proved using very enjoyable facts from number theory. In fact, the following characterization of primes can be used. An integer $p > 1$ is prime if and only if there exists an integer r so that $1 < r < p$ and

(i) $r^{p-1} - 1$ is divisible by p, and
(ii) If q is a prime so that $qd = p - 1$ for some integer d, then $r^d - 1$ is not divisible by p.

Given p, a witness $W(p)$ can be an integer r and all the prime divisors q of $p - 1$. A Turing machine could then verify in polynomial time that r satisfies the requirements with each $d = (p-1)/q$. Note that the number of distinct prime divisors q of $p-1$ is less than $\log_2 p$, so $W(p)$ is of polynomial size in terms of the size of the input, which is $\log p$.

It is even more difficult to prove that *PRIMES* is in **P**. That is perhaps the most celebrated recent result in complexity theory. The proof, by Agrawal, Kayal, and Saxena [1] was published in 2004, and takes only 12 pages! It is also worth pointing out that two of the three authors were undergraduate students at the time the proof was found.

The known containment relations between the three complexity classes that we have defined so far are shown in Figure 20.1.

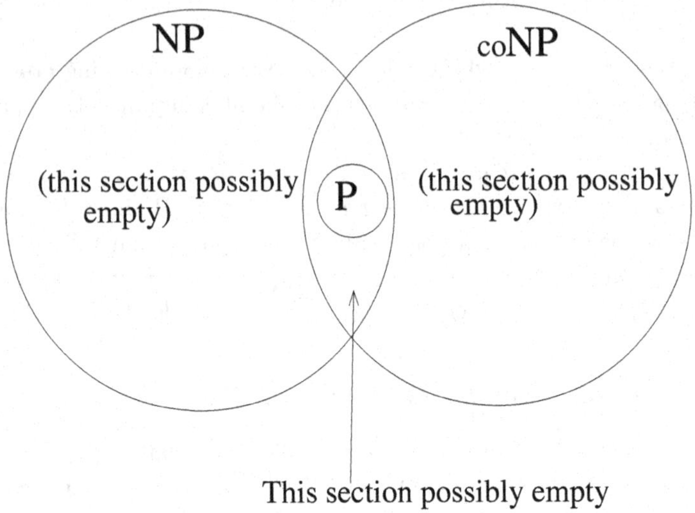

Fig. 20.1 The known inclusions between the three complexity classes defined so far.

At this point, you are asked to test your understanding of the concepts of this section by proving the following proposition.

Proposition 20.13. *The following two statements are equivalent.*

(1) $\mathbf{P} = \mathbf{NP}$.
(2) $\mathbf{P} = \mathbf{coNP}$.

We end the section by noting that it is not even known whether $\mathbf{NP} = \mathbf{coNP}$. It is widely believed that these two classes are different, just as it is widely believed that **P** and **NP** are different.

20.2.2.2 *Nondeterministic Turing Machines*

You may have wondered where the "**N**" comes in the name of the complexity class **NP**. After all, the definition of the class says that certain things have to be done in polynomial time, not in "non-polynomial" time.

The answer to that question comes from the following version of Turing machines, called *nondeterministic Turing machines.* For nondeterministic Turing machines, the first three parameters, that is, the tape, the head, and the set of states, are defined exactly as they were for the classic (deterministic) Turing machines. The difference lies in f, which we called the *transition function* or *program* in the case of deterministic Turing machines. This

$$f : S \times A \rightarrow (S \cup \text{“}Yes\text{''} \cup \text{“}No\text{''}) \times A \times \{\leftarrow, \rightarrow, stay\}$$

was a function, that is, given a certain input consisting of a state and a symbol at the cell currently read, it sent the machine into a uniquely determined state. This is why those Turing machines were called deterministic. In an undeterministic Turing machine, the function f is replaced by the *relation*

$$g : S \times A \subset [S \times A] \times [(S \cup \text{“}Yes\text{''} \cup \text{“}No\text{''}) \times A \times \{\leftarrow, \rightarrow, stay\}] .$$

In other words, a nondeterministic Turing machine has several legal courses of action in a generic step. Given a symbol in a cell and a state of the machine when reading that symbol, there are several ways in which the machine can continue.

Fine, you will say, but when will we say that such a nondeterministic Turing machine T accepts the input string x? What if a certain sequence of legal choices will result in T halting in the "Yes" state and some other sequence of legal choices will result in T halting in the "No" state? Will we take a majority vote?

It turns out that we will have a very weak notion of acceptance. We will say that T accepts x if there is at least one sequence of legal choices of action for T that results in T halting in the "Yes" state. If there is no such sequence, we will say that T rejects x.

With the acceptance of an input string now defined, we can define acceptance of a language L by a nondeterministic machine T. This definition is not surprising. We simply say that T accepts L if T accepts x if and only if $x \in L$.

How do we measure the running time of a nondeterministic Turing machine? We will *not* add up the running times it takes to carry out each computation that arises from a legal sequence of choices. Instead, we will define the running time of the nondeterministic Turing machine as the *maximum* running time among the running times of the possible computations. See Figure 20.2 for an illustration. We could interpret this definition by saying that in a nondeterministic Turing machine, all possible choices are followed up *concurrently*, so the total running time will indeed be the maximum individual running time.

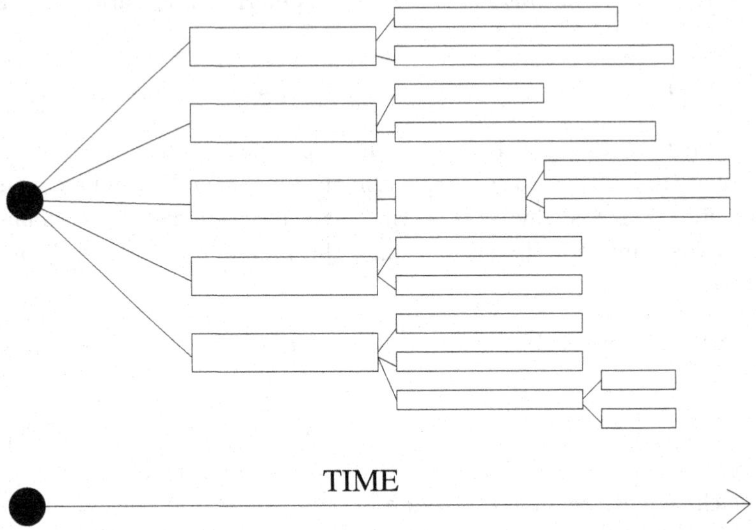

Fig. 20.2 Measuring the running time of a nondeterministic machine.

Finally, we are in a position to explain the name of **NP**. The class **NP** is the class of languages that can be accepted by a *nondeterministic* Turing machine in $O(n^k)$ time, for some positive integer k, where n denotes the size of the input. That is, **NP** stands for *nondeterministically polynomial.* Indeed, if a language L is in **NP**, then for $x \in L$, a witness $W(x)$ can be verified in polynomial time by a *deterministic* Turing machine. A nondeterministic Turing machine could then just go through all possible witnesses for x, and decide whether any of them are valid. As verifying a witness takes polynomial time, this nondeterministic machine would finish in polynomial time. If, on the other hand, no nondeterministic Turing machine could finish the task of checking all witnesses in polynomial time, then at least one possible witness could not be checked in polynomial time, implying that L is not in **NP**.

Note that even this alternative definition of **NP** makes it clear that $\mathbf{P} \subseteq \mathbf{NP}$ since a deterministic Turing machine T is just a special case of a nondeterministic one. That is, it is a nondeterministic Turing machine whose defining relation g happens to be a function. In other words, in each step, T happens to have only one legal choice. The fact that we cannot decide whether $\mathbf{P} = \mathbf{NP}$ can be expressed by saying that in some sense, we cannot decide whether nondeterministic machines are *really* stronger than deterministic ones.

Example 20.14. Let $HAMCYC$ be the language of all undirected graphs G that have a Hamiltonian cycle. Then $HAMCYC$ can be decided by a nondeterministic Turing machine T in polynomial time as follows. Let n be the number of vertices of G. Then there are $(n-1)!/2$ ways to arrange the vertices in a cycle. These will be the legal choices of T. No matter what choice T makes, T can then check whether that arrangement of vertices is a Hamiltonian cycle or not. In this stage, T can act as a deterministic machine, and will still only need $O(n)$ steps. So we find again that $HAMCYC \in \mathbf{NP}$.

20.2.3 NP-*complete Problems*

With a slight abuse of language, in this subsection we identify the language L with the problem of deciding whether $x \in L$.

Let us assume that we have a computer program that computes the prime factorization of any positive integer less than one billion. Let us further assume that for some purpose, we need to compute not the prime factorization of n, but the *number of its positive divisors*. If this is the case, we cannot simply ask the program to do all the work for us, but we will see that the program will in fact do *almost* all the work. Indeed, note that if $n = p_1^{k_1} p_2^{k_2} \cdots p_t^{k_t}$ where the p_i are different primes, then the number of positive divisors of n is precisely $\prod_{i=1}^{t}(k_i + 1)$ since m divides n if and only if $m = p_1^{a_1} p_2^{a_2} \cdots p_t^{a_t}$, with $0 \leq a_i \leq k_i$ for all i. Therefore, all we need to do is to run the program, take its output, and do something very simple with it, namely compute the product of certain numbers determined by the output.

The above example is a special case of a very general phenomenon in the theory of computation (or, in mathematics in general), namely the *reduction* of a problem to another one. Indeed, the above argument shows that if we can find the prime factorization of an integer, then we can also find the number of its positive divisors. In other words, the latter problem can be *reduced* to the former. Furthermore, the reduction did not take long when compared to the original algorithm, (think about this!), so it was "worth it". Of course, if the reduction had taken too long, we might try to solve the new problem directly, instead of reducing it to the old one.

Another example of reduction, one in which a *decision problem* is reduced to another one, will be presented shortly, in the proof of Theorem 20.22.

If decision problem A can be reduced to decision problem B in a short time, then it is natural to think that "B is at least as hard as A" in some sense. If every problem of a complexity class C can be reduced to a problem $B \in C$, then it is natural to think that B has some kind of a special role in C. The following definition is the most important example for this.

Definition 20.15. We say that the problem L is **NP**-complete if

(1) $L \in \mathbf{NP}$, and
(2) each $L' \in \mathbf{NP}$ can be reduced to L by a deterministic Turing machine in polynomial time.

You may be thinking now that the above requirement is rather strong, and therefore, it is usually rather hard to prove that a problem is **NP**-complete. Then you might be thinking that therefore, the number of **NP**-complete problems must be small, and so their class might be a rather restricted one. The first of these concerns is partly true, namely, it was difficult to find the *first* **NP**-complete problem. However, once an **NP**-complete problem is found, others are much easier to find, because of the following simple fact.

Proposition 20.16. *If L is an* **NP**-*complete language and L' is a language so that L is reducible to L', then L' is* **NP**-*complete.*

Proof. If $A \in \mathbf{NP}$, then A is reducible to L in polynomial time, and then L is reducible to L' in polynomial time. Therefore, A is reducible to L' in polynomial time. (Just run the two reducing Turing machines consecutively.) □

So once *one* **NP**-complete problem is found, others can be found by showing that the first one is reducible to them in polynomial time. The more **NP**-complete problems we have, the easier it is to find new ones, since there are more problems to play the role of L in Proposition 20.16.

The notion of **NP**-completeness provides a strategy for those who want to prove that $\mathbf{P} = \mathbf{NP}$. This is the content of the following corollary.

Corollary 20.17. *If there exists an* **NP**-*complete language* $L \in \mathbf{P}$, *then* $\mathbf{P} = \mathbf{NP}$.

Proof. If the **NP**-complete language L is in **P**, then any language $L' \in$ **NP** is also in **P**. Indeed, first reduce L' to L in polynomial time by a deterministic Turing machine, and then decide L in polynomial time by another deterministic Turing machine. □

As we said, it was not easy to find the first **NP**-complete language. We will now describe this language, without proving that it is **NP**-complete.

Let $x_1, x_2, \cdots, x_n$ be *Boolean variables*, which means that they can take two values, **true** and **false**. These variables will be called *literals*. We introduce the operations $\wedge$, $\vee$, and $^-$ on the set of literals as follows.

(1) $x_i \vee x_j =$ **true** if at least one of x_i and x_j is **true**. Otherwise, $x_i \vee x_j =$ **false**. This can be thought of as the "or" operation.
(2) $x_i \wedge x_j =$ **true** if both x_i and x_j are **true**. Otherwise, $x_i \wedge x_j =$ **false**. This can be thought of as the "and" operation.
(3) $\bar{x}_i =$ **true** if $x_i =$ **false** and $\bar{x}_i =$ **false** if $x_i =$ **true**. This can be thought of as the *negation* operation.

A Boolean expression is just a sequence of operations on literals, such as $(x_1 \wedge x_2) \vee \bar{x}_3$, or $(x_1 \wedge x_2) \vee x_1$. A Boolean expression is called *satisfiable* if we can assign the values **true** and **false** to its literals so that the expression evaluates to **true**.

Example 20.18. The Boolean expression

$$(x_1 \wedge x_2) \vee \bar{x}_3$$

is satisfiable. Indeed, setting $x_1 =$ **true**, $x_2 =$ **true**, and $x_3 =$ **false**, the expression evaluates to **true**.

Example 20.19. The Boolean expression

$$(x_1 \wedge \bar{x}_2) \wedge (\bar{x}_1 \vee x_2)$$

is not satisfiable. Indeed, the first parentheses will only evaluate to **true** if $x_1 =$ **true** and $x_2 =$ **false**, while in that case, the second parentheses will evaluate to **false**.

A Boolean expression in *conjunctive normal form* is a Boolean expression in which there are only $\wedge$ operations among the parentheses (the latter are called the *clauses*), and there are only $\vee$ operations within the parentheses.

Example 20.20. The Boolean expression

$$(x_1 \vee x_2) \wedge (x_1 \vee \bar{x}_3 \vee x_4) \wedge x_2 \wedge (x_1 \vee \bar{x}_4)$$

is in conjunctive normal form.

It can be proved that each Boolean expression is equivalent to one in conjunctive normal form. So restricting our attention to Boolean expressions in this form will not result in loss of generality, but it will simplify the handling of the expressions.

We are now in a position to announce *Cook's theorem*, the first result showing that a certain language is **NP**-complete.

Theorem 20.21 (Cook's theorem). [15] *Let SAT be the language of satisfiable Boolean expressions in conjunctive normal form. Then SAT is* **NP***-complete.*

It is easy to see that *SAT* is in **NP**. Indeed, the witness $W(x)$ for a given Boolean expression x is just an assignment of values to the literals of x. It then takes $O(n)$ time to verify that each clause indeed contains at least one literal with value 1. It is also easy to see that the total number of possible assignments is 2^m if we have m literals, so checking all possible assignments would take more than a polynomial amount of time.

The proof of Cook's theorem can be found in any textbook on Complexity Theory. For a reader-friendly presentation, we recommend [50]. We point out that even if we only consider Boolean expressions in conjunctive normal form so that each clause contains only three literals, the corresponding language *3SAT* is still **NP**-complete. This is because *SAT* is reducible to *3SAT* in polynomial time as we will see in the proof of the next theorem.

Theorem 20.22. *Let 3SAT be the language of Boolean expressions in conjunctive normal form so that each clause contains exactly three literals. Then 3SAT is* **NP***-complete.*

Proof. It goes without saying that *3SAT* $\in$ **NP** since an assignment of variables can play the role of the witness. We will now show how to reduce *SAT* to *3SAT* in polynomial time. That is, for each Boolean expression X

in conjunctive normal form, we will construct a Boolean expression $f(X)$ in which each clause contains exactly three literals so that X is satisfiable if and only if $f(X)$ is satisfiable.

We will construct $f(X)$ clause by clause. Say one of the clauses of X is $(x_1 \vee x_2 \vee \cdots \vee x_m)$. We will break this clause up into $m-2$ smaller clauses, which will also contain some new variables. In fact, let us replace $X_c = (x_1 \vee x_2 \vee \cdots \vee x_m)$ by the clause
$f(X_c) = (x_1 \vee x_2 \vee y_1) \wedge (x_3 \vee \bar{y}_1 \vee y_2) \wedge (x_4 \vee \bar{y}_2 \vee y_3) \wedge \cdots \wedge (x_{m-1} \vee x_m \vee \bar{y}_{m-3})$. That is, the first and last clauses are different from the rest. Other than that, the ith clause is $(x_{i+1} \vee \bar{y}_{i-1} \vee y_i)$ if $2 \leq i \leq m-3$.

Let us replace each clause X_d of X by the clause $f(X_d)$ defined this way. Let us now assume that X is satisfiable; that happens exactly when each clause of X is satisfiable. As X_c is satisfiable by a certain true-false assignment, there is at least one index $i \in [m]$ so that $x_i =$ **true** in that assignment. Choose the smallest such i. Now assign $y_j =$ **true** if $j < i-1$ and $y_j =$ **false** if $j \geq i-1$. This assignment will satisfy $f(X_c)$, since the first $i-2$ clauses will evaluate to **true** since the unbarred y_j variable in them will be **true**, the $(i-1)$st clause will evaluate to **true** since it will contain x_i, and the remaining clauses will evaluate to **true** since the variable $\bar{y}_j$ in them will be **true**.

This argument works for each clause of X, so we have proved that $f(X)$ is satisfiable if X is satisfiable. We still have to prove the converse.

Let us assume that $f(X) = \vee (f(X_c))$ is satisfiable, but X is not satisfiable. That means that there is an assignment of values to all variables x_i and y_j that satisfies each clause of $f(X)$, but not each clause of X. Let c be such that this assignment does not satisfy X_c, but satisfies $f(X_c)$. As $X_c = (x_1 \vee x_2 \vee \cdots \vee x_m)$, this means that in the assignment satisfying $f(X)$, the equality $x_i =$ **false** holds for all $i \in [m]$. Then, crucially, we can remove all the x_i from $f(X_c)$ and the obtained clause $f(X_C)$ will still evaluate to **true** (since no x_i is barred in $f(X_C)$). This implies that

$$y_1 \wedge (\bar{y}_1 \vee y_2) \wedge (\bar{y}_2 \vee y_3) \wedge \cdots \wedge (\bar{y}_{m-4} \vee y_{m-3}) \wedge (\bar{y}_{m-3})$$

is satisfied by the assignment satisfying $f(X)$. However, the last displayed expression is unsatisfiable. Indeed, to satisfy its first clause, we would have to set $y_1 =$ **true**, then to satisfy its second clause, we would have to set $y_2 =$ **true**, and so on. The next-to-last clause would force $y_{m-3} =$ **true**, and then the last clause would not be satisfiable.

So we have seen that X is satisfiable if and only if $f(X)$ is. As the creation of $f(X)$ takes only polynomial (in fact, linear) time, this shows that *SAT* is reducible to *3SAT* in polynomial time, proving our claim. □

The result of Theorem 20.22 is probably optimal in the following sense. If we restrict our attention to Boolean expressions which consist of clauses of exactly two literals, and define *2SAT* to be the language of those that are satisfiable, then *2SAT* is *very unlikely* to be **NP**-complete. This is because, as it is proved in Exercise 4, the language *2SAT* is in **P**! So if *2SAT* is **NP**-complete, then **P** = **NP**. The reader should wait until the end of this chapter before attempting to solve Exercise 4 as some additional definitions will be needed.

Many **NP**-complete problems involve graphs, and the proof of their **NP**-completeness often involve the reduction of *3SAT* to these problems. The reader is strongly encourage to attempt the solution of Exercise 6 for an elegant example.

The following are three examples of **NP**-complete problems. We point out that [19] is an entire book totally devoted to this complexity class!

Example 20.23. Let *HAMPATH* be the language of graphs that have a Hamiltonian path. Then *HAMPATH* is **NP**-complete.

Example 20.24. Let *SUBSETSUM* be the set of finite multisets of real numbers that have a non-empty submultiset whose sum of elements is equal to 0. Then *SUBSETSUM* is **NP**-complete.

See Exercise 5 for a variation of this problem.

Example 20.25. Let L be the set of pairs (p, q) so that p is a permutation that contains q as a pattern. Then L is **NP**-complete.

Note that it is very important in the above example that q is part of the input, that is, that the length of q is not given in advance. If the length of q were a given constant, then the corresponding language would be in **P**, as you will be asked to prove in Supplementary Exercise 14. This is an example of an important distinction which often decides whether a problem can be proved to be in **P** or to be **NP**-complete.

A special case of this example is the famous *traveling salesman problem.* See Supplementary Exercise 17.

Corollary 20.17 implies that if someone could find an efficient (read "contained in **P**") algorithm for the Hamiltonian cycle problem, or the subset sum problem, or the pattern avoidance problem, then we would know that there also exists an efficient algorithm for the several hundred other known **NP**-complete problems.

20.2.4 *Other Complexity Classes*

Instead of defining complexity classes based on how much *time* it takes for a Turing machine to solve the corresponding decision problems, one could look at the *space*, that is, the number of cells, the Turing machine will need.

Definition 20.26. We say that the language L belongs to the complexity class **PSPACE** if there exists a Turing machine T and a positive integer k and a constant c so that T accepts x if and only if $x \in L$ and the number of cells T will use when given input x is at most cn^k.

In other words, T needs only $O(n^k)$ cells to decide if $x \in L$, where n is the size of the input x.

As a Turing machine takes a unit of time to access each cell, the following proposition is immediate.

Proposition 20.27. *We have* $\mathbf{P} \subseteq \mathbf{PSPACE}$.

It is not known whether this inclusion is strict or not. The following containment relation is a little bit less obvious.

Lemma 20.28. *We have* $\mathbf{NP} \subseteq \mathbf{PSPACE}$.

Proof. Let $L \in \mathbf{NP}$. Note that as far as membership in **PSPACE** is concerned, the running time of the machines is not important. Therefore, if T is the nondeterministic Turing machine that accepts L in polynomial time, we could modify T to get the machine T' as follows. Let T' be the deterministic Turing machine that carries out each computation resulting from a legal sequence of choices by T, but it does so *consecutively* in some specified order, instead of *concurrently*, and so that each sequence *overwrites* the previous one. Then this T' is a deterministic machine. Indeed, in each stage, T' takes a uniquely defined step since it takes the next step of the currently selected sequence, and the order in which the sequences are processed is determined. Furthermore, T' uses polynomial space only, since each sequence, including the longest one, uses polynomial space only. Indeed, if a sequence s would take more than polynomial space to process, then T could not process that sequence in polynomial time. □

As it is not even known whether **PSPACE** is actually larger than **P**, it is not surprising that it is not known whether **PSPACE** is actually larger than **NP**.

So far, every complexity class we considered contained **P**. How about classes *contained* in **P**? In order to be able to introduce two interesting classes of that kind, we need the notion of *logarithmic space.* That is, we want to consider languages that can be accepted using $O(\log n)$ space only. "Nonsense", you could say at this point, since n is the size of the input given to the Turing machine, so just taking the input needs $n > O(\log n)$ steps. Therefore, when considering these complexity classes, we will not count the part of the tape that contains the input as part of the needed space. We will only count the space needed for the actual computation.

Now we are ready for the definition of two new complexity classes.

Definition 20.29. We say that the language H is in $\mathbf{L}$ if there exists a positive integer k and a deterministic Turing machine T so that for any input x of length $|x|$, the machine T can decide whether $x \in H$ using at most $k \log |x|$ cells.

A spectacular recent result in this regard is the following.

Theorem 20.30. *Let UST be the language of triples (G, s, t) so that G is an* undirected *graph, and s and t are two of its vertices so that there is a path from s to t in G. Then $UST \in \mathbf{L}$.*

Theorem 20.30 was proved by Omer Reingold in 2004 [35].

Definition 20.31. We say that the language H is in $\mathbf{NL}$ if there exists a positive integer k and a nondeterministic Turing machine T so that for any input x of length $|x|$, the machine T can decide whether $x \in H$ using at most $k \log |x|$ cells.

A famous example of a decision problem that is in $\mathbf{NL}$ is *REACHABILITY.* That is, given input (G, x, k), where G is a *directed* graph, x is a vertex of G, and k is a positive integer, a Turing machine must decide whether G has at least k vertices that are reachable from x by a directed path. The fact that this problem is in $\mathbf{NL}$ is the celebrated *Immerman-Szelepcsényi* theorem. Note that if an algorithm can decide *REACHABILITY*, then it can decide UST since we can set $k = 1$, and we can replace each edge of the undirected graph by two directed edges going in opposite directions.

It is not known whether *REACHABILITY* is in $\mathbf{L}$ or not, but it is known that there exists a deterministic Turing machine that can solve *REACHABILITY* using $O(\log^2 n)$ cells. Note that unlike $\mathbf{P}$ or $\mathbf{NP}$, the complexity class $\mathbf{L}$ does *not* allow for taking squares that way.

It is clear from the definitions that $\mathbf{L} \subseteq \mathbf{NL}$. Whether that inclusion is strict is not known. The following inclusion is a little bit more difficult to prove.

Lemma 20.32. *We have* $\mathbf{L} \subseteq \mathbf{P}$.

The reader is asked to make an effort to prove this lemma on his own. A proof is provided in the solution of Exercise 3. An enhanced version of the argument given in that solution (see [32]) proves the inclusion $\mathbf{NL} \subseteq \mathbf{P}$. Again, it is not known if this inclusion is strict.

Finally, for a change, we mention one inclusion that is known to be strict. It is known that $\mathbf{L} \neq \mathbf{PSPACE}$ (see [32]).

The following chain of inequalities summarizes the weak containment relations we mentioned in this chapter.

$$\mathbf{L} \subseteq \mathbf{NL} \subseteq \mathbf{P} \subseteq \mathbf{NP} \subseteq \mathbf{PSPACE}. \tag{20.1}$$

What is amazing about this chain of inclusions is that none of the inclusions in (20.1) is *known* to be strict. On the other hand, as we said, $\mathbf{L} \neq \mathbf{PSPACE}$. Therefore, at least one of the inclusions in (20.1) is *strict*. So there is at least one strict inclusion between *consecutive* expressions to be proved in this chain. Is there just one? If not, which one will be proved first?

Notes

A list of the seven Millennium Prize Problems can be found at the website of the Clay Institute, at http://www.claymath.org/millennium/. When this book goes to press, in 2013, thirteen years after the announcement of the million-dollar offers for these problems, only one of these problems has been solved.

A reader-friendly introduction to the topic of this chapter, just as to the topic of the previous chapter, is Herb Wilf's book *Algorithms and Complexity* [50]. Two very enjoyable and fairly comprehensive graduate-level textbooks are *Computational Complexity* by Christos Papadimitriou [32] and *Introduction to the Theory of Computation* by Michael Sipser [38].

Exercises

Note: in solving some of the Exercises of this chapter, the reader may use certain theorems or examples that were mentioned in the text without proof.

(1) Let L be the language of all *connected* graphs. Prove or disprove that $L \in \mathbf{P}$.

(2) Let L be the language of all *bipartite* graphs.

  (a) Prove that $L \in \mathbf{NP}$.

  (b) Prove that $L \in \mathbf{P}$.

(3) Prove Lemma 20.32.

(4) (+) Let *2SAT* be the language of Boolean expressions in conjunctive normal form so that each clause contains only two literals. Prove that *2SAT* $\in \mathbf{NL}$. Note that this implies that *2SAT* $\in \mathbf{P}$.

(5) Let *BIGSUBSETSUM* be the language of all finite multisets S of real numbers that have a submultiset T so that

  (a) the sum of all elements of T is 0, and

  (b) $|T| > 0.9 \cdot |S|$.

  Prove that *BIGSUBSETSUM* is **NP**-complete.

(6) Let *INDEPENDENT-SET* be the language of pairs (G, k) so that G is a simple graph and k is a positive integer so that G has an induced subgraph on k vertices that has no edges. Prove that *INDEPENDENT-SET* is **NP**-complete.

(7) A decision problem is called **NP**-hard if all problems in **NP** are reducible to it in polynomial time, by a deterministic Turing machine. Prove that the halting problem, discussed in Chapter 17, is **NP**-hard.

(8) It follows from the definition given in the previous exercise that the set of **NP**-hard problems contains the set of **NP**-complete problems. Is this containment strict?

(9) A problem is called **coNP** – *complete* if every problem in **coNP** is reducible to it in polynomial time by a deterministic Turing machine. A *tautology* is a finite Boolean expression that is satisfied by every assignment of its variables. For instance, $x_1 \vee \bar{x}_1$ is a tautology. Let *TAUT* be the language of all tautologies. Prove that *TAUT* is **coNP**-complete.

(10) Let *HAMCYC* be the language of graphs that contain a Hamiltonian cycle. Prove that *HAMCYC* is **NP**-complete.

(11) A *dominating set* in a graph G is a subset D of vertices so that any vertex that is not in D has a neighbor in D. Let *DOMINATING-SET* be the language of pairs (G, k), where G is a graph that has a dominating set consisting of k or less vertices. Prove that *DOMINATING-SET* is **NP**-complete.

(12) Let *SPANNING-TREE* be the language of ordered pairs (G, k) where

G is a simple graph that has a spanning tree in which each vertex has degree at most k. Prove that *SPANNING-TREE* is **NP**-complete.

Supplementary Exercises

(13) Explain, using the formal definition of (deterministic) Turing machines, that once a Turing machine entered the accepting state or rejecting state, it will stop.

(14) Let q be a given permutation pattern. Let L be the set of all permutations that contain q. Prove that $L \in \mathbf{P}$.

(15) Let L be the language of graphs containing a matching that consists of at least 10 edges. Prove or disprove that $L \in \mathbf{P}$.

(16) Prove Proposition 20.13.

(17) A salesman has to travel to each of n cities, visiting each of them exactly once, and ending in the same city where he started. The cost of travel between any two cities is given in advance. Prove that the problem of deciding whether this can be done at a cost less than a given C is **NP**-complete.

(18) Prove that if an **NP**-complete problem is in **coNP**, then $\mathbf{NP} = \mathbf{coNP}$.

(19) Let L be the language of finite multisets of real numbers such that L can be partitioned into two blocks A and B so that the elements of A and the elements of B have the same sum. Prove that L is **NP**-complete.

(20) Prove that if $\mathbf{NP} \subseteq \mathbf{coNP}$, then $\mathbf{NP} = \mathbf{coNP}$.

(21) Let *CLIQUE* be the language of pairs (G, k) where G is a graph and k is a positive integer so that G contains a subgraph isomorphic to K_k. Prove that *CLIQUE* is **NP**-complete.

(22) Recall that a *vertex cover* of a graph G is a subset C of the vertex set of G so that each edge of G has at least one endpoint in C. Let *VERTEX-COVER* be the language of pairs (G, k) where G is a graph and k is a positive integer so that G has a vertex cover of k elements or less. Prove that *VERTEX-COVER* is **NP**-complete.

(23) Let $SHORTESTPATH$ be the language of 4-tuples (G, k, a, b) where G is an undirected graph, and a and b are vertices of G so that there is a path between a and b that consists of at most k edges. Prove that $SHORTESTPATH \in \mathbf{P}$.

(24) Let $SUBGRAPH$ be the language of pairs (G, H) where G and H are graphs such that G has a subgraph isomorphic to H. Prove that $SUBGRAPH$ is **NP**-complete.
(25) Let $HITSET$ be the language of pairs (F, k), where F is a finite family of finite sets so that there is a k-element set that has a non-empty intersection with each set in F. Prove that $HITSET$ is **NP**-complete.

Solutions to Exercises

(1) Yes, $L \in \mathbf{P}$. Just run breadth first search starting at any vertex s. When the algorithm stops, check whether all vertices have been reached.
(2) (a) A witness $W(G)$ for a graph G could simply be a partition of the vertex set of G into two blocks. It can then be verified in $O(n^2)$ steps that there are no edges within the same block.
(b) Do breadth first search on the vertex set of G starting from some vertex s. Vertices at an even distance from s get colored red, and vertices at an odd distance from s get colored blue. If this algorithm never reassigns the color of a vertex, then $G \in L$, otherwise it is not.
(3) If at a given point of time, we are told the content of each cell of the tape of the deterministic Turing machine T, the position of its head, and state in which the machine is in, then using the transition function of T, we can compute all future moves of T. If T uses at most $k \log n$ cells, then there are at most $|A|^{k \log n}$ possibilities for the content of the tape, $k \log n$ possibilities for the position of the head, and at most $|S|$ states in which T can be. Therefore, the total number of *configurations* described by the parameters above is at most

$$\begin{aligned}
|A|^{k \log n} \cdot k \log n \cdot |S| &= e^{k \log n} \left(\frac{A}{e}\right)^{k \log n} k \log n \cdot |S| \\
&= n^k c^{\log n} \cdot C \cdot \log n \\
&= n^k \cdot n^{\log c} C \cdot \log n \\
&= \leq C n^{k + \log c + 1}.
\end{aligned}$$

Here $C = |S| \cdot k$ and $c = (A/e)^k$. Now T processes each configuration in unit time, so in Cn^{c_1} time, it will process all configurations. A configuration cannot occur twice, since that would put T into an infinite loop. This proves our claim.

(4) Let B be a Boolean expression in conjunctive normal form in which each clause contains exactly two literals. We are going to construct a *directed* graph G_B from B. The vertices of G_B are the literals of B and their negations, each occurring once. There is an edge from vertex x to vertex y if one of the clauses of B is $\bar{x} \vee y$. See Figure 20.3 for an example. Note that this clause is equivalent to the implication "if $x =$ **true**, then $y =$ **true**". That is, if an assignment satisfies B, and the value of a vertex v in that assignment is **true**, then the value of all vertices reachable from v by a directed path must also be **true**.

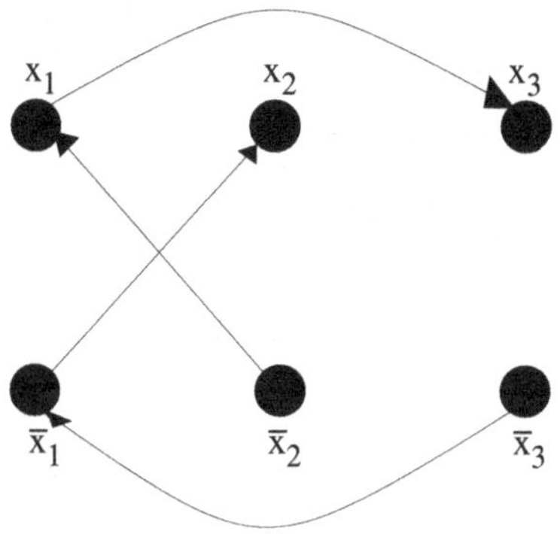

Fig. 20.3 The graph G_B defined by $B = (x_1 \vee x_2) \wedge (\bar{x}_1 \vee x_3)$.

We now claim that $B \in 2SAT$ if and only if there is no literal $x \in B$ so that there is a path from x to $\bar{x}$ in G_B, and also, a path from $\bar{x}$ to x in G_B. As the latter problem is in **NL** (it is an instance of *REACHABILITY*), our statement will then follow.

In order to prove that claim, let us assume that such a literal x exists, and assume without loss of generality that in an assignment satisfying B, the value of x is **true**. As there is a path in G_B from x to $\bar{x}$, and $\bar{x}$ is **false**, this contradicts to the property of G_B we just proved, that is, that all literals reachable from a **true** literal must also be **true**.

Conversely, if no such literal x exists, then we will define an assignment satisfying B. (Intuitively, "no literal will cause any trouble".) Note that by the definition of G_B, if there is an edge from x to y, then there is also an edge from $\bar{y}$ to $\bar{x}$. Start at a vertex u for which there is no path from u to $\bar{u}$. Assign **true** to all vertices reachable from u by a directed path, including u itself, and assign **false** to their negations. If this does not exhaust all vertices, then pick another vertex v whose value is not yet assigned, then repeat the procedure. We claim that this procedure will never cause a conflict at the assignment of any vertex. Indeed, if both z and $\bar{z}$ were reachable from u, then, by the symmetric property of G_B mentioned earlier in this paragraph, there would be paths from both $\bar{z}$ and z to $\bar{u}$. That would, by concatenation, yield a path from u to $\bar{u}$, contradicting our hypothesis.

Finally, the assignment defined in the previous paragraph will satisfy B. Indeed, in each step of the above procedure, we ensure that if $x =$ **true**, then in all clauses in which $\bar{x}$ occurs, the *other* literal is set to be **true**. Therefore, each clause will contain at least one literal that is **true** in the assignment.

(5) We are going to prove the statement by reducing *SUBSETSUM* to BIG*SUBSETSUM*. On any input multiset S, just add $9 \cdot |S|$ copies of 0 to S to get the new multiset S'. Then $S' \in$ *BIGSUBSETSUM* if and only if $S \in$ *SUBSETSUM*, and the statement follows since $|S'|$ is only ten times larger than S, so the Turing machine deciding if $S' \in$ *BIGSUBSETSUM* runs in polynomial time in the size of S as well.

(6) We show that *SAT* is reducible to *INDEPENDENT-SET*. Let B be a Boolean expression in conjunctive normal form that has k clauses. We will define a graph G_B that has an empty induced subgraph with k vertices if and only if B is satisfiable. The vertices of the graph are labeled by the literals of B. If a literal x_i occurs m times in B, then there are m vertices in G labeled by x_i, one for each occurrence. Now connect each vertex labeled x_i to each vertex labeled by $\bar{x}_i$. In addition, connect vertices if the corresponding literals appear in the same clause. (This does not mean that if x_6 and x_7 appear together in one clause C, then we connect all vertices labeled x_6 to all vertices labeled x_y; just the vertices corresponding to literals in C.)

See Figure 20.4 for an example.

If G_B contains an empty subgraph H on k vertices, then each vertex of H must correspond to a literal from a different clause, since each clause contributes a complete subgraph to G_B. Furthermore, none of these k vertices could correspond to a literal that is a negation of another literal corresponding to a vertex of H, since x_i and $\bar{x}_i$ are always adjacent. Therefore, assigning **true** to all literals represented in H by

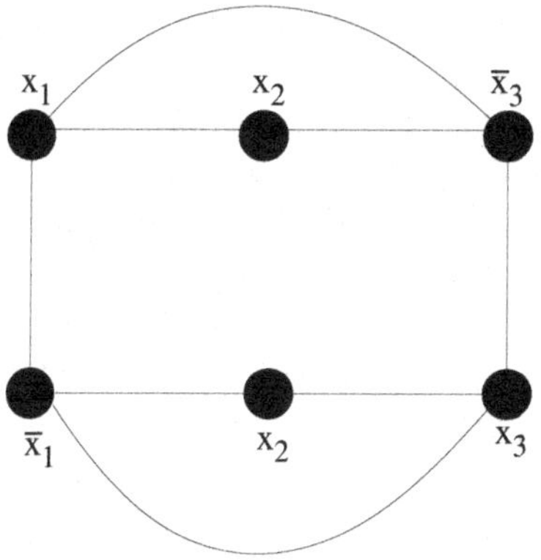

Fig. 20.4 The graph G_B of $B = (x_1 \vee x_2 \vee \bar{x}_3) \wedge (\bar{x}_1 \vee x_2 \vee x_3)$.

a vertex will satisfy each of the k clauses of B, and consequently, B. Conversely, if B is satisfiable by an assignment, then that assignment assigns **true** to at least one literal in each clause. Choosing such a literal from each clause will result in an empty subgraph with k vertices, since it will never happen that two adjacent vertices in separate clauses are assigned **true**, since such vertices correspond to pairs of literals that are negation of each other.
As the creation of G_B takes only polynomial time, the proof is complete.

(7) We claim that SAT is reducible to the halting problem. Indeed, if we could solve the halting problem by a Turing machine T, then we could input the pair $(G, (B, x))$ to T, where B is a Boolean expression in conjunctive normal form, x is an assignment of **true** and **false** values to the variables of B, and G is a Turing machine that halts if x satisfies B and stops otherwise. The assumption that T decides the halting problem would then imply that T decides SAT.

(8) Yes, this containment is strict. We have seen in the solution of the previous exercise that the halting problem is **NP**-hard. On the other hand, the halting problem is not in **NP**, and so it is not **NP**-complete. Indeed, if it were, then it would also be in **PSPACE**, and that would mean that it is decidable by a deterministic algorithm, and we saw in Chapter 17 that it is not the case.

(9) It is easy to see that $TAUT \in$ **coNP** since for a Boolean expression B, the role of the witness $W(B)$ can be played by an assignment that

does not satisfy B, and that can be checked in polynomial (in fact, linear) time.

In order to prove that $TAUT$ is **coNP**-complete, note that the complement SAT^c of SAT (that is the language of Boolean expressions that are not satisfiable) is **coNP**-complete, and is reducible to $TAUT$. Indeed, $B \in SAT^c$ if and only if $\bar{B} \in TAUT$.

(10) We show that *HAMPATH* is reducible to *HAMCYC*. Let G be a graph, add a new vertex v to G, and let v be adjacent to all other vertices of G. Then the new graph has a Hamiltonian cycle if and only if G had a Hamiltonian path.

(11) We prove that the language *VERTEX-COVER* is reducible to *DOMINATING-SET* in polynomial time. This suffices, since Exercise 22 shows that *VERTEX-COVER* is **NP**-complete. Let G be a graph. Construct the graph G' by first doubling each edge xy of G, and then splitting each new xy edge by a new vertex v_{xy}. We claim that G' has a dominating set of size k if and only if G' has a vertex cover of size k. First, it is obvious that if S is a vertex cover of G, then S is also a dominating set of G', since the vertices of G' are along the edges of G (old or new).

If G' has a dominating set D of size k, then we can replace the vertices of D that are of the form v_{xy} by one of x or y without losing the dominating property (why?). This leads to a dominating set D' of G that has k vertices. This set D' is a vertex cover of G (since it still dominates all vertices v_{xy}, and hence it covers all edges of G). This reduction algorithm clearly takes polynomial time only.

(12) We show that *HAMPATH* is reducible to *SPANNING-TREE*. Indeed, to decide if G has a Hamiltonian path, it suffices to decide if G has a spanning tree in which every vertex has degree at most two.

Bibliography

[1] Agrawal, M., Kayal, N., Saxena, N. (2004) "PRIMES is in **P**." *Annals of Mathematics*, **160**, 781–793.

[2] Albert, M., Elder, M. , Rechnitzer, A., Westcott, P., and Zabrocki, M. (2006) "A lower bound on the growth rate of the class of 4231 avoiding permutations", *Adv. Appl. Math.* **36** vol. 2, 96–105.

[3] Alon, N., Spencer, J. (2000) "The Probabilistic Method", second edition, Wiley-Interscience Series in Discrete Mathematics and Optimization. A Wiley-Interscience Publication. John Wiley and Sons, Inc.

[4] Andrews, G. E. (1976) "The Theory of Partitions", Encyclopedia of Mathematics and Its Applications, vol. 2, Addison-Wesley.

[5] Biggs, N. L. (2008) "Codes. An Introduction to Information Communication and Cryptography." Springer Verlag, London.

[6] Bollobás, B. (1978) "Extremal Graph Theory". London Mathematical Society Monographs, 11. Academic Press, Inc.

[7] Bóna, M. (2005) "Introduction to Enumerative Combinatorics" McGraw-Hill.

[8] Bóna, M. (2012) "Combinatorics of Permutations", second edition, CRC Press – Chapman Hall.

[9] Bóna, M. (1998) "The permutation classes equinumerous to the smooth class". *Electron. J. Combin.* **5** (1998), Research Paper R31.

[10] Bóna, M., Simion, R. (2000) "A self-dual poset on objects counted by the Catalan numbers and a type-B analogue", *Discrete Math.* **220**, no. 1-3, 35–49.

[11] Bóna M., MacLennan, A., White, D., (2000) "Permutations with roots", *Random Structures and Algorithms,* **17**, no. 2, 157–167.

[12] Bóna, M (1997) "A new proof of the formula for the number of the 3×3 magic squares", *Math. Magazine* **70**, 201–203.

[13] Brenti, F. (1997) "Unimodal, log-concave and Pólya frequency sequences in combinatorics," *Mem. Amer. Math. Soc.* **81**, no. 413.

[14] Colburn, C. J., Dinitz, J. H. (2007) "Handbook of Combinatorial Designs," Chapman Hall - CRC Press, Boca Raton, FL.

[15] Cook, S. (1971) "The complexity of theorem proving procedures" Proceed-

ings of the third ACM Symposium, ACM, 151–158.

[16] Cormen, T, Leiserson, C., Rivest, R., Stein, C. (2002) "Introduction to Algorithms", McGraw-Hill, second edition.

[17] R. Donaghey (1975) Alternating permutations and binary increasing trees. *J. Combinatorial Theory Ser. A* **18**, 141–148.

[18] P. Flajolet, R. Sedgewick (2009) "Analytic Combinatorics", Cambridge University Press.

[19] Garey, M. D., Johnson, S. (1979) "Computers and Intractability; A guide to the theory of **NP**-completeness. W. H. Freeman and Co.

[20] Gordon, H. (1997) "Discrete Probability", Springer.

[21] Graham, L., Rothschild, B., Spencer, J., (1990) "Ramsey Theory", Wiley Interscience, second edition.

[22] Greene, C., Bogarth, K. P., Kung, J. (1990) "The impact of the chain decomposition theorem on classical combinatorics. The Dilworth theorems", 19–29, Contemp. Mathematics, Birkhäuser.

[23] Harary, F., Palmer, E. M. (1973) "Graphical Enumeration", Academic Press, 1973.

[24] Kaiser, T., Klazar, M. (2003) On growth rates of hereditary permutation classes, *Electr. J. Combinatorics* **9** vol. 2, R10.

[25] Krattenthaler, C. (1995) "The major counting of nonintersecting lattice paths and generating functions for tableaux", *Mem. Amer. Math. Soc.* **115**, no. 552.

[26] Landman, B., Robertson, A., (2004) "Ramsey Theory on the Integers", American Mathematical Society.

[27] Lovász, L. (1994) "Combinatorial Problems and Exercises", Elsevier; Akadémiai Kiadó (Publishing House of the Hungarian Academy of Sciences), second edition.

[28] Lovász, L., Plummer, M. D. (1986) "Matching Theory", Akadémiai Kiadó (Publishing House of the Hungarian Academy of Sciences).

[29] MacMahon, P. A. (1916) "Combinatorial Analysis", vols. 1-2, Cambridge, 1916. (reprinted by Chelsea, New York, 1960).

[30] Marcus, A., Tardos, G. (2004) Excluded Permutation Matrices and the Stanley-Wilf conjecture, *J. Combin. Theory Ser. A* **107** no. 1, 153–160.

[31] Pach, J., Agarwal, P., (1995) "Combinatorial geometry" Wiley Interscience.

[32] Papadimitriou, C. H. (1994) "Computational Complexity", Addison-Wesley.

[33] Pittel, B. (1999) Confirming two conjectures about the integer partitions. *J. Combin. Theory Ser. A* **88** no. 1, 123–135.

[34] Recski, A. (1989) "Matroid Theory and Its Applications in Electric Network Theory and in Statics." Springer-Verlag; Akadémiai Kiadó (Publishing House of the Hungarian Academy of Sciences).

[35] Reingold, O. (2005) "Undirected ST-connectivity in log-space." STOC'05: Proceedings of the 37th Annual ACM Symposium on Theory of Computing, 376–385, ACM.

[36] Sagan, B. E. (1998) "Unimodality and the reflection principle", *Ars Combin.* **48**, 65–72.

[37] Simion, R., Ullman, D. (1991) "On the structure of the lattice of noncrossing

partitions", *Discrete Math.* **98**, no. 3, 193–206.

[38] Sipser, M. (2005) "Introduction to the Theory of Computation" Course Technology; second edition.

[39] Stanley, R. (1983) "Combinatorics and Commutative Algebra", Progress in Mathematics 41, Birkhäuser.

[40] Stanley, R. (1973) "Acyclic orientations of graphs", *Discrete Math.* **5** , 171–178.

[41] Stanley, R. (1997) "Enumerative Combinatorics", vol. 1, Cambridge University Press, second edition.

[42] Stanley, R. (1999) "Enumerative Combinatorics", vol. 2, Cambridge University Press.

[43] Tóth, G., Valtr, P. (2005) The Erdős-Szekeres theorem, upper bounds and generalizations, *Discrete and Computational Geometry - Papers from the MSRI Special Program (J. E. Goodman et al. eds.)* **52**, 557–568.

[44] Trotter, W. T. (1992) "Combinatorics and Partially Ordered Sets. Dimension Theory", Johns Hopkins Series in the Mathematical Sciences. Johns Hopkins University Press.

[45] (2003) Vella, A. "Pattern avoidance in permutations: linear and cyclic orders". Permutation patterns (Otago, 2003) *Electron. J. Combin.* **9** no. 2, Research paper 18, 43 pp, electronic.

[46] W. D. Wallis (2007) "Introduction to Combinatorial Designs", Chapman Hall - CRC Press, Boca Raton, FL.

[47] West, D. (2001) "Introduction to Graph Theory", Prentice Hall, second edition.

[48] West, J. (1996) "Generating trees and forbidden subsequences". Proceedings of the 6th Conference on Formal Power Series and Algebraic Combinatorics (New Brunswick, NJ, 1994). *Discrete Math.* **157**, no. 1-3, 363–374.

[49] Wilf, H. (2005) "Generatingfunctionology", AK Peters, third edition.

[50] Wilf, H. (2002) "Algorithms and Complexity". AK Peters, second edition.

Index